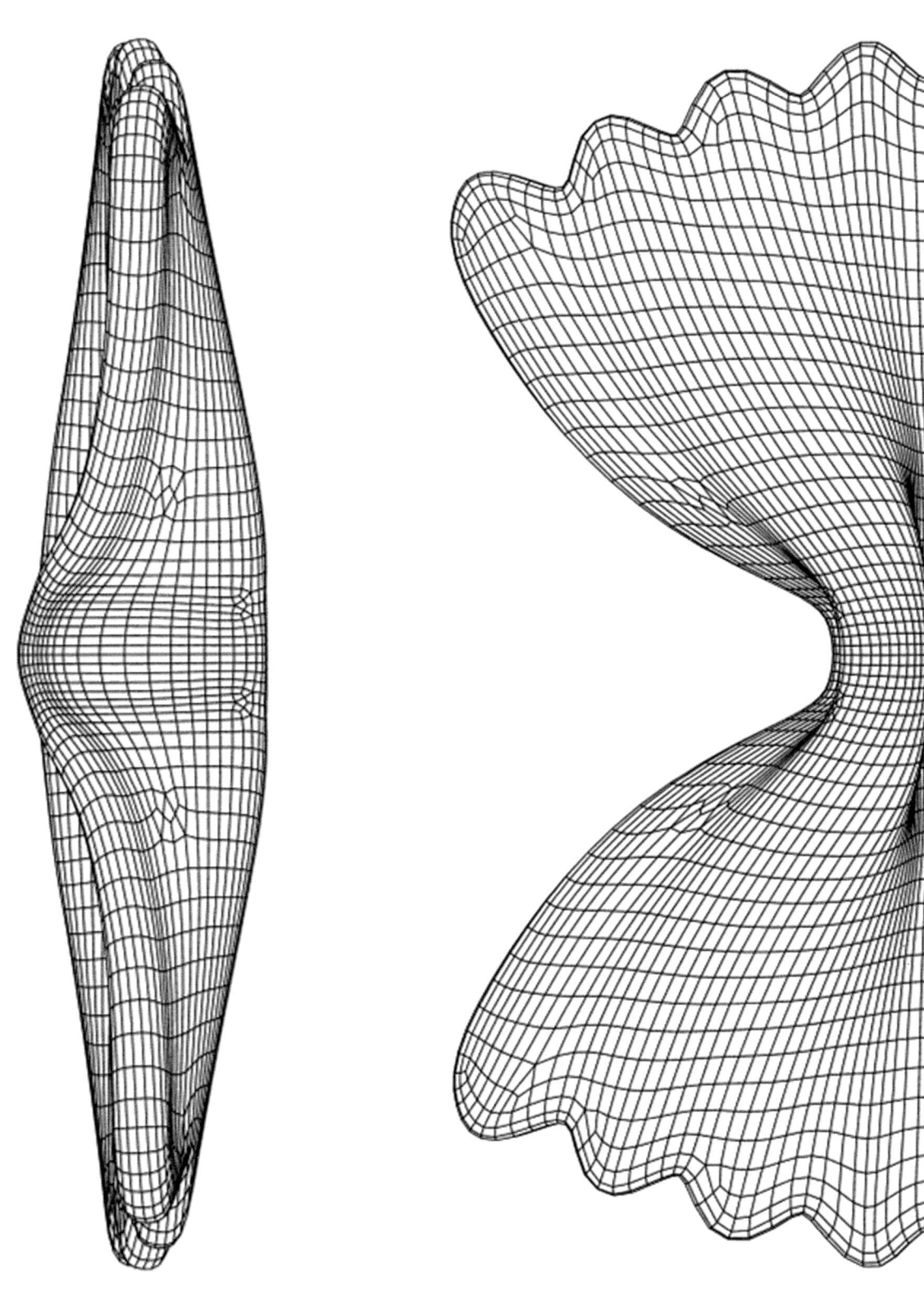

Sonja Stummerer & Martin Hablesreiter

fotos
Ulrike Köb, Ludwig Löckinger

food design XL

SpringerWienNewYork

Sonja Stummerer & Martin Hablesreiter
honey & bunny productions, Vienna, Austria
www.honeyandbunny.at

Gedruckt mit Unterstützung des Bundesministeriums für Unterricht, Kunst und Kultur und der Kulturabteilung der Stadt Wien, Wissenschafts- und Forschungsförderung.
Printed with the support of the Federal Ministry for Education, Arts and Culture, as well as of the Cultural Department of the City of Vienna, Promotion of Science and Research.

Das Werk ist urheberrechtlich geschützt. Die dadurch begründeten Rechte, insbesondere die der Übersetzung, des Nachdruckes, der Entnahme von Abbildungen, der Funksendung, der Wiedergabe auf photomechanischem oder ähnlichem Wege und der Speicherung in Datenverarbeitungsanlagen, bleiben, auch bei nur auszugsweiser Verwertung, vorbehalten.
This work is subject to copyright.
All rights are reserved, whether the whole or part of the material is concerned, specifically those of translation, reprinting, re-use of illustrations, broadcasting, reproduction by photocopying machines or similar means, and storage in data banks.

Die Wiedergabe von Gebrauchsnamen, Handelsnamen, Warenbezeichnungen usw. in diesem Buch berechtigt auch ohne besondere Kennzeichnung nicht zu der Annahme, dass solche Namen im Sinne der Warenzeichen- und Markenschutz-Gesetzgebung als frei zu betrachten wären und daher von jedermann benutzt werden dürfen.
Product Liability: The use of registered names, trademarks, etc. in this publication does not imply, even in the absence of specific statement, that such names are exempt from the relevant protective laws and regulations and therefore free for general use.

Alle nicht von Dritten beigestellten Abbildungen verstehen sich als künstlerisches Werk der Autoren. Die abgebildeten Objekte können von den Originalprodukten abweichen.
All illustrations not supplied by third parties are authors' artwork. Objects depicted in this book may differ from the original products.

© 2010 Springer-Verlag/Wien
Printed in Austria
SpringerWienNewYork is a part of
Springer Science + Business Media
springer.at

Graphische Gestaltung/Graphic Design: .... honey & bunny, Markus Hiden
Fotografien/ Photographs: .... Ulrike Köb, Ludwig Löckinger, Sonja Stummerer
Fotokonzept/ Photo Concept: .... honey & bunny
Fotoassistenz/ Photo assistant: .... Margarethe Götz
Lithographie/ Lithography: .... pixelstorm
Coverfoto/ Coverphoto: .... Ulrike Köb, Alessi

Lektorat deutsch/Proof reading German: Michael Walch, Vienna, Austria
Übersetzung D – E/Translation G – E:
Elisabeth Frank-Großebner, Vienna, Austria, Camilla Nielsen, Vienna, Austria
Druck/Printed by: Holzhausen Druck & Medien GmbH, 1140 Wien, Austria

Gedruckt auf säurefreiem, chlorfrei gebleichtem Papier – TCF
Printed on acid-free and chlorine-free bleached paper

SPIN: 12642373
Mit zahlreichen farbigen Abbildungen
With numerous figures in color

Bibliografische Information der Deutschen Nationalbibliothek
Die Deutsche Nationalbibliothek verzeichnet diese Publikation in der Deutschen Nationalbibliografie; detaillierte bibliografische Daten sind im Internet über http://dnb.d-nb.de abrufbar.

ISBN 978-3-211- 99230-2 SpringerWienNewYork

Inhalt | contents

Einleitung | introduction — 8
Vorwort | preface — 12

Die fünf Sinne - und wie sie die Gestaltung von Essen beeinflussen
the five senses - and their impact on the design of food — 18

Geruch & Geschmack | smell & taste — 24
Konsistenz | consistency — 36
Optik | optics — 76
Geräusch | sound — 100
Geschmacksforschung | taste research — 110

Form Follows Function - Die Funktion von Essen
form follows function - the function of food — 122

Konservierung | preservation — 126
Produktion | production — 144
Transport | transportation — 168
Verzehrssituation | where and how we eat — 188
Resteverwertung | recycling leftovers — 221
Abhängigkeit | dependency — 226
Oberflächendesign | surface design — 234
Teilen | sharing — 240
Functional Food | functional food — 258

Designprozess – wie entsteht unser Essen?
design process – how does our food come about? — 262

Du bist, was du isst – Der kulturelle Faktor von Food Design
you are what you eat – the cultural factor in food design — 274

Symbol | symbol — 278
Innen & Außen | inside & outside — 298
Abstraktion | abstraction — 306
Zeitgeist | zeitgeist — 314
Mythos | myth — 326

Zeittafel | timeline — 336

Danksagung | acknowledgment — 340
Fotoregister | photographic credits — 343
Literaturverzeichnis | bibliography — 346

Einleitung | In Europa kommen jährlich rund 10.000 neue Lebensmittel auf den Markt. Davon floppen 50 Prozent innerhalb von drei Monaten und nur eines von zwanzig hält sich länger als zwei Jahre auf dem Markt.

Nahrungsmittel sind Modeerscheinungen geworden und ihr Konsum hat oft mehr mit Lifestyle und Selbstverwirklichung zu tun als mit Ernährung. Essen ist in aller Munde, Schlagworte wie „Functional Food", „Molekulargastronomie" oder „Maillard-Reaktion" geistern durch die Medien. Dennoch leben wir in einer Gesellschaft, die über Herkunft, Inhalt und Bedeutung ihrer Nahrung so wenig Bescheid weiß wie keine zuvor. Dieses Buch möchte über einen wesentlichen, aber bisher wenig diskutierten Aspekt unseres Essens, jenen des Designs, informieren und gleichzeitig Interesse wecken, sich mehr mit dem, was wir täglich in uns hineinstopfen, auseinander zu setzen. Wir wollen die Frage stellen, welche Geschichten unsere Nahrung erzählt und was die alltäglichen Mahlzeiten über unsere Kultur aussagen.

Das Bedürfnis, Nahrung zu gestalten, ist so alt wie die Zivilisation. Von der Fülle an Nahrungsmitteln, mit denen uns die Natur versorgt, gelangen nur die wenigsten direkt, also roh und im Ganzen, auf den Teller. Die allermeisten werden in irgendeiner Weise verändert, ehe wir sie in den Mund stecken. Egal ob Früchte, Getreide oder Fleisch: Wir bearbeiten Grundprodukte, um sie genießbar und schmackhaft zu machen, sie zu konservieren oder zu transportieren. Mehr als eintausend Mal pro Jahr – vor jedem Essen – zerschneiden, kochen, verrühren und kombinieren, also gestalten wir das, was wir ernten oder schlachten, und betätigen uns somit als Food Designer.

Wenn wir essen, nehmen wir nicht einfach Kohlenhydrate, Eiweiße und Fette zu uns, sondern Produkte und Gerichte, die zubereitet und gestaltet wurden. Schon seit Jahrhunderten formen wir Teig nicht nur zu einfachen Brotlaiben, sondern verschränken ihn zu Bageln oder Brezeln und biegen ihn zu sichelförmigen Croissants. Wenn wir Käse zu zylindrischen Laiben formen und Schokolade in eckige Tafeln gießen, Fisch in geometrische Klötze schneiden und Maismehlbrei zu Erdnusslocken extrudieren, dann gestalten wir unser Essen – und das ganz bewusst.

Von Mode über Möbel bis zu Autos ist über andere Designsparten viel geschrieben worden, die Strukturen hinter Food Design sind aber überraschenderweise weitgehend unbekannt. Was aber ist der Ursprung des menschlichen Willens, Nahrung zu formen? Welche Beweggründe stecken hinter diesen Gestaltungsprozessen? Welche Wünsche, Sehnsüchte und Notwendigkeiten wollen wir dabei befriedigen?

Das vorliegende Buch recherchiert, woher die Formen unserer Nahrung kommen, welche Faktoren bei ihrer Gestaltung eine Rolle spielen und wer eigentlich darüber entscheidet, wie das, was wir täglich zu uns nehmen, beschaffen ist.

Sonja Stummerer & Martin Hablesreiter
Wien 2009

Introduction | In Europe about 10,000 new food products are launched every year. Out of these, 50 per cent fail within three months, and only one out of twenty will stay on the market for more than two years.

Food products have become subject to fads, and consumption is more a matter of lifestyle and self-fulfillment than nutrition. Food is literally on everybody's lips, with buzzwords such as "functional food", "molecular gastronomy" or "Maillard reaction" pervasive throughout the media. Nevertheless, we live in a society that knows less about the origin, content and meaning of food than any other before us. The goal of this book is to inform about a significant aspect of food which has hardly been discussed yet: the aspect of design. At the same time, it seeks to instill an interest in learning more about what we consume every day and in exploring the stories our food tells – about us and about our culture.

The wish to design food is as old as civilization. Out of the abundance of foods nature supplies us with, only very few land on our plates directly, as they are, raw and whole. Most of them will have changed one way or the other before we eat them. No matter whether it is fruits, grain or meat, we process the basic products to make them edible and tasty, to preserve them or transport them. More than a thousand times a year – before every meal – we cut, cook, mix and combine, i.e., we shape what we harvest or slaughter – so basically we are all food designers.

Whenever we eat, we do not just take in carbohydrates, proteins and fats but products and dishes that have been prepared and designed. For centuries we have given various shapes to bread, not just simple loaves but also folded bagels or soft pretzels or sickle-shaped croissants. When we pour cheese into cylindrical loaves and chocolate into rectangular bars, cut fish into geometrical slices and extrude cornmeal into peanut puffs, we design our food – and this in a deliberate fashion.

Much has been written about areas such as fashion, furniture and car design but surprisingly, the structures underlying food design have remained largely unknown. Why do humans wish to shape food? What are the motivations behind this design process? What are the desires, longings and necessities we seek to satisfy?

This book presents research about the origin of the shapes and forms our food comes in, the factors playing a role in its design and who it is that actually decides what the things we eat every day are like.

Sonja Stummerer & Martin Hablesreiter
Vienna 2009

FOOD DESIGN | Essen erhält uns am Leben; ohne zu essen, sterben wir. Um zu überleben, benötigen wir fortlaufend Proteine, Fette und Kohlenhydrate. Mittlerweile genügen eine Nadel, ein Schlauch und ein Infusionsbeutel, um einen Menschen mit diesen lebensnotwendigen Stoffen zu versorgen. Die künstliche Ernährung deckt den täglichen Kalorienbedarf, aber stillt sie auch den Hunger? Tatsächlich nutzen wir die Möglichkeit der künstlichen Ernährung nur in Notfällen: nach Unfällen, im Krankheitsfall, kurz vor dem Tod.

Essen bedeutet mehr als die bloße Zufuhr lebensnotwendiger Nährstoffe. Wir wollen keine Pillen oder Infusionsflaschen, sondern Nahrung, die wir sehen, angreifen, riechen, zerbeißen und schmecken können. Seit Anbeginn der Menschheit setzen wir ein unglaubliches Maß an Phantasie, Kreativität und Erfindungsgeist ein, um natürliche Grundprodukte nach unseren Vorstellungen zu verändern. Und so kommt Essen in einer scheinbar endlosen Fülle unterschiedlicher Rezepturen, Zubereitungsarten, Farben und Formen auf den Tisch.

Design? | Wir formen unsere Umwelt. Im Falle von Gebrauchsgegenständen spricht man von Design, ein Terminus, der zuerst in Großbritannien auftauchte. Seit der Gründung des Londoner „British Council of Industrial Design" im Jahre 1944 bezeichnet Design eine den Erfordernissen der Massenproduktion entsprechende Gestaltung von Gegenständen und Geräten aller Art.[1] Design erfüllt ästhetische, funktionale und kulturelle Anforderungen. Hotdogs, Cornflakes oder Schokoriegel sind bewusst gestaltete Massenartikel und damit genauso Industriedesignprodukte wie Autos, Kugelschreiber oder Sonnenbrillen.

Wir gestalten Essen, um den Genuss zu steigern und praktische Anforderungen wie Lagerfähigkeit zu erfüllen, aber auch um Werte zu transportieren und Mythen zu erzählen. Ähnlich wie in der Baukunst, wo Form, Nutzung und Stil zusammenwirken, bilden auch Gerichte und Speisen immer eine Einheit aus Genuss, Funktion und Kultur. Wenn wir Cola trinken oder Ketchup über unsere Pommes leeren, regt das nicht nur unseren Appetit, sondern auch unsere Phantasie an. Es stellt sich die Frage, wie Nahrung beschaffen sein muss, damit sie Erfolg hat.

Food Design ist eine alte Disziplin. Verfolgt man seine Geschichte, so zeichnen sich ganz klare Muster und Strategien ab, die zur Geschmacks- und Formenvielfalt unseres Essens geführt haben – und bis heute motivieren, das Spektrum der vorhandenen Nahrung ständig zu erweitern. Das vorliegende Buch zeigt, wie die Mechanismen hinter Food Design funktionieren und was dafür ausschlaggebend ist, dass ein Lebensmittel genau so und nicht anders auf den Tisch kommt.

FOOD DESIGN | Food keeps us alive: if we do not eat, we will die. We need a constant supply of proteins, fats and carbohydrates to survive. Nowadays, needles, tubes and infusion bags are enough to provide humans with these vital substances. Artificial nutrition may cover our daily calorie need, but can it still our hunger? The only time we actually take recourse to artificial nutrition is in emergencies: after accidents, in case of illness, or when death is imminent.

Eating means more than simply ingesting vital nutrients. We do not want pills or infusion bottles, we want food we can see, touch, smell, chew up and taste. Since the beginning of humankind, we have invested enormous amounts of imagination, creativity and invention in changing natural products according to our ideas. Thus, food is brought to our tables in a seemingly endless array of recipes, ways of cooking, colors and shapes.

Design? | We shape our environment. When we give form to objects for daily use, we call it design – a notion that was first used in the United Kingdom. Ever since the London-based "British Council of Industrial Design" was founded in 1944, the word "design" has been used to denote the shaping of objects and devices of all descriptions to cater to the needs of mass production.[1] Design meets esthetic, functional and cultural requirements. Hot dogs, cornflakes or chocolate bars are mass-produced articles that are designed with a specific purpose in mind, and thus, as much products of industrial design as cars, ball-point pens or sunglasses.

We design food for greater pleasure, and for practical purposes such as longer storage life, but also to convey values and tell myths. In architecture, form, use and style interact, and it is much in the same way that food always combines enjoyment, function and culture. When we drink Coke or put ketchup on our fries, it is not just to whet our appetite, but also to spur our imagination. The question is: what should food be like to be successful?

Food design is not a new discipline. Looking at its history, we can identify clear patterns and strategies which have led to diversity in the taste and forms of food – and have continued to motivate us as we keep expanding the range of available food up to the present day. This book will identify the mechanisms at work behind food design, and the reasons why something is put on the table in a certain shape and form, and no other.

Die Natur versorgt uns mit vielen unterschiedlichen Nahrungsmitteln: Früchte, Gemüse, Milch oder Fleisch stehen uns einfach so zur Verfügung. Doch das genügt uns nicht. Mehr als eintausend Mal pro Jahr, vor jedem Essen, zerschneiden, zerkochen, verrühren und kombinieren wir bewusst das essbare Angebot der Natur. Der menschliche Wille, Essbares zu gestalten, unterscheidet uns von allen anderen Lebewesen. Nahrungsmittel sind eben Genussmittel.

Nature supplies us with a rich diversity of foods. We all have ready access to an abundance of fruits, vegetables, milk and meat. But this does not seem to be enough. More than a thousand times a year, before each meal, we cut, cook, stir and combine the edible products of nature. The human will to design food is what distinguishes us from all other creatures. Foods are simply products to be enjoyed.

Food Design | Für uns bedeutet der Begriff „Food Design" die Entwicklung und Gestaltung von Lebensmitteln. Darunter verstehen wir die Summe aller Prozesse und Entscheidungen, die dazu dienen, Essen erfolgreich reproduzierbar und wiederholbar zu gestalten. Das betrifft längst nicht nur das optische Erscheinungsbild einer Speise oder eines Produktes, sondern auch die geschmackliche Gestaltung, die Konsistenz, die Textur, die Oberfläche, das Kaugeräusch, den Geruch und vieles mehr.

Da wir uns dem Thema Food Design als Architekten und Designer nähern, versteht sich das vorliegende Buch als Designbuch und nicht als Fachliteratur für Lebensmitteltechnologen. Themen wie Gentechnik, Novel oder Functional Food finden daher nur insofern Erwähnung, als sie uns gestalterisch relevant erscheinen. Ziel des Buches ist es, teils bekannte Tatsachen aus den unterschiedlichsten Wissensbereichen entsprechend unserem Zugang zum Thema Design zusammenzuführen, zu ordnen und zu vernetzen. Wir begeben uns auf die Suche nach dem Ursprung von Food Design und versuchen die Frage zu beantworten, warum unser Essen genau so gestaltet ist und nicht anders.

Kunsthandwerk und Design bilden einen integralen Bestandteil jeder Kultur. Die Gestaltung von Nahrungsmitteln beziehungsweise Food Design ist ein Teilbereich des herkömmlichen Industriedesigns und somit ein wesentlicher Faktor unserer Zivilisation. So gesehen, bedeutet Food Design wesentlich mehr als lediglich die synthetische Erzeugung oder die gentechnische Veränderung von Nahrung. Natürlich spielen der Gebrauch künstlicher Aromen, die chemische Synthese zuvor isolierter Grundsubstanzen und der Einsatz neuartiger Zubereitungsverfahren bei der Gestaltung mancher Nahrungsmittel eine Rolle, den Begriff „Food Design" aber allein auf diese Prozesse zu beschränken, empfinden wir als falsch. Food Design ist jegliche Gestaltung von Nahrung nach den Regeln der Reproduzierbarkeit und erfüllt – ebenso wie herkömmliches Design – sinnliche, funktionale und kulturelle Anforderungen.

Food Design | For us the notion 'food design' refers to the development and shaping of food. In our understanding this includes all the processes and decisions related to successfully designing food in a reproducible and recurring way. This now no longer applies just to the appearance of a dish or a product but also to the design of taste, consistency, texture, surface, the sound of chewing, smell and much more.

Since we are approaching the subject of food design as architects and designers, the present book sees itself as a design book and not as specialized literature for food engineers. Themes such as genetic engineering, novel or functional food are thus only addressed whenever we think they are relevant to design. The goal of the book is to bring together facts (in part familiar ones) from a wide diversity of areas of knowledge and to order and network them in keeping with our approach to the subject of design. We have embarked upon a search for the origins of food design and are trying to answer the question why our food is designed precisely the way it is and in no other way.

Artisanship and design constitute an integral part of every culture. The design of food – food design, for short – is a subfield of conventional industrial design and thus a crucial factor of our civilization. Seen in this light, food design means significantly more than just the synthetic manufacturing or the alteration of food on the basis of genetic engineering. To be sure, the use of artificial aromas, the chemical synthesis of hitherto isolated basic substances and the utilization of novel modes of preparation play a role in the design of some foods, but to limit the notion of food design to these processes alone, would, in our view, be misguided. Food design refers to all design of food based on the rules of reproducibility and as such fulfills sensual, functional and cultural demands – much like conventional design.

Styling und Dekoration | Food Design wird oft mit Food Styling oder Dekoration verwechselt. Das Anrichten von Speisen am Teller oder auf Buffets ist ein einmaliger Vorgang, der meist individuellen, spontanen und eher beliebigen Entscheidungen folgt. Design steht im Gegensatz zu bloßem Stylen immer in Zusammenhang mit Reproduzierbarkeit und Massenfertigung. Wobei Massenfertigung nicht unbedingt mit industrieller Fertigung gleichzusetzen ist: Pasta, verschiedenste Brotformen oder Süßspeisen wie der Gugelhupf wurden über Jahrhunderte hinweg händisch, aber nach dem immer gleichen Muster millionenfach von den unterschiedlichsten Personen zubereitet – und sind unserer Definition nach daher ganz klar Designobjekte.

Warum gestalten wir unser Essen? | Das Buch versucht der Motivation für die Gestaltung von Nahrung auf den Grund zu gehen und gliedert sich, den drei Hauptmotiven folgend, in die übergeordneten Kapitel: Sinnlicher Genuss, Funktion und Kultur. Der Frage, wer darüber entscheidet, wie unser Essen gestaltet ist, gehen wir im Kapitel Designprozess nach, in dem wir neben historischen Beispielen die unterschiedlichsten Food Designer und ihre Arbeitsweise vorstellen.

Lustgewinn | Essen ist eine sinnliche Erfahrung. Es bereitet uns Lust, Essbares anzusehen, anzugreifen, seinen Duft einzuatmen, es auf der Zunge zu fühlen und zu zerkauen. Schon ein einfaches Stück Brot bietet eine Vielzahl an sinnlichen Empfindungen. Wir lieben das Knacken beim Zerbeißen der knusprigen Rinde. Wir genießen das Kratzen der rauen Oberfläche am Gaumen und das flaumig weiche Volumen von porigem Teig. Erregungen, wie jene, wenn zartschmelzende Schokolade auf der Zunge zergeht und sich die Süße langsam in der Mundhöhle verteilt, sind durchaus mit sexuellen Empfindungen vergleichbar.

Neben der Sexualität ist der Vorgang des Essens jener mit dem größten körperlichen Lustgewinn. An keiner anderen Stelle des Körpers ist der Mensch so sensibel wie am oberen und am unteren Ende des Verdauungstraktes und an den Sexualorganen.[2] Seit Anbeginn der Menschheitsgeschichte investieren wir Kreativität und Zeit, um das Erlebnis Essen zu intensivieren. Hauchdünne Schokoblättchen, vielschichtige, flüssig gefüllte, knackig umhüllte oder in Krokant gewälzte Pralinen, luftig-leichte Sorbets, flaumige Cremen oder aus dünnen Teigblättern gerollte oder geschichtete Backwaren zeigen, dass die Steigerung des Lustgewinns eine der wesentlichsten Triebfedern von Food Design ist. Wir trachten danach, unsere Nahrung so zuzubereiten, darzureichen und zu gestalten, dass ihr Verzehr möglichst alle Sinne anregt und uns größtmöglichen Genuss bereitet.

Funktionale Aspekte | Essen muss aber nicht nur schmecken, es muss auch funktionieren. Vermutlich ist der menschliche Drang, Nahrung zu formen, genauso alt wie jener, Kleidung und Wohnraum zu gestalten und Kunst zu schaffen. Ähnlich wie Mode und Architektur erfüllt auch Food Design nicht nur sinnliche Anforderungen, sondern folgt auch ganz pragmatischen Überlegungen: Kleidung soll wärmen, Architektur vor Wind und Wetter schützen und Essen muss nicht nur verzehrtauglich, lager- und transportfähig sein, sondern auch einer ganze Reihe weiterer Ansprüche genügen.

Styling and Decoration | Food design is often confused with food styling or decoration. Arranging food on plates or buffet tables is a one-time process mostly based on individual, spontaneous and rather random decisions. Contrary to mere styling, design is always linked with reproducibility and mass production. In this context, mass production does not necessarily have to be industrial production. Pasta, various types of bread or pastries such as the "gugelhupf" have been made by hand by a variety of people over centuries, millions of times, but always according to the same pattern – they are thus clearly design objects according to our definition of the term.

Why Do We Design Our Food? | This book will try to identify the motivation for designing food. Its structure follows the three main motivations: sensory and sensual pleasure, function and culture. The chapter "Design Process" will look at the issue of who decides what our food is like; we will present historical examples as well as contemporary food designers and their working methods.

Pleasure Gain | Eating is a sensory and sensual experience. We enjoy looking at food, touching it, smelling its fragrance, feeling it on our tongues and chewing it. Even a simple slice of bread offers a multiplicity of sensations. We love it when the crispy crust cracks as we bite it. We relish the scratchy feel of the rough surface on the palate and the fluffy, soft mass of porous dough. The excitement caused by chocolate softly melting on the tongue, with the sweet taste slowly filling the mouth, is quite comparable with sexual arousement.

Apart from sexuality, no other process gives us more pleasure gain than eating. After all, the most sensitive parts of the human body are the upper and lower ends of the digestive tract and the sexual organs.[2] Since the beginning of civilization, we have invested creativity and time into intensifying the eating experience. Ultra-thin chocolate, multi-layered confectionery with a liquid center and a coat of crunchy chocolate or brittle, airy sherbet, light cream and pastry rolled or layered out of thin sheets of dough show that heightened pleasure gain is one of the main driving forces in food design. We want to prepare, serve and shape our food in such a way that consumption is a feast for all senses, so we can relish it as much as possible.

Functional Aspects | Food is, of course, not only expected to taste good, it also has to function properly. The human drive to shape food is presumably as old as the wish to design clothes and housing, and to create art. In much the same way as fashion and architecture, food design not only fulfills needs of the senses, it is also supposed to conform to very practical considerations: garments should keep us warm, architecture should protect us from the elements, and food is not only there to be edible, storable and transportable, it also has to meet an array of other requirements.

14 Vorwort | preface

Esswaren als Bewohner einer bunten Comicwelt: Gummibären, Salzfische, Hello-Kitty-Cracker, Pasta in Form von Tannenbäumen, Sternen und Blumen, Knabberbären und Minihamburger aus Fruchtgummi. Allein in Europa kommen jährlich rund 10 000 neue Nahrungsmittel auf den Markt.

Food inhabiting a colorful comic world: gummi bears, salt fish, Hello Kitty crackers, pasta in the form of fir trees, stars and flowers, cracker bears and mini-hamburgers made of fruit gum. In Europe alone 10,000 new food products are launched every year.

Sometimes, we prefer to wear comfortable shoes or decide to buy a family car, and for similar reasons, we opt for practical fish fingers or deep-frozen vegetables with a long life. The range of desired functions runs the gamut from efficient cultivation, harvesting and processing to ease of packaging, transport, storage and consumption, from bite-size dimensions, simple preparation, amenability to portioning and suitability for takeaways to wholesomeness, conduciveness to health and curative effect. The functional specifications have an impact on the properties of the final product and, along with pleasure gain, they represent the second central basis of food design.

Food as a Cultic Object | In a round wafer, there are hardly any calories; it is not especially beautiful, has no taste and could be manufactured much more efficiently if it were square. Nevertheless, round wafers are produced and consumed regularly as hosts.

Our eating habits mirror our cultural standards. Through what we eat, we define ourselves and dissociate ourselves from other civilizations. Acquired eating behavior means coherence – within the family, in our social environment, in our own culture.

Symbolically charged offering bread is one of the oldest examples of food design. Some of the ancient motives have survived until today, for example the plaited pastry or the croissant. Even if the contents and meanings of these forms have meanwhile been forgotten, they originally played a decisive role in the design of these foods.[3]

Successful food not only tastes good, it also has a story to tell. Food design turns simple ingredients into national dishes or tokens of love, sexual innuendos, offerings and religious feasts. The conveyance of symbolic contents and higher cultural values is the third fundamental driving force of food design. We eat things because of their high nutritional value, because we like their taste, because they are readily available and easy to consume. And we also eat things because they have cultural value, because they reinforce our identity and feel for life.

Sensory and Sensual Pleasure + Function + Culture = Food Design | For thousands of years, humankind has gone to great lengths to design food. We have invested creativity and inventive talent, we have mused about and experimented with products found in nature so as to adapt them to our needs. Food design is underpinned by numerous ideas, thoughts, causalities, coincidences and requirements. Being architects and designers, we claim that we have identified three main objectives: to heighten sensory and sensual pleasure in the consumption of food, to fulfil functional aspects of all descriptions, and to convey cultural values.

Food design is most likely as old a discipline as architecture or art, and it is at least as sophisticated. Food design is not only a part of industrial design and thus part of the applied arts, it is also a human need and an expression of civilization. Examples of good food design should be added to the design collections of international museums. When the Museum of Modern Art in New York shows design classics from everyday life such as Bic ballpoint pens, VW Beetles or Alessi jugs, our question is: What about fish fingers, Kinder Surprise Eggs, Pringles…?

1 | © Bibliographisches Institut & F. A. Brockhaus AG, 2003
2 | We thank Prof. Gisla Gniech for this information.
3 | see: Hansferdinand Döbler: Kochkünste und Tafelfreuden; Orbis, 2002

die 5 Sinne
the 5 senses

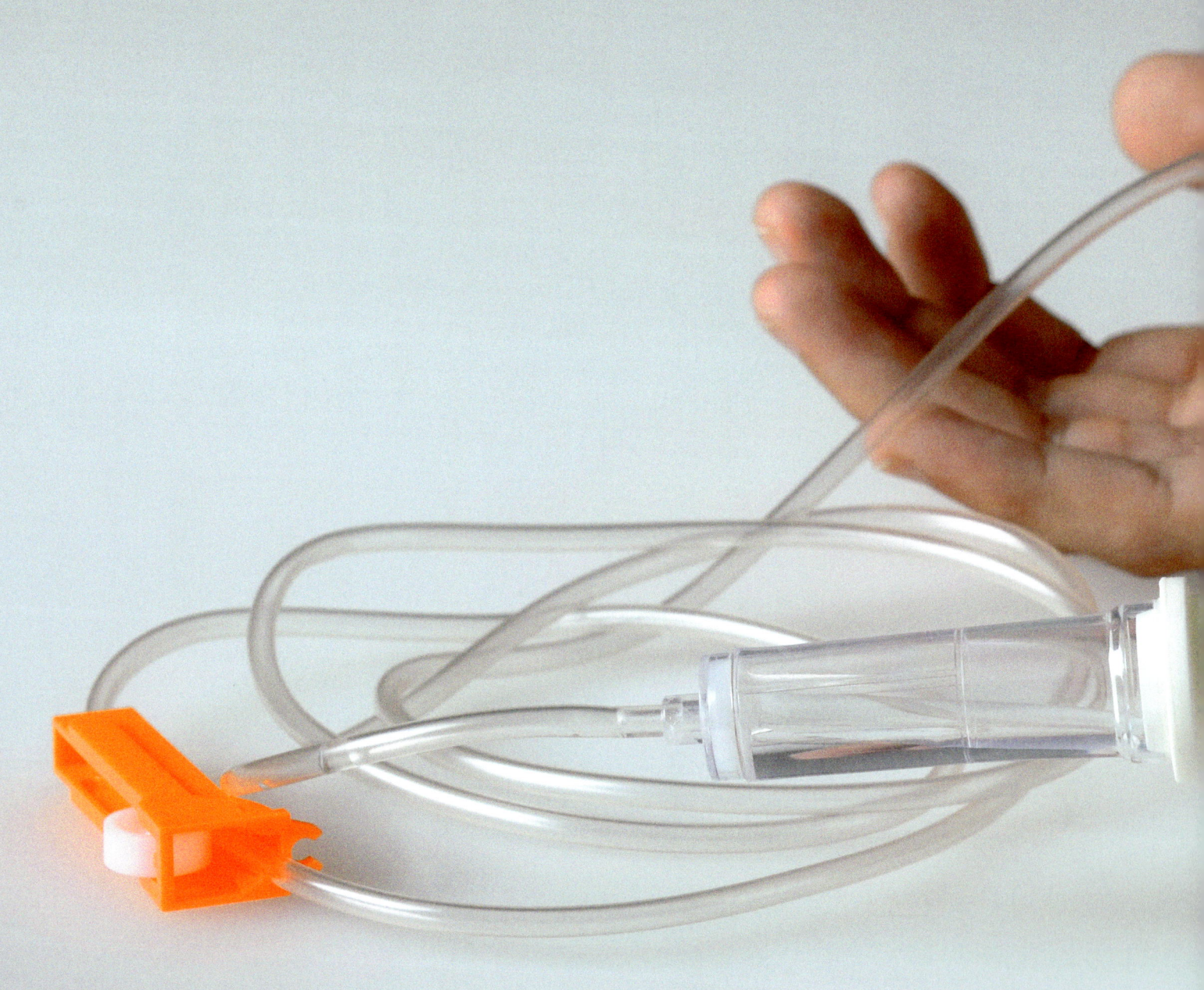

The Five Senses and Their Impact on the Design of Food

Here's the shape of things to come from the perspective of the 1960's: in the world of the future, food pills have come to replace the (time-consuming) act of eating. Science-fiction movies and TV series of those days had their protagonists live on colorful but otherwise nondescript lozenges and sugar-coated pills which they took from featureless contraptions and dispensers. Even today, rumor has it that astronauts in outer space live on nothing but pills. In fact, human beings are technically able to cover their daily calorie intake without eating. Parenteral nutrition, the administration of hypertonic solutions by means of infusions, can supply us with everything we need to survive. However, the dream of efficient calorie intake is doomed to fail because we want to eat – even in outer space. Good food is psychologically essential in everyday life - for everyone of us, including astronauts, and we do not want to do without.

Eating is enjoyment and pleasure. We do not just expect food to fill our tummies, it also has to indulge our senses; we want to relish it. It is no coincidence that words such as "relish" or "enjoyment" have been linked with eating and drinking for ages. The definitions of "relish" in the Merriam-Webster Dictionary include a reference to positive physical sensations and mental well-being: "enjoyment of or delight in something that satisfies one's tastes, inclinations or desires". Under the entry "genieszen" ("to enjoy"), the Brothers Grimm's German Dictionary quotes a definition by Immanuel Kant: "enjoyment [for] that is the word used to denote intensity of gratification".[1]

People spend much time and effort on inventing new foods and drinks, and on improving existing recipes in such a way that we will relish them even more. If eating were only about calorie intake, we could simply use infusion bags and pills. However, our food is tempting just because it comes in an abundance of different flavors, recipes and ways of cooking, with an enormous diversity of shapes, colors, surfaces and textures. After all, food does not only serve survival, but also provokes lust and temptation.

Enhancing the sensory pleasure of enjoying food is one of the main driving forces behind food design. To be exact, the sensation commonly known as "taste" is an interaction of all five senses, which the natural philosophers of Antiquity broke down into vision, hearing, smell, taste and touch.[2] When we say that something "tastes good", we also mean this to include that it looks good, smells good, feels good in the mouth and sounds good when we chew it. The following chapter is devoted to the factors which contribute to the sensory experience of "eating" and how they impact the design of our food.

last page | Food without design: In purely technical terms, infusion bags serve the same purpose as elaborately designed meals. They all provide nutrients necessary for survival.
this page | Food is a product to be enjoyed. One main motivation for designing food is to enhance the sensual pleasure of eating.

1 | See: Merriam-Webster Dictionary, Immanuel Kant "Critique of Judgment", trans. James Creed Meredith, 1911.
2 | See: Karl-Heinz Plattig, Spürnasen und Feinschmecker; Springer Verlag 1995

Geruch & Geschmack
smell & taste

Jede Geschmacksrichtung reizt die Geschmacksknospen auf unterschiedliche Art. Salz beispielsweise erzeugt eine leichte Spannung. Verantwortlich für die Wahrnehmung von Saurem ist ein Molekül namens PKD2L1, das Ionen in Zellen hinein- und hinaustransportiert. Die Bewertung von Geschmack – also die Summe der Geruchs- und Geschmackswahrnehmungen wasser- bzw. speichellöslicher chemischer Stoffe in der Mundhöhle sowie die Textureindrücke des Gegessenen – ist erlernt. Dieser Prozess beginnt bereits im Mutterleib und ist besonders in der Kindheit prägend. Süß ist positiv, bitter ist negativ, salzig und sauer sind – je nach Intensität – ambivalent. Die geschmacklichen Vorlieben verschieben sich aber auch im Laufe des Lebens: Kinder und Alte lieben süß intensiver, Kinder in der Wachstumsphase präferieren höhere Salzkonzentrationen. Erwachsene legen einen Teil ihrer Bitter-Aversion ab und schätzen bittere Getränke wie Bier oder Kaffee. Alle heutigen Nahrungsmittelprodukte sind geschmacklich gezielt komponiert. Die erfolgreichste Geschmackskombination ist jene von süß und fett, das derzeit als sechste Grundgeschmacksrichtung diskutiert wird. Saure Naschwaren wie saure Drops oder Schlangen beispielsweise werden speziell für Kinder erzeugt, die acidophil sind, also eine starke Vorliebe für sauer haben – das ist rund ein Drittel aller Kinder.[1]

Each taste stimulates the taste buds in a different way. Salt, for instance, generates a slight tension. The perception of a sour taste is the result of a molecule known as PKD2L1 which transports ions into and out of the cells. The assessment of taste – that is, the sum total of smell and taste perceptions of water- or saliva-soluble chemical substances in the mouth cavity as well as the texture impressions of something we have eaten – is acquired by learning. This process already begins in the womb and is particularly decisive in childhood. Sweet is positive, bitter negative, salty and sour are – depending on intensity – ambivalent. The preferences in taste shift in the course of a lifetime: Children and older people like a more intense degree of sweetness, while children in a growth phase prefer higher concentrations of salt. Adults give up part of the aversion to bitter tastes and appreciate bitter drinks such as beer or coffee. The most successful combination of taste is that of sweet and fat, which is presently being discussed as a sixth basic taste. Sour snacks such as sour drops or jelly snakes are especially created for children who are acidophilic, that is, have a strong preference for a sour taste – which applies to about a third of all children.[1]

GESCHMACK + GERUCH – ein perfektes Team

Auch wenn es uns Lust bereitet, unser Essen anzusehen, der volle Genuss einer Speise entfaltet sich erst im Mund. Entgegen der landläufigen Meinung findet die Geschmackswahrnehmung nur zu einem kleinen Teil tatsächlich auf der Zunge statt. Die Geschmackszellen sind über die gesamte Mundhöhle verteilt, sitzen also auch am Gaumen, im Kehlkopf und in der Speiseröhre. Diese „Schmeckzellen" sind relativ einfach aufgebaut und in Gruppen zu 15 bis 40 als so genannte Geschmacksknospen angeordnet. Neugeborene kommen mit etwa 10.000 Geschmackszellen zur Welt, danach verringert sich ihre Anzahl kontinuierlich bis zu 2.000 im Greisenalter.

Der Geschmackssinn kann nur vier Primärgeschmacksrichtungen erkennen: süß, sauer, salzig und bitter. In Ostasien kennt man mit „umami" noch eine fünfte Geschmacksqualität. Das aus dem Japanischen stammende Wort bedeutet so viel wie „köstlich schmeckend" und bezeichnete ursprünglich die Eigenschaft von Seetang, den Geschmack von Fleisch und Fleischprodukten zu verstärken. 1909 wurde erstmals eine Substanz namens „Glutamat" isoliert, die der Geschmacksrichtung „umami" entspricht und seitdem als Geschmacksverstärker Verwendung findet.

Evolutionär hat sich der Geschmackssinn entwickelt, um dem Menschen das Überleben zu sichern. „Süß" signalisiert einen hohen Kohlenhydratgehalt und damit einen hohen Nährwert, „umami" lässt auf gute Eiweißlieferanten schließen. Salziges ist wichtig für den Ionenhaushalt, Saures fördert die Verdauung. „Bitter" beinhaltet eine Warnfunktion, um uns vor dem Konsum ungenießbarer oder giftiger Substanzen zu schützen.[2] In der Praxis kommen die vier bzw. fünf Primärgeschmacksrichtungen kaum in reiner Form vor – so schmeckt zum Beispiel nur Kochsalz wirklich salzig –, die meisten Nahrungsmittel nimmt man als Kombination mehrerer Grundgeschmäcker wahr.[3]

Wie die Geschmacksarten zusammengestellt werden, um eine wohlschmeckende Speise beziehungsweise ein erfolgreiches Produkt zu bekommen, hängt mit der jeweiligen kulinarischen Tradition zusammen und ist von Kultur zu Kultur äußerst unterschiedlich. Die populärste Kombination ist jedenfalls jene von süß und fett, das derzeit als mögliche sechste Grundgeschmacksart diskutiert wird. Manche Aromen verstärken einander, andere schwächen einander ab: Eine Prise Salz im Kuchenteig intensiviert die Süßwahrnehmung, ein Stück Zucker im Kaffee dagegen reduziert dessen Bitterkeit. Auch trigeminale Empfindungen wie Chili, Pfeffer, Kohlensäure, Alkohol oder Kren, senken die Wahrnehmungsintensität aller anderen Geschmäcker.[4]

TASTE + SMELL – A Perfect Team

Even though we may enjoy looking at our food, we will only be able to fully relish it when its taste unfolds in our mouths. Contrary to popular belief, the tongue is not the entire taste organ but only a small part of it. Taste cells are spread all over the oral cavity and beyond – on the palate, in the larynx and in the esophagus. They are relatively simple in structure, arranged in groups of 15 to 40 cells to form the so-called taste buds. Newborn babies have about 10,000 taste cells; their number dwindles continuously as we grow older, leaving us with only 2,000 in old age.

Our senses can only identify four primary tastes: sweet, sour, salty and bitter. In East Asia, there is a fifth taste called "umami" – the word is of Japanese origin. It means "tasty" and initially referred to the way in which seaweed enhances the flavor of meat and meat products. In 1909 a substance known as monosodium glutamate was isolated for the first time; its taste corresponds to what "umami" is like, and it has been used as a food additive and flavor enhancer ever since.

Evolution has developed our sense of taste in such a way that it ensures our survival. "Sweet" indicates a high carbohydrate content, which means high nutritional value. "Umami" reflects high protein levels. Salty food is important for our electrolyte balance and sour stuff is good for digestion. "Bitter" has a warning function which protects us from taking in unfit and poisonous substances.[2] In practice, the four or five primary tastes hardly ever occur in their pure form – for example, only cooking salt has that really salty taste – in most foods, what we get is a combination of several primary tastes.[3]

The way in which the primary tastes are combined to produce a delicious dish or a successful product will depend on the culinary tradition of an area and thus vary greatly from culture to culture. The most popular combination of all is sweet and fatty – fatty being discussed as a potential sixth primary taste at present. Some flavors enhance each other, others attenuate each other: a pinch of salt in cake dough intensifies the sweetness whereas sugar in our coffee will reduce its bitter taste. Even foods which trigger trigeminal sensations – for example chili, pepper, carbonic acid, alcohol or horseradish – reduce the intensity of other tastes.[4]

Unsere Zunge kann nur 5 verschiedene Geschmäcker voneinander unterscheiden: süß, sauer, salzig, bitter und umami. Differenzierungen innerhalb dieser Geschmacksrichtungen nehmen wir mit dem Geruchssinn wahr.

Our tongue can only distinguish 5 different tastes – sweet, sour, salty, bitter and umami. We can perceive differences within these tastes by means of our sense of smell.

Zunge und Nase als perfektes Team | Beim Essen treffen das gustatorische und das olfaktorische System, also der Geschmacks- und der Geruchssinn aufeinander. Sie haben die Aufgabe, Riech- und Schmeckreize in Impulse umzuwandeln, die dann an das Gehirn weitergeleitet werden. Das wichtigste Organ für die Geschmackswahrnehmung ist nicht der Mund, sondern die Nase. Als so genannter Fernsinn liefert sie zunächst einen wesentlichen Beitrag zur Erstanalyse des Servierten. Wir riechen, wie frisch etwas ist, welche Zutaten es enthält, wir entwickeln eine Idee, wie es schmecken wird, und können beurteilen, ob etwas reif, noch genießbar oder schon verdorben ist.

Riechzellen sind etwa tausendmal empfindlicher als Geschmackszellen und dementsprechend bedeutsamer für den Genuss von Essen. Tatsächlich wird der „Geschmack" eines Lebensmittels wesentlich stärker über das „choanale Riechen" innerhalb der Mundhöhle wahrgenommen als über die Geschmackszellen selbst. Dabei gelangen Nahrungsmoleküle vom Schlund aus durch die hinteren Nasenöffnungen zu den Riechsensoren in der Nase.[5]

Die Nase reagiert mithilfe spezieller Rezeptorproteine auf chemische Wechselwirkungen von Riech- und Schmeckstoffen und kann auf diese Weise rund 10.000 verschiedene Gerüche identifizieren. Da die Geschmackszellen nur zwischen süß, salzig, sauer und bitter unterscheiden, brauchen wir die Nase für die Differenzierungen aller Nuancen innerhalb dieser vier Hauptgeschmacksrichtungen.

Erdbeere oder Kirsche? Zitrone oder saurer Apfel? Mit zugehaltener Nase lässt sich das nicht feststellen. Erst wenn die entsprechenden Geruchspartikel die Rezeptoren unserer Nase erreichen, nehmen wir die vielschichtigen Aromen des Essens wahr. Mit verstopfter Nase schmecken wir daher (fast) nichts, ein Phänomen, unter dem jeder von uns während einer Erkältung leidet.

Der volle Geschmack entfaltet sich meist auch nicht sofort, nachdem wir etwas in den Mund gesteckt haben, sondern erst wenn wir das Essen im Mund zwischen Zunge, Gaumen und Wangen hin und her bewegen. Vorgänge wie das Kauen, Lutschen, Befühlen und Zerdrücken der Nahrung steigern die Sinneswahrnehmung, indem sie sie erwärmen und zu stärkerer Abdampfung flüchtiger Moleküle führen, die wir dann riechen.[6] Der zeitliche Ablauf, wann sich welche Aromen im Mund entfalten, wird bei manchen Gerichten sogar als besonderes Gestaltungselement eingesetzt. Als besonders spannend empfinden wir Esswaren, die Aromen in ihrem Inneren verstecken, wie zum Beispiel gefüllte oder gerollte Lebensmittel. Der Alkohol oder die Fruchtcreme in einer hermetisch geschlossenen Praline entfaltet Geschmack und Geruch erst im Mund und liefert dadurch ein unmittelbareres und intensiveres Geschmackserlebnis, als wenn wir diese Aromen bereits vor dem Zubeißen wahrnehmen könnten.

Tongue and Nose – A Perfect Team | When we eat, the gustatorial and olfactory systems, our senses of taste and smell, get together. Their job is to turn olfactory and gustatorial stimuli into impulses which are then transported to our brains. In fact, the nose, not the mouth, is the most important organ of taste. Smell is a so-called distance sense contributing significantly to the initial analysis of what we are served. We smell if something put before us is fresh or not, which ingredients it contains, we get an idea of what it could taste like and we can judge if something is ripe or still edible, or if it has gone bad.

Olfactory nerve cells, or smell cells, are about a thousand times more sensitive than taste cells and thus more important for relishing food. As a matter of fact, food is not so much "tasted" via the taste buds but sensed via "choanal smell" within the oral cavity when molecules of food reach the olfactory sensors in the nose through the posterior nasal apertures from the throat.[5]

With the help of special receptor proteins, the nose reacts to chemical interactions of smell and taste substances, and is thus able to distinguish roughly 10,000 different smells. As the taste cells can only differentiate between sweet, salty, sour and bitter, we need the nose to draw distinctions between all nuances within these four primary tastes.

Strawberry or cherry? Lemon or sour apple? We are unable to tell when we hold our noses. We can only sense the diversity of aromas in food when the respective smell particles reach the receptors in our noses. With a clogged-up nose we will taste nothing or hardly anything – a phenomenon everyone who has ever had a cold will be familiar with.

In most cases, the full taste of food will not unfold immediately after we have put it into our mouth but only as we move it around between tongue, palate and cheeks. Processes like chewing, sucking, feeling and crushing of food enhance sensation because they warm the foods and lead to increased evaporation of volatile molecules which we then smell.[6] The chronology according to which certain aromas unfold in the mouth is even used as a special element in the composition of some dishes. We consider food hiding aromas inside to be particularly exciting, for example stuffed or rolled food products. Alcohol or fruit cream in a hermetically sealed piece of chocolate will only unfold its taste and scent inside the mouth; it will give us a more direct and intensive taste experience than would be the case if we had been able to sense the aromas before biting into it.

Unsere Nase kann rund 10.000 verschiedene Gerüche identifizieren. Selbst Lebensmittel wie Kartoffelchips, Butter oder rohes Fleisch haben einen ganz spezifischen Geruch, auch wenn wir diesen nicht immer bewusst wahrnehmen.

Our nose can identify about 10,000 different smells. Even foods such as potato chips, butter or raw meat have a very specific smell, even if we do not consciously perceive it.

Wie beeinflussen Geruch und Geschmack die Gestaltung von Essen? | Dass Fooddesigner mit der Entwicklung von Rezepturen und Geschmäckern arbeiten, liegt auf der Hand, dass auch der Geruch bei der Gestaltung von Nahrungsmitteln eine wichtige Rolle spielt, ist vielleicht weniger offensichtlich.

Gerüche, die beim Zubereiten und Servieren von Speisen und Getränken entstehen, lassen uns sprichwörtlich das Wasser im Mund zusammenlaufen. Den Duft eines frischen Bratens, eines noch warmen Brotes oder einer Tasse Kaffee einzuatmen, erzeugt ein großes Lustgefühl. Dabei kann der gezielte Einsatz bestimmter Zutaten, vor allem von Gewürzen, den olfaktorischen Genuss noch zusätzlich steigern. Safran, Zimt und Vanille verströmen durch die ätherischen Öle, die sie enthalten, intensive, betörende Gerüche. Schon die Aztekenkönige verfeinerten ihre Trinkschokolade deswegen mit Vanille und im alten Rom bestreute man das Hochzeitsbett mit Safranfäden.

Auch Rosenöl ist eine uralte, vermutlich persische Zutat, mit der Nahrung aromatisiert wird. Der Grieche Theophrast schreibt bereits um 320 vor Christus über die Gewinnung von Rosenöl, das dem Wein zugesetzt wurde, und durch Plinius den Älteren ist überliefert, dass die Römer ihre Nahrungsmittel mit Rosenöl beduftetem. Auch die Besonderheit im Winter gebackener Kekse und Gewürzbrote liegt nicht nur in ihrem Geschmack, sondern auch im speziellen, „typisch weihnachtlichen" Geruch, den sie verströmen.

How Do Smell and Taste Impact the Design of Food? | It comes as no surprise that food designers work in the development of recipes and tastes, but it may be less evident that the scent of food also plays an important role in the design of food.

Scents emerging when food and drinks are prepared and served literally make our mouths water. Inhaling the fragrance of fresh Sunday roast, hot bread or a cup of coffee will generate enormous pleasure. The targeted use of certain ingredients, especially spices, can add to olfactory enjoyment. Due to the ethereal oils they contain, saffron, cinnamon and vanilla exude enticing scents. This is why the Aztec kings used vanilla to refine the taste of their hot chocolate and wedding beds in ancient Rome were strewn with saffron threads.

Rose oil is another age-old ingredient, presumably from Persia, used to add flavor to food. As early as in 320 BCE the Greek philosopher Theophrastus wrote about the production of rose oil which was added to wine, and Pliny the Elder also reported that the ancient Romans scented their foods with rose oil. Something similar applies to cookies and gingerbread baked in the winter; they are not only special for their taste but also their "Christmasy" scent.

Gerüche bilden einen fixen Bestandteil jedes Geschmackserlebnisses, können es verstärken oder in eine bestimmte Richtung lenken. Im Restaurant „The Fat Duck" nahe London stimmt zum Beispiel der Geruch eines Stücks schottischen Moorbodens, der jedem Gast vor einem schottischen Moorhuhn serviert wird, auf den Genuss desselben ein.

In Japan werden im Frühjahr Kirschblüten eingesalzen und dann das restliche Jahr über als Tee getrunken. Mit heißem Wasser aufgegossen, riechen die trockenen Blüten auch noch im November intensiv nach Kirschen. Dass der Tee selbst dafür sehr salzig und eigentlich nach nichts schmeckt, wird großzügig in Kauf genommen. Der spezielle Geruch konserviert für Japaner das Gefühl einer bestimmten Jahreszeit und ist – da er nur salzig schmeckt und keinen Nährwert hat – der einzige Grund, weswegen Kirschblütentee überhaupt getrunken wird.

Heute kommen viele industriell hergestellte Nahrungsmittel mit natürlichen und / oder künstlichen Aroma- oder Duftstoffen auf den Markt. Der erste künstliche Aromastoff war Vanillin, welches 1874 vom Chemiker Wilhelm Haarmann in Zusammenarbeit mit Ferdinand Tiemann aus dem Rindensaft von Nadelhölzern hergestellt wurde. Bis heute ist Vanillin der meistverwendete Aromastoff der Welt. Von Vanillezucker bis zu Eiscreme wird Vanillin für eine breite Palette von Nahrungsmitteln verwendet. Gerichte mit Vanillegeschmack sind in weiterer Folge zu

Scents are a fixture in every taste experience, they can reinforce it or direct it into a certain direction. At "The Fat Duck" restaurant near London, a piece of highland peat served to each guest who orders red grouse will put him or her in the right mood for enjoying the Scottish specialty.

In Japan cherry blossoms are pickled in salt in the spring and used for making tea during the rest of the year. In hot water, the dried blossoms exude an intense cherry scent, even in November. The Japanese cavalierly neglect the fact that the tea is salty and has no particular other taste. Its special scent reminds them of a certain season, which is the only reason at all for the Japanese to drink cherry blossom tea – salty and devoid of nutritional value as it is.

Today, many industrially produced food products contain natural and/or artificial flavors or scents. The first artificial aroma was Vanillin, first produced from the bark oil of coniferous trees by the chemist Wilhelm Haarmann in cooperation with Ferdinand Tiemann in 1874. To this very day, Vanillin continues to be the most widely used aroma in the world. From vanilla sugar to icecream – Vanillin is used in a broad range of foods. Dishes with vanilla aroma have

Der gezielte Einsatz von Gewürzen kann den olfaktorischen Genuss beim Essen enorm steigern: Kurkuma, Basilikum, Rosmarin, Nelken, Kümmel, Rosen, Zimt, Minze, Kardamom, Safran, Muskat und Vanille.

The targeted use of herbs can greatly enhance the olfactory pleasure of eating: turmeric, basil, rosemary, cloves, caraway, roses, cinnamon, mint, cardamom, saffron, nutmeg and vanilla.

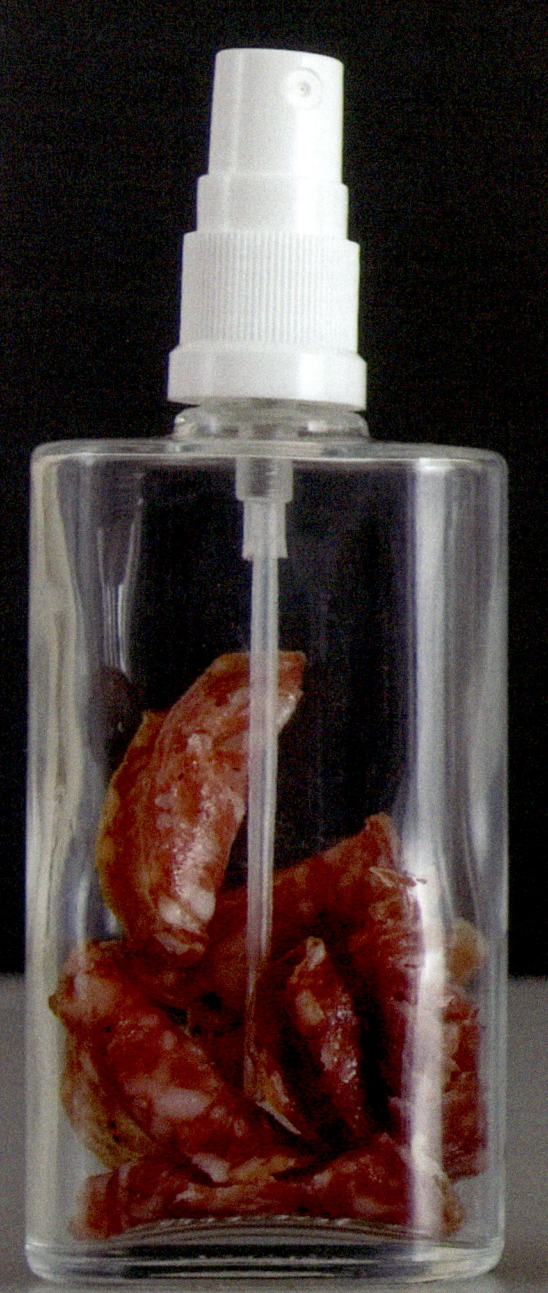

Massennahrungsmitteln geworden, Vanille zu einer der beliebtesten Geschmacksrichtungen, so dass die Nachfrage heute, da künstliche Aromastoffe zusehends abgelehnt werden, gar nicht mehr mit natürlicher Vanille gedeckt werden könnte.

Ähnlich ergeht es der beliebtesten Geschmacksrichtung Europas – der Erdbeere. Die weltweite Produktion von Erdbeeren würde gar nicht ausreichen, um die gesamte Herstellung von Erdbeerjoghurt zu decken.[7] Aromastoffe ersetzen oder unterstützen heute nicht nur teure Gewürze, sondern ganze Zutaten. Hersteller bieten Geruchsstoffe wie „Gekochtes Rindfleisch", „Emmentaler" oder „Butter" an, welche in Fertig- und Halbfertigprodukten zum Einsatz kommen. Künstliche Aromen bilden mittlerweile einen fixen Bestandteil unseres Alltags und haben auch unser Geschmacksverhalten nachhaltig verändert.

meanwhile become food for mass consumption, and vanilla ranks among the most popular flavors worldwide so that demand can no longer be met by the use of natural vanilla alone, even though artificial aromas are increasingly rejected.

The situation is similar when we look at Europe's most popular fruit flavor – strawberry. The worldwide production of strawberries would not be enough to meet demand from strawberry yoghurt producers.[7] In our day and age, artificial aromas not only replace or enhance expensive spices but even entire ingredients. Manufacturers offer aromatic substances such as "boiled beef", "gruyère cheese" or "butter" which are used in convenience food and semi-finished food products. Artificial flavors are part and parcel of our every-day life, they have changed the way we experience taste with lasting effect.

1 | Wir bedanken uns bei Dipl. Ing. Klaus Dürrschmid von der Universität für Bodenkultur für diese Information
2 | vgl.: Roswitha Muttenthaler, Geschmacksache, was Essen zum Genuss macht, 2008
3 | vgl.: Karl-Heinz Plattig, Spürnasen und Feinschmecker; Springer Verlag 1995
4 | Wir bedanken uns bei Dipl. Ing. Klaus Dürrschmid von der Universität für Bodenkultur für diese Information.
5 | vgl.: Karl-Heinz Plattig, Spürnasen und Feinschmecker; Springer Verlag 1995
6 | vgl.: Karl-Heinz Plattig, Spürnasen und Feinschmecker; Springer Verlag 1995
7 | vgl.: Hanne Tügel, Experiment Schlaraffenland, GEO Magazin 6/ 98

1 | A note of thanks for this item of information goes to Klaus Dürrschmid.
2 | See: Roswitha Muttenthaler, Geschmacksache, was Essen zum Genuss macht, 2008
3 | See: Karl-Heinz Plattig, Spürnasen und Feinschmecker; Springer Verlag 1995
4 | A note of thanks for this item of information goes to Klaus Dürrschmid.
5 | See: Karl-Heinz Plattig, Spürnasen und Feinschmecker; Springer Verlag 1995
6 | See: Karl-Heinz Plattig, Spürnasen und Feinschmecker; Springer Verlag 1995
7 | See: Hanne Tügel, Experiment Schlaraffenland, GEO Magazin 6/ 98

Ob Salami, Reis, Salzgebäck oder Käse: Der Geruch von Lebensmitteln bildet einen zentralen Bestandteil des Geschmackserlebnisses.

Whether it is salami, rice, salty snacks or cheese - the smell of foods constitutes a central aspect of the experience of taste.

Konsistenz
consistency

KONSISTENZ – Der mechanische Sinn – das Tasten | „*Gleichzeitig prüfen die taktilen Sensoren der Mundhöhle die sogenannten haptischen Qualitäten, über den Temperatursinn die Temperatur und – über den Schmerzsinn – die Frage, ob stechende bzw. leicht schmerzhafte Sensationen vorliegen.*" (Karl Heinz Plattig)[1]

Zusätzlich zur "klassischen" Geschmackswahrnehmung mit Mund und Nase erfahren wir Essen auch mit dem Tastsinn. Essen wir mit Messer und Gabel, können wir zum Beispiel anhand des Widerstandes beim Hineinstechen oder Durchschneiden feststellen, ob ein Kotelett weich und saftig oder zäh und strohig, ein Kuchen flaumig oder trocken ist. Mit Besteck zu essen ist allerdings eine vergleichsweise junge Erfindung. In vielen Teilen der Erde wird bis heute mit den Händen gegessen und auch bei uns nimmt Fingerfood einen besonderen Stellenwert ein.

Mit bloßen Fingern | Mit den Fingern zu essen macht Spaß, gerade weil es uns unsere Erziehung im Normalfall verbietet. Fingerfood erhöht aber auch den Genuss, denn es lässt uns Nahrungsmittel schon vor der Mundhöhle hautnah spüren. Es bereitet Lust, die poröse Oberfläche von Erdnussflocken, die gummiartig knusprige Textur von Popcorn oder die Salzkristalle auf der Oberfläche von Knabbergebäck zu ertasten. Auch die Sesamkörner an der Oberseite von Hamburgerbrötchen, die Kokosflocken auf Pralinen oder die spröde Fragilität von Kartoffelchips verschaffen dem Tastsinn einen gewissen Kick.

Fingerfood intensiviert den Kontakt zwischen Essen und Esser und steigert das sinnliche Erleben. Egal, ob britische Scones, französische Petits Fours oder spanische Tapas, der Mensch "greift" gerne nach kulinarischen Genüssen. Auch das Sushi verspeisen Japaner traditionell mit den Händen. Wir betasten die Oberfläche von Essen gerne mit den Fingerkuppen und können dabei bereits Rückschlüsse auf den zu erwartenden Geschmack ziehen. Wir fühlen, ob etwas brennheiß oder eiskalt, cremig oder hart, die Oberfläche porös, matschig oder pelzig ist. Der Tastsinn informiert uns vorab über Temperatur, Frische und Beschaffenheit eines Nahrungsmittels und kann so zum Beispiel auch vor zu Heißem oder Verdorbenem warnen.

Der japanische Lebensmittelhersteller Snow Brand hat sich von der japanischen Vorliebe für eine spezielle Art, Essen mit den Fingern zu befühlen, sogar zur Entwicklung eines neuen Produktes inspirieren lassen. Getrocknete Kalamari, ein beliebter, traditioneller Snack, haben eine besondere, gummiartige und zugleich faserige Textur, so dass sich mit den Fingern einzelne Fasern abziehen lassen. Snow Brand hat nun einen speziellen Käsestick entwickelt, dessen Konsistenz den haptischen Qualitäten der Kalamari nachempfunden ist und der sich in ähnlicher Weise mit den Fingern in Fasern zerrupfen lässt, bevor man ihn in den Mund steckt.

Die Konsistenz liefert 60 Prozent des Geschmacks | Noch sensibler reagiert der Tastsinn in der Mundhöhle. Wenn wir etwas in den Mund stecken, umschließen wir es mit unserem eigenen Körper und erleben es daher so intensiv wie kaum ein anderes Objekt. Laut der Bremer Wahrnehmungspsychologin Professor Gisla Gniech sind wir nur an den Sexualorganen und an der unteren Öffnung des Verdauungstraktes mit ähnlich vielen Druck-, Schmerz- und Temperatursensoren ausgestattet wie im Mund. Ob wir nun lutschen, saugen, beißen oder kauen – das Mundgefühl entscheidet letztlich über Erfolg oder Misserfolg, über Geschmack oder Nichtgeschmack eines Lebensmittels. Der Mund ist vor allem auch wegen seiner taktilen Fähigkeiten unser wichtigstes Geschmacksorgan.

CONSISTENCY – The Mechanical Sense – Touch | *"At the same time, the tactile sensors in the oral cavity check the so-called haptic qualities, whilst they deal with temperature through the temperature sense and via the pain sense, they feel for stinging or slightly painful sensations." (Karl Heinz Plattig)*[1]

In addition to "classic" taste through mouth and nose, we also experience food with our sense of touch. If we use a knife and fork, the resistance we feel when we stick the fork into the pork chops or cake and cut them will tell us if the porkchops are soft and juicy or tough and dry, and if the cake is fluffy or stale. Cutlery is a relatively new invention. In many parts of the world, people still eat using their hands only, and finger food is coming to the fore in the western world, too.

Eating with Your Hands | It is fun to eat with your hands, especially because our education normally does not permit it. Finger food is also more enjoyable because we can feel the food really close up before we put it into our mouths. We relish the porous surface of peanut puffs, the rubbery yet crunchy texture of popcorn and the salt crystals on the surface of pretzels. We also get a haptic kick out of the sesame seed on hamburger buns, the coconut flakes on chocolates and the brittle fragility of potato chips, or crisps, as they are called in the UK.

Finger food intensifies the contact between food and eater, thus enhancing sensory experience. No matter if it is English scones, French petits fours or Spanish tapas, people like "reaching out and touching" culinary tidbits. The Japanese traditionally eat sushi with their hands. We like touching food with out fingertips to draw conclusions about the taste to expect. We can feel if something is scalding hot or ice cold, creamy or hard, if the surface is porous, mushy or furry. Our sense of touch informs us about temperature, freshness and quality of food, and it can warn us of food that is too hot or rotten.

For the Japanese food producer Snow Brand, the fact that the Japanese like feeling food in a special way even inspired a new product. Dried calamaris, a popular traditional snack, have a particular, rubbery and stringy texture so that individual filaments can be peeled off for eating. Snow Brand developed a special cheese stick with a texture reminiscent of the haptic qualities of calamaris; you can likewise peel off strings with your fingers before putting them into your mouth.

Consistency is 60 Per Cent of Taste | The sense of touch in our oral cavities responds even more sensitively. Putting something into the mouth means enclosing it in the body and thus experiencing it more intensely. According to Professor Gisla Gniech, a Bremen-based perceptual psychologist, the only other parts of the body where the number of pressure, pain and temperature sensors is similarly large as in the mouth are the sexual organs and the orifice at the end of the digestive tract. No matter whether we suck, suckle, bite or chew – eventually, the feeling inside the mouth will decide if food is a success or failure for us, if it is tasty or not. Thus, the mouth is also our most important organ of taste because of its many tactile abilities.

Links außen | Mit dem Tastsinn „begreifen" wir Oberfläche und Konsistenz von Nahrungsmitteln und können ihre Genießbarkeit überprüfen. Eine feste, knackige Tomate oder Erdbeere wirkt frisch und appetitlich, während uns eine weiche, matschige Frucht abstößt.
Unten | Rumfingern und Zerrupfen erlaubt: japanischer Käsesnack mit gezielt komponierter Textur, sodass sich mit den Fingern einzelne Fasern abziehen lassen.

left | With the sense of touch we "grasp" the surface and consistency of foods and can assess their palatability. A solid, crisp tomato or strawberry appears fresh and appetizing whereas a soft, mushy fruit is off-putting.
bottom | Fumbling and pulling to pieces allowed: Japanese cheese snack with a specifically composed texture so that individual fibers can be pulled off with the fingers.

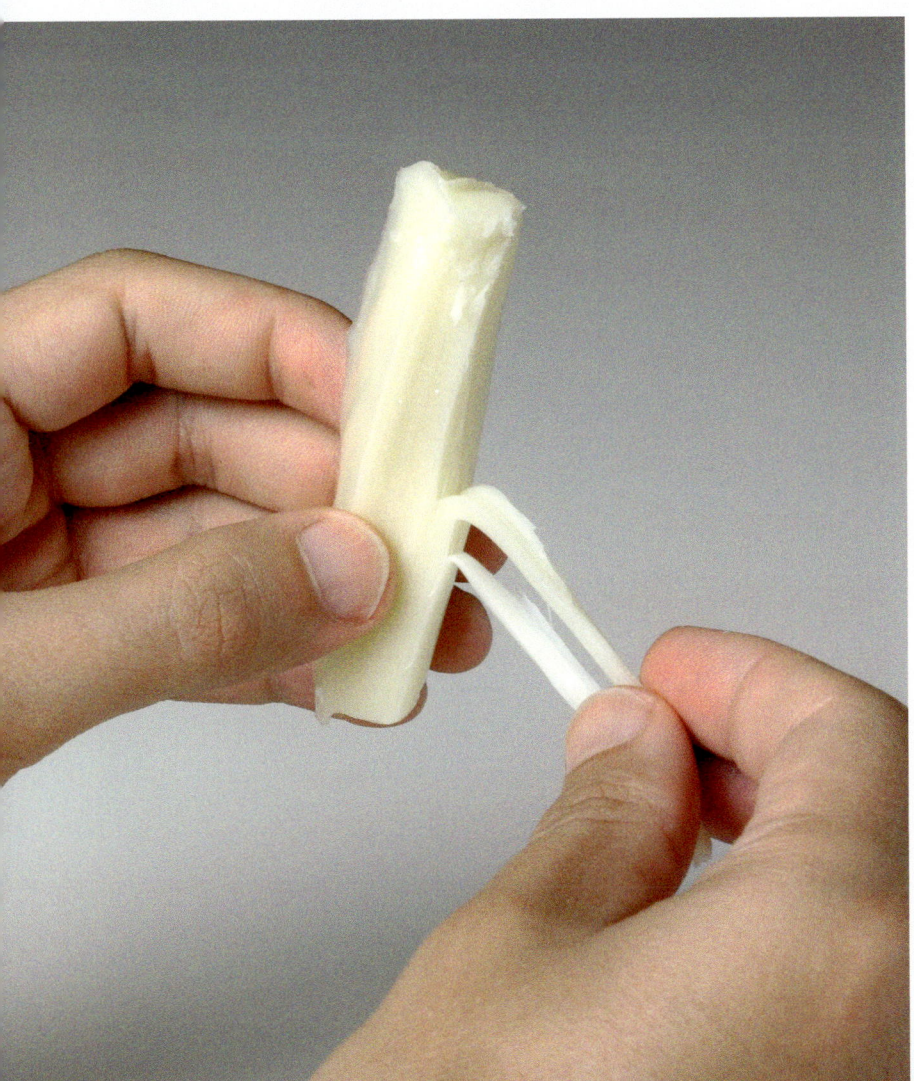

Der Tastsinn überhäuft uns mit einer Fülle von Informationen. Wir erleben Konsistenz, Textur und Temperatur einer Speise. Ist sie rau oder glatt, elastisch oder bröselig, knackig oder weich – oder eine Kombination von alledem? Die Art und Weise, wie sich die Oberfläche anfühlt, ob sie strukturiert ist wie bei einem Knäckebrot oder feucht und glitschig wie die Haut eines Würstchens, wie viel Beißwiderstand Zähne und Kiefer beim Kauen überwinden müssen und wie sich das Essen im Mund verteilt, entscheidet ganz wesentlich darüber, wie sehr uns etwas schmeckt. Denn Geschmack ist nicht nur die Wahrnehmung der Inhaltsstoffe des Gegessenen, sondern besteht zu einem Großteil – laut Gisla Gniech sogar zum überwiegenden Teil – aus den taktilen Empfindungen, die im Mund erlebt werden. Demnach macht die Konsistenz, also die Anfühlqualität auf Zunge, Zähnen, Gaumen und im Rachen, zirka 60 Prozent des Geschmackserlebnisses aus und ist daher wichtiger als der „eigentliche" Geschmack.

Mithilfe des Tastsinns entscheiden wir über Genuss und Ekel, über Frische und Nahrhaftigkeit einer Speise. Instinktiv ertasten wir, ob ein Nahrungsmittel reif oder unreif, frisch oder vergammelt ist. Wir wissen, dass eine genießbare Frucht nicht matschig, frisches Brot nicht hart und gutes Fleisch nicht zäh sein darf. Die Textur signalisiert dem Gehirn: Hier ist ein knackiges, frisches Produkt; oder aber: Achtung, hier ist ein altes, ein weiches, also ein schlechtes Produkt. Weiche Kartoffelchips beispielsweise werden sofort abgelehnt. Mit dem Tastsinn prüfen wir, ob das Gegessene genießbar ist und wir es tatsächlich hinunterschlucken können.

Es bereitet uns aber auch große Lust, herzhaft in unser Essen hineinzubeißen und es sorgfältig zu zerkauen. Fällt der Widerstand im Mund weg, so verlieren wir die Freude am Essen. Menschen mit Kau- und Schluckproblemen, die nur noch Brei zu sich nehmen können, büßen mit der Zeit sogar ihren Lebenswillen ein.

Our sense of touch provides us with an abundance of information. We experience consistency, texture and temperature of food. Is it rough or smooth, elastic or crumbly, crispy or soft – or a combination of all these qualities? The way a surface feels, if it is structured like cracker bread, or humid and slick like the skin of a sausage, the resistance our teeth and jaws have to overcome when chewing, and how the food spreads in the mouth, all of this has an impact on how well we like it. After all, taste not only comprises the way we sense the ingredients in what we eat but also and mostly the haptic sensations in the mouth – according to Gisla Gniech, the haptic element is even far more significant. Hence, the consistency, that is to say what we feel on our tongues, teeth and palates as well as in our throats makes for about 60 per cent of the taste experience and is thus more important than what food does for our taste buds.

With the help of touch, we decide if we will relish something or find it disgusting, if the food is fresh and nutritious or not. Instinctively, we feel whether something is ripe or green, fresh or rotten. We know that eating-quality fruits should not be mushy, fresh bread should not be hard, and good meat should not be tough. The texture tells the brain: This is a fresh, crispy product, so enjoy! Conversely, it also indicates: This is soft and old, so it has gone bad – watch out! For example, soft potato chips will be turned down immediately. We use our sense of touch to check if food is safe to eat and swallow.

Moreover, taking a hearty bite and carefully chewing food also involves pleasure. If there is no resistance to our teeth, the pleasure of eating is lost. People who have problems chewing and swallowing food and therefore have to stick to mash will even lose their will to live if this condition persists.

„Sie kennen sicherlich die Geschichte, dass eine Frau zum Sozialamt geht und sagt, sie brauche ein neues Gebiss, weil ihre Zähne kaputt seien und sie nicht mehr richtig essen könne – und das Sozialamt bewilligt ihr sieben Euro für einen Mixer. Damit kann sie natürlich ihr Essen zerkleinern, aber der Lustgewinn des Kauens und Beißens ist weg." (Gisla Gniech)[2]

High and Low | Die Art und Weise, wie ein Mensch einem Nahrungsmittel gegenübertritt, lässt sogar Rückschlüsse auf seinen psychischen Zustand zu. Gisla Gniech meint, dass sich die Psyche eines Menschen unter anderem in seinen Vorlieben für bestimmte Konsistenzen beim Essen zeigt. Mit den vier Entwicklungsstufen des Kindes vom Sauger über den Lutscher und Beißer zum Kauer geht auch eine psychologische Entwicklung des Lustempfindens und des sozialen Verhaltens einher. Das Kind entwickelt sich vom Baby, das an der Brust saugt, zum Erwachsenen, der nicht nur Nahrung, sondern auch Probleme des Lebens beißen und zerkauen, also verarbeiten muss. Wer nur zubeißt und unzerkaut hinunterschluckt, bekommt nicht nur Magenbeschwerden, sondern hat auch Schwierigkeiten bei der Bewältigung von Problemen.

Gisla Gniech teilt Esser grundsätzlich in zwei Kategorien: die Low- und die High-Sensation-Seeker, also jene, die Sanftes, Feines und Softes lieben, und jene, die es gerne farbig, knackig und exotisch haben. Psychologisch lassen sich den zwei Esstypen – aggressiver Lustesser und stiller Genießer – auch bestimmte generelle Verhaltensmuster zuordnen. Wer beim Essen nach high sensations sucht, agiert auch im Leben angespannter, interessierter, aktiver, kreativer, non-konformer und hektischer. Low-Sensation-Seeker sind im Gegensatz dazu gelassenere, phlegmatische Menschen.

"I'm sure you know the story of the woman who went to the social welfare office and asked for a denture because her teeth were all rotten and she was no longer able to chew food – and the social welfare office granted her a few dollars so she could buy a blender. Of course, she can chop up her food now but the pleasure of biting and chewing is gone." *(Gisla Gniech)*[2]

High and Low | The way in which people approach food actually allows for conclusions concerning his or her mental state. According to Gisla Gniech, the psyche of a person is reflected in his or her predilection for certain types of food, amongst other things. The four stages of development in children from suckling to sucking, biting and chewing go hand in hand with the psychological development of pleasure and social behavior. From the baby suckling at the mother's breast, a child turns into an adult who has to bite off, chew and swallow food as well as the problems of life he or she has to cope with. If we bite off more than we can chew, we will not only get indigestion but also difficulties in mastering problems.

For Gisla Gniech, eaters can, as a rule, be broken down into two categories: low- and high-sensation seekers, those who are into the soft and sophisticated stuff and those who like their food colorful, crisp and exotic. Psychologically, the two types of eaters – the aggressive pleasure-hunter and the person who enjoys in his or her own quiet way – also act according to certain general behavioral patterns. Those who seek high sensations in food are more tense, interested, active,

Ob eine Frucht unreif, reif oder verdorben ist, erkennen wir nicht nur an der Farbe, sondern auch an der Konsistenz.

We can judge if a fruit is unripe or spoiled – not just from the color but also from its consistency.

Wie sich ein Lebensmittel im Mund anfühlt, ob es cremig und weich ist wie geschlagene Sahne oder rau und porig wie ein Erdnussflip, wie kräftig wir zubeißen müssen und wie es sich beim Kauen in der Mundhöhle verteilt, macht bis zu 60 Prozent des Geschmackserlebnisses aus.

Up to 60 % of the taste experience depends on whether food is creamy and soft like whipped cream or rough and porous like a peanut flip, on how hard we have to bite into something and how it distributes in the mouth cavity when we chew it.

Studien zufolge gibt es etwa 30 Prozent High-Sensation-Seeker, wobei diese Zuordnung auch situativ ist, das heißt jeder Mensch kann einmal so und ein andermal anders agieren. Grob vereinfacht, produziert die Industrie 50 Prozent weiche, mild schmeckende und 50 Prozent harte, deftige Nahrungsmittel, um den gesamten Markt abzudecken.

Was als „richtige" oder „falsche" Konsistenz gilt, ist natürlich auch kulturell erlernt. Die Meinungen über die optimale Festigkeit eines Gerichts können daher je nach Region stark variieren. Man denke nur an das berühmt-berüchtigte englische Gemüse, das andere Gesellschaften als matschig verabscheuen. Oder an Teigwaren: Sie müssen in Italien extrem bissfest sein, in den USA dagegen weich gekocht. Auch bei Schokolade gehen die Meinungen über das perfekte Mundgefühl stark auseinander. Im angelsächsischen Raum liebt man sie eher „bröselig", während die Mitteleuropäer ihre Schokolade cremiger oder, wie Kritiker sagen, „schleimiger" mögen.

In unserem Gehirn ist eine ganze Reihe von Texturen abgespeichert. Wir wissen, wie sich Nahrungsmittel anfühlen sollen. Entspricht das Gegessene nicht diesem Erfahrungsschatz, lehnen wir es sofort ab. Ein Joghurt zum Beispiel muss eine ganz bestimmte Viskosität haben, damit er schmeckt. Wenn er zu dünn ist, schmeckt er nicht, und wenn er zu dick ist, auch nicht.[3] Eine Zutat, die für die Konsistenz vieler Produkte entscheidend ist, ist Zucker. Er macht zum Beispiel einen Teig nicht nur süß, sondern auch dichter und knuspriger. Die texturbildenden Eigenschaften von Zucker stellen Lebensmittelhersteller im Zeitalter von Übergewicht und wachsendem Ernährungsbewusstsein vor ein unlösbares Problem. Zuckerersatzstoffe bringen zwar oft die geschmacklich gewünschte Süße, aber nicht das entsprechende Mundgefühl. Bestimmte Lebensmittel schmecken ohne Zucker einfach nicht, weil sie nicht die gewünschte Konsistenz haben.

Was wir lieben … | Viele Lebensmittel lieben wir wegen ihrer ungewöhnlichen Konsistenz, nicht wegen ihres Geschmacks. Insalata Caprese zum Beispiel schmeckt genau genommen vorrangig nach Tomaten, Basilikum und Olivenöl, der Mozzarella ist kaum wahrnehmbar. Auch der Eigengeschmack von Weinbergschnecken wird von Zutaten wie Knoblauchbutter überdeckt und Kaviar schmeckt sehr salzig und eigentlich nicht besonders gut. Trotzdem sind alle diese Gerichte Spezialitäten, weil sie mit ganz speziellen Texturen aufwarten. Die schwammig-weiche Konsistenz von Mozzarella, das gummiartige Gefühl beim Zerbeißen einer Schnecke oder das Zerplatzen der einzelnen Kaviarperlen im Mund erzeugt sensorische Eindrücke, die einzigartig sind und daher

creative and hectic than others in real life and tend to be non-conformist. By contrast, low-sensation seekers are calm and phlegmatic persons. According to surveys, about 30 per cent of people belong to the category of high-sensation seekers although this can depend on the situation, i.e. people may act this way on one occasion and behave differently in the other. In grossly simplified terms, this means that the food industry produces 50 per cent soft and mild products and 50 per cent hard and hearty foods to cover the entire market.

The "right" or "wrong" consistency is also a matter of cultural learning. Opinions on the optimum firmness of food may vary strongly from region to region. Just think of the notoriety of English vegetables which are considered overcooked and mushy by many other societies. Or pasta, which has to be extremely firm (al dente) in Italy but should be soft in the USA. Chocolate is another case in point – again, opinions on the perfect mouth feel diverge: People in the English-speaking countries like their chocolate "crumbly" whereas Central Europeans go for the creamy kind (or slimy, as critics would say).

Our brain has stored a whole range of textures; we know what food should feel like. If something we eat does not correspond to our repository of experiences, we reject it immediately. For example, yoghurt has to have a specific viscosity to taste good. It should neither be too watery nor too thick.[3] Sugar is an ingredient with a decisive impact on the consistency of many products. Not only does it sweeten dough, it also makes it more dense and crunchy. In an age of overweight and growing nutritional awareness, the food industry is in a quandary when it comes to sugar and its texture-shaping properties. Sugar substitutes are sweet but they do not give rise to the same mouth feel as sugar, and certain foods simply do not taste good without sugar because they do not have the desired consistency.

What We Love … | There are many foods we love for their unusual consistency, not their taste. Insalata Caprese is a good example: it actually tastes primarily of tomatoes, basil and olive oil, we hardly perceive the mozzarella cheese. Edible snails have almost no taste of their own, it is covered up by the garlic butter, and caviar tastes salty and not particularly nice at all. Nevertheless, all these dishes are specialties because of their very specific textures. The spongy-soft consistency of mozzarella, the rubbery touch of snails or the feel of caviar grains

Oben | Zuckerwatte wurde 1897 von den Konditoren William Morrison und John C. Wharton in Nashville, Tennessee, erfunden. Geschmacklich hat sie – da sie zu 100 Prozent aus Zucker besteht – nicht viel zu bieten. Faszinierend ist aber ihre luftig-leichte und zugleich kratzig-raue Struktur.
Unten | Nahrungsmittel mit „erfolgreicher" Konsistenz: Pringles sind auf Grund ihrer zweiseitigen Krümmung dünn, knusprig und zerbrechen dennoch nicht in der Packung; Fischstäbchen faszinieren durch den Unterschied zwischen weichem, saftigem Fischfleisch und knuspriger Panier, die je nach regionaler Vorliebe von den Herstellern mehr oder weniger grob gekörnt und mehr oder weniger stark geröstet wird; die weiche Textur von Streichkäse entsteht bei der Herstellung von Schmelzkäse aus eingeschmolzenem Hartkäse, welche den Schweizern Walter Gerber und Fritz Stettler 1911 erstmals gelang. Sie wollten allerdings nicht die Konsistenz von Käse verändern, sondern die Haltbarkeit verlängern; die Rezeptur der pickig-süßen Marshmallows entwickelte sich aus einer Arznei, die schon im 11. Jahrhundert aus Eibischwurzeln hergestellt wurde. In Frankreich nutzte man den klebrigen Eibischwurzelsaft später auch kulinarisch und vermengte ihn mit aufgeschlagenem Eiweiß und Zucker zur so genannten „pâte de guimauve", einem Vorläufer heutiger Marshmallows.

top | Cotton candy was invented by confectioners William Morrison and John C. Wharton in Nashville, Tennessee, in 1897. In terms of taste it does not have much to offer – since it consists of 100 per cent sugar. What is fascinating, however, is its airy-light but also scratchy-rough structure.

bottom | Success stories in food consistency: As they are shaped like a hyperbolic paraboloid, Pringles are thin, crunchy but they still don't break in the package; fish fingers are fascinating because of the contrast between soft, juicy fish meat and crunchy breading which depending on re-gional preferences is more or less coarsely grained or more or less strongly roasted; the soft texture of cream cheese is created when processed cheese is produced from molten hard cheese: The first to succeed with this technique were the Swiss Walter Gerber and Fritz Stettler in 1911. Originally, they did not intend to change the consistency of cheese but to stop the process of ripening; the recipe of sticky-sweet marshmallows developed out of a medicine which was produced from white mallow root in the eleventh century already. In France the sticky white mallow root juice was later also used for culinary purposes and mixed with stiff egg whites and sugar to create the so-called "pâte de guimauve", a precursor of today's marshmallows.

besonders geschätzt werden. Das Mundgefühl kann für Vor-liebe oder Ablehnung von Gerichten den Ausschlag geben, der Geschmack ist dabei oft zweitrangig.

Flaumig | Ob Speisen bissfest oder weich sein sollen, ent-scheiden jeweils Geschmack und kulturelles Verständnis. Da-bei kann eine ausgesprochen weiche oder zart- schmelzende Konsistenz ebenso faszinieren wie eine knackige, spröde Zusammensetzung. Unter anderem gilt besondere Flaumig-keit als Konsistenzideal. Die „Salzburger Nockerln", eine süße, im Ofen überbackene Creme aus Eischnee, verdanken ihren Erfolg einer überaus luftigen Beschaffenheit. Kleine Frucht-stückchen scheinen in dem leichten Schaum geradewegs zu schweben und beinahe von der Gabel zu fließen. Das berühmte Dessert aus der Mozartstadt zeigt exemplarisch, dass nicht unbedingt der Geschmack den Ausschlag für kuli-narischen Erfolg geben muss.

bursting in the mouth, they all create sensory impres-sions that are unique and thus especially appreciated. Mouth feel can be crucial in liking or disliking food, the actual taste is often secondary in importance.

Fluffy | Taste and cultural approach decide if food should be firm to the bite or soft. In this context, soft or mouth-melting consistency can be as fascinat-ing as a crisp feel or brittle texture. Special fluffiness can be particularly ideal consistency. The Austrian soufflé known as "Salzburger Nockerln", sweetened egg white beaten into shape and baked in the oven, owes its success to its light and creamy qualities. Small pieces of fruit seem to be hovering in the tender foam, nearly gliding off the fork. The famous dessert from the town where Mozart was born shows that taste is not necessarily decisive for culi-nary success.

Ähnliches gilt auch für Zuckerwatte. Das Jahrmarktprodukt, das 1897 von den Konditoren William Morrison und John C. Wharton in Nashville, Tennessee, erfunden und 1904 bei der Weltausstellung in St. Louis vorgestellt wurde, hat geschmacklich nicht viel zu bieten. Es besteht zu 100 Prozent aus Zucker. Fasziniert reagiert der Konsument aber auf die ungewöhnliche, luftig-leichte und gleichzeitig kratzig-raue Textur. Die besondere, faserige Struktur entsteht durch ein spezielles Herstellungsverfahren. Dabei wird Zucker bis zum Fließpunkt von rund 150 Grad Celsius erhitzt und mittels Zentrifugalkraft vom so genannten Spinnkopf weggeschleudert. An der kühlen Luft erstarrt der Zucker als Karamellfaden und wird auf einem runden Stab aufgewickelt. Das Endprodukt ist nicht kristallin wie Zucker, sondern amorph und daher weich wie Watte.

Nicht zuletzt durch die so genannte Molekulargastronomie sind besonders lockere Konsistenzen heute sehr begehrt und Gerichte wie Parmesanluft, Petersilienschaum oder Bonbonwolken ergänzen unser gewohntes Wahrnehmungsspektrum. Zutaten werden dafür mit Distickstoffmonoxid (Lachgas) mithilfe von ISI-Flaschen aufgeschäumt oder mit Lecithin vermischt und mit dem Stabmixer aufgeschlagen, um süße und salzige „Espumas", also Schäume aller Art, herzustellen.[4] Die Idee ist nicht neu, denn schon seit Jahrhunderten schlägt man mit Schneebesen Luft in Sahne oder Eiweiß, um besondere Leichtigkeit zu erreichen. Was man der Molekularküche allerdings zugute halten muss, ist das systematische Experimentieren mit unterschiedlichen Technologien zur Erzielung neuartiger Konsistenzen und das Hinterfragen von klassischen Geschmack-Textur-Mustern. Wenn wir bisher rohe Karotten, gekochte Karotten, Karottenpüree, Karottensuppe und Karottensaft gekannt haben, so tun sich nun auch die Möglichkeiten von Karottenkrokant, Karottengel, Karottenstaub oder Karottenkaviar auf.

Cremig | Auch Joghurts, Pürees, Muse oder Speiseeis sind Produkte, deren Akzeptanz stark von ihrer Konsistenz abhängt. Generell gilt: Je cremiger, desto besser. Bei Speiseeis hängt die Cremigkeit vor allem mit der Kristallstruktur der Eismoleküle zusammen, die wiederum temperaturabhängig ist. Bei der händischen Herstellung wird die Masse beim Frieren immer wieder aufgeschlagen, um die entstehenden Kristalle klein und die Textur cremig zu halten. Die industrielle Herstellung bewegt sich im Spannungsfeld zwischen möglichst niedriger Verarbeitungstemperatur, um die Cremigkeit zu gewährleisten, und höheren Verarbeitungstemperaturen, um die Masse verarbeitungsfähig zu halten. Auf industriellen Fertigungsstraßen wird die Eismasse zunächst durch Zylinder mit -40 Grad Celsius kalten Außenwänden geschickt. Daran bilden sich Eiskristalle, die kontinuierlich von

We find the same phenomenon in cotton candy, called candy floss in the UK. The carnival staple was invented by the confectioners William Morrison and John C. Wharton in Nashville, Tennessee, in 1897 and first presented at the St. Louis World's Fair in 1904 (See: Wikipedia). In terms of taste, it is nothing to write home about, save that it consists of sugar and nothing but sugar. Nevertheless, we are fascinated by the unusual, airy yet rough texture. It takes a special procedure to make it, which also accounts for its specific thready structure. Sugar is heating to the yield point of 150 degrees Centigrade and slung out of the so-called spinning head by means of centrifugal force. As soon as the thin threads of sugar hit the air, they cool and solidify, to be collected on a stick by the operator of the machine. The final product is not crystal-shaped like sugar but amorphous and thus as soft as cotton wool.

So-called molecular gastronomy has made fluffy consistencies widely sought after, and dishes involving parmesan air, parsley froth or candy clouds now complement our accustomed spectrum of perception. The ingredients are either frothed up using nitrous oxide (laughing gas) from ISI whipped-cream canisters or mixed with lecithin and foamed up with a hand-held blender to end up as "espumas", foams of all descriptions.[4] The idea is nothing new because air has been added to cream and egg white for centuries to achieve special lightness, the only difference being the use of a whisk. We have to give molecular cuisine credit for the systematic approach to experimenting with different technologies so as to come up with new consistencies, and for questioning conventional patterns of taste and texture. So far, we have had raw carrots, boiled carrots, carrot puree, cream of carrot soup and carrot juice. Now we have new options, such as carrot brittle, carrot gel, carrot dust and carrot caviar.

Creamy | Yoghurts, purees, pulp or ice cream are products the acceptance of which depends strongly on their consistency. The general rule is "the creamier, the better". In ice cream, creaminess is contingent on the crystal structure of the ice molecules, which in turn depends on temperature. When ice cream is produced manually, the body is whisked again and again during the freezing process to keep the emerging crystals small and texture creamy. In industrial production, there is a constant trade-off between the lowest possible processing temperature to ensure creaminess and higher temperatures so that the body can still be processed. In industrial ice cream production lines, the ice cream body is first sent through cylinders with exterior walls cooled to

Messern abgeschabt und zur weiteren Verwendung bereitgestellt werden. Gießt man diese danach in Formen, muss die Verarbeitungstemperatur erhöht werden, um Lufteinschlüsse zu verhindern. Dadurch verliert das Eis allerdings wieder an Cremigkeit. Eissorten, die besonders cremig sein sollen, wie zum Beispiel das Magnum, werden daher nicht gegossen, sondern im so genannten Extrusionsverfahren hergestellt. Dabei wird die Grundmasse bei sehr niedrigen Temperaturen durch eine Düse gepresst und geschnitten. Dadurch bleibt die kleinteilige Kristallstruktur des Eises erhalten und fühlt sich anschließend im Mund softer an als bei gegossenen Eislutschern.[5]

Konsistenz und Cremigkeit ist bei der industriellen Herstellung von Eis ein derart zentrales Thema, dass sich eigene Forschungsprojekte damit befassen. Auf diese Weise hat man zum Beispiel bei einem arktischen Tiefseefisch ein Protein entdeckt, das bei der Erzeugung von Eis verwendet werden kann, um die Bildung großer Eiskristalle zu verhindern. Mittlerweile wird das Protein aus Hefe hergestellt und in den Vereinigten Staaten bei der Produktion von Eis bereits eingesetzt.

Emulsionen | Von vielen Esswaren wie Joghurts, Dressings, Senf, Mayonnaise oder Saucen erwarten wir, dass sie cremig sind. Physikalisch betrachtet, sind diese Nahrungsmittel Wasser-Fett-Emulsionen, also mehr oder weniger stabile Mischungen zweier nicht ineinander löslicher Komponenten. Wir empfinden sie dann als besonders cremig und angenehm, wenn die Emulsion gut gemischt ist, das heißt die Fetttröpfchen möglichst klein und gut verteilt sind und sich die einzelnen Zutaten nicht absetzen. Wenn zum Beispiel Wasser aus der Senftube tropft, so erscheint uns das unappetitlich.

Um Emulsionen langfristig möglichst temperatur- und pH-Wert-unabhängig stabil zu halten, werden eigene Stoffe beigemengt, die so genannten Emulgatoren. Die Idee, Emulgatoren zur Herstellung sämiger Saucen zu verwenden, ist nicht neu, sondern stammt bereits aus früheren Jahrhunderten. Die Sauce Hollandaise aus dem 17. Jahrhundert, die ihren Namen den Butterexporten von Holland nach Frankreich verdankt, und die Mayonnaise, die in Frankreich im 18. Jahrhundert erfunden wurde, nutzen die Fähigkeit von Eidotter, um Wasser-Fett-Gemische stabil zu halten. Eidotter enthalten viel Lecithin, einen Emulgator, der mittlerweile künstlich hergestellt wird und in der Nahrungsmittelindustrie breite Verwendung findet.

- 40 degrees Centigrade. The ice crystals form on the walls and are continuously scratched off by means of blades for further processing stages. If they are filled into molds, the processing temperature has to be raised to avoid air being trapped. However, higher temperatures cause the ice cream to become less creamy. This is the reason why especially creamy varieties such as the Magnum are not poured into molds but extruded through a nozzle at very low temperatures and cut up. The crystals remain small and the ice cream feels softer in the mouth than poured ice cream on a stick.[5]

Consistence and creaminess are central to the industrial production of ice cream – to such an extent that special research projects were launched. This way, anti-freeze proteins were discovered in ocean pout, a fish living in arctic waters; it can be used in the production of ice cream to prevent the development of large ice crystals. This ingredient is now produced from yeast and used in ice cream production in the United States.

Emulsions | Many foods such as yoghurts, salad dressings, mustard, mayonnaise or sauces are expected to be creamy. In terms of physics, they are emulsions of water and fat, more or less stable mixtures of two components which are basically unblendable. Whenever emulsions are well mixed, i.e. when the droplets of fat are as small as possible and well distributed, without any flocculation of individual ingredients, the emulsion is perceived as pleasant and creamy. By contrast, when water starts dripping from the mustard tube, it looks gross to us.

Special substances, so-called emulsifiers, are mixed into emulsions to keep them stable in the long run, irrespective of temperature and pH values. The idea to use emulsifiers for thick sauces is nothing new but came up centuries ago. Sauce hollandaise, a creation of the seventeenth century which owes its name to the butter exports from Holland to France, and mayonnaise, which was invented in eighteenth-century France, make use of the fact that egg yolk keeps water-fat mixtures stable. Egg yolk contains a lot of lecithin, an emulsifier which is nowadays produced artificially and widely used in the food industry.

Kremser Senf, Dijon-Senf, Mayonnaise, Cocktail-Sauce, Erdbeerjoghurt und Ketchup: In unserem Gehirn sind Texturen in ihren feinsten Nuancen abgespeichert. Entspricht das Gegessene nicht diesem Erfahrungsschatz, lehnen wir es ab. Joghurt beispielsweise darf weder zu dünn- noch zu dickflüssig sein, damit es uns schmeckt.

Kremser mustard, Dijon mustard, mayonnaise, cocktail sauce, strawberry yoghurt and ketchup: Textures are stored in our brain in their most subtle nuances. If what is eaten does not correspond to this trove of experience, we will immediately reject it. Yoghurt, for instance, must neither be too thin nor too thick to taste good.

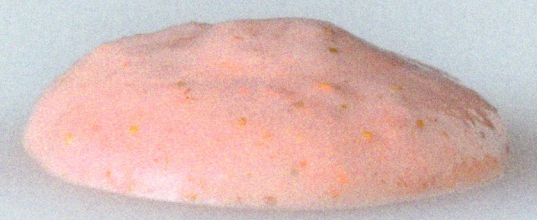

Qimiq – das Wunderprodukt
Eine technologische Neuentwicklung der letzten Jahre in diesem Zusammenhang ist „Qimiq", ein Milchprodukt, welches zu neunundneunzig Prozent aus fettreduzierter Sahne und zu einem Prozent aus Gelatine besteht. Die Gelatine wird in einem patentierten Herstellungsverfahren mit der Sahne vermischt, was Qimiq die Eigenschaft verleiht, seine Konsistenz immer wieder zu verändern – auch wenn man weitere Zutaten beifügt. Beim Erwärmen wird Qimiq flüssig, beim Kühlen wieder fest. So wird zum Beispiel ein bei Zimmertemperatur streichfähiger Thunfischaufstrich auf warmen Nudeln zu einer Thunfischsauce.

Knusprig
Von Cornflakes bis zu Keksen, von Toastbrot bis zu Knabbergebäck verdankt eine Vielzahl von essbaren Produkten ihren Erfolg ihrer knusprigen Konsistenz. Bereits im 3. Jahrhundert vor Christus erzeugten die Römer krosse

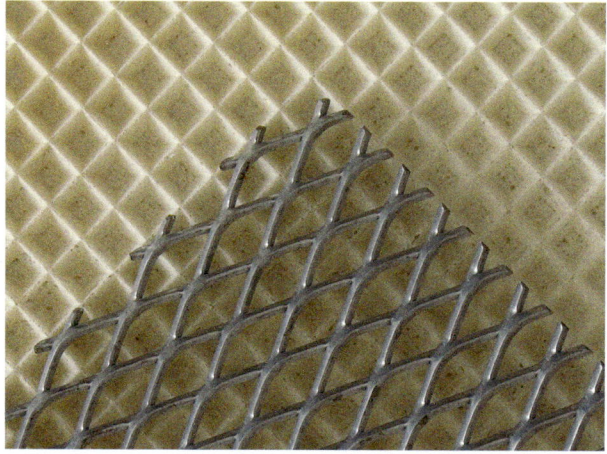

Teigwaren, die man nach dem Herstellungsverfahren „bis coctum", zweimal Gebackenes, nannte. Bezeichnungen wie Zwieback oder Biscuit erinnern noch heute an das Produktionsverfahren, das den ursprünglich ungesüßten Teigwaren ihre knackig-resche Textur verlieh. Knusprige Kekse und andere scharf gebackene Produkte wurden ursprünglich wegen der langen Haltbarkeit geschätzt und fanden unter anderem als Seemannsnahrung Verwendung. Ende des 19. Jahrhunderts importierte Hermann Bahlsen die Tradition der Teekuchen und -backwaren von den britischen Inseln nach Deutschland und machte krosse Backwaren auch in Kontinentaleuropa populär. Aus der Not der Lagerfähigkeit wurde eine Tugend – die spröde Konsistenz zum geschätzten Markenzeichen von Keksen.

Knusprigkeit und Stabilität
Sind Knusprigkeit und Stabilität gleichermaßen gefragt, bietet die klassische Form der Waffel eine überzeugend einfache und ebenso wirksame Lösung. Stege im Quadratraster stabilisieren eine hauchdünne und daher sehr knusprige Teighaut. Die ältesten Waffeleisen wurden in Belgien und Frankreich gefunden und stammen bereits aus dem 9. Jahrhundert. In Frankreich existierte schon im 13. Jahrhundert nachweislich eine eigene Zunft der Waffelbäcker. Spätestens im 15. Jahrhundert verbreitete sich die Rezeptur auch in den Niederlanden und im norddeutschen Raum. Schon auf dem 1559 entstandenen Gemälde „Kampf zwischen Fasching und Fasten" von Pieter Brueghel dem Älteren sind Waffeln mit der typischen, quadratisch strukturierten Oberfläche zu sehen.

Links von oben nach unten | Gummibären, Zuckerwatte, Süßer Speck oder Marshmallows, japanisches Yuba aus getrockneter Sojamilchhaut, Knabbersticks, an der Oberfläche strukturierte Schokolade
Reihe von links nach rechts | Waffel, Crumpets, asiatische Reisnudelblätter, hausgemachte Waffelrolle, Milchbrot
Großes Bild rechts | Waffelproduktion bei der Firma Manner in Wien. Neben besonderer Cremigkeit ist auch Knusprigkeit ein Konsistenzideal. Bereits die Römer erzeugten krosse Teigwaren, die man nach dem Herstellungsverfahren „bis coctum", zweimal Gebackenes, nannte.

left, top to bottom | Gummi bears, cotton candy, marshmallows, Japanese Yuba made of dried soy milk skin, snack sticks, chocolate with a structured surface
left to right | wafers, crumpets, Asian rice noodle sheets, home-made waffle roll, milk bread
large photo on right | Wafer production at Manner co. in Vienna. In addition to special creaminess, crunchiness is also an ideal consistency. Already the Romans created crisp pastries named "bis coctum" (baked twice) after the production technique.

Qimiq – The Miracle Product | One technological development of the recent past is "Qimiq", a milk product consisting of 99% low-fat cream and 1% gelatin. Gelatin is blended with cream in a patented production process which gives Qimiq the welcome ability of changing consistency again and again – even if further ingredients are added. Qimiq turns liquid when heated and solidifies again when cooling down. For example, what is tuna bread spread at room temperature becomes tuna sauce on hot pasta.

Crunchy | From cornflakes to cookies, from bread for toasting to pretzels, there is a wide range of food products which owe their success to their crunchy consistency. As early as in the third century BCE, the Romans baked a crusty type of bread they

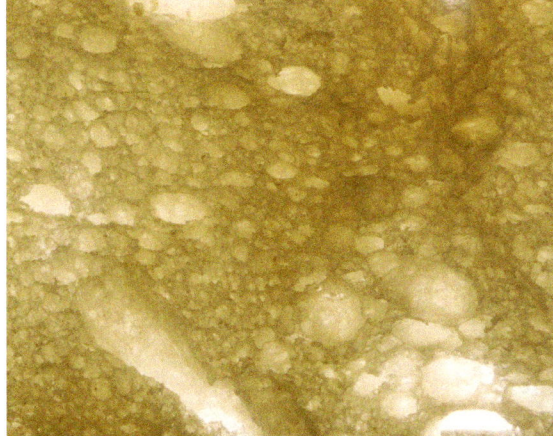

called "bis coctum", ("twice baked") after the baking process. Words in other languages, such as the German "Zwieback" or the French "biscuit", still hark back on the production technique which gives the originally unsweetened rusk its crunchy texture. In the beginning, crispy biscuits and other products baked at high temperatures were popular due to their long life and primarily used on sea journeys. At the end of the nineteenth century Hermann Bahlsen brought the tradition to Germany from the British Isles and crispy biscuits also became well-liked in mainland Europe. That way, what started as a necessity – long life – turned into a virtue, and the brittle consistency of biscuits was from now on their signature feature.

Crispy and Stable | When both crispiness and stability are desired, the classic waffle shape is a convincingly simple yet effective solution. The dividers in the square grids stabilize a thin, and thus very crispy, skin of dough. The oldest waffle irons were found in Belgium and France, dating back to the ninth century. In France, documents prove that the guild of waffle bakers existed as early as in the thirteenth century. The recipe had found its way to the Netherlands and northern Germany by the fifteenth century. The painting "The Battle between Carnival and Lent" by Pieter Brueghel the Elder, dating from 1559, is a case in point here, as it includes an early depiction of waffles and their typical honeycomb structure.

Products without stabilizing longitudinal or transversal dividers – which practically consist of nothing but surface – are also extremely fascinating. Paper-thin fried slices of potato – the forerunners of today's potato chips – are said to have been invented by George Crum, a chef of the "Moon Lake Lodge" hotel in Saratoga Springs, New York, in 1853. Allegedly, a guest had complained about French fries too thick for his liking, so Crum served him extra-thin fried chips. According to a second version of the story, it was a matter of coincidence because Crum's sister accidentally dropped a thin potato slice into hot frying fat. No matter which version is true, the new side dish rose to fame under the name "Saratoga chips". In the 1920's Herman Lay, a traveling salesman from the American south, invented a potato peeling machine and launched the industrial production of potato chips. Originally, they were unseasoned and moderately successful. After World War II, the Irish Tayto company found a technique to add seasoning and flavors to the fried potato slices. In 1954 the "Cheese and Onion" as well as "Salt'n'Vinegar" varieties were first marketed. Since then, chips (or crisps, depending on which side of the Atlantic you are on) have become the most popular snack product worldwide. In some countries they are even served with sandwiches or burgers as a regular side dish. 72,000 tons of potato chips per year are eaten in Germany alone.[6]

Potato chips are not just well-liked for their rich taste but also, and especially for their crunchiness. Production requires sound know-how because a squashy raw potato slice has to be turned into a thin and crisp fried chip which is not suppose to crumble. The starch content of the potato is decisive. Optimum thickness is selected depending on the type of potato supplied, so not all potato chips are the same in thickness even though they may come from the same factory.

Successful on the Surface | Paper-thin slices or leaves differ from compact food in that the surface from which we can best take in the flavor without chewing is bigger. Thinly sliced dry-cured ham, cheese shavings or the famous carpaccio actually melt in the mouth and taste better than thick slices of the same product. We prefer thinly sliced foods because the larger surface enhances contact with the taste cells and the emanation of scent particles. Even industrially produced food exploits the image-related benefit and taste advantages of paper-thin units, offering chocolate leaves like the wafer-thin Swiss Lindt chocolates or "Eclipse" strips which are mere macromillimeters thick, a mixture of candy and edible paper. "After Eight", invented by the English producer "Rowntree's" in 1962, combines the consistency benefit of thin chocolate layers with the British tradition of having a mint leaf after a meal.

left | Lebanese confectionery
right | Yoghurt with colored chocolate buttons

Aufregende Struktur | Den Genuss steigern auch speziell geformte oder gekrümmte, glatte Oberflächen wie zum Beispiel jene des zweiseitig gewölbten Dixi-Bonbons oder die konkaven Löcher im Emmentaler, welche die Zunge zum sanften Darüberstreichen und Hineinfühlen anregen. Andere Esswaren setzen bewusst auf strukturierte Oberflächen. Das sanfte Kratzen der Panier gehört zum Geschmack von Hühnernuggets, Wiener Schnitzel und Fischstäbchen einfach dazu. Brot wird mit Salz, Mohn, Sesam oder anderen Körnern bestreut, um die Oberfläche für Zunge und Gaumen interessanter zu machen. Kristallzucker, Mandelsplitter oder Kokosflocken auf Süßigkeiten dienen nicht nur als Dekoration, sondern im wahrsten Sinn des Wortes als besonderer „Gaumenkitzel", grobkörniges Meersalz auf Fleisch- oder Fischspeisen erzielt einen ähnlichen Effekt.

Raue, pelzige oder gerillte Strukturen sorgen für ein abwechslungsreiches kulinarisches Vergnügen mit höherem Wiedererkennungseffekt. Die Oberflächen vieler Schokoriegel sind daher geriffelt, gerillt oder anders texturiert und erzeugen ebenso wie Cornflakes oder saure Drops aufgrund ihrer rauen Beschaffenheit ein angenehmes Kribbeln auf der Zunge. Eine Extremvariante dieser Oberflächeneffekte sind Brausepulver und süße oder saure Panaden, die im Mund „explodieren". Chemisch funktioniert das durch Hydrogenkarbonate und Milch-, Wein- oder Zitronensäuren, die miteinander reagieren, sobald sie mit Speichel in Berührung kommen. Dabei wird Kohlendioxid freigesetzten, das die Kristalle zersprengt und für das eigenartige Prickeln auf der Zunge sorgt. Einen ähnlichen Effekt haben auch Krabbenchips: Der Speichel führt zu mechanischen Veränderungen, die die Struktur des Chips plötzlich zusammenbrechen lassen, im Mund fühlt sich das dann wie kleine Explosionen an.[7]

Gelees | Die Liste von Texturen und Konsistenzen, die wir bei Esswaren lieben, ist beinahe endlos. Neben flaumig, cremig und knusprig üben auch weiche, aber dennoch schnittfeste Gelees, wie wir sie von Terrinen, Sülzen, roter Grütze oder englischen Jellys kennen, eine besondere Faszination auf uns aus. Seit dem 16. Jahrhundert werden in Großbritannien Früchte mithilfe von Stärke und Gelatine zu elastisch-flexiblen, aber dennoch formstabilen Desserts verarbeitet.[8] Die Formenvielfalt dieser Jellys blickt auf eine lange kulinarische Tradition zurück. Modelle aus viktorianischer Zeit werden mittlerweile als teure Antiquitäten gehandelt.

Auch die japanische Küche kennt ein traditionelles Geleegericht. „Yokan" ist ein Mus aus süßen Azuki-Bohnen und Yam, das mit Agar, einem Geliermittel aus den Zellwänden von Algen, zu einem bissfesten rechteckigen Riegel geformt und in Scheiben geschnitten zum Tee serviert wird. Heute verwenden Hersteller vor allem Agar, den mittels Mikroorganismen hergestellten Vielfachzucker Gellan oder das Verdickungsmittel Xanthan, um aus Cremen, Suppen oder Fonds gelartige Objekte zu formen.[9]

Exciting Structure | Food products with particular shapes or curved smooth surfaces also enhance enjoyment, such as the convex dextrose candy known as "Dixi", very popular with children in Austria, or the concaves holes in Swiss cheese which invite the tongue to gently feel the surface before biting into it. Other foods rely on structured surfaces. The slightly raspy feel of the breading is part and parcel of chicken nuggets, Wiener Schnitzel and fish fingers. Bread is covered with salt, poppy seeds, sesame seeds and other grains to make the surface more interesting for the tongue and palate. Crystal sugar, almond slivers and coconut rasps on sweets are more than just decoration, they literally "titillate the palate", along similar lines as coarse sea salt grains on fish or meat.

Rough, furry or grooved structures account for diversified culinary pleasure with more recognition effect. Many chocolate bars have surfaces with ribs, grooves or other textures to create a pleasant tingle on the tongue, just like cornflakes or acid drops. Sherbet powder as well as sweet and sour coatings "exploding" in the mouth are the most extreme variants of surface effect. They are due to a chemical reaction of hydrogen carbonate, lactic acid, tartaric acid or citric acid when they come into contact with saliva. The carbon dioxide released at this point causes the crystals to burst, which in turn leads to the tingling sensation on the tongue. The same goes for shrimp chips a.k.a. prawn crackers: saliva causes mechanical changes which make the chip structure break down, resulting in little explosions in the mouth.[7]

Jellies | The list of textures and consistencies we love in our food is seemingly endless. We like it fluffy, creamy and crunchy, and we should not forget the soft yet firm jellies we find so fascinating in terrines, brawn, red berry compote, English jelly or Jell-O. Since the sixteenth century, fruit, starch and gelatin have been turned into elastic yet stable desserts.[8] The diversity of shapes and forms which jellies come in looks back on a long tradition, and Victorian jelly molds have meanwhile become expensive antiques.

The traditional jelly dish in Japanese cuisine is called "Yokan", a puree of sweet azuki beans and yam, mixed with agar agar, a gelling agent produced from the cell walls of algae. It is shaped in firm rectangular bars and cut into slices to be served with tea. Nowadays, manufacturers mainly use agar agar, gellan (multiple sugar produced from micro-organisms) or the thickening agent xanthan to turn creams, soups or stocks into jellied products.[9]

Elastisch, plastisch, klebrig

Selbst extrem klebrige Nahrungsmittel wie Türkischer Honig, die beim Essen an den Zähnen haften, lieben wir gerade ihrer Konsistenz wegen. „Toffee", eine gehärtete Mischung aus Zuckersirup und Butter, ist zum Beispiel im Vereinigten Königreich extrem beliebt. Das eingedeutschte „Karamellbonbon" ist eine Abwandlung der vermutlich aus dem frühen 19. Jahrhundert stammenden englischen Erfindung.

Medizinische Zwecke erfüllte ursprünglich eine klebrige, weiße Substanz aus den Wurzeln des Echten Eibisch. Kandierte Stücke dieser Eibischwurzel sollen schon im 11. Jahrhundert als Mittel gegen Erkältung eingesetzt worden sein. Die Franzosen nutzten die spezielle Konsistenz der klebrigen Masse zuerst für kulinarische anstatt für medizinische Zwecke. Aus aufgeschlagenem Eiweiß, Zucker und den Inhaltsstoffen der Eibischwurzeln produzierten sie „pâte de guimauve". Daraus entstanden Marshmallows, der pickig-süße Inbegriff jeder US-amerikanischen Thanksgiving-Feier.

Aufblasbar und unverdaulich – der Kaugummi

Extrembeispiel eines klebrigen Nahrungsmittels mit unendlich großem Beißwiderstand ist der Kaugummi. Er kommt in Streifen, als rosarote Schnur, in Zigarettenform und als bunte Drageekugel auf den Markt und wird beim Kauen niemals weich. Ein weiterer Konsistenzbonus von unverdaulichen Kaumassen sind die bis zu 50 Zentimeter großen Blasen, die sich mit gewissen Kauprodukten erzeugen lassen.

Kaugummikauen ist ein uralter Brauch, der sich durch archäologische Funde knapp 10.000 Jahre zurückverfolgen lässt. Bereits die alten Ägypter kauten im Rahmen gewisser Kulthandlungen gummiartige Substanzen. Später war im antiken Griechenland die Sitte, an Harzstücken des so genannten Mastixbaumes zu knabbern, weit verbreitet. Auf dem nordamerikanischen Kontinent entdeckte die indianische Urbevölkerung die Vorzüge des Harzes der Rottanne, das letztendlich auch die Siedler überzeugte. Zu Beginn des neunzehnten Jahrhunderts galt diese Frühform des Kaugummis als begehrte Handelsware entlang der amerikanischen Ostküste. 1848 gebar der Seemann John Curtis die Idee, Kaugummis mit Geschmacksstoffen zu aromatisieren, und landete damit einen Riesenerfolg. Er war auch der Erste, der ein Kauprodukt auf Basis des aus Rohöl gewonnenen Paraffins auf den Markt brachte.

Der Ursprung der Kaugummiherstellung im heutigen Sinn geht auf den Fotografen und Hobbyerfinder Thomas Adams zurück, der etwa um 1870 den aus Mittelamerika stammenden Naturgummi Chicle in die Vereinigten Staaten importierte. 1893 brachte der Seifenhersteller William Wrigley Jr. die bis heute erfolgreichen Sorten „Spearmint" und „Juicy Fruit" auf den Markt, 1914 dann den „Doublemint", der mit Pfefferminzöl aromatisiert ist.

Für einen uneingeschränkten Kaugenuss sorgte damals wie heute ein unverdaulicher Basisgummi, die eigentliche Seele jedes Kauprodukts. Die Qualität eben dieser Grundsubstanz ist für ein elastisches, geschmeidiges Kauverhalten ohne Bröckeln sowie für ein möglichst intensives und gleichmäßiges Geschmackserlebnis verantwortlich. Mittlerweile haben synthetische Polymere, die neben verminderter Klebrigkeit auch für länger anhaltenden Geschmack sorgen, den Naturgummi Chicle als Kaubasis weitgehend verdrängt.

Elastic, Plastic, Sticky

In spite of the fact that it clings to our teeth, we even love extremely sticky food such as Turkish delight for its consistency. Toffees, a mixture of hardened sugar syrup and butter, are very popular in the United Kingdom. The English invention which probably dates back to the early nineteenth century, found its way to Germany, where it continues to be sold in a slightly modified form as caramel candy.

A sticky white substance from the roots of althaea officinalis originally served medical purposes. Candied pieces of the root are said to have been used against colds as early as in the eleventh century. The French were the first to use the special consistency of the gummy substance for culinary purposes. They made "pâte de guimauve" from beaten egg white, sugar and an extract from the root. Marshmallows, the sticky-sweet epitome of all American Thanksgiving parties, were derived from it; the English name of the confection actually reflects the English name of the plant.

Inflatable and Indigestible – Chewing Gum

Chewing gum is the most extreme example of a sticky food product with infinite resistance to biting. It comes in all shapes and forms: strips, pink ropes, cigarettes and sugar-coated spheres, and no matter how long we chew it, it will never really go soft. Another bonus of this indigestible material for chewing is that certain varieties can be inflated to form bubbles of up to 20 inches in diameter.

Chewing gum is an age-old custom; archeological finds corroborate that it already existed about 10,000 years ago. The ancient Egyptians chew gum-like substances during certain rituals. Later on, the ancient Greeks liked nibbling on pieces of resin from the mastic tree. The indigenous population of North America discovered the merits of chewing the resin of the common spruce, which eventually also convinced the white settlers. In the early nineteenth century, this early type of chewing gum was much in demand as a commodity traded along the east coast of the United States. In 1848 the sailor John Curtis came up with the brilliant idea of flavoring chewing gum with aromas. He was also the first to successfully market gum based on parrafin, a crude-oil based substance.

The origins of chewing gum as we know it today harks back on the efforts of photographer and hobby inventor Thomas Adams, who imported chicle gum from Central America around 1870. In 1893 the soap manufacturer William Wrigley Jr. launched "Spearmint" and "Juicy Fruit", followed in 1914 by "Doublemint", chewing gum flavored with peppermint oil; all three varieties are still on the market and continue to sell well.

What was then still applies now: the actual heart and soul of every gum is an indigestible basis which accounts for unrestricted chewing enjoyment. The quality of the gum base ensures elasticity and suppleness without crumbling as well as the optimum in flavor intensity and homogeneity. Meanwhile, synthetic polymers have come to replace natural chicle gum because they are less sticky and produce longer lasting flavor.

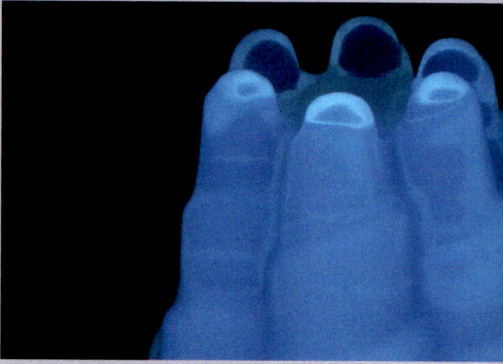

Seit dem 16. Jahrhundert werden in Großbritannien Früchte mithilfe von Stärke und Gelatine zu elastisch-flexiblen, aber dennoch formstabilen Desserts, so genannten Jellies verarbeitet. Auf die zeitgemäße Gestaltung von Jellies haben sich die Londoner Designer Sam Bompas und Harry Parr spezialisiert.

Since the 16th century, fruits have been processed with starch and gelatin in Great Britain to produce so-called jellies– elastic-flexible but still form-stable desserts. The London designers Sam Bompas and Harry Parr have specialized in the contemporary design of jellies.

Extrembeispiel eines plastischen Nahrungsmittels mit unendlich großem Beißwiderstand ist der Kaugummi. Als sein Erfinder gilt der Fotograph Thomas Adams, der um 1870 den Naturgummi Chicle aus Mittelamerika in die Vereinigten Staaten importierte. 1893 brachte der Seifenhersteller William Wrigley Jr. die bis heute erfolgreichen Sorten „Spearmint" und „Juicy Fruit" auf den Markt, 1914 dann den „Doublemint", der mit Pfefferminzöl aromatisiert ist.

Chewing gum is an extreme example of a plastic food with infinite resistance to biting. The photographer Thomas Adams is considered to have invented it. Around 1870 he imported the Chicle natural gum from Central America to the United States. In 1893 the soap manufacturer William Wrigley Jr. launched the "Spearmint" and "Juicy Fruit" brands that remain successful to the present day. In 1914 "Doublemint", which is aromatized with peppermint oil, was launched.

Um aus dem Grundgummi verkaufsfähige Kauprodukte zu generieren, wird die Kaumasse vorerst erhitzt, geknetet und mit weiteren Stoffen verfeinert. Glycerin und andere Pflanzenöle sorgen für zusätzliche Weichheit und Homogenität. Bei einem entsprechend hohen Anteil weichmachender Zusätze entsteht jene hohe Elastizität, mit deren Hilfe und einigem Geschick sich die Gummis zu dünnen Membranen dehnen lassen, ohne vorschnell zu zerplatzen. An den so genannten „Bubble Gums" schätzen Kinder neben der Blasenfähigkeit vor allem auch die Süße. Der beigefügte Zucker dient dabei aber nicht nur der geschmacklichen Optimierung, sondern hält die Masse gleichzeitig frisch und elastisch. Eine ähnliche Aufgabe erfüllt auch die Drageekruste zahlreicher Kaugummiprodukte. Hierfür wird die gewalkte und auf das gewünschte Format zugeschnittene Masse mit einer harten Schutzschicht aus Zucker, Farb- und Aromastoffen versehen, die dem Kaugummi nicht nur Formstabilität und einen knackigen ersten Biss verleiht, sondern ihn auch vor dem Austrocknen schützt.

Veränderungen im Mund | Als besonders verführerisch empfinden wir auch Esswaren, die während des Verzehrs ihren Aggregatzustand verändern. Speiseeis beispielsweise wird als fester Stoff in den Mund gesteckt, schmilzt auf der Zunge und wird als Flüssigkeit geschluckt. Auch Schokolade zergeht langsam auf der Zunge. Verantwortlich dafür ist ihr Schmelzpunkt, der ungefähr der menschlichen Körpertemperatur entspricht. Verstärkt wird dieser Effekt noch durch gewisse Formgebungen, indem Schokolade beispielsweise zu hauchdünnen Blättchen gegossen wird, die im Mund besonders schnell schmelzen. Ähnlich wirken auch die Lufteinschlüsse in der so genannten Luftschokolade. Die poröse Struktur kratzt auf der Zunge und vergrößert die Oberfläche und damit die Kontaktpunkte zwischen Essen und Esser. Mit luftiger Schokomousse gefüllte Schokokugeln, die Aero Bubbles, werben sogar mit dem speziellen Texturerlebnis von Lufteinschlüssen, wenn auf der Packung zu lesen ist: „Feel the bubbles melt" – „Fühl, wie die Bläschen schmelzen"

Symbiose von hart und weich | Besonders aufregend sind aber nicht nur Texturen an sich, sondern vor allem auch die Kombination verschiedener Konsistenzen in ein und demselben Lebensmittel. Das Zusammenspiel von Knusprigem, Cremigem und Flüssigem in einer Speise ermöglicht uns vom herzhaften Reinbeißen bis zum lustvollen Lutschen eine breite Palette an genüsslichen Erfahrungen in einem einzigen Gericht. Auch Industrieprodukte wie die „Planets" von Mars spielen ganz bewusst mit einer möglichst breiten Palette an Texturen. „Soft, crispy and chewy", liest der Käufer auf der Packung und wird damit auf ein besonderes Konsistenzerlebnis hingewiesen. Wonach die Kugeln schmecken, bleibt im Dunkeln.

Bereits 1922 forderte der französische Küchenchef J. Maincave ausdrücklich „die Verbindung von Weichem und Knusprigem" und trug damit dem menschlichen Verlangen nach Abwechslung Rechnung. Kontraste innerhalb derselben Speise begeistern Menschen in allen Kulturkreisen. So werden in der Schweiz Brotstücke in geschmolzenen Käse getunkt, in Italien wird bissfeste Pasta mit flüssigen Saucen kombiniert und in Mexiko werden Nachos in sämige Guacamole getaucht. Lebensmittel werden bewusst so gestaltet, dass sie möglichst viele und möglichst unterschiedliche

To turn the basic substance into a marketable product, gum is heated, kneaded and refined with additives. Glycerine and other vegetable oils make it softer and more homogeneous. A high content of softeners gives the gum the elasticity needed (in addition to some skill) to inflate it so it forms a thin membrane without bursting all too quickly. Children not only like the ability to blow bubbles, bubble gums are also and primarily appealing to them because of their sweetness. The added sugar serves two purposes: it accounts for optimum taste whilst keeping the gum fresh and elastic at the same time. The sugar coating covering many chewing gums is applied to a similar end. Gum is rolled, cut into the desired shape and then covered with a hard protective coat of sugar, coloring and flavoring agents which not only helps it keep its shape and makes it crunchy when we first bite into it – it also protects the gum from drying out.

Changes in the Mouth | Food which changes its aggregate state as it is eaten is especially appealing to us. Ice cream is eaten in its solid state, melts on the tongue and is swallowed as a liquid. Chocolate also slowly melts in the mouth. This is due to the fact that its melting point is close to human body temperature. The effect is enhanced by certain shapes, e.g. paper-thin leaves which dissolve particularly fast. Air trapped in so-called aerated chocolate works along similar lines. The porous structure tickles the tongue and increases the surface, and thus the number of contact points between eater and food. Little chocolate balls filled with light chocolate mousse called Aero Bubbles even advertise the special experience with texture. The package says: "Feel the bubbles melt".

A Symbiosis of Hardness and Softness | Not only textures as such but also combinations of various consistencies in one and the same food can be especially exciting. The interaction of crunchy, creamy and liquid elements gives us a broad range of enjoyable experiences, from taking a hearty bite to the pleasure of sucking, in one go. Some industrial products such as Mars Planets deliberately play with the wide scope of textures. The package reads "Soft, crispy and chewy", indicating to the buyer that a special experience with consistency is in the offing. The actual taste of the chocolates remains in the dark.

As early as in 1922 the French chef J. Maincave expressly called for "the combination of soft and crispy elements", thus paying tribute to the human desire for variety. People of all cultural backgrounds enjoy contrasts brought together in one dish. In Switzerland, they dip pieces of bread into molten cheese, in Italy, firm pasta is combined with liquid sauces, and in Mexico, nachos are eaten with creamy guacamole.

Die kleinen rosaroten Kügelchen in diesem Trinkhalm verwandeln normale Milch, sobald sie durch ihn hindurch gesaugt wird, in einen süßen, zartrosa Erdbeerdrink.

The small pink-red balls in this drinking straw transform milk into a sweet, light-pink strawberry drink as soon as it is sucked through it.

Konsistenzen in sich vereinen. Deswegen streut man Rosinen in den Kuchen, mischt Schokosplitter ins Vanilleeis oder Smarties ins Joghurt. Cashew-Nüsse in indischen Currys erfüllen dieselbe Aufgabe wie Sonnenblumenkerne im Schwarzbrot oder Keksteigstücke und Karamellklumpen im Speiseeis. Als Vorreiter der Idee, „chunks" – also klein geschnittene Stückchen aller möglichen anderen Esswaren – in Eiscreme zu mischen, gelten die beiden New Yorker Aussteiger Ben Cohen und Jerry Greenfield, die 1978 im US-Bundesstaat Vermont eine Speiseeisfabrik gründeten. Unter dem Motto „anything goes" entwickelten sie Sorten, die unter anderem Cookie-, Bonbon-, Salzbrezel- oder sogar ungebackene Mürbteigbrocken enthalten.

Süße Kombinationen | Unzählige Desserts gründen ihren Erfolg auf dem Zusammenspiel verschiedener Konsistenzen, sei es in Form von Schichten, Füllungen oder als Rouladen. Sinnliche Kontraste bietet zum Beispiel die Crème Brûlée: Ein beinahe glasharter Karamellfilm bedeckt eine cremig weiche Mousse aus Ei und Zucker.

Mit ähnlichen Texturextremen spielt die Cremeschnitte, auch bekannt unter dem französischen Namen millefeuille (wörtlich: Tausendblatt). Die Idee, Teig in hauchdünnen, knusprigen Blättern zu backen, stammt ursprünglich aus Zentralasien. Schon von den mittelalterlichen nomadisierenden Turkvölkern weiß man, dass sie ein Faible für Brot aus feinen, gefalteten Schichten hatten. Aus diesem Brot entwickelten sich im Laufe der Zeit der Blätterteig und in Folge das salzige Börek und das süße Baklava. Der französische Koch Antonin Carême beschrieb Ende des 18. Jahrhunderts eine Süßspeise aus gefaltetem Blätterteig namens „Szegedinertorte", die mit Karamell überzogen war. Daraus hat sich dann – der Legende nach – die Creme-Teig-Schichtspeise namens Millefeuille entwickelt.

Erotische Geschmackserlebnisse | Wenn die Zunge zum ersten Mal die Schokoladehaut einer Praline durchdringt und im Inneren auf einen völlig unerwarteten Geschmack in Form einer süßen Creme stößt, so ist das ein sinnlicher Moment der Sonderklasse. Das Spiel unterschiedlicher Bissfestigkeiten ist das Erfolgsgeheimnis vieler Süßwaren. In Nusssplittern gewälzt, mit Schokolade überzogen, cremig gefüllt und in der Mitte eine knackige Haselnuss – manche Pralinen vereinen bis zu fünf verschiedene Konsistenzen, erlebbar mit einem einzigen Biss. Die Art des Verzehrs solcher vielschichtiger Süßigkeiten verrät auch manches über den Charakter des Essers. Während die einen herzhaft zubeißen, um sogleich in den Genuss der Fülle zu geraten, lösen die anderen vorsichtig Schicht für Schicht ab, um so die Vorfreude auf den süßen Kern zu zelebrieren.

Eine Ohrfeige und ihre Folgen | Gefüllte Lebensmittel sind ein kulinarisches Abenteuer. Die Ungewissheit über die zu erwartenden Konsistenzen und Geschmäcker im Inneren üben einen besonderen Reiz auf uns aus. Eines der diesbezüglich erfolgreichsten Desserts der letzten Jahre stammt vom französischen Sternekoch Michel Bras: Der „Chocolat Coulant" ist ein heiß servierter Schokoladekuchen, aus dessen Innerem beim Anschneiden flüssige Schokolade rinnt.

Inbegriff des gefüllten Desserts ist die Praline. Der Legende nach beruht ihre Erfindung auf einem Missgeschick. Demnach rutschte einem französischem Küchenjungen im Jahr 1671 eine Schüssel mit Mandeln aus den Händen und fiel zu Boden. Der wütende

Food is deliberately designed to combine as many different consistencies as possible. This is why we put raisins into cake dough, mix chocolate rasps into vanilla ice cream and lace our yoghurt with chocolate buttons – no matter whether they are Smarties or M&Ms. Cashew nuts in Indian curries fulfil the same task as sunflower seeds in brown bread or cookie crumbles and caramel in ice cream. The idea of putting "chunks" – pieces of candy bars, confections or dough – into ice cream was pioneered by the two New York dropouts Ben Cohen and Jerry Greenfield, who founded an ice cream factory in the US state of Vermont in 1978. Their motto was "anything goes" and they developed varieties containing pieces of cookies, candies, pretzels and even unbaked shortcrust dough.

Sweet Combinations | Innumerable desserts are well liked because they bring together various consistencies, for instance in fillings or in layer cakes and sponge rolls. Crème Brûlée is a fine example of sensory contrasts: a layer of hard caramel covers creamy and soft custard.

The cream slice, also known by its French name millefeuille (which literally translates as "thousand sheets") plays with similar texture contrasts. Originally, the idea of baking dough in thin, crusty sheets came from Central Asia. We know that the Turkic nomads of the Middle Ages liked bread made of thin folded sheets. Puff pastry as well as savory borek and sweet baklava developed from this bread. At the end of the eighteenth century, the French cook Antonin Carême described a puff pastry cake with caramel icing called after the Hungarian town of Szeged. Legend has it that mille-feuille, the slice consisting of cream and dough layers, was derived from it.

Erotic Taste Experiences | The moment when the tongue first pierces the skin of a piece of chocolate, hitting the completely unexpected taste of sweet cream filling inside, is a sensory and sensual experience beyond compare. The interaction of different degrees of firmness is actually the secret of success for many types of confections. Rolled in nut slivers, coated with chocolate, filled with cream or with a crunchy hazelnut inside – some confections or chocolates combine up to five different consistencies that can be experienced in one bite. The way in which people eat such multi-layered sweets also tells us a lot about the personality of the eater. Whilst some take a hearty bite to get to the filling rightaway, others carefully nibble off one layer after the other to celebrate the feeling of anticipation until they get to the sweet core.

A Slap in the Face and the Consequences | Foods with stuffing or filling are culinary adventures. The uncertainty about the consistencies and flavors to be relished when we get to the interior has a special appeal. One of the most successful desserts of recent years was French cook Michel Bras' "Chocolat Coulant". The Michelin three-star chef created a chocolate cake served hot with a liquid chocolate center revealed when the cake is cut.

The bite-size confectionery known as praline is the epitome of dessert. Allegedly, it was invented due to a mishap. In 1671 a French kitchen boy dropped a bowl with almonds.

Selbst extrem klebrige Nahrungsmittel wie weißen Nougat, Torrone, Toffee, Karamell oder Türkischen Honig, die beim Essen an den Zähnen haften, lieben wir gerade wegen ihrer Konsistenz.

We even love highly sticky foods such as white nougat, Torrone, toffee, caramel or Turkish delight that stick to our teeth when we eat them just because of their consistency.

In Mandelsplitter gewälzt, mit Schokolade überzogen, danach eine knusprige Waffelschicht, innen eine flaumige Creme und in der Mitte eine knackige Nuss: Manche Pralinen vereinen bis zu fünf verschiedenen Konsistenzen in sich – alles erlebbar bei einem einzigen Biss.

Rolled in almond slivers, covered with chocolate, followed by a crispy waffle layer, fluffy cream on the inside and a crunchy nut in the center. Some chocolates unite up to five different consistencies – all to be experienced in one single bite.

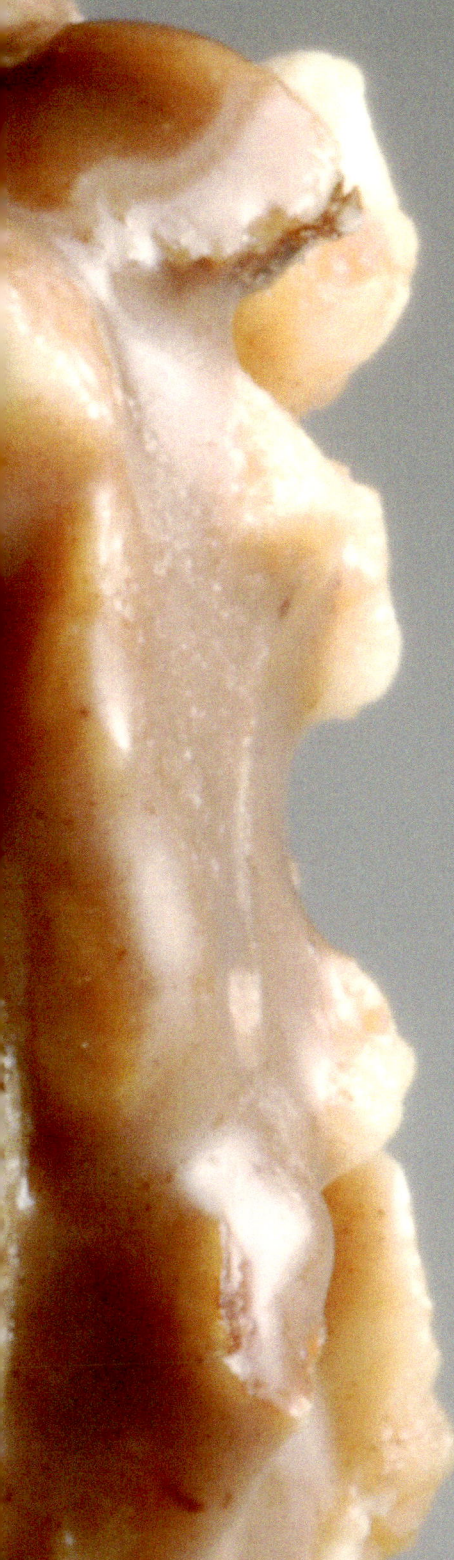

Küchenchef holte zu einer Ohrfeige aus und verschüttete einen Topf mit gebranntem Zucker über den verstreuten Mandeln. Das Malheur war perfekt, zumal der Herzog von Plessis-Praslin auf sein Dessert wartete. Notgedrungen servierten die verzweifelten Köche das Produkt des kulinarischen Debakels und lösten damit unerwartete Begeisterung aus. Monsieur Plessis-Praslin zeigte sich von der neuartigen Näscherei so beeindruckt, dass er der Süßspeise kurzerhand seinen Namen gab. Im Laufe der Zeit wandelte sich Praslin zur heute üblichen Bezeichnung Praline, der Karamellüberzug wich der wesentlich stabileren Schokoladeglasur.

Flüssig gefüllt | Von Marzipan über Nougat bis zu Sirup und Alkohol sind der Auswahl der Zutaten von Hülle und Fülle von Konfekt wenig Grenzen gesetzt. Die Frage der Konsistenz spielt aber immer eine wesentliche Rolle dabei. Pralinen werden bewusst so gestaltet, dass sie möglichst viele Texturen in sich vereinen: außen in Nusskrokant gewälzt, in Schokolade getunkt, darunter vielleicht noch eine knusprige Waffelschicht, dann eine weiche Creme oder ein zähflüssiger Likör und in der Mitte noch eine knackige Kirsche oder eine Nuss als Kern.

Sensorisch am interessantesten sind natürlich flüssige Füllungen, die allerdings auch technisch am aufwändigsten sind. Haben Sie sich beispielsweise schon einmal gefragt, wie der Rum in die Kugel kommt? Hersteller verrühren dazu Alkohol und Zucker zu einer zähflüssigen Lösung, spritzen die Masse in die jeweils passende Form – und warten. Während dieser Ruhephase weicht der Zucker nach außen und bildet dort eine kristalline Hülle um den flüssigen Alkoholkern. Da diese Ummantelung allerdings zu spröde ist, um den Weg in die Verkaufsregale schadlos zu überstehen, werden die Kugeln anschließend mit Schokolade dragiert – und fertig ist die Rumkugel, die beim Zerbeißen wegen der innen liegenden Zuckerkruste auch immer so schön knirscht.

Eine andere Rezeptur, die fasziniert, weil feste Zutaten flüssige umhüllen, stammt aus der Molekulargastronomie. Mithilfe von Natriumcitrat, Alginsäure und Calciumchlorid werden unterschiedliche Flüssigkeiten wie Säfte, Saucen und Getränke zu einige Millimeter bis mehrere Zentimeter großen Tropfen verkapselt oder „sphärisiert". Das Ergebnis sind so genannte „liquid drops", die man sich so ähnlich wie wassergefüllte Luftballons vorstellen kann: Sie haben eine zähe Haut und einen flüssigen Kern, fühlen sich so ähnlich an wie Kaviarperlen und platzen im Mund beim Zerdrücken oder Draufbeißen.

The angry chef made a swipe at him to slap his face and spilled a pot with caramelized sugar, which landed on the almonds scattered on the floor. A perfect disaster – after all, the Duke of Plessis-Praslin was waiting for his dessert. For want of a better solution, the desperate cooks served him the outcome of the unfortunate incident. Unexpectedly, Monsieur Plessis-Praslin was so impressed by the new variety of sweets that he gave it his name. As time went by, "Praslin" turned into today's "praline", and the caramel coating came to be replaced by the more stable chocolate exterior. The word "praline" is still widely used in the South of the US whilst "chocolates" is more common in the other English-speaking countries.

A Heart of Liquid | There is almost no limitation to the choice of materials for the center of chocolates – from marzipan to gianduia, from syrup to alcohol, almost anything is possible. Nevertheless, the issue of consistency is always important. Chocolates are designed in such a way that they invariably combine as many textures as possible: hazelnut brittle on the outside, a coat of chocolate which hides a crispy wafer layer, then soft cream or thick liqueur, with a cherry or a nut in the center.

Liquid fillings are most interesting for our senses; however, they are also most difficult to produce. Have you ever asked yourself how rum chocolate dragees are made? Producers mix alcohol and sugar into a gooey substance, inject it into the desired mold and – wait. It takes some time for the sugar to move to the outside and form a crystalline coat around the liquid alcohol center. However, the sugar coat is too brittle and would not survive transport to the shop. This is why the spherical sweets are coated with chocolate – which makes the first bite particularly delightful as the sugar layer beneath it is so nice and crunchy.

Another fascinating way of producing specialties in which solid ingredients surround a liquid center comes from the realm of molecular gastronomy. With the help of sodium citrate, alginic acid and calcium chloride, various liquids such as juices, sauces and beverages can be encapsulated in drops ranging in size from 1/8 to half an inch or an inch. This is called "liquid drops" and we can imagine them to look like water-filled balloons. With a tough skin and a liquid center, they feel like caviar grains, exploding when we crush or bite them. The unusual

Die ungewöhnliche Tasterfahrung im Mund potenziert sich mit der unerwarteten Geschmacksgebung: Grüner Kaviar schmeckt nach Erbsen, knallrote Kugeln, die in einer orangen Flüssigkeit schweben, entpuppen sich als Campariblasen in Orangensaft.

Von weich bis knusprig, von porig bis fluffy: Brot |

Für den Konsumenten ist Brot ein vergleichsweise einfaches Alltagsprodukt. Für Fooddesigner dagegen ist vor allem seine Konsistenz ein komplexes Thema. Die Erwartungen an die Textur von Brot und Backwaren divergieren je nach Kultur. Während viele Kontinentaleuropäer knuspriges, grobporiges Brot lieben, bevorzugt der angloamerikanische Raum weiches, elastisches Brot, dessen Rinde sich kaum vom Innenleben unterscheidet. In Europa haben sich eigene Zubereitungsmethoden und Formen entwickelt, welche die Bildung einer knusprigen Kruste fördern. So werden zum Beispiel Croissants und Salzstangen aus dünnen Teigstücken gerollt. Diese Methode schafft eine strukturierte Oberfläche, die beim Backen für dunkle Hochpunkte und weniger stark gebackene, weichere Einkerbungen sorgt. Außerdem lockern Lufteinschlüsse, die beim Rollen entstehen, die Konsistenz auf und verhindern schlecht durchgebackene Teigklumpen in der Mitte.[10]

Ähnlich funktioniert auch die Wiener Kaisersemmel, die traditionellerweise „geschlagen" wird. Der Bäcker faltet dabei ein flaches, rundes Teigstück von fünfzig Gramm viermal über seinen Daumen zur Mitte hin ein, bevor er den letzten Zipfel in den Hohlraum des ersten steckt und somit die Semmel schließt. Das Zentrum bleibt hohl, so dass der Teig im Inneren stets gut gebacken aus dem Ofen kommt, die fünf Segmente für eine strukturierte, knusprige Kruste.[11] Bei der maschinellen Erzeugung hingegen wird der charakteristische fünfteilige Stern in einen kompakten Teigklumpen gedrückt. Das Innenleben von Maschinensemmeln schmeckt daher oft lasch und teigig. Die ursprüngliche Idee, ein möglichst knuspriges Gebäck zu erschaffen, wich der Industrialisierung, seine optische Auswirkung, der berühmte Stern, ist jedoch geblieben.

From Soft to Crusty, from Porous to Fluffy: Bread |

For consumers, bread is a relatively simple product for every-day life. For food designers its qualities, and especially its consistency, are a complex matter. Expectations in respect of bread and bakery products vary according to cultural background. People from mainland Europe like their bread crusty and large-pored while the English-speaking world expects bread to be soft, elastic, with a crust that hardly differs from the center. Methods of baking and shapes that have developed in Europe are conducive to a crispy crust. Croissants and pretzel rolls consist of thin pieces of dough rolled up for a surface structure in which grooves remain soft whereas peaks are darker and crustier. Air is trapped due to the process of rolling the dough up, which helps avoid undercooked clusters of dough in the center.[10]

The Kaiser roll, also known as a Vienna roll or hard roll, comes about according to similar principles. Traditionally, the dough has to be "folded". The baker folds four corners of a flat piece of dough towards the center over his thumb, finally closing the roll by placing the fifth corner in the hollow space underneath the first one. The hollow center makes sure that the dough will always be thoroughly baked, the five segments form a structured crust.[11] By contrast, hard rolls produced by machines are compact pieces of dough imprinted with the star-shaped pattern. This is why the inside of machine-made rolls is often doughy and limp. The original idea of the crusty bakery product was lost but its visual appearance, the famous star shape, has been preserved.

Runder, nach außen gewölbter, japanischer Cracker mit schwarzem Sesam an der Außen- und dreidimensionalem Pinien-Motiv an der Innenseite.

Round curved Japanese cracker with black sesame on the outside and three-dimensional pine motive on the inside.

Für den europäischen Geschmack soll das optimale Brot eine krosse Rinde haben, die Röstaromen beinhaltet, das Innenleben weich und porig sein. Die beiden Parameter sind herstellungstechnisch jedoch schwierig zu vereinbaren. Das optimale Ergebnis liefert ein klassisches Sauerteigbrot, das bei niedriger Temperatur sehr lange fermentiert wird. Der wirtschaftliche Nachteil für die Hersteller ist, dass solche Brote eine lange Produktionsdauer haben und daher sehr teuer zu produzieren sind. Deswegen sucht man seit Jahren nach Techniken, die trotz kürzeren Fermentationszeiten möglichst flaumige Brotteige erzeugen. Einerseits experimentiert man mit chemischen Zusatzstoffen, andererseits mit physikalischen Methoden, die mithilfe von neuen Knet- und Backtechniken oder durch unterschiedliche Temperaturführung ein Brot herstellen, das wie traditionelles Brot schmeckt, aber in wesentlich kürzerer Zeit produziert werden kann.[12]

The type of bread best suited for Europeans should have a crunchy crust with a roasted flavor, with a soft and porous interior. In terms of production, these two parameters are hard to reconcile. Classic leavened dough which is fermented for a long time at a low temperature will yield the best result. The economic downside for manufacturers is that this type of bread takes long to produce and is therefore expensive. For years, producers have been looking for techniques to make dough that is fluffy in spite of shorter fermentation periods. They continue to experiment with chemical additives and physical methods, such as new kneading and baking technologies as well as various baking temperatures to come up with bread with a traditional taste yet shorter production times.[12]

Saucen – die flüssige Komponente | Im Alltag kommt unsere Vorliebe für Konsistenzunterschiede schon bei der klassischen Kombination von Fleisch und Sauce zum Ausdruck. Auch wenn ein gut gebratenes Steak mit einer bissfesten Kruste und einem weichen Inneren an sich schon mehrere Texturen bietet, rundet eine sämige Sauce dieses Gericht noch zusätzlich ab. Sirupartige Dips, cremige Saucen oder flüssige Dressings werden in allen Kulturen als Geschmacksverstärker zu Fleisch und Fisch geschätzt. Stark eingekochte Fonds sind dabei ebenso begehrt wie dickflüssige Chutneys oder spezielle Öle und Essige. In Norditalien, im Gebiet um die Stadt Modena, wird der Saft von spätgelesenen Trebbiano-Trauben durch Kochen eingedickt, gefiltert und vergoren. Langjährige Lagerung in verschiedenen Holzfässern führt schließlich zu einem fruchtigen Essig, der fast die zähflüssige Konsistenz von Öl hat. Unter dem Namen „Aceto Balsamico" wird er als Delikatesse gehandelt und unter anderem als Sauce für Früchte und Vanilleeis verwendet.

Im Gegensatz zur festen Nahrung bilden Saucen die flüssige Komponente vieler Gerichte und werden sowohl bei salzigen als auch bei süßen Speisen als Gimmick gesehen, der besonders gut schmeckt. Erdbeersauce zum Eis, der Bratensaft zum Fleisch oder das Ketchup zu den Pommes haben eines gemeinsam: dass ohne sie dasselbe Gericht nur halb so gut wäre. Pommes Frites werden nicht nur des Geschmacks wegen in Ketchup getunkt, sondern auch aufgrund seiner sämigen Textur, die sich von den knusprigen Pommes stark unterscheidet und diese „saftiger" macht. Ketchup entwickelte sich ursprünglich aus einer südostasiatischen Fischsauce namens „ketsiap", die mit Tomaten rein gar nichts zu tun hatte. Im 18. Jahrhundert brachten Seefahrer die Rezeptur über Europa nach Amerika, wo Henry Heinz, Sohn deutscher Einwanderer, 1878 das heute weltbekannte Tomatenketchup herausbrachte. Nicht ganz so berühmt, dafür wesentlich älter ist der Senf. Senfsamen wurden in China schon vor 3000 Jahren in der Küche verwendet, im antiken Rom soll dann die Idee entstanden sein, sie gemeinsam mit Essig, Salz und Traubenmost zu einer Würzpaste zu verarbeiten. Das erste überlieferte Rezept für Senf stammt von dem Römer Columella und wurde im ersten Jahrhundert nach Christus verfasst.

Sauces – the Liquid Component | In everyday life our predilection for differences in consistency can already be seen in the classical combination of meat and gravy. Even if a well-roasted steak that has a crispy crust and is soft on the inside offers several textures by itself, a viscid sauce really rounds off this dish. Syrupy dips, creamy sauces or liquid dressings are popular in all cultures as elements that enhance the taste of meat and fish. Thickened gravies are just as popular as gooey chutneys or special oils and vinegars. In northern Italy, in the area around the town of Modena, the juice of late-vintage Trebbiano grapes is thickened by cooking, filtered and fermented. After they have been stored for many years in various wooden casks, they become fruity vinegar that almost has the thick consistency of oil. It is sold as a delicacy under the name "aceto balsamico" and is even used as a sauce for fruit and vanilla ice cream.

As opposed to solid food, sauces constitute the fluid component of many dishes and are seen as a gimmick, a clever add-on, in both salty and sweet dishes, as something that is particularly tasty. Strawberry sauce with ice cream, gravy with meat or French fries with ketchup have one thing in common: without the sauce the dish would only be half as good. French fries are dipped into ketchup not just because of the taste but also because of the viscid texture that differs dramatically from the crunchy French fries, making the latter more "juicy". Ketchup was originally a Southeast-Asian fish sauce known as 'ketsiap' that had absolutely nothing to do with tomatoes. In the eighteenth century, seafarers brought the recipe from Europe to America where Henry Heinz, the son of German immigrants, launched the meanwhile world-famous tomato ketchup in 1876. Mustard is not so famous but it is considerably older. Mustard seeds were already used in cooking in China 3,000 years ago. In ancient Rome the idea emerged to mix them with vinegar, salt and grape must to create an herbal paste. The first mustard recipe to be transmitted came from the Roman Columella and was written in the first century after Christ.

Die neben Ketchup weltweit meistverbreitete Sauce ist die Sojasauce, welche bereits vor mehr als 2500 Jahren in China entstand. Von dort aus wurde die Rezeptur im 6. Jahrhundert von einer buddhistischen, vegetarisch lebenden Glaubensgemeinschaft nach Japan gebracht, wo man sie heute zur Zubereitung fast aller Gerichte verwendet. Bei der traditionellen Herstellungsmethode werden Sojabohnen gemahlen, gedünstet und mit geröstetem Reis- oder Weizenschrot gemischt. Danach wird die Masse mithilfe von Mikroorganismen in eine Trockenmaische verwandelt, mit Wasser angereichert und mit Salz gewürzt. Bei der anschließenden Fermentation spaltet sich das Sojaeiweiß in Aminosäuren auf, welche die Farbe und das Aroma der Sojasauce definieren. Durch holländische Händler gelangte die Sojasauce im 17. Jahrhundert nach Europa.

Eine gestalterisch raffiniertere Methode, Fleisch oder Fisch mit flüssigen Komponenten zu kombinieren, als bloßes Eintunken sind gefüllte Rezepturen. Das Hühnchen Kiew zum Beispiel, eine panierte Hühnerfleischrolle, ist mit Kräuterbutter gefüllt, die beim Frittieren schmilzt und herausfließt, sobald man die Roulade anschneidet. Mit flüssiger Butter, bissfestem Fleisch und knuspriger Panade bietet das Hühnchen Kiew immerhin drei sehr unterschiedliche Texturen. Praktisch, handlich und relativ sauber schafft es auch die österreichische Käsekrainer, flüssige und feste Komponenten zu kombinieren. Die Käsekrainer ist eine Wurst, die mit Käsestückchen gefüllt ist, welche beim Braten schmelzen. Die Abwandlung der slowenischen Wurstspezialität Kranjska Klobasa aus dem Gebiet Krain wurde angeblich in den 1980er Jahren in Graz entwickelt.

The most widespread sauce after ketchup is soy sauce which was created in China more than 2,500 years ago. From there, a group of vegetarian Buddhists brought it to Japan in the sixth century where it is still used to prepare almost all dishes. In its traditional production mode the soybeans are ground, steamed and mixed with roasted rice or shredded wheat. Then the mass is transformed into a dry mash with the help of microorganisms. Water is added and salt is used for taste. In the subsequent fermentation the soy protein splits into amino acids which are what define the color and aroma of the soy sauce. Thanks to Dutch merchants, soy sauce reached Europe in the seventeenth century.

Recipes with fillings are an elegantly creative way to combine meat or fish with liquid components, as a mere dipping. Chicken Kiev for instance is a breaded chicken roll filled with herbal butter which melts when the meat is fried. Once the chicken roll is cut the butter flows out. With melted butter, meat cooked al dente and a crispy breading, Kiev chicken features three very different textures. The Austrian "Käsekrainer" (cheese sausage) is a practical and relatively clean way to combine liquid and solid components. The Käsekrainer is a sausage filled with pieces of cheese that melt when the sausage is roasted. A variation of the Slovenian sausage specialty Kranjska Klobasa from the Krain region was allegedly developed in Graz in the 1980s.

Links | Säfte, Saucen und Dressings bilden die flüssige Komponente von Nahrung. Die neben Ketchup weltweit meistverbreitete Sauce ist Sojasauce, deren Rezeptur bereits vor mehr als 2500 Jahren in China entstand. Die japanische Firma Kikkoman wurde 1917 von den Familien Mogi und Takanashi als Noda Shoyu Co. Ltd. gegründet.
Mitte | Für die Erzeugung von traditionellem Balsamessig aus Modena wird der Saft von spätgelesenen Trebbiano-Trauben vergoren und mehrere Jahre in Holzfässern gelagert. Balsamico ist extrem fruchtig, hat fast die Konsistenz von Öl und wird mittlerweile von der steirischen Firma Gölles auch aus Äpfeln und anderen Früchten hergestellt.

left | Juices, sauces and dressings constitute the liquid component of food. The most widespread sauce apart from ketchup is soy sauce whose recipe was created more than 2,500 years ago in China. The Japanese Kikkoman co. was founded as Noda Shoyu Co. Ltd. by the Mogi and Takanashi families in 1917.
center | To create traditional balsamico vinegar from Modena, the juice of Trebbiano grapes picked late in the year is fermented and kept in wooden casks for several years. Balsamic is extremely fruity, almost has the consistency of oil and today, it is also made from apples and other fruits by the Styrian Gölles company.

Dem Wurstbrät aus Speck, Schweine- und Rindfleisch wird 10 bis20 Prozent Käse beigemengt. Dieser ist in kleine Würfel geschnitten, die sich verflüssigen, sobald die Wurst heiß wird. Damit treffen bei jedem Biss drei verschiedene Konsistenzen – zähflüssiger Käse, weich gegartes Wurstbrät und knackiger Schweinedarm – aufeinander. Steigern lässt sich das Texturspiel noch zusätzlich, indem man die Käsekrainer vor dem Braten mit einer zweizackigen „Stupfgabel" mehrfach ansticht. Durch die kleinen Löcher treten dann geringe Käsemengen aus und bilden an der Oberfläche einen knusprigen Käsefilm. Auch bei gewissen Fleisch- und Leberkäsesorten werden Emmentaler-, Paprika- oder andere Gemüsestücke ins Wurstbrät gemischt, um den Genuss durch Konsistenzunterschiede zu erhöhen.

Getränke – Mundgefühl, Textur, Prickeln | Nicht nur Esswaren punkten mit Textur. Gerade für Getränke ist das Mundgefühl oft der alles entscheidende Erfolgs- oder Misserfolgsfaktor. Auch Flüssiges kann mit unterschiedlichen Konsistenzen aufwarten: zum Beispiel Bier, das aus gestalterischer Sicht eine Kombination aus flaumigem Schaum, perlender Kohlensäure und wässriger Flüssigkeit darstellt. Einen besonderen Stellenwert bei Biertrinkern hat vor allem der Schaum, nicht nur weil er angenehm auf den Lippen kitzelt, sondern auch im Vergleich zu den meisten anderen Getränken etwas Einzigartiges ist. Technisch betrachtet, ist Bierschaum ein wenig stabiles Gefüge aus kleinen Flüssigkeitsperlen und Luft. Aufrechterhalten wird er durch bestimmte Proteine, welche die Spannung an der Oberfläche der winzigen Biertröpfchen erhöhen und in Schwebe halten. Da die Menge und die Beständigkeit von Bierschaum von den meisten Konsumenten als Qualitätsmerkmal wahrgenommen werden, erforschen große Bierkonzerne dessen Beschaffenheit und Stabilität.

Ten to twenty per cent of cheese are added to the sausage meat made of bacon, pork and beef. The diced cheese melts as soon as the sausage is heated. As a result three different consistencies come together in every bite: softly cooked, viscously thick cheese, sausage meat that has been cooked until soft, and crisp swine gut. This play of textures can be intensified by using a special fork (a so-called 'Stupfgabel') to pierce the cheese sausage several times before roasting. A small amount of cheese leaks through the tiny holes, leaving a crunchy cheese layer on the surface. There are also different types of meat and Leberkäse where pieces of Gruyère cheese, pepper or other vegetables are added to the sausage meat to enhance the taste by means of different consistencies.

Drinks – Texture and a Tingling Sensation | Not just food gains extra points with texture. For drinks in particular, the mouth feel is often crucial for their being a success or total failure. Also, liquids can offer a large variety of consistencies. Take beer as an example: in terms of design it represents a combination of fuzzy foam, bubbly carbonic acid and watery liquid. For beer drinkers the foam plays a special role not just for the way it gently tickles the lips but also because it is something truly unique compared to most other drinks. Technically speaking, beer foam is a hardly stable structure composed of small pearls of liquid and air. It is held together by certain proteins which enhance the tension on the surface of the tiny drops of beer and keep them suspended. Since most drinkers are able to perceive the amount and consistency of beer foam, large corporations study its consistency and stability.

Raffinierte Kombination von flüssig, fest und knusprig: Die Käsekrainer ist eine Wurst, deren Fülle mit Käsestückchen vermengt ist, welche beim Braten schmelzen. Steigern lässt sich das Texturspiel noch zusätzlich, indem man die Käsekrainer vor dem Braten mit einer zweizackigen „Stupfgabel" mehrfach ansticht. Durch die kleinen Löcher treten dann geringe Käsemengen aus und bilden an der Oberfläche einen knusprigen Käsefilm.

A clever combination of liquid, solid and crunchy: The 'Käsekrainer' is a sausage whose filling is mixed with pieces of cheese that melt when the sausage is roasted. The play of textures can be further enhanced when the Käsekrainer is pierced several times with a special fork ("Stupfgabel") before roasting. Small amounts of cheese come through the surface, forming a crunchy film of cheese.

Milchkaffee, Bier, Mineralwasser oder Tee: Nicht nur Esswaren punkten mit Textur, auch Flüssiges kann mit unterschiedlichen Konsistenzen aufwarten. Bier beispielsweise bietet drei verschiedene taktile Qualitäten: wässrige Flüssigkeit, perlende Kohlensäure und flaumigen Schaum. Technisch betrachtet, ist Bierschaum ein instabiles Gefüge aus kleinen Flüssigkeitsperlen und Luft. Aufrechterhalten wird er durch bestimmte Proteine, welche die Spannung an der Oberfläche der winzigen Biertröpfchen erhöhen und sie in Schwebe halten.

Milk coffee, beer, mineral water or tea – not just foods collect points with texture, liquids can also offer a variety of consistencyconsistencies. Beer, for instance, shows three different tactile qualities: watery liquid, bubbling carbonic acid and soft foam. In technical terms, beer foam is an instable formation of small bubbles of liquid and air. It remains intact due to certain proteins which increase the tension on the surface of the tiny drops of beer and keep them in suspense.

Links | Ein in Europa (noch) ungewohntes Texturerlebnis erzeugt koreanischer AloeVera-Saft mit Fruchtanteil.
Rechts | Um das Mundgefühl von Getränken zu verändern, werden unterschiedliche Zusatzstoffe beigemengt, unter anderem Kohlensäure, welche ursprünglich der Konservierung diente. Der deutsche Uhrmacher Johann Jacob Schweppe ließ sein Verfahren, Wasser mit Kohlensäure zu versetzen, 1783 für medizinische Zwecke patentieren. Das „Blubberwasser" wurde zunächst Patienten befreundeter Ärzte verabreicht.

left | Korean AloeVera juice containing fruit offers an experience of texture (still) unusual in Europe.
right | To change the mouth feel of drinks, various substances are added, including carbonic acid, which was originally used for conservation. The German clockmaker Johann Jacob Schweppe had his method of adding carbonic acid to water patented for medical purposes in 1783. The "bubbly water" was first administered to patients of befriended doctors.

Bierschaum hat aber auch eine geschmackliche Funktion, indem er die Bitterstoffe neutralisiert. Ohne den festen Schaum, der während des Brauens mithilfe von Stickstoff entsteht, wäre das irische Guiness noch viel bitterer. In Lokalen wird über spezielle Zapfhähne Stickstoff eingeblasen und Bierdosen enthalten immer öfter eine Metallkapsel mit flüssigem Stickstoff. Beim Öffnen reduziert sich der Druck in der Dose, der Stickstoff wird freigesetzt und sorgt für den erwünschten, feinen Schaum.[13]

Um das Mundgefühl von Getränken zu verändern, werden unterschiedliche Zusatzstoffe beigemengt, unter anderem Zucker und Kohlensäure. Die Beigabe von Kohlensäure diente ursprünglich der Konservierung von Wasser. Der Silberschmied und Uhrmacher Johann Jacob Schweppe aus Witzenhausen im deutschen Hessen entwickelte im späten 18. Jahrhundert ein Verfahren, um Wasser mit Kohlensäure zu versetzen, und ließ es 1783 für medizinische Zwecke patentieren. Das „Blubberwasser" wurde zunächst Patienten befreundeter Ärzte verabreicht. In der Folge gründete Schweppe mehrere Sodawasserfabriken und expandierte 1831 nach London, wo 1834 das Produktangebot um aromatisierte Getränke wie die „Schweppes Aerated Lemonade" erweitert wurde. Die Idee der kohlensäurehaltigen, angenehm auf der Zunge prickelnden Limonade war geboren.

Der Urtyp der heutigen Limonade ist der englische „Lemon Squash", ein Produkt aus Wasser, Zucker und Zitronensaft, das ursprünglich aus Indien stammt und dort noch immer unter dem Namen „lemon soda" erhältlich ist. Um 1870 begannen englische Offiziere in Indien malariavorbeugende Chinintabletten mit Limetten, Limonen und Wasser einzunehmen. Schweppe griff die Idee des Lemonsodas mit Chinin auf, steigerte das Prickeln des Zitronensaftes durch die Beigabe von Kohlensäure und kreierte das „Indian Tonic Water". Um 1885 entwickelte der Apotheker Dr. John Stith Pemberton in Atlanta mit dem „Pemberton's French Wine Coca" den Vorläufer von Coca-Cola. Erst 55 Jahre später entstand auch Fanta: Da die Produktion von Coca-Cola von 1942 bis 1949 in Deutschland völlig eingestellt war, entwickelte der Chemiker Dr. Schetelig im deutschen Essen 1940 als Ersatzprodukt eine Limonade namens Fanta.

Cremiger Instantkaffee | Auch bei Getränken wie Kaffee spielt das Mundgefühl eine wichtige Rolle. Neben Geschmack und Farbe bürgt die cremige Oberfläche einer frischen Tasse Espresso für Qualität. Die so genannte „Crema" sollte eine geschlossene Decke ohne Luftbläschen bilden, zwei Minuten lang erhalten bleiben und beim Schwenken der Tasse am Tassenrand haften. Um auch Instantkaffee, vor allem Instant-Cappuccino, eine besonders cremige Textur zu verpassen, haben Lebensmittelkonzerne ein eigenes Zubereitungsverfahren entwickelt. Dabei wird das Kaffeepulver sprühgetrocknet, das heißt, während der Trocknung wird unter Druck Luft eingeblasen. Dadurch bilden sich im Instantpulver mikroskopisch kleine Lufteinschlüsse, die beim Auflösen des Kaffees leicht schäumen und auch bei Instantkaffee die besagte Crema erzeugen.[14]

Yet beer foam also has a taste-related function in the sense that it neutralizes the bitter substances. Without the firm foam that is created while brewing beer by means of nitrogen, Irish Guinness beer would have a much more bitter taste. In taverns special taps are used to blow in nitrogen and beer cans now more often contain a metal capsule with liquid nitrogen. When the can is opened, the pressure in the can is reduced and the nitrogen is released, which creates the desired fine foam.[13]

In order to alter the sensation of drinks in the mouth, various substances are added, including sugar and carbonic acid. The original purpose of adding carbonic acid was to preserve water. The silversmith and clockmaker Johann Jacob Schweppe from Witzenhausen in German Hessen developed a procedure in the late 18th century for adding carbonic acid to water and had this technique patented in 1783 for medical purposes. The "bubbly water" was first administered to the patients of befriended doctors. Subsequently, Schweppe founded several soda water factories and in 1831 expanded production to London where in 1834 aromatic drinks such as "Schweppes Aerated Lemonade" were added to the product range. The idea of a carbonated lemonade with a pleasant tingling sensation on the tongue was born.

The original type of today's lemonade is the English "Lemon Squash", a product made of water, sugar and lemonade, which originally came from India and is still available there under the name "lemon soda". Around 1870 English officers in India began taking quinine tablets with limes, lemons and water to prevent malaria. Schweppe took up the idea, increasing the tingling sensation of the lemonade by adding carbonic acid and creating the "India Tonic Water". Around 1885 Atlanta-based pharmacist Dr. John Stith Pemberton created a precursor of Coca-Cola with his "Pemberton's French Wine Coca". It was not until 55 years later that Fanta was also created. Since production of Coca-Cola was completely halted in Germany from 1942 to 1949, chemist Dr. Schetelig created an alternative drink in Essen in 1940 – a lemonade with the name of Fanta.

Creamy Instant Coffee | Also in drinks like coffee the response of taste buds plays an important role. In addition to taste and color the creamy surface of a cup of freshly made espresso guarantees quality. The so-called 'crema' is supposed to create a closed blanket without any air bubbles, to stay intact for two minutes and to stick to the edge of the cup when the cup is tilted. To also give instant coffee, in particular instant cappuccino, an especially creamy texture, food corporations have developed a special processing technique. Coffee powder is spray-dried, that is to say during the drying process air is injected under pressure. This way, microscopically tiny air bubbles create a light foam when the coffee is dissolved.[14]

Auch der Brauch, Tee oder Kaffee mit Milch zu trinken, ist eine Methode, um die Textur dieser wässrigen Getränke zu verändern beziehungsweise cremiger zu machen. Historisch hat sich die Tradition, Tee mit Milch zu mischen, allerdings aus anderen Gründen entwickelt. Aus Angst, das wertvolle Porzellan, aus dem man das sündhaft teure Modegetränk schlürfte, könnte beim Eingießen des heißen Tees zerspringen, füllte man vorsichtshalber zuerst etwas kalte Milch in die Tasse. In Großbritannien, wo rund 90 Prozent ihren Tee mit Milch mischen, gießt man bis heute zuerst die Milch und erst danach den Tee ein.[15]

The custom of drinking coffee with milk is also a way of altering the texture of this watery drink or to make it creamier. Historically, the tradition of mixing tea with milk has, however, developed for other reasons. Out of fear that the one might break the precious porcelain from which one sipped the sinfully pricey fashionable drink, one took the precaution of first pouring some cool milk into the cup. In Great Britain where 90 percent of tea drinkers mix their tea with milk, one still first pours milk into a cup before adding tea.[15]

1 | vgl.: Karl-Heinz Plattig, Spürnasen und Feinschmecker; Springer Verlag 1995
2 | Interview für den Dokumentarfilm: „Food Design"; Geyrhalter Film GmbH Wien
3 | Wir bedanken uns bei Werner Mlodzianowski für diese Information.
4 | Wir bedanken uns bei Heinz Hanner für diese Information.
5 | Wir bedanken uns bei Frank Förster für diese Information.
6 | vgl.: Christoph Wagner, Fast schon Food; Lübbe Verlag 2001; und Wikipedia
7 | Wir danken Dipl. Ing. Klaus Dürrschmid für diese Information.
8 | Wir danken den „Jelly Mongers" für diese Information.
9 | Wir bedanken uns bei Heinz Hanner für diese Information.
10 | Wir bedanken uns bei Bäckermeister Hoheneder für diese Information.
11 | Wir bedanken uns bei Bäckermeister Hoheneder für diese Information.
12 | Wir bedanken uns bei Werner Mlodzianowski für diese Information.
13 | Wir danken Dr. Peter Wilde vom Institute of Food Research in Norwich für diese Information.
14 | Wir danken Dr. Peter Wilde vom Institute of Food Research in Norwich für diesen Hinweis.
15 | Wir danken Steven Twining für diese Information.

1 | See: Karl-Heinz Plattig, Spürnasen und Feinschmecker; Springer Verlag 1995
2 | Interview for the documentary "Food Design"; Geyrhalter Film GmbH Vienna
3 | A note of thanks for this information goes to Werner Mlodzianowski.
4 | We thank Heinz Hanner for this information.
5 | A note of thanks for this piece of information goes to Frank Förster.
6 | See: Christoph Wagner, Fast schon Food; Lübbe Verlag 2001; and Wikipedia
7 | A note of thanks for this piece of information goes to Klaus Dürrschmid.
8 | A note of thanks for this item of information goes to "Jelly Mongers".
9 | We thank Heinz Hanner for this piece of information.
10 | We thank master baker Hoheneder for this item of information.
11 | We thank master baker Hoheneder for this item of information.
12 | A note of thanks for this item of information goes to Werner Mlodzianowski.
13 | We thank Dr. Peter Wilde from the Institute of Food Research in Norwich for this information.
14 | We thank Dr. Peter Wilde from the Institute of Food Research in Norwich for this information.
15 | We thank Steven Twining for this information.

Der Brauch, Tee oder Kaffee mit Milch zu mischen, ist eine Methode, um die Textur dieser wässrigen Getränke cremiger zu machen. Historisch hat sich die Tradition, Tee mit Milch zu trinken, allerdings aus anderen Gründen entwickelt: Aus Angst, das wertvolle Porzellan könnte beim Eingießen des brennheißen Tees zerspringen, füllte man vorsichtshalber zuerst etwas kalte Milch in die Tasse. In Großbritannien, wo rund 90 Prozent ihren Tee mit Milch mixen, gießt man bis heute zuerst die Milch und erst danach den Tee ein.

The custom of mixing tea or coffee with milk is a method of making the texture of these watery drinks creamier. Historically, this tradition was developed out of the fear of breaking the precious porcelain with the hot tea, some cold milk was first poured into the cup. In Great Britain, where about 90 per cent of the people have their tea with milk, the tradition of pouring milk first, then tea, has survived until today.

Optik
optics

DIE OPTIK – Das Auge isst mit

Meistens sagt uns bereits der Anblick einer Speise, ob sie uns schmecken wird oder nicht. Schon von weitem kann allein der optische Eindruck den Appetit anregen oder Ekel auslösen. Wir sehen, ob wir etwas überhaupt in den Mund stecken wollen, und malen uns bereits im Geiste den Geschmack aus, der uns erwartet.

Wie muss Essen gestaltet sein, damit uns schon beim Anblick das Wasser im Mund zusammenläuft? Die Proportionen von Fleisch-, Fisch- und Gemüsestücken auf dem Teller, die Anzahl der gereichten Zutaten, das Zusammenspiel unterschiedlicher Formen, Farben, Proportionen und Texturen auf dem Teller entscheiden, ob uns ein Gericht harmonisch, also appetitlich erscheint. Wird unsere Vorstellung von Ästhetik optisch befriedigt, so steigert das den Genuss, das Essen „schmeckt" besser, auch wenn es in Wahrheit nur anders aussieht.

Die alles entscheidende Frage für Fooddesigner ist: Was empfinden wir bei Esswaren als optisch ansprechend? Gibt es zum Beispiel allgemein gültige Gestaltungsvorgaben für möglichst appetitlich wirkende belegte Brötchen? Erschwerend wirken sich dabei individuelle und kulturelle Unterschiede aus. Ästhetisches Empfinden ist sozial erlernt und variiert von Gesellschaft zu Gesellschaft.

Außerdem ist die Bewertung von Geschmack und Stil beim Essen genauso Moden unterworfen wie in anderen Designsparten: Symmetrische, asymmetrische, minimalistische und überladen manierierte Kompositionen wechseln einander auf dem Teller ab. Auch die Vorliebe für bestimmte Formen, Farben, Portionsgrößen oder Zutaten sind vom jeweiligen Zeitgeist abhängig. Heute belächelt man grüne Kiwi- und gachgelbe Ananasscheiben als Dekor auf gebratenem Fleisch oder rosarotem Schinken. Klein gewürfelte, orange-gelb-grüne Gemüsestückchen tragen den Zeitstempel der 1960er und 70er Jahre. Ästhetik ist eben nicht zeitlos.

Die Ästhetik des Essens

Generell lässt sich vielleicht sagen, dass wir infolge der instinktiven Prägung, uns möglichst breitgefächert zu ernähren, Speisen vorziehen, die eine größtmögliche optische Vielfalt bieten. Wobei Vielfalt, historisch betrachtet, auch für Überfluss und Luxus steht, denn je mehr Zutaten, desto aufwändiger sind deren Beschaffung und die Zubereitung des Gerichts. Auch heute empfinden wir eine Speise als besonders appetitanregend, wenn sie aus vielen unterschiedlich großen, unterschiedlich geformten Elementen mit unterschiedlichen Texturen in unterschiedlichen Aggregatzuständen und Farben besteht. Ein ovales, flaches Stück Fleisch von solider, faseriger Textur mit einer transparenten Flüssigkeit als Sauce, schmalen, langen, grünen Bohnen und gelbem, cremigem Kartoffelpüree bietet eine relativ breite Palette optischer Eindrücke. Auch belegte Brötchen wirken dann besonders lecker, wenn kleine und große, runde und eckige, feste, knackige und weiche, farblich kontrastierende Objekte auf ihnen vereint sind.

OPTICS – You also eat with your eyes

Usually, just the look of a dish tells us whether it will taste good or not. Even from a distance the optical impression can whet our appetite or trigger disgust. We can see whether we want to stick something in our mouth and can already mentally picture the taste.

How does food have to be styled so that just looking at it makes our mouth water? The proportions of pieces of meat, fish and vegetables on the plate, the number of ingredients used, the interplay of various shapes, colors, proportions and textures on a plate already decide whether a dish appears harmonious, that is to say appetizing to us. If our idea of esthetics is satisfied optically, this intensifies the pleasure, food then "tastes" better, even if, in reality, it only looks different.

The all-decisive question for food designers is: what do we perceive as optically appealing in food? Are there, for example, generally valid design parameters for open-face sandwiches that are made as appetizing as possible? Individual and cultural differences make things even more difficult. Esthetic perception is something that is socially acquired and varies from society to society.

Moreover, the assessment of taste and style in food is just as much subject to fashions as other design areas – symmetric, asymmetric, minimalist and overly stylized compositions alternate on a plate. Also the predilection for certain shapes, colors, proportions or ingredients are dependent on a given zeitgeist. Today one laughs at green kiwi and bright yellow pineapple slices as decoration on roasted meat or pink-red ham. Small diced orange-yellow-green pieces of vegetables bear the stamp of the 1960s and 1970s. Esthetics is simply not timeless.

The Esthetics of Food

Generally speaking, we could perhaps say that our nutrition encompasses as large a spectrum as possible in keeping with our instinctual disposition. In other words, we prefer dishes that offer as much optical diversity as possible. From a historical perspective, diversity also stands for surplus and luxury, since the more ingredients the more complicated it is to procure them and to put together the dish. Even today we perceive a dish as being particularly appetizing when it consists of many different-sized, different-shaped elements with different textures in different aggregate states and colors. An oval, flat piece of meat with a solid, fiber-rich texture with a transparent liquid as a sauce, thin, long green beans and yellow, creamy mashed potatoes offers a relatively broad range of optical impressions. Even sandwiches with cold cuts then appear particularly appealing when small and big, round and rectangular, solid, crisp and soft objects of contrasting colors are assembled on them.

Belegte Brötchen, frisches Gemüse, Fruchttorten oder traditionelle, japanische Süßigkeiten: Auf Grund der instinktiven Prägung, uns möglichst vielfältig zu ernähren, machen uns Speisen dann besonders Gusto, wenn sie aus möglichst vielen verschiedenen, großen und kleinen, runden und eckigen, festen und flüssigen, knackigen und weichen, farblich kontrastierenden Elementen bestehen.

Sandwiches, fresh vegetables, fruit cakes or traditional Japanese sweets: Because of the instinctive disposition of keeping our nutrition as diverse as possible, foods appear especially appetizing to us when they are made up of as many different large and small, round and square, solid and liquid, crunchy and soft, multi-colored elements as possible.

A look at Japan shows that other cultures have relatively clear guidelines regarding the design of dishes. In the traditional Kaiseki cuisine the preparation of meals follows the esthetic principles of Ikebana, the Japanese art of flower arrangement. Until several years ago Japanese cooks even had to learn Ikebana in the course of their training, but in the meantime the curricula are no longer so rigid. The fundamental principles have, however, remained the same. The arrangement basically adheres to the shape of a triangle. Three main elements are placed in the shape of an equilateral triangle, the biggest or highest element always being farthest away from the eye of the person eating. In terms of color, the meal is supposed to have only one main accent, which is to be framed by two more subtle shades.[1]

Flowers as a Model | Even in Europe parallels can be drawn between the arrangement of flowers and that of food. A bouquet of flowers always consists of one, three, five, seven or any other uneven number of flowers. Four roses in a vase is something we see as optically unsatisfying, an idea we also find reflected in many foods. The Viennese Kaiser roll or the Italian Rosetta consists of five and not of four or six parts. Traditional, multi-level wedding cakes always have an uneven number of "stories". Even 'Linzer Augen' or other filled, round cookies always have a hole in the middle or three holes that are evenly distributed over the surface. According to Konrad Paul Liessmann from the Department of Philosophy at the University of Vienna, the symbolic-esthetic appeal of uneven numbers lies in the fact that they are indivisible. An uneven number of elements thus comes to bear in objects and acts with which something inseparable, common is to be conjured, as for instance, marriage, partnerships, promises of fortune and the like.

That this esthetic sensation is not a generally valid phenomenon but a culturally acquired one can be seen for instance in Japan where round crackers are decorated with two equally big dabs of fruit syrup on the surface. The arrangement of two identical objects next to each other can be found daily in Japan. Contrary to Europe or America, it is seen as optically appealing. Sushis are also always served in pairs. Europeans, by contrast, show little sympathy for combinations of equal pairs. In France, Germany or Italy there are always an uneven number of decorations arranged in the middle of round objects such as cakes or cookies. A cake with two or four symmetrical dabs of whipped cream looks strange, unbalanced and simply not "nice".

Geometric vs. Natural | Not just the arrangement of a dish but also the design of ingredients and food contributes to making it a real treat. An appearance that is as natural as possible can be just as appealing as perfect geometric bodies. Traditional English finger sandwiches, contemporary French dessert combinations or Japanese sushis are artificially rectangular and are appealing for

Round objects such as cakes or cookies are always decorated with 1, 3 or more than 5 equally distributed elements. A cookie with 2 holes or a cake with two cream toppings appears optically unbalanced and disrupts our aesthetic feelings, just like a Bundt cake whose hole is not centered. However, this esthetic approach is not commonly valid but culturally acquired – Japan is unlike Europe and the US, in that people consider two adjacent equivalent objects to be visually appealing.

künstlich rechteckig und gefallen gerade wegen der perfekten Bearbeitung des Naturproduktes durch den Menschen. Auch kreisrunde Torten, Pizzen und kugelförmige Pralinen oder Lollipops faszinieren durch ihre perfekte Geometrie.

Interessant bei der formalen Analyse von Esswaren ist auch, dass die japanische Kultur im Vergleich zur westlichen extrem viele rechteckige Formen kennt. Für Anthropologen spiegelt das die stark hierarchische Struktur der japanischen Gesellschaft wider. Demokratische Gesellschaften neigen dazu, ihr Essen rund zu formen, da der Symbolik der Gemeinschaft und der gleichmäßigen Verteilung der Nahrung innerhalb der Gruppe große Bedeutung beigemessen wird. In Japan dagegen spielt die gerechte Aufteilung bei Tisch eine untergeordnete Rolle, im Vordergrund stehen die Harmonie der Form und hierarchische Regeln, die Ablauf und Zeremonie des Essens prägen.[2]

Im Gegensatz zu geometrischen Formen steht die ästhetische Wirkung besonderer Natürlichkeit und scheinbar zufälliger Anordnungen wie jener von Salzkristallen auf Broten und Salzstangen. Ein fast philosophischer Zugang zum Thema Natürlichkeit findet sich wiederum in der japanischen Esskultur, wo Formen auch dann als besonders harmonisch gelten, wenn sie so wenig wie möglich durch Menschenhand oder Werkzeuge „verfälscht" wurden.

Links | Japanisches Yuba ist ein Nebenprodukt der Tofuproduktion, für das die Haut auf der gekochten Sojamilch abgeschöpft und getrocknet wird. Zu kleinen Paketen gefaltet und mit einer Schleife zusammengebunden, wird es zum Beispiel in der Suppe serviert.
Rechts | Kelp-Kombu ist ein Meeresgewächs, das für die Zubereitung von Suppe verwendet wird. Die besondere Gestaltung in der Form des Mosubu („Zusammenbringens"), einem Motiv aus dem Shinto-Symbolismus, weist auf die Verwendung bei Tisch hin, etwa in einer Schale Suppe, die dem Gast serviert wird.

left | Japanese Yuba is a side product of tofu production for which the skin on the cooked soy milk is skimmed and dried. It is folded to form small parcels, tied with a ribbon and then served, for instance, in soup.
right | Kelp-Kombu is a sea plant that is used for preparing soup. The special design in the form of the mosubu ("bringing together"), a motive from Shinto symbolism, refers to the use at the table, for instance, in a bowl of soup that is served to the guest.

just the way the natural product is perfectly processed by man. Even circular cakes, pizzas and ball-shaped chocolate truffles or lollipops are fascinating because of their perfect geometry.

An interesting aspect in the formal analysis of food is also the fact that Japanese culture knows an extreme number of rectangular shapes as opposed to western culture. For anthropologists this reflects the strongly hierarchical structure of Japanese society. Democratic societies have the tendency to shape their food in a round way since the symbolism of community and the equal distribution of food within the group is attributed great meaning. In Japan, by contrast, the equal distribution at the table plays a lesser role, while the focus is on the harmony of form and hierarchical rules that define the sequence and ceremony of eating.[2]

Different from geometric shapes, there is the esthetic effect of special naturalness and seemingly coincidental arrangements like those of salt crystals on bread and salt rolls. In Japanese culture we find an almost philosophical approach to the theme of naturalness where forms are also seen as particularly harmonious when they have been "adulterated" as little as possible by human hand or tools.

Das japanische Konzept, ein Naturprodukt in Geschmack und Aussehen möglichst wenig zu verändern, wird zum Beispiel bei traditionellen, halbrunden Reiscrackern ersichtlich. Im Zuge ihrer Herstellung wird der Teig zu einer Rolle geformt und dann auf ein Holzbrett gelegt, wo er sich durch seine weiche Konsistenz zu einem halbrunden Querschnitt verformt. Die so entstandene Form gilt in Japan als besonders schön, da sie sich völlig „natürlich", ohne Zutun durch Menschenhand, ergibt.[3]

Was wir beim Essen als „schön" empfinden, ist also stark kulturell geprägt und hängt untrennbar mit den philosophischen Konzepten und symbolischen Werten, die wir mit bestimmten Gestaltungsprinzipien verbinden, zusammen. Jeder Mensch hat durch seine Erziehung und durch sein gesellschaftliches Umfeld gelernt, gewisse Dinge als schön und andere als hässlich zu empfinden. Die Frage nach der Ästhetik des Essens ist daher in Wahrheit eine kulturelle und keine optische.

Das Auge als Scanner der Verträglichkeit | Das Auge entscheidet aber nicht nur über „appetitlich" oder „unappetitlich", sondern auch über „genießbar" oder „ungenießbar". Wenn wir unser Essen betrachten, steigert das nicht nur den Lustgewinn, sondern dient auch als eine Art „Security Check", der vor dem Verzehr verdorbener Zutaten warnt. Mit dem Gesichtssinn ist der Mensch in der Lage, Alter und Genießbarkeit einer Mahlzeit abzuschätzen. Ist die Oberfläche eines Schinkens matt und rosa oder schillert sie weißlich oder gar grünlich? Glänzt die Kruste einer Brezel oder wirkt sie bereits stumpf? Befinden sich Flüssigkeitsablagerungen an den Rändern von Cremen, Joghurts oder Ketchup? Hersteller helfen beim optischen Eindruck etwas nach, indem sie ihre Produkte zum Beispiel nachträglich glänzen und so frischer erscheinen lassen. Äpfel werden gewachst, aber auch Gummibären erhalten eine feine Schicht aus Bienenwachs, damit sie schön farbig leuchten. Schokolade wird mit Schellack, Smarties werden mit Baumharz aufpoliert, Häppchen und Desserts mit Gelatine überzogen.

Anhand des optischen Eindrucks von Farbe, Konsistenz und Oberfläche können wir beurteilen, wie frisch ein Lebensmittel ist. Die große Kunst bei der industriellen Produktion von Desserts und Puddings zum Beispiel liegt weniger im Geschmack als darin, eine Rezeptur zu finden, die sowohl die Creme als auch das Sahnehäubchen über Wochen in der exakt gleichen Konsistenz hält, ohne dass sich die Zutaten entmischen oder sich Flüssigkeit absetzt. Hersteller schaffen es unter anderem durch die Beigabe von Hydrokolloiden und langkettigen Zuckern, Emulsionen über Wochen und Monate hinweg zu stabilisieren.[4]

Neben Oberfläche und Textur liefert das Auge als Sicherheitsscanner auch wertvolle Informationen zum Thema Farbe. Aufgrund von Erfahrungswerten können wir mithilfe der Farbe die Beschaffenheit des Essens überprüfen: Wir haben gelernt, den Reifegrad von Obst und Früchten oder den Fettgehalt von Milch anhand der Farbe abzuschätzen. Viele dieser Analysen erfolgen unterbewusst. Gelblicher Käse zum Beispiel macht uns mehr Gusto als weißlicher, weil wir uns einen höheren Fettgehalt und damit mehr Nährwert und einen intensiveren Geschmack erwarten.

The Japanese idea of changing the taste or look of a natural product as little as possible is visible for instance in traditional, semi-circular rice crackers. During production the dough is shaped to form a roll and then it is placed on a wooden board where, thanks to its soft consistency, it forms a semi-round cross-section. The resulting shape is seen in Japan as being particularly appealing since it is completely "natural" without any human intervention.[3]

What we view as "appealing" in food is thus strongly culturally informed and inextricably linked to philosophical concepts and symbolic values that we associate with certain design principles. Each individual has learned through upbringing and social environment to see certain things as nice and others as ugly. The question of culinary esthetics thus in reality boils down to one of culture and not of optics.

The Eye as a Means for Scanning Agreeable Food | The eye not only determines whether something is 'appetizing' or 'unappetizing' but also whether it is 'agreeable' or 'disagreeable'. Looking at food not only increases the pleasure gained but also serves as a sort of 'security check' that warns us against eating food with spoiled ingredients. The eyes enable a person to assess the age and agreeability of a meal. Is the surface of ham dull and pink or does it have a whitish or even a green shimmer? Does the crust of a pretzel shine or is it already dull? Is there residual liquid on the edges of creams, yoghurts or ketchup? Manufacturers try to enhance the optical impression by for instance making their products shine so that they look fresher. Apples are waxed and even gummi bears are covered with a fine layer of bee wax so that their colors glow. Chocolate is polished with shellac, smarties with tree resin, and canapés and desserts are covered with gelatin.

Thanks to the optical impression of color, consistency and surface we are able to assess how fresh food is. The great art in the industrial production of desserts and puddings, for instance, lies less in taste than in finding a recipe that keeps the consistency of both cream and the layer of whipped cream exactly the same for weeks without ingredients dissolving or fluid settling at the bottom. By adding hydrocolloids and long-chain sugar compounds, manufacturers succeed in stabilizing emulsions for weeks or even months.[4]

In addition to surface and texture, the eye serves as a precautionary scanner by also providing valuable information about color. Based on empirical values, we know that color can be used to determine the condition of food. We have learned to assess the ripeness of fruit or the fat content of milk on the basis of color. Many of these analyses are unconscious. Yellowish cheese for instance appeals more to us than whitish cheese since we expect it to have a higher nutritional value and a more intensive taste.

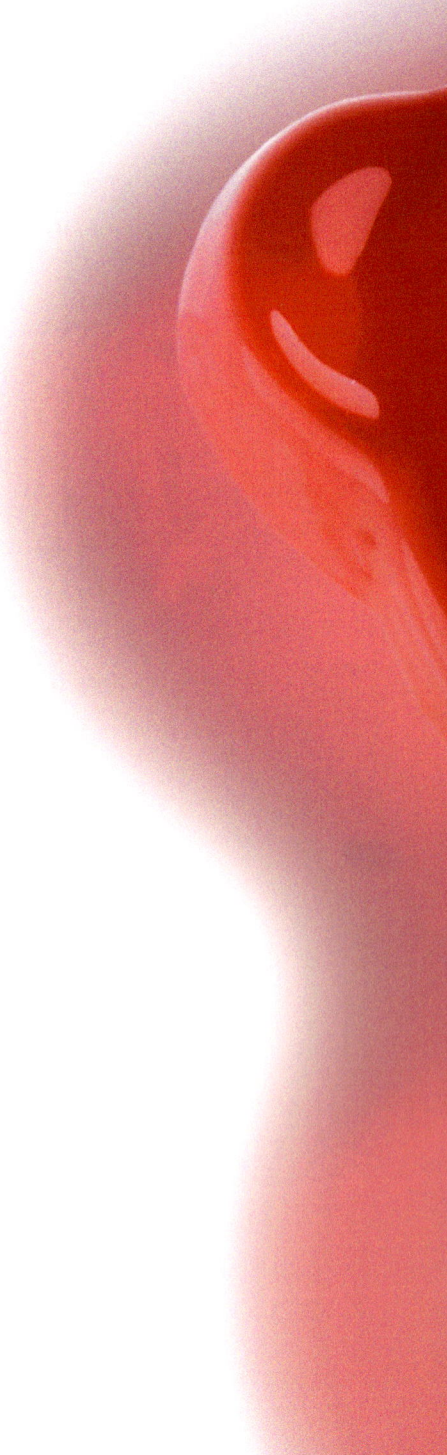

Die Farbe erweckt bestimmte Geschmackserwartungen. Besonders gerne greifen wir zu Rotem, weil es uns an reife Früchte erinnert und wir uns einen süßen Geschmack erhoffen.

The color triggers certain taste expectations. We are particularly attracted to red because it reminds us of ripe fruits and we hope for a sweet taste.

Der Geschmack von Farbe | „*Wir haben an der Universität Experimente durchgeführt, bei denen wir Lebensmittel eingefärbt haben. Zum Beispiel haben wir Kartoffeln schwarz, Blumenkohl grün und Spargel rot gefärbt. Und obwohl der Geschmack – darauf haben wir wirklich sehr geachtet – gleich geblieben ist, haben die Leute das nicht gegessen, weil sie mit der Farbe einen ganz anderen Geschmack assoziierten als den, wonach das Lebensmittel dann tatsächlich schmeckte.*" Gisla Gniech[5]

Farben machen uns Appetit und erwecken bestimmte Geschmackserwartungen. Sind rote Speisen also süß und schwarze bitter? Die Wahrnehmung der Farbe gibt bei der Entscheidung, ob wir etwas überhaupt in den Mund stecken oder nicht, oft den Ausschlag. Die deutsche Sozialwissenschaftlerin Eva Heller ließ bei einem Experiment 1.888 Personen den vier Hauptgeschmacksrichtungen bitter, sauer, salzig und süß Farben zuordnen. Grün und Gelb wurden dabei vorwiegend mit „sauer" assoziiert, Rosa, Orange und Rot mit „süß". Bei Weiß, Grau und Blau erwarteten die Testpersonen einen salzigen Geschmack, Violett, Schwarz und Braun ließ auf Bitteres schließen.[6]

Der Mensch liebt Essbares in Farbtönen, die an natürliche Esswaren erinnern: Farben wie Rot, Orange, Gelb oder Grün assoziieren wir mit reifen Früchten oder knackigem Gemüse. Sie signalisieren uns Bekömmlichkeit und Wohlgeschmack und regen unseren Appetit an. Hingegen besteht eine instinktive Hemmschwelle vor dem Verzehr dunkler, violetter oder unnatürlich intensiver Farbschattierungen. Tatsächlich gibt es auch nicht so viele natürliche Esswaren, die blau oder schwarz sind, weswegen uns ihr Anblick am Teller skeptisch macht und warnt, dass das Dargebotene bitter, ungenießbar oder sogar giftig sein könnte. Es ist sicher kein Zufall, dass es viele Menschen gerade vor Blutwurst und Sepiatinte graut, viele keine schwarzen Oliven, Kaviar oder Lakritze mögen. Die dunkle Farbe spielt bei der Ablehnung dieser Esswaren eine zentrale Rolle.

Tatsächlich schmecken viele dunkle Lebensmittel bitter, wie zum Beispiel Kaffee, der ja genau genommen kein Nahrungs-, sondern ein Genussmittel ist. Selbst Schokolade schmeckt ohne Zuckerzusatz sehr herb. Die allseits beliebte Milchschokolade dagegen ist mittelbraun. Hellere Brauntöne interpretieren wir als kohlenhydrathaltig, nahrhaft und sanft schmeckend. Nicht zuletzt wegen der unzähligen Nahrungsmittel auf Getreidebasis erscheinen die meisten Esswaren in Braun- und Beigetönen.

Zudem übermittelt die Farbe dem Gehirn klare Botschaften darüber, welcher Geschmack zu erwarten ist, wie bekömmlich und wie frisch das Dargebotene ist. Wie das Experiment von Eva Heller bestätigt, lassen uns sanfte Gelb- und Brauntöne einen milden Geschmack erwarten, dunkles Goldbraun Röstaromen. Bei Dunklem und Violettem wie Radicchio oder Kaffee stellen wir uns auf bittere Geschmäcker ein. Weißliches wird mit einem sanften, milchigen Geschmack, Gelbgrünes mit „säuerlich" verbunden. Kräftiges Grün steht für natürlich und saftig, satte Gelb-, Orange- und Rottöne für fruchtig. Instinktiv greift der Mensch am liebsten nach Rotem. Die Farbe erinnert uns an süße Früchte wie vollreife Tomaten, Erdbeeren oder Himbeeren. Rot steht für reif, süß und höchste Erregung.

The Taste of Color | "*We conducted experiments at the university in which we dyed food. For instance, we dyed potatoes black, cauliflower green and asparagus red. And even though the taste remained the same – something we went to pains to ensure – people wouldn't eat it because they would associate a very different taste with the color, one that had nothing to do with the actual taste of the food in question.*" Gisla Gniech[5]

Colors whet our appetite and trigger certain taste expectations. Is something red and sweet or black and bitter? The perception of food is often decisive in determining whether we put something in our mouth or not. German sociologist Eva Heller did an experiment in which she had 1,888 persons assign colors to the four main tastes – bitter, sour, salty and sweet. Green and yellow were predominantly associated with "sour" whereas pink, orange and red were associated with "sweet". With white, gray and blue, respondents expect a salty taste. Violet, black and brown make people think of something bitter.[6]

People like edible things in colors that remind them of natural foods. We associate colors such as red, orange, yellow or green with ripe fruits or crunchy vegetables. They signal something agreeable and with a good taste and whet our appetite. By contrast, people generally have an instinctive inhibition against the consumption of dark, violet or artificially intensive shades of color. In reality, there are also not so many natural foods that are blue or black which is why seeing them on a plate makes us skeptical and warns us that what is being served could be bitter, unpalatable or even poisonous. It is certainly not a coincidence that many people find blood sausage and cuttlefish ink disgusting and many people do not like black olives, caviar or licorice. The dark color certainly plays a central role in people's dislike of these foods.

A number of dark food products actually taste bitter, as for instance coffee, which is not considered to be food strictly speaking, but a natural stimulant. Even chocolate tastes very bitter without any sugar. By contrast, the widely loved milk chocolate is middle brown. We interpret lighter shades of brown as containing carbohydrates, being nutritious and subtle in taste. Because of the many cereal-based food products, the appearance of most foods is in shades of brown and beige.

Color clearly conveys to the brain what taste is to be expected, how agreeable and fresh the presented food is. As Eva Heller's experiment confirms soft shades of yellow and brown lead us to expect a mild taste, while dark golden browns makes us expect roast aromas. In dark and violet foods such as radicchio or coffee we expect bitter tastes. Something whitish is associated with a subtle milk-like taste whereas yellow-green is related to something "acerbic". An intense green stands for natural and succulent, rich shades of yellow, orange and red stand for fruity. Instinctively, people prefer to reach for something red. The color reminds them of sweet fruits such as fully ripe tomatoes, strawberries or raspberries. Red stands for ripe, sweet and for utmost excitation.

Natürliche Farben wie Orange, Gelb oder Grün signalisieren uns Bekömmlichkeit. Bei dunklen, violetten oder unnatürlich intensiv gefärbten Nahrungsmitteln sind wir dagegen skeptisch: Sie könnten ungenießbar oder gar giftig sein.

Natural colors such as orange, yellow or green, signal something appetizing. By contrast, dark, purple or artificially dyed foods with an intense hue make us skeptical: they could be inedible or even poisonous.

88 Optik | optics

Lebensmittel, die in Kontrastfarben zu ihren natürlichen Zutaten eingefärbt sind, wirken irritierend und abstoßend. „Falsche" Farben unterbinden die optische Kontrolle auf Genusstauglichkeit und lassen keine Rückschlüsse darauf zu, wie das Angebotene schmeckt.

Foods that are dyed in colors contrasting with their natural ingredients appear irritating or repulsive. "Off" colors suppress the optical control of edibility and do not allow us to conclude how the food on offer might taste.

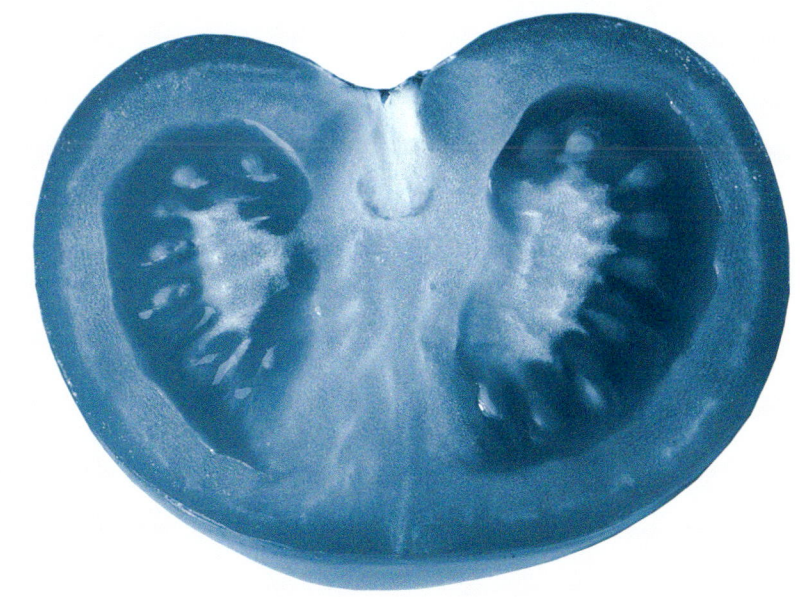

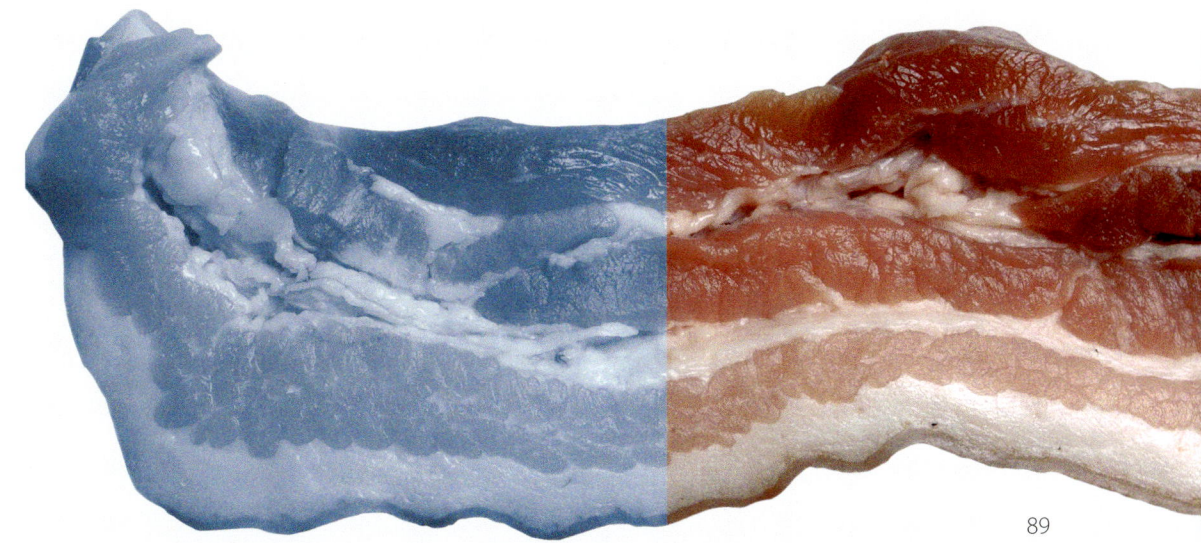

Ein Umstand, den sich vor allem die Süßwarenindustrie, wie die deutsche Firma Haribo, zunutze macht. Laut Umfragen greift die Mehrzahl der Konsumenten am liebsten zu roten Gummibärchen. Der Konzern reagiert auf diese Vorliebe, indem er doppelt so viele rot gefärbte Bären in die Packungen füllt wie weiße, gelbe, orange oder grüne. Unter den Gummibärchen haben die roten eindeutig die Mehrheit. Dabei hat die Leuchtkraft der roten Bären in den letzten Jahren deutlich nachgelassen: Im Zuge der Gesundheitswelle wurde in Europa auf pflanzliche und somit mattere Farbstoffe umgestellt.

Weißer Genuss | Die Interpretation von Farben ist aber nicht nur evolutionär vorgegeben, sondern auch kulturell erlernt. In unserem Kulturkreis gilt beim Essen gemeinhin Rot als anziehendste Farbe, weil wir uns von ihr ein extrasüßes und intensives Aroma erwarten. Ganz anders wird Geschmack dagegen in Japan bewertet, wo man helle Farben und möglichst sanft Schmeckendes bevorzugt. Für den Japaner soll der Geschmack einer Speise den Esser nicht überwältigen, sondern zurückhaltend, möglichst sanft, ja sogar neutral sein und „sich im Unendlichen verlieren". Besondere Faszination übt daher ein Aroma aus, das sich genau an der Schnittstelle zwischen Geschmack und Nichtgeschmack bewegt und gerade noch wahrzunehmen ist. Ähnlich verhält es sich in Japan auch mit der Farbe. Käse, Brot und Milchprodukte werden als appetitlicher empfunden, wenn sie sehr hell bzw. fast weiß sind.

Zurück nach Europa, wo die rote Farbe für den Erfolg vieler Gerichte und Produkte bürgt, so auch für das in den 1950ern von Arrigo Cipriani, dem Besitzer des berühmten Restaurants „Harry's Bar" in Venedig, kreierte Carpaccio. Die tiefrote Farbe der rohen Rinderfiletscheiben erinnerte Cipriani an Gemälde des italienischen Renaissancemalers Vittore Carpaccio, nach dem er sein Gericht benannte. Auch profanere Produkte werden gern nach ihrer Farbe benannt. In Australien bezeichnet man Ketchup als „red sauce" und eine ursprünglich französische, kalte Kräutersauce heißt in Deutschland „Grüne Sauce" und in Italien „salsa verde".

Färben und verführen | „*Geschmack ist erlernt. Das heißt, wir halten das für natürlich, was wir gewohnt sind. Bei Tests mit Erdbeerjoghurt etwa werden jene Produkte, die am rötesten und am intensivsten aromatisiert sind, am besten gewertet. Mit dem Geschmack von echten Erdbeeren hat das nichts zu tun.*" Werner Mlodzianowski[7]

Die Geschmacksrichtungen von Süßwaren werden ohne die zugehörige Farbe meist nicht erkannt. Erdbeerbonbons zum Beispiel werden nur dann als solche identifiziert, wenn sie rot sind. Bei Blindverkostungen von Gummibären können nur drei Prozent der Testesser die fünf Geschmacksrichtungen den jeweiligen Farben richtig zuordnen. Auch werden die fünf verschiedenen Sorten der Gummibären in Blindtests ganz anders bewertet, als wenn der Testesser bei der Verkostung die Farbe des jeweiligen Bären sehen kann. Dann nämlich gewinnen immer die roten.

Rotes erweckt Appetit, Schwarzes dagegen Ekel: Rote Esswaren wie Fleisch, Rohschinken oder reife Früchte bereiten uns Lust, weil wir instinktiv einen hohen Nährwert erwarten. Dunkles dagegen stößt uns ab. Es ist kein Zufall, dass viele Menschen schwarze Oliven, Sepiatinte, Kaviar, Blutwurst oder Lakritze ablehnen.

This is a factor that the candy industry, and in particular the German company Haribo, is capitalizing on. According to surveys, the majority of consumers prefer red gummi bears. The corporation has responded to this predilection by filling packages with twice as many red dyed bears than white, yellow, orange or green ones. The red bears are clearly in the majority. At the same time the radiance of the red bears has visibly declined in recent years. As a result of the health wave, European manufacturers have switched to plant-based and thus paler dyes.

Relish in White | The interpretation of colors is not just a product of evolution, it is also acquired through culture. In our culture, red is generally regarded as an appealing color in food because we expect it to have an extra-sweet and intensive aroma. In Japan, by contrast, taste is assessed very differently: subdued colors and things with a highly subtle taste are preferred. For the Japanese, the taste of food should not overwhelm the eater but be reserved, as subtle as possible, even neutral and "become lost in infinity". An aroma exerts particular fascination when it moves on the interface between taste and non-taste and can barely be perceived. Something similar is true for color as well in Japan. Cheese, bread and dairy products are seen as more appetizing when they are very light, almost white.

Back to Europe where the red color guarantees the success of many dishes and products, as for instance Carpaccio created in the 1950s by Arrigo Cipriani, the owner of the famous Venice restaurant "Harry's Bar". The deep red color of the slices of raw beef reminded Cipriani of the paintings by Italian Renaissance painter Vittore Carpaccio, after whom he then named his dish. Even more profane products are often named after their color. In Australia ketchup is called "red sauce" and a cold, originally French, herb sauce is called "green sauce" in Germany and "salsa verde" in Italy.

Coloring and Seducing | *"Taste is something that is acquired. That is, we believe that something is natural when we are accustomed to it. In tests with strawberry yoghurt those products that were the reddest and had the strongest aroma received the best grades. This has nothing to do with the taste of real strawberries." Werner Mlodzianowski*[7]

The tastes of sweets are usually not recognized without the corresponding color. Strawberry flavored sweets for instance are only identified as such when they are red. In the blind tasting of gummi bears only three percent of the testers can correctly ascribe the five tastes to the right colors. The five different types of gummi bears are also assessed completely differently in random tests when the tester is able to see the bears in question when sampling them. Then the red ones always win.

Red things arouse appetite whereas black things are disgusting. Red foods such as meat, ham or ripe fruits provide pleasure since we instinctively expect a high nutritional value. By contrast, dark things turn us off. It is no coincidence that many people refuse to eat black olives, squid ink, caviar, blood sausage or licorice.

Die Farbpsychologie spielt bei der Bewertung und insofern auch beim Kauf von Lebensmitteln eine entscheidende Rolle. Natürlich schließt die Gestaltung von Nahrungsmitteln die Veränderung ihrer natürlichen Farbe mit ein, wobei die Idee des Färbens wesentlich älter ist als die Nahrungsmittelindustrie. Viele traditionelle Gerichte spielen mit „künstlichen" Farbeffekten. In Nordindien wird das berühmte Tandoori-Huhn traditionellerweise mit Chilipulver rot gefärbt, heute greift man meist zu simpler Lebensmittelfarbe, die denselben Effekt erzielt. In Persien oder Spanien wird weißer Reis durch die Zugabe von Safran oder Kurkuma optisch aufpoliert, beide Gewürze färben leuchtend gelb. Fettreicher Cheddar aus der Grafschaft Leicester wird bei der Lagerung saftig orange. Schon früh begann man mit Karottensaft nachzuhelfen, um dem Esser einen vermeintlich höheren Fettgehalt vorzutäuschen. Und nach jahrelangem Verzehr von gelbem Vanilleeis nimmt mancher Konsument überrascht zur Kenntnis, dass Vanille eigentlich schwarz ist.

Dass unser Auge derart sensibel auf den Zusammenhang von Fettgehalt und Farbe reagiert, führt sogar zu Problemen bei der Entwicklung fettarmer Produkte. Nanotechnologe David Julian McClements vom Food Science Institute der University of Massachusetts beschäftigt sich mit der Frage, wie die interne Struktur von Emulsionen wie Milch, Sahne, Softdrinks, Eiscreme, Salatdressings, Dips, Mayonnaise, Saucen oder Suppen deren Farbe beeinflusst. Da die kleinen Fettpartikel, die in diesen fettigen Lösungen verteilt sind, das Licht streuen (ein ähnlicher optischer Effekt, wie man ihn von Wassertröpfchen kennt, die Wolken bilden), erscheinen sie milchig. Je weniger Fett eine Emulsion enthält, desto klarer und damit farbintensiver wirkt sie trotz gleichem Farbstoffgehalt. Dieses Phänomen führt bei fettreduzierten Milchprodukten zu Akzeptanzschwierigkeiten. Um dem Problem Abhilfe zu verschaffen, versucht man die Fettteilchen durch Nanopartikel, etwa TiO2, Titandioxid, mit einem Durchmesser von 100 Nanometern, zu ersetzen, die das Licht in ähnlicher Weise streuen.[8]

Von den instinktiven oder erlernten Assoziationen, die Farben in uns hervorrufen, können wir uns nur schwer lösen. Gelbe Butter, mit Beta-Carotin aufgemotzter Käse oder knallorange Fruchtsäfte erscheinen uns geschmacksintensiver, selbst wenn wir wissen, dass diese Produkte künstlich gefärbt sind. Viele Eislutscher, Kaugummis und Süßwaren bestechen durch ihre fruchtige Buntheit, ihre Farbintensität bewegt sich allerdings oft im Grenzbereich der Akzeptanz. Zu grelle Töne suggerieren Künstlichkeit, ein Attribut, das im Zeitalter der Biowelle nicht unbedingt positiv bewertet wird.

Blau als Rebellion | Farben sprechen das Unterbewusstsein an. Hersteller von Kaugummis, Zuckerln oder Frühstückscerealien wissen diesen Umstand gewinnbringend zu nutzen. Dabei wird der gewählte Farbton auch dem Alter der jeweiligen Zielgruppe angepasst. Wir haben uns zum Beispiel gefragt, warum es blaue Smarties, aber keine blauen Gummibärchen gibt. Letztere sprechen mittlerweile alle Altersgruppen an und werden auch von Konsumenten über vierzig geschätzt, die sich vor blauen oder schwarzen Farbtönen ekeln. Lebensmittel, die in Kontrastfarben zu ihren natürlichen Zutaten eingefärbt sind, wirken auf Erwachsene irritierend und abstoßend. Ein hellblauer Emmentaler, grün gefärbte Erdbeersauce oder violett-blaue Tomaten lösen

Color psychology plays a crucial role in assessing and thus also in purchasing food. Of course, the design of food also includes the alteration of their natural color, with the idea of adding color being older than the food industry. Many traditional dishes play with "artificial" color effects. In North India the Tandoori chicken was dyed red with chili powder. Today a simpler food dye is used that has the same effect. In Persia or Spain saffron or curcuma is added to white rice to enhance it visually – both spices make it radiantly yellow. Fatty cheddar from the shire of Leicester becomes a juicy orange when it is stored. Already in earlier times, carrot juice was used to give the consumer the impression of a higher fat content. And after having eaten pale yellow vanilla ice cream for years, some consumers are surprised to discover that vanilla is actually black.

Nano-engineer David Julian McClements from the Department of Food Science at the University of Massachusetts focuses on the question of how the internal structure of food emulsions influences their color. These food emulsions include products such as milk, cream, soft drinks, ice-cream, salad dressings, dips, mayonnaise, sauces, and soups. The tiny fat droplets present in these foods scatter light (like the tiny water droplets that make up clouds), making the products appear cloudy. The less fat an emulsion contains the less cloudy it appears and the more color-intensive it seems, despite having the same dye content. This loss of cloudiness leads to problems in the production of low fat products with desirable appearances. To counteract this problem attempts have been made to replace the fat particles with nano-particles, e.g., TiO2, titanium dioxide, with a diameter of 100 nanometers, that disperse light in a similar way.[8]

It is difficult for us to overcome instinctive or acquired associations that colors trigger in us. Yellow butter, cheese that has been spiffed up with beta carotene or bright orange fruit juices all appear more intensive in terms of taste, even if we know that these products are artificially dyed. Many popsicles, chewing gums and sweets are so appealing because of their fruity brightness; the intensity of colors however often moves towards the limits of acceptance. Shades that are too garishly bright suggest artificiality, an attribute that is not necessarily seen as positive in the age of healthy life style...

Blue as a Rebellion | Colors address the unconscious. Manufacturers of chewing gum, candies or breakfast cereals know how to capitalize on this phenomenon. Here the color selected is also adapted to a given target group. For example, we asked ourselves: why there are blue smarties but no blue gummi bears? The latter meanwhile appeal to all age groups and are also popular with customers over forty who are taken aback by blue and black shades. Food that has been dyed colors contrasting with their original ingredients has an irritating and repel-

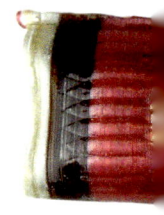

Seidenbonbons, japanische Cracker mit Blumen- beziehungsweise Blattdekor, gedrehte Zuckerstange, sauer panierte Fruchtgummizungen, zweifärbige Farfalle rigate, Lakritze-Röllchen, transparenter Lollipop mit Kaugummifülle, Farbkreis nach dem Schweizer Maler Johannes Itten, geschichtetes Lakritze-Bonbon: Die Interpretation von Farben ist auch kulturell erlernt. Wirkt bei uns Buntes appetitanregend, so liebt man in Japan weißes Essen und empfindet Käse, Brot oder Milchprodukte dann als besonders lecker, wenn sie sehr hell beziehungsweise fast weiß sind.

Silk candies, Japanese crackers with flower and leaf décor, twisted rock candy, fruit gum sticks with a sour sugar coat, two-color Farfalle rigate, licorice rolls, transparent lollipops with chewing gum filling, color circles after the Swiss painter Johannes Itten, layered licorice candy. The interpretation of colors is also culturally acquired. While something in bright colors has an appetizing effect for us, people in Japan love white food and consider cheese, bread or milk products as being particularly tasty when they are very light or almost white.

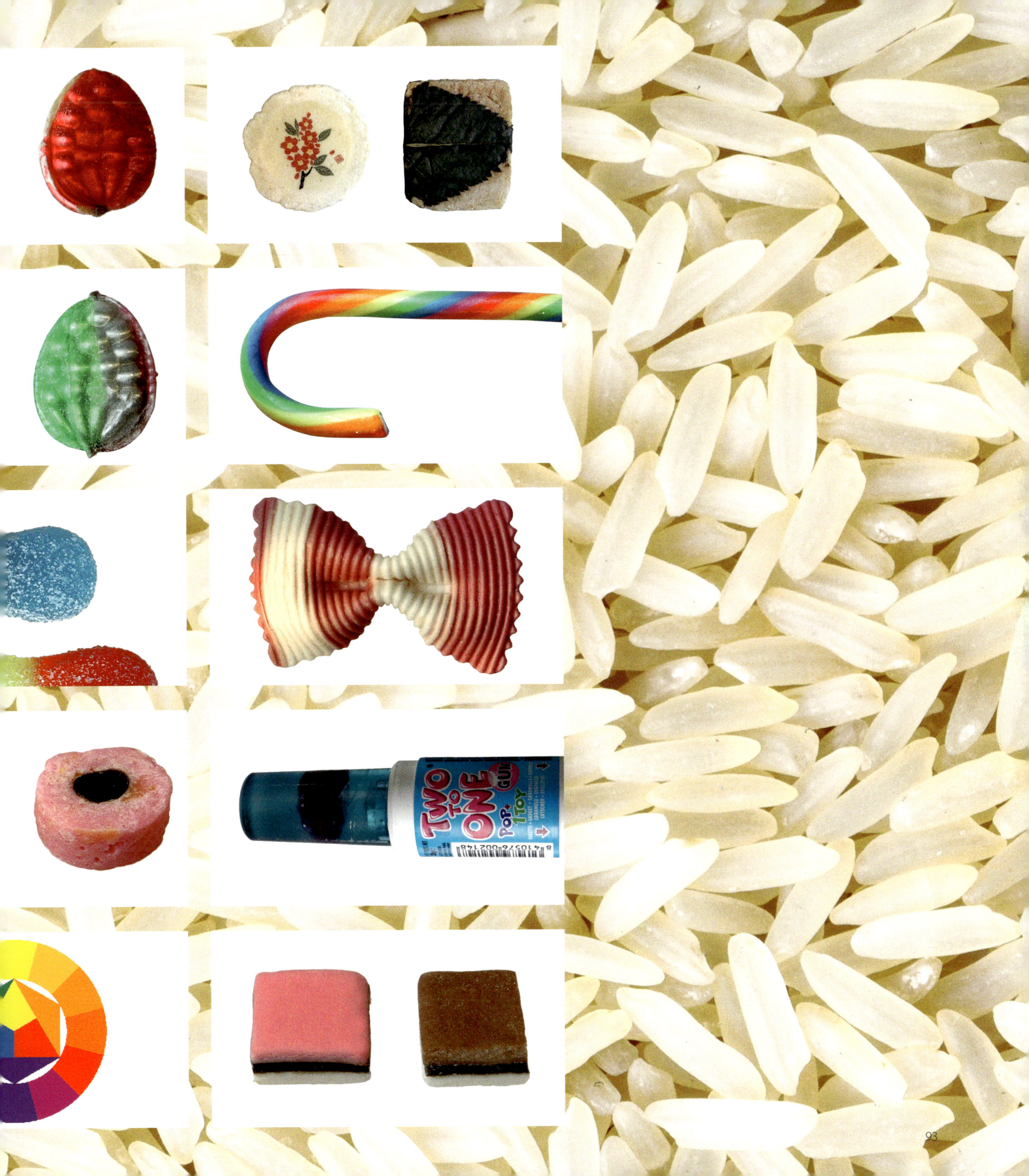

Verwunderung und Ekel aus und werden zumeist auch nicht gegessen. Smarties hingegen sind bis heute ein Kinderprodukt. Junge Konsumenten verzehren mit Vorliebe, was unnatürliche Violett- und Blautöne hat, wie auch schwarze Bonbons oder blaues (!) Ketchup. Frühstückskekse, die beim Eintauchen die Milch blau färben, oder pinkfarbene Margarine sind nicht zuletzt deswegen bei den Jüngsten so begehrt, weil den Erwachsenen davor graut. Mit dem Konsum von unnatürlich gefärbtem Essen können Kinder ihre Eltern provozieren und vor Altersgenossen prahlen.

Auch der US-Kultdrink Coca-Cola spielt mit dem rebellisch-gefährlichen Touch dunkler Töne. Wenn unter Kindern immer wieder Gerüchte grassieren, wonach Coca-Cola ungesund oder gar giftig sei, so unterstützt das nur seinen Erfolg. Tatsächlich blickt die bekannteste Limonade der Welt auf eine bemerkenswerte Geschichte im Spannungsfeld zwischen Genuss, Berauschung und Gesundheitstherapie zurück. Zur Zeit seiner Erfindung war Coca-Cola alles andere als ein Kindergetränk. Um 1885 entwickelte der Apotheker Dr. John Stith Pemberton in Atlanta ein kokainhaltiges Rauschgetränk, das er „Pemberton's French Wine Coca" nannte. Als Vorbild diente ein französisches Modegetränk namens „Vin Mariani", bei dem Bordeauxwein mit Kokablättern versetzt wurde. Im Zuge der Prohibition, die in den Vereinigten Staaten ab 1886 galt, entfernte Pemberton zunächst den Alkohol aus der Rezeptur und bewarb seine Erfindung als „alkoholfreies Getränk für Intellektuelle". Coca-Cola galt als „gut für die Gesundheit und lebensverlängernd". Es wurde gegen Müdigkeit, Verdauungsstörungen und Nervenkrankheiten eingesetzt. 1902 schließlich musste auch die kokainhaltige Substanz weichen, womit der Entwicklung zur Kinder- und Jugenderfrischung nichts mehr im Wege stand.[9]

ling effect on adults. A light blue Gruyère cheese, strawberries colored green or violet-blue tomatoes trigger surprise and disgust and will not be eaten. Smarties by contrast are a children's product to this very day. Young consumers eat things in artificial shades of violet and blue like black candies or blue (!) ketchup and they do so with a vengeance. Breakfast cookies that turn the milk blue when dipped in or pink-colored margarine are so popular with young people because adults find them repulsive. Children can provoke their parents by eating artificially dyed food and brag to their peers.

Even the cult drink Coca-Cola plays with the rebellious dangerous touch of dark shades. Rumors that continue to spread among children that Coca-Cola is unhealthy or even poisonous, basically only bring home its success. In reality, the world's best-known lemonade looks back on a remarkable history going from pleasure, intoxication to health therapy. At the time it was invented, Coca-Cola was anything but child's play. Around 1885 pharmacist Dr. John Smith Pemberton from Atlanta developed a drink containing cocaine which he dubbed "Pemberton's French Wine Coca". This was modeled after a fashionable French drink called "Vin Mariani" – Bordeaux wine mixed with coca leaves. With Prohibition, which began in the United States in 1886, Pemberton first removed the alcohol from his drink and advertised his concoction as a "non-alcoholic drink for intellectuals". Coca-Cola was seen as being "healthy and conducive to longevity". It was used against exhaustion, digestive disorders and nervous illnesses. In 1902 the substance containing cocaine was also removed so that the development of a refreshment for children and young people was possible.[9]

Getränke und ihre Farbe von links nach rechts | Milch, Mineralwasser, Espresso, Coca-Cola, Campari Soda, Red Bull, Orangensaft, Fanta, Blue Curaçao

Beverages and their color, from left to right | milk, mineral water, espresso, Coca Cola, Campari soda, Red Bull, orange juice, Fanta, Blue Curaçao

Warum ist Cola so erfolgreich? Ein Grund dafür könnte in seiner Farbe liegen. Cola ist dunkel, fast schwarz, undurchsichtig und in seinem Inneren perlt es. Der optische Eindruck lässt keinerlei Rückschlüsse auf Inhaltsstoffe oder Geschmack zu, weswegen sich jemand, der es nicht kennt, eigentlich davor ekeln müsste. Für die sinnliche Wahrnehmung ist Cola ein Mysterium, denn obwohl es fast schwarz ist, schmeckt es nicht bitter, sondern im Gegenteil „picksüß". Cola stellt damit in jeder Hinsicht unseren kulinarischen Erfahrungsschatz auf den Kopf – und fasziniert.

Ampel-Kombinationen | Appetitanregend oder abstoßend ist aber nicht nur die Farbe selbst, sondern auch die Farbkombination innerhalb eines Gerichts. Der Instinkt sagt uns unterbewusst, dass Gerichte mit möglichst vielen unterschiedlichen, als bekömmlich eingestuften Farben auch viele unterschiedliche Nährstoffe enthalten. Der farbliche Kontrast lässt ein Gericht geschmackvoller, interessanter und frischer erscheinen: grüner Schnittlauch auf der gelb-braunen Suppe, eine gelbe Zitronenscheibe zum schwarz-braunen Cola, grünes Basilikum auf weißem Mozzarella, weißes Schlagobers zum dunkelbraunen Schokokuchen, rote Erdbeeren auf weißem Rahm, dunkelrote Heidelbeeren im weißen Stilton. Eine Extremvariante dieser farblichen Akzentuierung war die in den äußerst farbenverspielten Achtzigerjahren zu allen nur erdenklichen Gerichten servierte, giftgrüne Kiwischeibe.

Besonders die Dreierkombination rot-grün-gelb lässt uns das Wasser im Mund zusammenlaufen. Eine Pizza, auf der gelber Mais, grüne Paprika-, rote Tomaten- und weiße Mozzarellastückchen liegen, spricht wesentlich mehr an als eine rein rote Pizza mit Tomaten, rotem Paprika und Speck. Auch die so genannte „bunte" Salatschüssel

What accounts for the great success of Coca-Cola? One reason might be its color. It is dark, almost black, opaque and bubbles on the inside. The outer appearance reveals nothing about the ingredients or taste, so someone who does not know it should actually find it disgusting. In terms of sensory perception, "Coke" is a mystery. Even though it is almost black, it does not taste bitter, but is, quite the contrary, "sickly sweet". "Coke" thus turns our repertory of culinary experiences upside down – and is thus so fascinating.

Traffic-Light Combinations | Not just the color but also the combination of colors in food can be appetizing or disgusting. Instinct tells us unconsciously that food with as many different colors as possible that are categorized as agreeable also contains many different nutrients. The colorful contrast makes a dish appear tastier, more interesting and fresher. Chives on yellow-brown soup, a yellow slice of lemon with black-brown cola, green basil on white mozzarella, white whipped cream on dark-brown chocolate cake, red strawberries on white cream, dark-red berries against a white background. An extreme variant of this color-based accentuation was the poison-green kiwi slice that could be found with all manner of conceivable dishes in the 1980's – a period of extremely obsessive play with color.

In particular the combination of the three colors red-yellow-green makes our mouth water. A pizza on which there is yellow corn, green peppers, red tomatoes and white pieces of mozzarella cheese is much more eloquent than a purely red pizza with tomatoes, red pepper and

nutzt den positiven Effekt farblicher Vielfalt und weist auf diesen Vorteil auch gleich in ihrem Namen hin. Ein typisches Beispiel für die appetitanregende Dreifarbenkombination rot-grün-gelb ist auch das klassische Dekor von Fleischspeisen mit einer gelben Zitronenscheibe, einem grünen Petersilienbüschel und roten, eingelegten Paprikaschoten. In vielen Gaststätten wird diese Art des Anrichtens so wichtig genommen, dass sie sogar eigens in der Speisekarte als „fein garniert" verzeichnet ist.

Auch oder gerade bei der Gestaltung von Fertiggerichten wird auf die farbliche Abstimmung der Zutaten geachtet. Der japanische Lebensmittelriese Nissin, bei uns vor allem durch Instantnudelsuppen im Becher bekannt, entwickelt etwa die Rezepturen seiner Instantsuppen immer auch nach optischen Kriterien. Die Kombination der Zutaten muss nicht nur gut schmecken, sondern auch möglichst vielfältig sein, was Größe und Farbe betrifft. Für Buntheit im Becher sorgen unter anderem grüner Lauch und rote Schrimps, die extra nach Farbe sortiert werden.

Als Nissin-Gründer Momofuku Ando 1971 die Cup-Noodles auf den Markt brachte, vereinte das Konzept der neuen Instantsuppe im Wegwerfbecher einige Ideen, die vor allem auch die Optik des Produktes betrafen. Eine davon war der Einsatz der damals noch jungen Technologie des Gefriertrocknens. Neben dem guten Erhalt der Inhalts- und Geschmacksstoffe hat das Gefriertrocknen auch einen entscheidenden optischen Vorteil: Die Farben der Zutaten bleiben in voller Frische erhalten. Während sich Gemüse bei normalem Tiefkühlen oder beim Trocknen gräulich-bräunlich verfärbt, bleiben Lauchstücke nach dem Gefriertrocknen grün, Tomaten rot und Mais dottergelb.

Ein weiterer Trick von Cup-Noodles ist die gekräuselte Struktur der Nudeln, die zu einem Laibchen geformt werden, das gerade um eine Spur kleiner ist als der Becherquerschnitt. Gießt man das Fertiggericht mit heißem Wasser auf, quillt das Nudellaibchen auf und drückt die gefriergetrockneten Zutaten an die Oberfläche. Die bunten Zutaten schwimmen oben auf, was natürlich appetitlicher aussieht, als wenn sie unten im Becher verteilt wären.[10]

ham. Even the so-called 'mixed-color' salad bowl makes use of the positive effect of colorful diversity, even signaling this advantage in its name. A typical example of this appetizing triple-color combination of red-green-yellow is also the classical decoration of meat dishes with a yellow slice of lemon, a green bunch of parsley and red, pickled sweet peppers. In many restaurants this type of preparation is taken so seriously that it is even noted on the menu as "garnish".

The coordination of colors in the ingredients is also taken into account in designing instant meals – here perhaps to a higher degree than in other meals. Nissin, the Japanese food megacorporation known here mainly for its instant noodle soups served in a cup, develops the recipes of its instant soups according to visual criteria. The combination of ingredients must taste good but it must also have the greatest diversity in terms of size and color. Green leeks and red shrimps, specially selected for color, also brighten up the soup in the cup.

When, in 1971, Nissin founder Momofuku Ando introduced the cup noodles on the market, the concept of the new instant soup assembled several ideas in the throw-away cup version which also relate to the visual appearance of the product. One was the technology of freeze-drying, which was still new at the time. In addition to preserving the ingredients' aroma, freeze-drying has one further advantage. It fully keeps the colors of the ingredients. While vegetables can turn grey-brown when stored in normal freezers or dried, leeks remain green, tomatoes red and corn yolk-yellow after they are freeze-dried.

A further trick of cup noodles is the wavy curled structure of the noodles that have a block-like form which is just a bit smaller than the cross-section of the cup. When you pour boiling water over the instant meal, the noodle block swells, pushing the freeze-dried ingredients to the surface. The colorful ingredients float on the top, which, of course, looks more appetizing than if they were spread on the bottom of the cup.[10]

Neben bunt glaciertem Popcorn werden noch viele andere Lebensmittel gefärbt, um uns Appetit zu machen. Im alten England täuschte mit Karottensaft gefärbter Cheddarkäse einen höheren Fettgehalt und damit einen intensiveren Geschmack vor. Auch gelbe Butter oder gefärbter Orangensaft erscheinen uns geschmacksintensiver. Und nach jahrelangem Verzehr von gelbem Vanilleeis nimmt manch europäischer Konsument überrascht zur Kenntnis, dass Vanille eigentlich schwarz ist.

In addition to brightly glazed popcorn many other foods are dyed to whet our appetite. In old England cheddar cheese dyed with carrot juice feigned a higher fat content and thus a more intensive taste. Yellow butter or dyed fruit juices also strike us as more intensive in taste. And after having eaten yellow vanilla icecream for many years, some European consumers are surprised to find that vanilla is actually black.

Cup-Noodles wurden von Momofuku Ando 1971 in Japan auf den Markt gebracht. Das Design der Instantsuppe im Wegwerfbecher nimmt speziell auf die optische Wirkung Bedacht. Die Nudeln sind zu einem Laibchen geformt, das gerade um eine Spur kleiner ist als der Becherquerschnitt. Gießt man heißes Wasser darauf, quillt es auf und drückt die anderen Zutaten an die Oberfläche, welche dann wie frisch garniert wirken.

Cup noodles were launched by Momofuku Ando in Japan in 1971. The design of the soup in a throwaway cup was particularly geared to making a visual impression. The noodles are formed as a patty which is just a tiny bit smaller than the cross-section of the cup. When hot water is poured on it, it swells and presses all the other ingredients to the surface which then look freshly garnished.

1 | Wir bedanken uns beim Restaurant „Nada Man" in Tokio/ Japan für diese Information.
2 | Wir bedanken uns bei Elisabeth Andoh für diese Information.
3 | Wir bedanken uns bei Keiji Toyozumi von Kakiyama in Tokio für diese Information.
4 | Wir bedanken uns bei Werner Mlodzianowski für diese Information.
5 | Interview für den Dokumentarfilm: „Food Design"; Geyrhalter Film GmbH Wien
6 | vgl.: Wie Farben wirken. Farbpsychologie, Farbsymbolik, kreative Farbgestaltung. Reinbek bei Hamburg 1999
7 | Werner Mlodzianowski
8 | Wir danken David Julian McClements für diese Information
9 | vgl.: Christoph Wagner; Fast Schon Food; Lübbe Verlag 2001
10 | Wir bedanken uns bei President Koki Ando für diese Informationen.

1 | We thank the restaurant "Nada Man"/ Tokyo/ Japan for this information.
2 | We thank Elisabeth Andoh for this information.
3 | We thank Keiji Toyozumi from Kakiyama/ Tokyo/ Japan for this information.
4 | We thank Werner Mlodzianowski for this item of information.
5 | Interview for the documentary "Food Design"; Geyrhalter Film GmbH Vienna
6 | See Wie Farben wirken. Farbpsychologie, Farbsymbolik, kreative Farbgestaltung. Reinbek bei Hamburg 1999
7 | Werner Mlodzianowski
8 | We thank David Julian McClements for this Information.
9 | See: Christoph Wagner, Fast Schon Food; Lübbe Verlag 2001
10 | We thank President Koki Ando for this information.

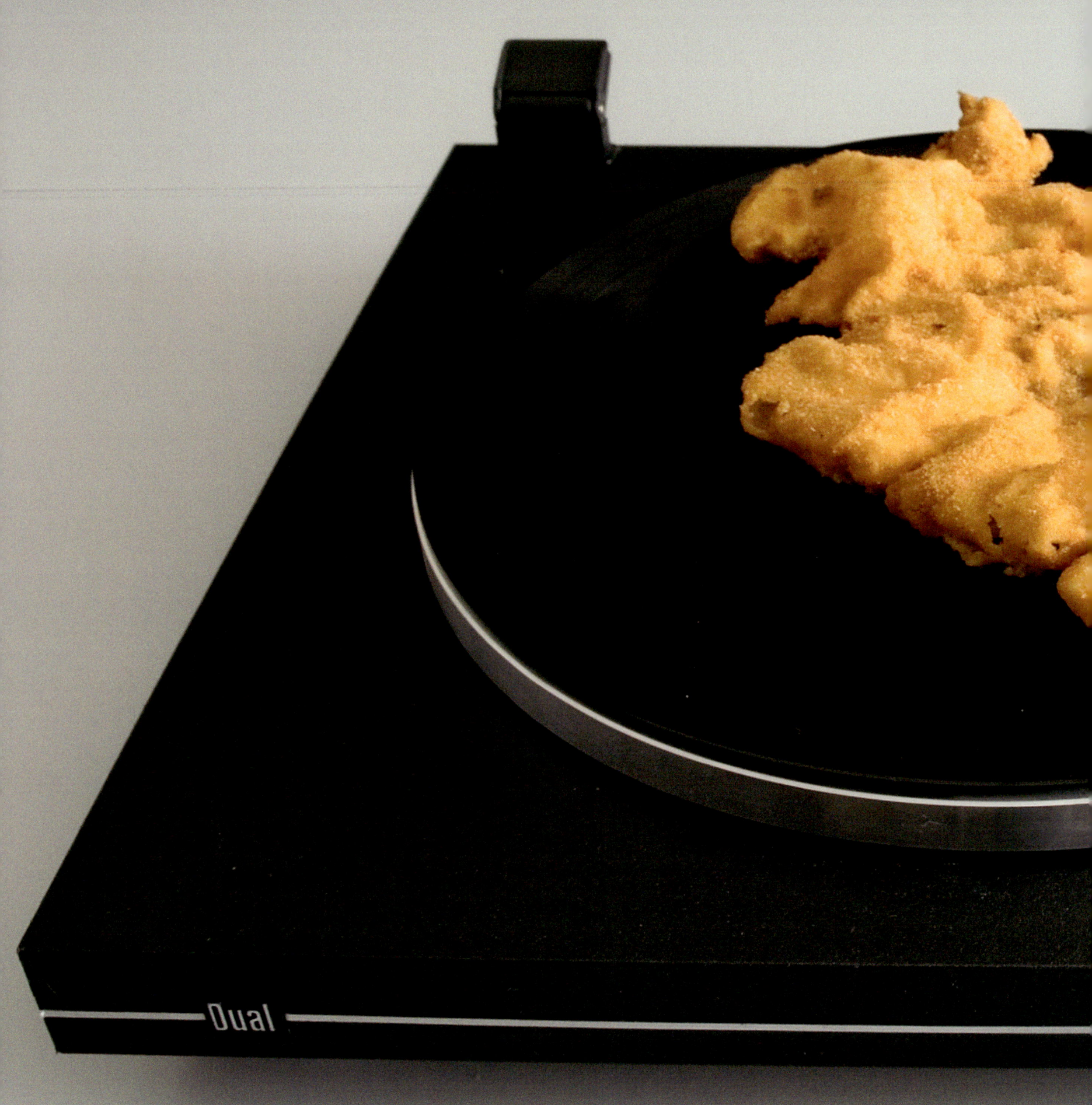

Geräusch
sound

GERÄUSCH – auch das Ohr isst mit

„Wenn man einen Zwieback kaut, dann gibt das zunächst ein Geräusch, das sehr knusprig, crunchy, laut, markant und auffällig ist. Wenn man den Zwieback dann zerkleinert, wird das Geräusch immer wärmer, weicher, angenehmer und schließlich merkt man, wenn das Geräusch gerade noch da ist, jetzt kann man die Nahrung runterschlucken. Das heißt, das akustische Signal ist gleichzeitig eine Kontrolle für den Fortschritt des Kauprozesses. Und dann beginnt der Zyklus von vorne." Friedrich Blutner[1]

„Kauen besteht aus 30–40 eindeutig differenzierbaren Klangmustern. Je differenzierter diese sind, umso interessanter, und umso besser schmeckt es uns." Friedrich Blutner[2]

Auch unsere Ohren entscheiden darüber, ob uns etwas schmeckt oder nicht. Cornflakes oder Kartoffelchips schmecken dann besonders gut, wenn sie beim Reinbeißen laut krachen. Einen „leisen" Kartoffelchip lehnen wir hingegen sofort ab. Entspricht das Geräusch beim Zubeißen nicht unseren instinktiv gesteuerten oder erlernten Erwartungen, spucken wir das Gegessene schnell wieder aus. Es könnte verdorben sein. Das Ohr kontrolliert, ob ein Lebensmittel frisch ist, oder stellt fest, dass etwas mulmig oder schimmlig klingt und besser nicht geschluckt werden sollte.

Natürlich hören nicht alle Menschen gleich. Grob unterscheidet man zwei akustische Typen: Menschen, die eher aktive, kernige und helle Geräuschkulissen lieben, und Menschen, die sonore, dunkle und weiche Töne bevorzugen, also – sehr vereinfacht – Obertonhörer und Grundtonhörer. Das Klangmuster erfolgreicher Lebensmittel muss für beide Gruppen funktionieren und eine möglichst ausgewogene Harmonie zwischen hellen und dunklen, sonoren Tönen bieten. Der deutsche Sounddesigner Friedrich Blutner bezeichnet das als Kombination von Ferrari- und Händelklängen einerseits mit Porsche- und Brahmsnoten andererseits.

Kauen und Sprechen, Hören und Schlucken

Jeder Mensch hat seinen eigenen Kautakt. Für geschulte Ohren ähnelt dieser Rhythmus einem Musikstück, einer Art Symphonie des Essens. Die „Kauistik", ein Teilbereich der Sprachakustik, geht sogar so weit, im Kaugeräusch den Ursprung der Sprachentwicklung zu sehen. In einer Studie des Wiener Phonetik-Professors Felix Trojan (veröffentlicht 1975) wurden zehn Versuchspersonen angewiesen, einen Bissen Schinkenbrötchen mit offenen Lippen zu kauen und dabei nach Herzenslust zu schmatzen. Die entstandenen Laute wurden aufgezeichnet und mit dem Ergebnis ausgewertet, dass sich stimmhaftes Kauen mit den ersten Lauten der Sprachentwicklung deckt.

Das Kaugeräusch signalisiert uns auch, wann die Nahrung fein genug zerkleinert ist, dass sie geschluckt werden kann. Der natürliche Kaurhythmus lässt Geräusche entstehen, die im Verlauf des Kauens mehr und mehr abnehmen. Rutscht der wahrgenommene Geräuschpegel unter einen bestimmten Grenzwert, nimmt das Gehirn an, dass das Gegessene weich und breiig ist, der Schluckreflex wird ausgelöst. Hier ortet Friedrich Blutner noch großes Potenzial für zukünftiges Food Design. Was etwa, wenn man künstlich eine Substanz erzeugen könnte, die im Fortlauf des Kauens immer lautere statt leisere Klänge auslöst?

Frisch oder süß?

Für alles Knackige und Knusprige, von Würstchen bis zu Knabbergebäck, gilt: Je lauter und heller es beim Zubeißen klingt, desto frischer erscheint es uns. Appetit macht uns das Frischesignal allerdings nur in Kombination mit dumpfen Tönen, welche nahrhafte Substanzen wie Zucker auslösen. Denn zu helle Beißgeräusche erinnern uns an nicht essbare Substanzen wie Styropor. Das Zubeiß- und Kaugeräusch eines optimalen Keks beispielsweise verbindet frische Töne mit dunklen, warmen Lauten, die uns Nahrhaftigkeit erkennen lassen. Er klingt dann nicht nur knusprig und frisch, sondern auch lecker und süß. Selbst in Zeiten von Überernährung sind wir – was das Kaugeräusch anlangt – „Opfer" unserer Instinkte, denn Kalorienreiches klingt für unser Ohr einfach besser als Kalorienarmes.

Auch der Klang von Röstaromen ist Musik für unsere Ohren. Friedrich Blutner hat dazu ein Experiment mit unterschiedlich lang getoasteten Toastbrotscheiben durchgeführt. Die Testpersonen sollten die Scheiben auseinander brechen und nur am Klang entscheiden, welcher Toast am besten schmeckt. Das Ergebnis war eindeutig: Je mehr das Brot geröstet war, desto eher wollten die Versuchspersonen auch hineinbeißen. Wir können also allein am Klang feststellen, ob etwas knusprig gebraten ist, ohne es dabei zu sehen oder zu riechen![3]

Knallige Würstchen

Auch Würste zählen zu den akustisch markanten Lebensmitteln. Ein unverwechselbares Knackgeräusch ist das Markenzeichen vieler Würste, unter anderem auch der 1805 von Johann Georg Lahner in Wien erfundenen Frankfurter bzw. Wiener Würstchen. Zwar bieten Hersteller mittlerweile haut- und damit geräuschlose Würstchen an, doch wird der Genussfaktor dadurch unweigerlich reduziert. Denn der Knalleffekt

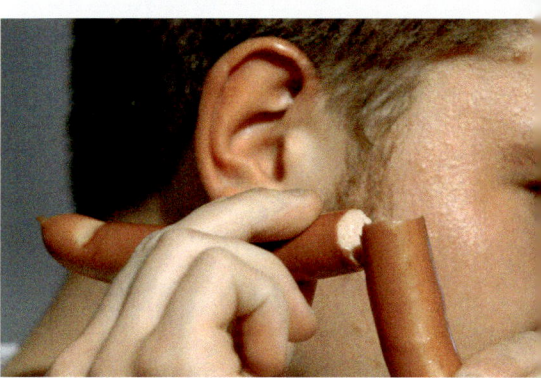

Oben | Blindverkostung von Würstchen, durchgeführt von der deutschen Akustikfirma Synotech. Geschmacklich am besten bewertet werden immer jene, die am lautesten knacken.
Unten | Wenn Konzerne akustisch markante Lebensmittel wie Kekse oder Cornflakes entwickeln, werden eigene Sounddesigner beigezogen.

top | Blind tasting of sausages, carried out by the German acoustics company Synotech. The sausages which made the loudest sound when broken were the ones that got the best evaluations in terms of taste.
bottom | When companies develop acoustically striking food products such as cookies or cornflakes, they enlist the services of special sound designers.

SOUND – We Also Eat With our Ears |
"When you chew a piece of rusk then this first produces a sound that can be described as very crunchy, loud, striking. If you crush the rusk, the sound becomes warmer, softer, more pleasant and finally, when you notice that the sound is fading, you can swallow the food. That is to say, the acoustic signal is also a way to check the progress of a chewing process. And then the cycle starts again." Friedrich Blutner[1]

"Chewing consists of 30 to 40 sound patterns that can be clearly differentiated. The more differentiated they are, the more interesting and the better the taste is for us." Friedrich Blutner[2]

Our ears, too, decide whether we find something tasty or not. Cornflakes or potato chips taste particularly good if they have a loud crunch when you bite into them. We immediately reject a "soft" potato chip. If the sound does not match our instinctive or acquired expectations, we spit out what we just bit into. It could be spoiled. Our ear checks the food to see if it is fresh or it determines that something sounds queasy or moldy and should thus be avoided.

To be sure, not everyone hears the same way. We can make a general distinction between two types of acoustic individuals. People who love active, powerful and light sound backdrops and people who prefer sonorous, dark and soft tones – very simplified, those who hear overtones and those who hear key tones. The sound pattern of successful food must function for both groups and offer as much harmony as possible between light, dark and sonorous tones. The German sound designer Friedrich Blutner describes this as a combination of the sounds of Ferrari and Händel on the one hand, and Porsche and Brahms notes on the other.

Chewing and Speaking, Listening and Swallowing |
Each person has his/her own chewing rhythm. For trained ears this resembles the rhythm of a piece of music, a sort of symphony of eating. "Chewology", a subrealm of linguistic acoustics, even goes so far as to claim that the origins of language are to be found in the sound of chewing. In a study by Viennese phonetics professor Felix Trojan (published in 1975) ten subjects were asked to chew a bite of ham sandwich with open lips and to smack to their heart's content. The resulting sounds were taped and then evaluated with the finding being that the sound of chewing corresponds to the first sounds developed in language.

The sound of chewing also signals to us when food is crushed enough so that it can be swallowed. The natural rhythm of chewing produces sounds that can increasingly fade in the course of chewing. If the perceived noise level drops below a certain limit value, the brain will assume that what has been eaten is soft and mushy, and the swallowing reflex is activated. Here Friedrich Blutner identifies great potential for future food design. For instance, what would happen if one could create a substance that triggers ever louder, instead of softer, sounds while being chewed?

Fresh or Sweet? |
For all crunchy and crispy foods, from sausages to snacks, the following holds: The louder and lighter they sound when eaten, the fresher they appear to be. However, our appetite is only whetted when they are combined with duller sounds which are triggered by nutritious substances such as sugar. Overly light sounds on biting remind us of inedible substances such as styrofoam. The sounds of biting and chewing the ideal cookie, for instance, link fresh sounds with dark, warm tones that enable us to recognize the nutritional value. Then it not only sounds crispy and fresh, but also tasty and sweet. Even in times of over-nutrition we are 'victims' of our instincts – at least with regard to the sound of chewing, since something that is rich in calories still sounds better to our ear than something low in calories.

Even the sound of roast aromas is music to our ears. Friedrich Blutner did an experiment with slices of toast that had been toasted for different lengths of time. The test persons were asked to break the slices of toast and only decide on the basis of sound which toast would taste best. The result was clear. The more toasted the bread was the more the test persons wanted to bite into it. We can determine alone on the basis of sound whether something is roasted to the point of being crunchy without seeing or smelling it![3]

The Bright Sound of Popping Sausages |
Sausages stand also out among foods in an acoustical sense. An unmistakable crunching sound is the telltale sign of many sausages, including the Frankfurter or Wiener sausages invented in 1805 by Johann Georg Lahner of Vienna. Manufacturers meanwhile also offer sausages without skins and thus without sounds but this invariably reduces the pleasure factor. The popping effect that occurs when biting into the sausage is so typical that this is even indicated in the names of certain sausages, such as "Knackwurst". In the case of the Frankfurter or Wiener

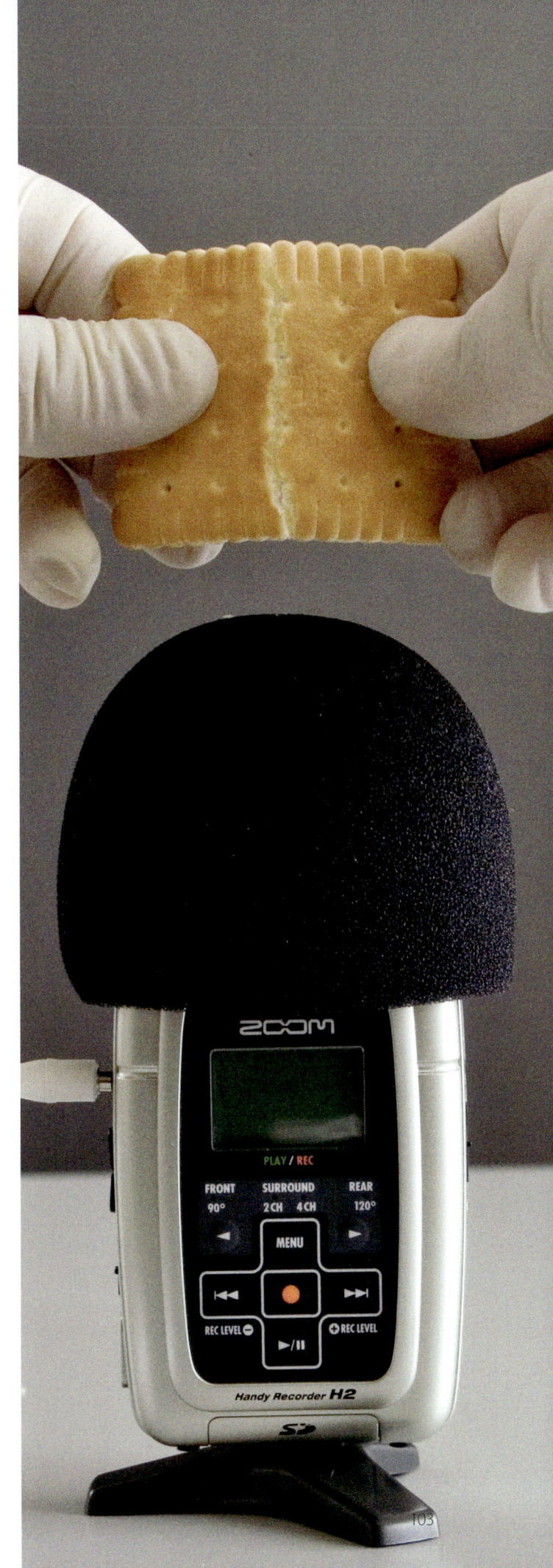

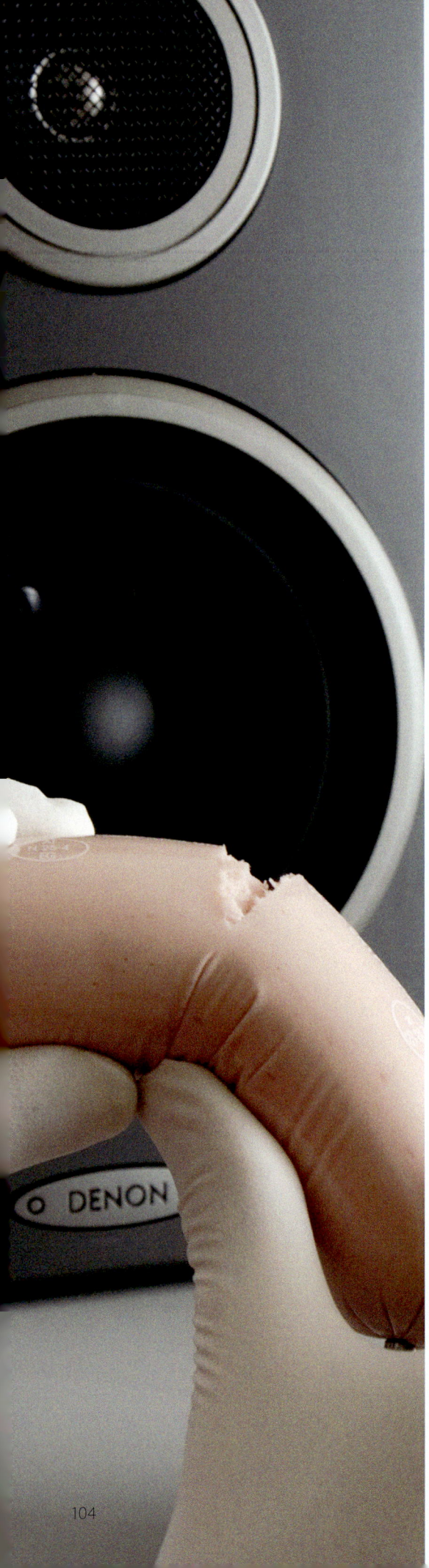

beim Zubeißen ist derart charakteristisch, dass ihn gewisse Wurstsorten wie die Knackwurst sogar im Namen führen. Im Fall der Frankfurter respektive Wiener entsteht das Knackgeräusch, wenn der dünne Schafsdarm, in den das Brät aus Rind- und Schweinefleisch gefüllt wird, zerplatzt. Und das passiert bei jedem Abbeißen genau genommen zweimal: Mithilfe einer akustischen Kamera, die Bild und Ton synchron in extremer Zeitlupe aufzeichnet, zeigt sich, dass das Knacken von Würsten eigentlich aus zwei unmittelbar aufeinander folgenden, lauten Knallgeräuschen besteht. Einmal knackt es, wenn der Oberkiefer die Wursthaut durchbeißt, ein zweites Mal, wenn die Zähne des Unterkiefers die Haut zum Platzen bringen.

Der Knalleffekt beim Zerbeißen einer Wurst macht bis zu 70 Prozent des erlebten Geschmacks aus. Bei Verkostungen verschiedener Würste gewinnen immer jene, die am lautesten knacken. Sounddesigner Friedrich Blutner führte in seiner Heimat, dem sächsischen Erzgebirge, einen Test durch, bei dem alle Fleischer der Umgebung dasselbe Wurstbrät in ihre jeweils bevorzugten Därme füllten. Die Würste wurden gekocht und einer ausgewählten Runde von Testessern serviert, die nicht wussten, dass alle Wurstfüllungen gleich waren und sich nur die Häute voneinander unterschieden. Und tatsächlich „schmeckte" das lauteste Würstchen am besten und gewann. Warum das Siegerwürstchen so viel lauter knackte als alle anderen, ist übrigens eine nette Anekdote. Der verwendete Schafsdarm stammte aus der Mongolei, wo die Tiere mangels Alternativen harte Gräser und sogar Disteln fressen, die den Verdauungsapparat offensichtlich besonders widerstandsfähig machen, wodurch ihr Darm dann besonders laut knackt.

Zu warmes Bier hören | Das Ohr ist eines unserer sensibelsten Organe. Allein am Klang können wir sauberes von schmutzigem Wasser und kaltes von warmem Bier unterscheiden. Am Geräusch, das entsteht, wenn Flüssigkeiten auf einer Oberfläche, also zum Beispiel am Boden eines Bierglases, auftreffen, lässt sich ihre Temperatur abschätzen. Warmes Bier klingt schal und lasch, kaltes hingegen hell und knacksig. Nach dieser Ouvertüre folgt die Durchführung:

Wurst, the popping sound is produced when the thin sheep's gut filled with the meat of beef and pork bursts. And this happens twice, to be exact, when someone bites into the sausage: with the help of an acoustic camera which synchronously documents picture and sound in extreme slow motion, it can be shown that the bursting of sausages consists of two immediately successive loud popping sounds. Once when the upper jaw bites through the sausage skin and the second time when the teeth on the lower jaw make the skin burst.

The popping effect that occurs when someone bites into a sausage accounts for 70 per cent of the taste that is perceived. When various types of sausage are sampled, the ones that always win are those that have the loudest popping sound. In his home region, the Saxon Erzgebirge, sound designer Friedrich Blutner conducted a test in which all butchers of the area filled the same sausage meat in their favorite guts. The sausages were cooked and then served to a selected group of samplers who were unaware of the fact that all sausage fillings were the same and that only the skins varied. And in fact, the loudest sausage "tasted" the best and won. Why the winning sausage popped so much louder than all of the rest is, incidentally, a nice anecdote. The sheep's gut that was used came from Mongolia where the animals for lack of alternatives ate hard grasses and even thistles that make their digestive system particularly resistant so that their guts made an especially loud popping sound.

Hearing Beer That Is too Warm | The ear is one of our most sensitive organs. On the basis of sound alone, we are able to distinguish clean water from dirty water and cold beer from warm beer. Its temperature can be recognized from the sound produced when fluids hit on a surface, for instance, on the floor of a beer glass. Warm beer sounds hollow and lifeless, cold beer, by contrast, bright and crisp.

Links | Bei Würsten macht das Knackgeräusch beim Zubeißen bis zu 70% des Geschmacks aus.
Unten | Allein am Klang können wir sauberes von schmutzigem Wasser und kaltes von warmem Bier unterscheiden.

left | In sausages the snapping sound when someone is taking a bite accounts for up to 70% of the taste.
bottom | Just on the basis of sound we can distinguish clean water from dirty water and cold from warm beer.

This overture is followed by implementation: the gushing sound of liquid when the glass is being filled. Acoustically trained persons can distinguish up to 100 different beer brands from this sound. A prerequisite for this is of course that all beers have exactly the same temperature. The finale of the sound composition is the foam crown that rises up with an unbelievably fine and diverse texture of sounds. Any connoisseur is, for instance, able to distinguish light beer from dark beer.

The acoustic impressions produced while beer is being poured are related to the structure of frequency and time of the fine foam bubbles bursting and the pearls of carbonated acid, and bubbling sounds, when the beer flows out of the bottle. This rhythm lets us to learn about the taste. If it bubbles too fast, the beer strikes us as light and tasteless, and if it does this too slowly, the beer appears stale. Friedrich Blutner even draws parallels between the most stimulating beer bubbling frequency and the rhythm of opera vibratos. "The rhythm of 5 to 6 hertz makes us happy and a beer which gurgles from the bottle in this rhythm simply sounds happy," as he observes. The sounds that beer makes can be influenced by the composition of the drink on the one hand and by the shape of the bottle on the other hand. Clever breweries design the bottleneck — consciously or intuitively — so that a striking gurgling sound between 5 and 6 hertz is produced when the beer is poured.

Pleasure Derived from Sound Patterns | Similar to the combination of consistencies, various sound nuances can also enhance the pleasure of eating. The more varied it sounds, the better it also tastes. Also because of the sounds that are created by chewing — a steak, for instance, tastes particularly good if it has been seared on the outside and is still bloody on the inside. In this case, there is a relatively loud crunching sound when one bites it. Then one hears flowing sounds in the mouth, when the juice exits the meat fibers through chewing. 'Dry' and 'juicy' are two central acoustic concepts (a dry space or a juicy sound as opposed to a rich sound), and it is thus no surprise that an allusion to "juicy meat" is often a question of acoustics. According to Friedrich Blutner, meat that has not been thoroughly cooked is thus so good

auch deshalb so gut, weil wir uns durch die Quietsch- und Gluckergeräusche beim Zerkauen an archaische Zeiten erinnert fühlen, in denen wir das Fleisch noch roh von der Jagdbeute nagten.

Ziel der Arbeit von Sounddesignern ist unter anderem die Differenzierung von Geräuschmustern, denn das lässt das Gegessene interessant erscheinen. Friedrich Blutner arbeitet derzeit zum Beispiel an einem Speiseeis, das wie ein Plattenspieler funktioniert: Auf einer harten Trägerschicht befinden sich Zutaten, die beim Hineinbeißen unterschiedlichste Klänge erzeugen. Laut Blutner lassen sich dabei 30 bis 40 unterschiedliche, eindeutig differenzierbare Klangmuster erzielen.[4]

Sounddesign | „Guter Geschmack ist wissenschaftlich nicht fassbar. Essen ist eine multisensuelle Wahrnehmung, die äußerst subjektiv ist. Mit Subjektivität kann unsere Gesellschaft jedoch schlecht umgehen, da sie immer davon ausgeht, alles müsse objektiv beschreibbar sein. Wenn ich einen Sound designe, so ist das durchaus mit der Kreation eines Duftes vergleichbar: man greift zwar auf technische Hilfsmittel zurück, letztendlich entscheiden aber mein Talent, meine Vorstellungskraft und meine Differenzierungs- und Merkfähigkeit der unterschiedlichen Klänge darüber, wie ich die einzelnen Komponenten zusammenfüge und ob daraus eine ausgewogene Geräuschkomposition entsteht oder nicht." Friedrich Blutner[5]

Wenn große Konzerne neue Gerichte oder Produkte entwickeln, werden auch deren Geräusche gestaltet. Sounddesign ist ein Teilbereich von Fooddesign, für große Projekte werden dazu eigene Sounddesigner hinzugezogen. Einer von ihnen ist Friedrich Blutner. Ausschlaggebend für die Beschäftigung mit dem Klang von Nahrung war für den gelernten Psychoakustiker eine Tomatenverkostung bei einem Geigenseminar in Italien: „Wir haben die Tomaten zerschnitten und festgestellt, dass jede Sorte anders klingt." Seitdem erforscht Friedrich Blutner, wie wir zu Geräuschen, die bei der Konsumation von Nahrung entstehen, emotionale Beziehungen aufbauen und wie man bestimmte Nahrungsmittel so gestalten kann, dass sie zum „richtigen" Geräusch fähig werden.

because the bubbling and squeaking sounds produced by chewing also remind us of archaic times in which we would still gnaw the meat from the hunter's bounty.

One goal of sound designers' work is to differentiate sound patterns since this makes what we eat appear interesting. Friedrich Blutner is presently working on an ice cream that functions like a record player. He places ingredients that create the most different sounds when biting on hard surfaces such as chocolate According to Blutner 30 to 40 different sound patterns that can be clearly distinguished can be produced.[4]

Sound Design | "Good taste cannot be grasped by scientific means. Eating is a matter of multi-sensory perception that is highly subjective. Our society, however, finds it difficult to deal with subjectivity, since it always assumes that everything must be described objectively. When I design sound, then this can be compared with the creation of a perfume. You resort to technical means but ultimately, however, my talent, my imagination and my ability to differentiate and retain the different sounds decide how I put together the individual components and whether this results in a balanced sound composition or not." Friedrich Blutner[5]

When large corporations develop new dishes or products, they also design their sounds. Sound design is a subfield of food design and for large projects sound designers are recruited. One of them is Friedrich Blutner. The trained psychoacoustic expert's interest in the sound of food was triggered when he attended a tomato sampling at a violin seminar in Italy. "We sliced the tomatoes and discovered that each type sounded different." Ever since Friedrich Blutner has been investigating how we establish emotional relations to sounds that are produced when we eat food and how we can design certain foods so that they are able to create the 'right' sound.

Der deutsche Akustiker Friedrich Blutner hat sich auf das Sounddesign von Lebensmitteln spezialisiert. Diagrammatische Darstellung von Frequenz und Zeit des Kaugeräusches von Brezeln, Boskoop Äpfeln, Kartoffelchips und Sprühsahne.

The German acoustics expert Friedrich Blutner has specialized in the sound design of food products. Diagrams used to depict the frequency and time of chewing sounds of pretzels, Boskoop apples, potato chips and canned whipped cream.

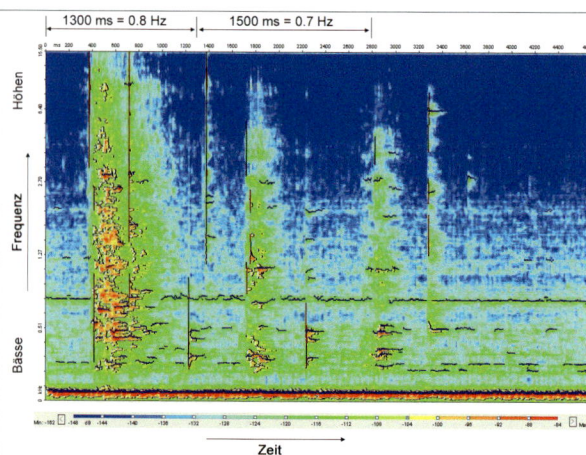
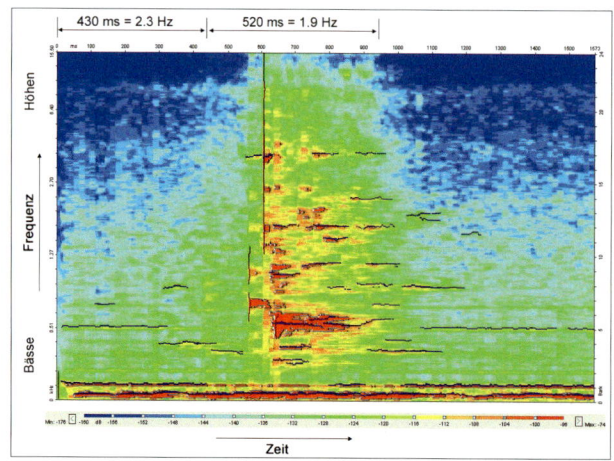
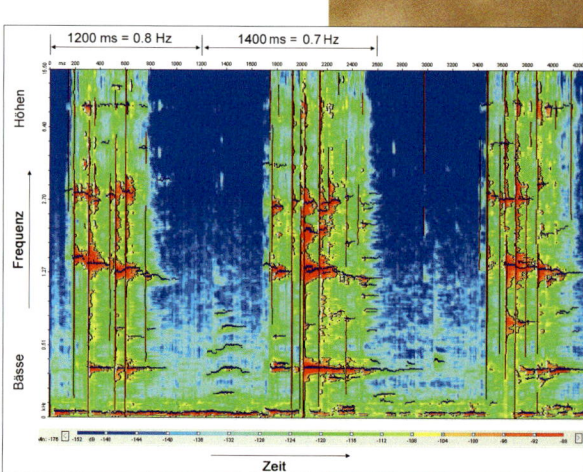

Am Beginn von Blutners Studien steht eine möglichst authentische Aufzeichnung und Wiedergabe von Beiß- und Kaugeräuschen der fraglichen Produktgruppe. Die Schwierigkeit dabei liegt einerseits darin, dass sich im Mund während des Essens keine Mikrophone platzieren lassen, und andererseits daran, dass Kaugeräusche auch über die Knochen übertragen werden. Ähnlich wie uns die eigene Stimme unnatürlich und verfremdet anmutet, wenn wir sie vom Tonband hören, decken sich auch Kaugeräusche, die über kleine Mikrophone in den Ohren der Testesser aufgezeichnet werden, nie hundertprozentig mit den tatsächlichen Wahrnehmungen während des Essens.
(Anmerkung: Ludwig van Beethoven (1770 - 1827) war am Ende seines Lebens fast taub. Trotz dieses Leidens war es ihm möglich Musik zu hören. Er setzte seinen Taktstock an das Klavier und biss darauf. So konnte er die Musik über seinen Kieferknochen hören. Die Knochenleitung ist im Zusammenhang mit Essgeräuschen deshalb so wichtig, weil sie Kaugeräusche direkt und zeitlich sehr präzise an das Innenohr leitet und darüber hinaus einen hohen Störabstand sicherstellt. Steht man beispielsweise neben einem Presslufthammer, so kann man sich nicht mehr unterhalten, aber sehr wohl die Geräusche beim Biss in einen Apfel wahrnehmen. Trotzdem bekommen wir vom Apfelessen keinen Hörschaden!)

Ein weiteres Problem ist, dass der Mundraum ein derart intimer Körperbereich ist, dass wir fremde Kaugeräusche als ekelhaft empfinden. Wir wollen nicht wissen, wie es im Mund klingt, wenn jemand Cornflakes, Würstchen oder Chips isst. Im Tonstudio von Synotech erfolgt die Aufnahme daher mit einem so genannten Kunstkopfmikrophon, einem stilisierten Kopf, in dessen Ohren zwei hochsensible Mikrophone angebracht sind. Die Versuchsperson beißt direkt vor dem Kunstkopf in das entsprechende Lebensmittel. Das aufgezeichnete Geräusch entspricht jenem, das wir hören, wenn jemand direkt neben uns isst. Diese Sounddaten werden dann mithilfe von sehr empfindlichen Kopfhörern von Testpersonen abgehört, verglichen und bewertet.

Ein anderes Testverfahren für die akustischen Eindrücke während des Essvorgangs sind so genannte „Krustimeter". Eine zylindrische Maschine simuliert das Kaugeräusch, indem ein Kolben das Testgut systematisch zerstampft. Mikrophone zeichnen die dabei erzeugten Geräusche auf, die dann mithilfe eines Computerprogramms ausgewertet werden. Ein weiteres Messgerät, das bei der Erforschung von Essgeräuschen zum Einsatz kommt, ist die akustische Kamera. Sie zeichnet parallel die Geräusche und die Mundbewegungen des Essvorganges in extremer Zeitlupe auf. Frequenz und Amplitude werden beim Abspielen optisch durch entsprechende Diagramme sichtbar gemacht und mit den Aufnahmen der Kieferbewegungen überlagert. Mit diesem Verfahren lässt sich zum Beispiel optisch darstellen, wann die Haut eines Würstchens tatsächlich platzt.

Aus allen Daten wird dann für eine konkrete Produktentwicklung ein so genannter „Zielsound" definiert, das heißt es wird aufgrund der Bewertungen ähnlicher, bereits bestehender Produkte festgelegt, wie der neue Cracker optimalerweise klingen sollte. Nun folgt der letzte Schritt, üblicherweise der schwierigste: die Herstellungsparameter so zu modifizieren, dass das Geräusch beim Abbeißen dem Zielsound möglichst nahe kommt. Bei einem Keks zum Beispiel lässt sich der Sound durch die Geometrie beeinflussen. Ein dicker Keks klingt anders als ein dünner Keks, ein kleiner Keks höher als ein großer. Ein anderer Parameter ist die Zusammensetzung des Kekses. Mehr Zucker beispielsweise klingt crunchyer und knuspriger. Schließlich spielt auch der Backvorgang eine Rolle. Ziel ist die Erschaffung sinnlich ausgewogener, standardisierter Geräuschkulissen. Letztlich ist Sounddesign ein oft langwieriger Prozess von trial and error, bei dem an allen möglichen Parametern geschraubt und probiert wird, bis der Klang dem Zielsound tatsächlich entspricht.[6]

Umgebungsgeräusche | Ob und wie uns etwas schmeckt, wird aber nicht nur durch die Geräusche beeinflusst, die wir selbst beim Essen erzeugen, sondern auch von der Essumgebung. Der britische Avantgardekoch Heston Blumenthal experimentiert mit unterschiedlichen Geräuschkulissen, die er seinen Gästen zu einzelnen Gerichten vorspielt. So hat er herausgefunden, dass dasselbe Ei intensiver und besser schmeckt, wenn seine Gäste dabei Hühnergegacker hören. Bei einem anderen Experiment halbiert Blumenthal eine Auster und serviert die eine Hälfte zu Tiergeräuschen, die andere zu Meeresrauschen. Die Gäste schwören danach, dass jene Hälfte, die sie zum Meeresrauschen gegessen haben, intensiver und salziger geschmeckt hat als die andere. Während uns fremde, unerwünschte Geräusche vom Essen ablenken und die Konzentration auf den Geschmack stören, können passende Geräusche das Geschmackserlebnis psychologisch unterstützen, ja sogar physisch intensivieren.

In his research Blutner first creates an authentic documentation and rendering of biting and chewing sounds of the product group in question. The difficulty is that it is not possible to place microphones in one's mouth while eating and that the sounds of chewing are also transmitted by the bones. Just as one's own voice can sound artificial and strange when we hear a recording of it, the chewing sounds that are taped by means of small microphones in the ears of the test eater, never converge one hundred percent with the actual perceptions while eating. (Note: Ludwig van Beethoven (1770-1827) was almost deaf at the end of his life. In spite of this he was able to listen to music. He would place his baton on the piano and bite on it. This way he was able to hear music by way of his jawbone. The bone connection is so important in connection with eating sounds because it conducts chewing sounds directly and temporally very precisely to the inner ear and also secures a high signal-to-noise ratio. If, for instance, one stands next to a pneumatic drill, it is impossible to converse but one can hear the sounds created when someone bites into an apple. Nevertheless, our hearing is not damaged by eating apples!)

Another problem is that the mouth is such an intimate part of the body that we find the chewing sounds made by someone else disgusting. We do not want to know what it sounds like in someone's mouth when this person is eating cornflakes, sausages or chips. Therefore, the Synotech sound studio uses a so-called plastic microphone, a stylized head with two highly sensitive microphones attached to the ears for recordings. The test person bites into a piece of food directly in front of the stylized head. The recorded sound corresponds to what we hear when someone is eating right next to us. This sound data is then heard, compared and assessed by test persons by means of very sensitive headphones.

Another test method for assessing acoustic impressions during eating is the so-called "crustymeter". A cylindrical machine simulates a chewing sound as a piston systematically pounds test material. Microphones record the resulting sounds which are then evaluated by means of a computer program. A further testing device that is used to study eating sounds is the acoustic camera. It simultaneously records the sounds and the mouth movements of a person eating in extreme slow motion. Frequency and amplitude are visualized when the recordings are played back by means of diagrams and overlapped with pictures of chewing movements. By means of this method it is for example possible to visually document when the skin of a sausage bursts.

All of the data is then used to define a so-called "target sound" for developing a specific project, that is, on the basis of the evaluations of similar, already existing products, it is specified how the new cracker should, ideally, sound. This is then followed by the last step, which is generally the most difficult one: modifying the production parameters in such a way that the sound heard when biting into, say, a cracker comes as close as possible to the target sound. In a cookie, for instance, the sound can be influenced by geometry. A thick cookie sounds different than a thin cookie, a small cookie sounds higher than a large one. Another parameter is the composition of a cookie. More sugar, for instance, sounds crunchier and crustier. Baking also plays a role. The goal is to create sensually balanced, standardized sound backdrops. Ultimately, sound design is often a tedious process of trial and error in which all sorts of parameters are adjusted and readjusted until the sound matches the one aimed at.[6]

Surrounding Sounds | Whether and how something tastes for us is not just influenced by sounds that we create while eating but also by the surroundings. British avant-garde cook Heston Blumenthal experiments with different sound backdrops which he plays for his guests to go along with individual courses. He discovered that the same egg tastes more intensive and better when his guests hear chickens cackling. In a different experiment Blumenthal halves an oyster and serves one half with animal sounds, the other with the sound of the ocean. The guests later swear that the half that they ate to the sound of the ocean tasted more intensive and saltier than the other one. While alien, undesired sounds distract us from eating and disrupt our focus on taste, the fitting sounds can psychologically back the experience of taste and even make it more intensive in physical terms.

1 | Interview für den Dokumentarfilm: „Food Design"; Geyrhalter Film GmbH Wien
2 | Interview für den Dokumentarfilm: „Food Design"; Geyrhalter Film GmbH Wien
3 | Wir bedanken uns bei Friedrich Blutner für alle diese Informationen und Hinweise.
4 | Wir bedanken uns bei Friedrich Blutner für alle diese Informationen und Hinweise.
5 | Interview für den Dokumentarfilm: „Food Design"; Geyrhalter Film GmbH Wien
6 | Wir bedanken uns bei Friedrich Blutner für alle diese Informationen und Hinweise.

1 | Interview for the documentary "Food Design"; Geyrhalter Film GmbH Vienna
2 | Interview for the documentary "Food Design"; Geyrhalter Film GmbH Vienna
3 | We thank Friedrich Blutner for this information.
4 | We thank Friedrich Blutner for this information.
5 | Interview for the documentary "Food Design"; Geyrhalter Film GmbH Vienna
6 | We thank Friedrich Blutner for all these pieces of information.

Laut Friedrich Blutner lassen sich bei Nahrungsmitteln 30 bis 40 unterschiedliche, eindeutig differenzierbare Klangmuster erzielen. Und je vielfältiger etwas klingt, desto besser schmeckt es auch.

According to Friedrich Blutner, 30 to 40 clearly distinguishable different sound patterns can be obtained in foods. And the more diverse something sounds, the better it tastes.

Geschmacksforschung
taste research

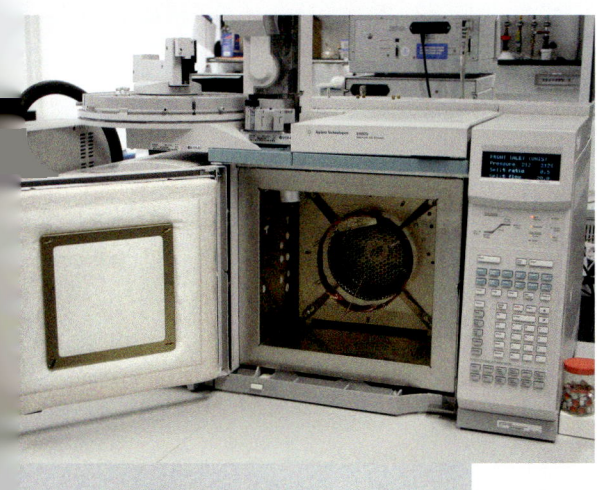  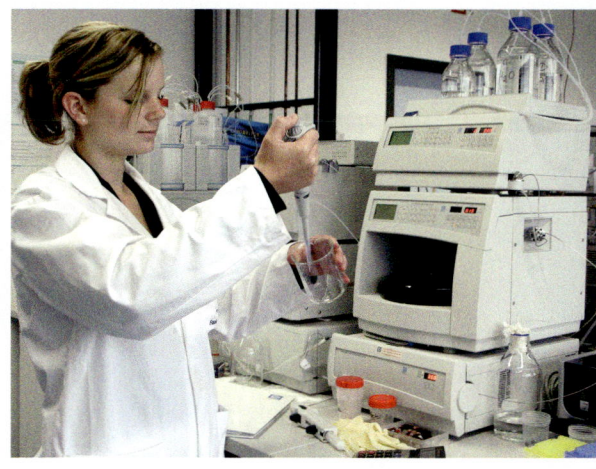

Geschmacksforschung | Wie kommt „guter" Geschmack eigentlich zustande? Welche Zutaten harmonieren besonders gut miteinander? Welche Substanzen schmecken gut, welche schlecht?

Wenn uns etwas gut „schmeckt", meinen wir damit nicht nur die Wahrnehmung des Gegessenen auf Zunge und Gaumen. Das Erlebnis „Geschmack" ist immer ein Zusammenspiel aller sinnlichen Erfahrungen, die wir beim Essen machen. Um die Vorgänge des Riechens, Schmeckens, Tastens, Sehens und Hörens für die Entwicklung neuer Nahrungsprodukte zu nutzen, wird Geschmack wissenschaftlich erforscht. Die so genannte Geschmacksforschung, die sich mit der wissenschaftlichen Analyse von Geschmack befasst, ist allerdings eine junge Wissenschaft und steckt derzeit erst in den Kinderschuhen. Wie die Geschmacksinformationen von Mund und Nase aufgenommen, ans Gehirn weitergeleitet und dort in den Geschmackseindruck umgewandelt werden, ist größtenteils ungeklärt. Niemand weiß also genau, wie in unserem Körper „guter" Geschmack eigentlich zustande kommt.

Historisch betrachtet, wurde die Analyse des Geschmacks als leibnahen Sinns über Jahrhunderte hinweg den Geisteswissenschaften zugewiesen. Themen wie Ästhetik oder Geschmacksaussagen fielen in den Bereich der Philosophie. Erst seit der Moderne beschäftigt sich auch die Naturwissenschaft mit Geschmacksfragen. Die Mediziner waren die Ersten, die sich für die Erforschung der menschlichen Wahrnehmung interessierten, zunächst allerdings vor allem im negativen Sinn, etwa mit dem Verlust des Geruchs- oder Geschmackssinnes nach einem Unfall.

Taste Research | How does "good" taste evolve? Which ingredients harmonize particularly well? Which substances taste good, which ones bad?

If something "tastes" good, we are not just referring to the perception of what we have eaten on our tongue and gums. The "taste" experience always involves an interplay between all sensual experiences we have while eating. In order to use the processes of smelling, tasting, touching, seeing and listening to develop new food products, taste is being scientifically studied. So-called taste research that deals with the scientific study of taste is, however, a young science and is still in an embryonic stage. It is still largely unexplained how taste information is taken in by the mouth and the nose, passed on to the brain and transformed into a taste impression there. No one really knows how "good" taste actually evolves in our body.

Viewed historically, the analysis of taste as a physical sense has been ascribed to the humanities for centuries. Themes such as esthetics or assertions regarding taste fell into the range of philosophy. It is only in modern times that science is also dealing with issues related to taste. Medical doctors were the first to be interested in studying human perception, though first mainly in a negative sense, for instance with the loss of the sense of smell or taste following an accident.

Links | Testflüssigkeit zum Kalibrieren.
Kleine Bilder von links nach rechts | Pipettieren einer Flüssigkeit unter einer Abzugshaube, HPLC (high performance liquid chromatography) Hauptgerät geöffnet, Sprühtrockner klein, Laborantin Frauke Kramer beim Pipettieren einer klaren Flüssigkeit im Labor

left | Test liquid for calibration
small photos from left to right | sampling liquid under a ventilation hood, HPLC (high performance liquid chromatography) equipment; spray dryer; lab technician Frauke Kramer sampling clear liquid in lab

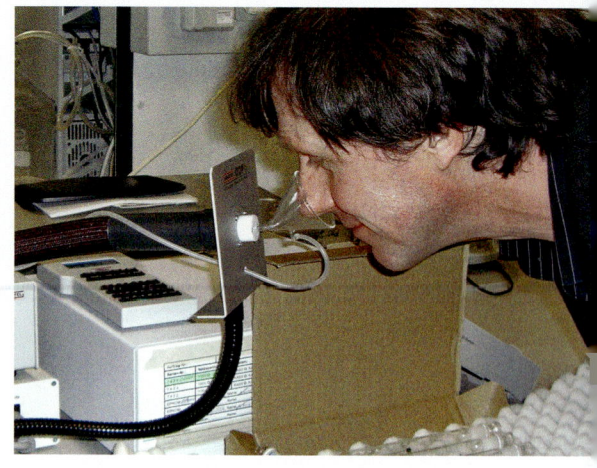

Mittlerweile ist Geschmacksforschung als naturwissenschaftliches Fachgebiet anerkannt. Forschungseinrichtungen wie das Technologietransferzentrum Bremerhaven und Universitäten wie die Pennsylvania State University oder die Universität für Bodenkultur in Wien untersuchen, welche chemischen Reaktionen während des Kochens, Bratens oder Backens stattfinden und wie diese den Geschmack im fertigen Lebensmittel „erzeugen".

Maillard-Reaktion | Am Institut für Food Science an der Pennsylvania State University in den Vereinigten Staaten beschäftigt man sich unter anderem mit der sensorischen und chemischen Analyse jener Substanzen, die den Geschmack von Esswaren beeinflussen, und der Frage, wie Geschmack auf molekularer Ebene entsteht. Ein Forschungsschwerpunkt ist die so genannte Maillard-Reaktion, eine Bräunungsreaktion, bei der sich Aminosäuren und Zucker unter Hitzeeinwirkung zu neuen Verbindungen umwandeln. Sie ist dafür verantwortlich, dass Fleisch beim Braten eine braune Kruste annimmt, Brot beim Backen dunkel wird und sich Pommes im Öl goldgelb färben. Wenn ein Schweinebraten mit einer schönen braunen Kruste aus dem Ofen kommt, dann istdiese durch die Maillard-Reaktion enstanden. Da wir geschmacklich gerade auf diese gebräunten Oberflächen und ihre Röstaromen ungemein „abfahren", spielt die Maillard-Reaktion bei der Entstehung „guten" Geschmacks eine zentrale Rolle. Von der Klärung der genauen Umstände, wann, warum und wie die Maillard-Reaktion erfolgt, erhofft man sich Erkenntnisse über den Zusammenhang von Zutaten, Rezepturen und Zubereitungsverfahren, die heute rein empirisch erfolgen, sowie die gezielte Erzeugung der beliebten Röstaromen unter kontrollierten Bedingungen mit immer gleichem, also standardisierbarem Ergebnis. Bedeutend ist das zum Beispiel bei der Herstellung immer gleich schmeckender Fertiggerichte.

Aromastoffe | Eine andere Frage, mit der sich die Geschmacksforschung beschäftigt, ist, welche chemische Zusammensetzung ein Nahrungsmittel hat, das gut schmeckt, und warum gerade die Kombination dieser Substanzen besser schmeckt als eine andere. Im Technologietransferzentrum Bremerhaven wollte man zum Beispiel wissen, ob sich ein relativ billiger Wein chemisch von einem teuren (wie zum Beispiel aus dem Gebiet um Bordeaux) unterscheidet, also ob sich die Qualität eines Lebensmittels chemisch nachweisen lässt.

Meanwhile the study of taste as a scientific field has become recognized. Research institutions such as the Technology Center of Bremerhaven, Pennsylvania State University or the Vienna University of Agriculture study which chemical reactions take place while cooking, roasting or baking and how these "generate" taste in the finished food.

The Maillard Reaction | The Institute for Food Science at the Pennsylvania State University in the United States carries out sensory and chemical analyses of those substances that influence the taste of food and the question of how taste evolves on a molecular level. One research focus is the so-called Maillard reaction, a browning reaction in which amino acids and sugar become transformed into new compounds under the influence of heat. It is responsible for meat taking on a brown crust when it is being roasted, for bread becoming dark when baking and for French fries turning golden-yellow when they are fried in oil; when a pork roast comes out of the oven with a nice brown crust, then a Maillard reaction has taken place. Since we really get off on the taste of these browned surfaces and their roast aromas, the Maillard reaction plays a central role in the emergence of "good" taste. It is hoped that clarifying the exact conditions, explaining when, why and how the Maillard reaction takes places will give insight into the composition of ingredients, recipes and modes of preparation that work purely empirically today. Moreover, it is expected that the deliberate creation of the popular roast aromas will become possible under specific conditions, always leading to the same, that is to say standardized result. This is important, for instance, for manufacturing ready-to-serve meals that always have the same taste.

Aromatic Substances | Another question addressed by taste research is the composition of food that tastes good and why precisely the combination of these substances tastes better than another. For example, researchers at the Bremerhaven Technology Transfer Center (TTZ) wanted to know whether a relatively inexpensive wine was chemically different from an expensive one (as for example one from the Bordeaux region), that is, whether the quality of food can be chemically proven.

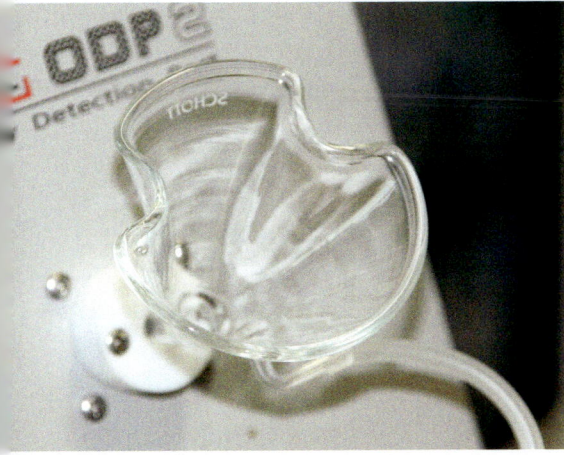
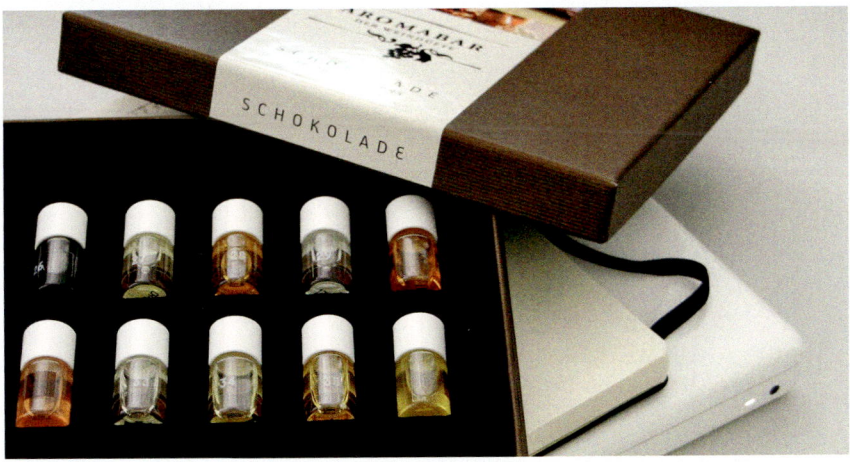

Reihe von links nach rechts | Laborantin Julia Disqué im TTZ Bremerhaven beim Abmessen einer Kalibrierungsflüssigkeit, Gaschromatograph mit Sniffing-Port, das Gerät wird zur Analyse der chemischen Zusammensetzung von Nahrungsmittelaromen verwendet.
Ganz rechts | 10 unterschiedliche Schokoladearomen.

top left to right | lab technician Julia Disqué measuring calibration liquid; gas chromatographer with sniffing port for the analysis of chemical compositions in food flavours.
far right | 10 different chocolate aromas

Das Aroma von Rotwein wird von rund tausend verschiedenen Substanzen gebildet. Etwa 50 davon sind besonders geschmacksprägend und – wie das TTZ nachweisen konnte – in so genannten „Qualitätsweinen" in größerer Konzentration vorhanden als in „Billigwein". Aber erst nach einer umfassenden Analyse aller, also auch der restlichen 950 Aromastoffe, wird man in der Lage sein, das Rätsel rund um „guten" und „schlechten" Rotwein wissenschaftlich zu lösen.[1]

Geschmack ist – chemisch betrachtet – äußerst komplex, das heißt er wird von unzähligen, unterschiedlichen Substanzen gebildet, die oft nur in sehr geringen, schwer nachweisbaren Konzentrationen enthalten sind. Die moderne Messtechnik ermöglicht erst seit wenigen Jahren, dass man „Wohlgeschmack" überhaupt feststellen kann.

Eine Erdbeere zum Beispiel enthält über dreihundert Aroma gebende Substanzen, davon sind vielleicht zwei oder drei – im konkreten Fall Furaneol, 4-Decanolid und Linalool - wirklich maßgeblich, das heißt zu neunzig oder fünfundneunzig Prozent am Geschmack der Erdbeere beteiligt. Aber auch die übrigen 297 Stoffe tragen zum harmonischen, runden Geschmack der Frucht bei. Wenn nun künstliche Erdbeeraromen auf Basis von Furaneol, 4-Decanolid und Linalool entwickelt werden, so schmeckt das Ergebnis zwar nach der echten Pflanze, wird aber im Vergleich dazu immer als künstlich, zu intensiv und unausgewogen wahrgenommen.[2]

Geschmack als wissenschaftliche Größe | Ein wesentlicher Bestandteil der Geschmacksforschung ist die wissenschaftliche Erfassung der Vielfalt aller sinnlichen Wahrnehmungen, die wir während des Essens erleben. Die Umwandlung von Geschmack, Geruch, Kaugeräusch, haptischen und optischen Eindrücken in messbare Einheiten und reproduzierbare Zahlenwerte und die Verknüpfung dieser Daten mit den emotionalen Beliebtheitsurteilen von Konsumenten ist von zentraler Bedeutung für die Optimierung bestehender und die Entwicklung erfolgreicher neuer Lebensmittel.

Sensorik | Konkret erfolgt die Erforschung von Geschmack durch die kombinierte Auswertung von chemischen Analyseverfahren einerseits und der Bewertungsergebnisse professioneller sensorischer Prüfpersonen (Testesser) andererseits. Sie beurteilen Speisen und Getränke hinsichtlich Geschmack, Geruch, Aussehen, Textur und Geräusche in möglichst objektivierter Weise. Obwohl

The aroma of red wine is the product of about a thousand different substances. About fifty of them have a particularly influential effect on taste and – as TTZ was able to prove – are present in greater concentration in so-called "quality wines" than in cheap ones. But it is only after an extensive analysis of all, that is the fifty plus the remaining 950 aromatic substances that researchers will be able to scientifically solve the mystery shrouding "good" and "bad" red wine.[1]

In chemical terms, taste is highly complex, that is to say, it is the product of countless different substances often only contained in very small concentrations for which it is difficult to find evidence. In recent years, modern measuring technology has finally made it possible to determine "good flavor".

A strawberry, for instance, contains more than three hundred aromatic substances, of which two or three – more specifically furaneol, 4-decanolid and linalool – are really decisive, that is to say, they contribute ninety or ninety-five per cent to the taste of strawberries. However, all the remaining 297 substances contribute to the round harmonious taste of fruit. When artificial strawberry aromas are being developed now on the basis of furaneol, 4-decanolid and linalool, the result still tastes like the real plant but compared with it, it is always perceived as being artificial, too intensive and unbalanced.[2]

Taste as a Scientific Parameter | A crucial part of taste research is the scientific study of the diversity of all sensory perceptions that we have while eating. The transformation of taste, smell, chewing sound, tactile and optical impressions in measurable entities and reproducible numeric values and the linking of these data with assessments of consumer's emotional satisfaction is essential for optimizing existing food and developing successful new food.

The Study of Sensory Stimuli | The study of taste actually takes place through the combined procedures of chemical analysis on the one hand and the evaluation results of professional sensory

In so-called sensorics labs foods are tested under controlled conditions. The usual criteria for assessing foods are appearance, taste, smell and general impression. The foods are given a numeric code for this purpose. To change the color impression, the test booths can also be illuminated with red light.

wir im digitalen Zeitalter leben, werden Lebensmittel immer noch primär von Menschen bewertet. Testgeräte können zwar einzelne Teilbereiche wie die Viskosität oder die chemische Zusammensetzung eines Lebensmittels feststellen, sind aber nicht in der Lage, Esswaren in ihrer „geschmacklichen" Gesamtheit zu erfassen. Die Schwierigkeit besteht darin, dass Geschmack keine objektive, wissenschaftliche Größe ist und der Konsument nicht nur die technisch messbaren Parameter, also die physikalischen oder chemischen Eigenschaften eines Produktes beurteilt, sondern auch das Gefühl, das ein Gericht in ihm auslöst.[3]

Eine wichtige Grundlage der Geschmacksforschung ist daher die Sensorik, jenes Wissensgebiet, das sich mit der sinnlichen Wahrnehmung und Bewertung von Produkten beschäftigt. Sie dient sowohl zur Qualitätskontrolle vorhandener als auch zur Entwicklung neuer Produkte. In so genannten Sensoriklabors werden Lebensmittel unter kontrollierten Bedingungen von Testpersonen bewertet. Die Esswaren werden dem jeweiligen Tester in anonymisierter Form, also nur mit einem Zahlencode versehen, in seine Testkabine gereicht. Danach wird die Kabine geschlossen und der Testesser kann völlig abgeschlossen von allen Umwelteinflüssen seine Bewertung abgeben. Die gängigen Kriterien zur Beurteilung sind Aussehen, Geschmack, Geruch und Gesamteindruck. Je nach Zielsetzung kann auch die Lichtsituation in den Kabinen zwischen Tageslicht, Kunstlicht oder Rotlicht variiert werden oder die Verkostung überhaupt im Dunkeln stattfinden.[4]

testers (test eaters) on the other hand. They assess food and drinks with regard to taste, smell, appearance, texture and sounds as objectively as possible. Even though we are living in a digital age, food is still primarily judged by people themselves. Testing devices can be used to determine certain factors such as the viscosity or the chemical composition of food but they are still unable to capture the entire "taste" of food. The difficulty has to do with the fact that taste does not constitute an objective, scientific parameter. Moreover, the consumer not only assesses the technically measurable parameters, i.e., the physical or chemical qualities of a product, but also the sensation that a food triggers in him/her.[3]

The study of sensory stimuli – the area of knowledge that deals with the sensual perception and evaluation of products – is thus an important foundation of taste research. It serves both quality control of existing food and the development of new products. In so-called sensory labs food is tested by respondents under controlled conditions. The food is served to the tester in a cubicle in an anonymous form, that is to say, only identified by a number code. Then the cubicle is closed and the test eater can assess the food in complete isolation from all environmental influences. The usual criteria for an appraisal are appearance, taste, smell and general impression. Depending on the goal, the light situation in the cabin can be varied between daylight, artificial light or red light or the tasting can take place in total darkness.[4]

Testgeräte | Parallel zur sensorischen Bewertung wird dasselbe Produkt von unterschiedlichen Testgeräten physikalisch und chemisch untersucht. So lässt sich zum Beispiel die Konsistenz einer gekochten Nudel testen, indem man sie so lange dehnt, bis sie reißt, und den dafür nötigen Kraftaufwand misst. Stimmt der Wert nicht mit der von menschlichen Testessern als optimal eingestuften Nudelkonsistenz überein, können Parameter wie Rezeptur, Trocknungszeit, Herstellungstemperatur et cetera so lange adaptiert werden, bis der gewünschte Wert erreicht ist.[5]

Die Analyse der chemischen Zusammensetzung von Nahrungsmittelaromen erfolgt mit dem so genannten Gaschromatographen, der eine Substanz in das Spektrum ihrer einzelnen Inhaltsstoffe zerlegt. Diese werden dann an einem so genannten Sniffing Port von einer menschlichen Testperson sensorisch, also hinsichtlich ihres Geruches, bewertet und protokolliert. Kennt man sowohl die chemische Zusammensetzung eines Produktes als auch die sensorische Wahrnehmung seiner Einzelkomponenten, so lassen sich Rückschlüsse ziehen, welche Substanzen für spezifische Geschmacksbilder verantwortlich sind.[6]

Andere Testgeräte messen die Textur, die Oberflächenspannung, die Zähigkeit von Teig, den Beißwiderstand oder das Kaugeräusch. Akustische Messungen an Cornflakes, Knabbergebäck und anderen spezifische Geräusche verursachenden Lebensmitteln werden mit dem so genannten Acoustic Analyzer durchgeführt. Durch mechanische Belastungen wie Penetration, Schneiden oder Scherung wird zum Beispiel das Kaugeräusch im Mund simuliert und von Mikrophonen aufgezeichnet. Die Testergebnisse werden ausgewertet, mit den von Konsumenten als appetitlich und passend empfundenen Zielsounds verglichen und die Rezepturen hinsichtlich ihrer akustischen Auswirkungen optimiert.[7]

Resümee | Angesichts des Feuerwerks an lustvollen Wahrnehmungen, die wir genießen, wenn wir eine Speise mit allen Sinnen erfahren, scheint es durchaus verständlich, dass der Mensch rund um die Gestaltung seiner Nahrung so viel Aufwand treibt. Die Steigerung des sinnlichen Genusses ist eine der wesentlichsten Motivationen, Food Design zu betreiben. Der Mensch reagiert auf seine Nahrung aber nicht nur sensuell, sondern immer auch kulturell. Alle Sinneseindrücke, die wir beim Essen wahrnehmen, werden in das so genannte limbische System unseres Gehirns geleitet. Dieser Teil ist sowohl für das Gedächtnis, als auch für das Affekt- und Triebverhalten verantwortlich. Ein Nahrungsmittel kann nur dann positive Emotionen auslösen, wenn diese den erlernten, kulturellen Standards entsprechen. Werden zum Beispiel Nahrungstabus verletzt, können wir keinen Genuss verspüren. Geschmack ist somit letztlich eine subjektive Angelegenheit, abhängig von äußeren Einflüssen wie dem gesellschaftlichen und kulturellen Umfeld. Jeder Mensch, jede Kultur, jede Epoche „schmeckt" anders, hat andere Vorlieben und andere Ansprüche und nimmt dasselbe Essen anders wahr.

Testing Devices | Parallel to the sensory evaluation, the same product is chemically and physically tested with the help of various devices. For instance, the consistency of cooked noodles is tested by pulling them until they break and measuring the force this requires. If the value does not coincide with the noodle consistency seen as ideal by the test eaters, parameters such as recipe, drying time, manufacturing temperature etc. can be adapted until the desired value is obtained.[5]

The analysis of the chemical composition of food aromas is carried out by means of so-called gas chromatography which decomposes a substance into the spectrum of its individual components. These are then evaluated and documented by a human tester at a so-called sniffing port, that is, in terms of smell. If we know both the chemical composition of a product and the sensory perception of individual components, we will be able to determine which substances are responsible for specific taste impressions.[6]

Other testing devices can be used to measure the texture, surface tension, the viscosity of dough, the biting resistance or the chewing sound. Acoustic measurements of cornflakes, snacks and other foods producing specific sounds are carried out using a so-called acoustic analyzer. Mechanical stress such as penetration, cutting or shearing is used to simulate a chewing sound in the mouth, which is taped via microphones. The test results are evaluated, compared with the target sounds that the consumer has perceived as appealing and fitting. The recipes are then optimized, taking into account their acoustic effects.[7]

Conclusion | In light of this panoply of pleasurable sensations that we enjoy when we experience food with all of our senses, it seems understandable that people go to such efforts to design food. Enhancing the sensual pleasure is one of the greatest motives for pursuing food design. An individual responds to food not only in a sensory but also in a cultural way. All sensory impressions that we have while eating are conducted to the so-called limbic system of the brain. This region is responsible not just for memory but also for affective and instinctive behavior. Food can only trigger positive emotions when it corresponds to the acquired, cultural standards. If food taboos are violated, we will be unable to feel pleasure. Taste is thus ultimately a subjective matter, dependent on external influences such as our social and cultural environments. Each individual, each culture, each period in time "tastes" different, has different predilections and different aspirations, and thus perceives the same food in a different way.

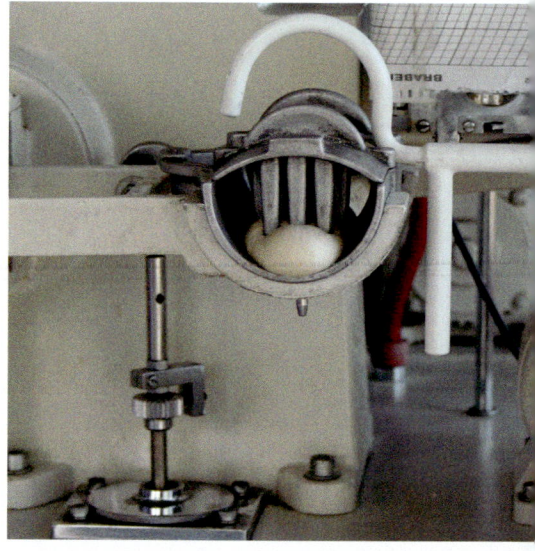

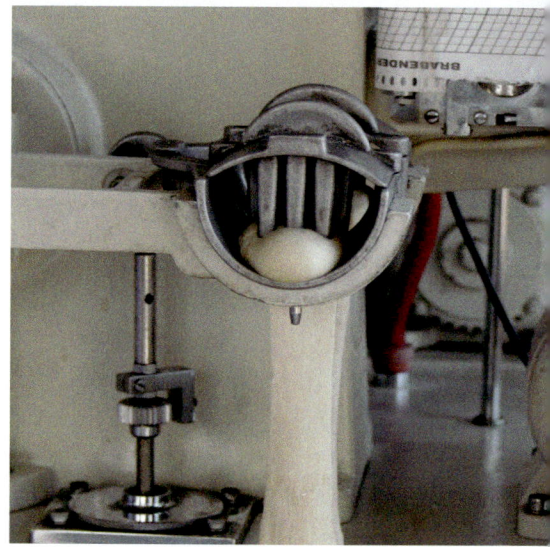

Geräte zur Überprüfung von Texturen: Die Konsistenz von Brot oder Schokolade wird getestet, indem man den Teig so lange dehnt, bis er reißt – bzw. die Schokolade punktuell belastet, bis sie bricht – und den dafür nötigen Kraftaufwand misst. Stimmt dieser Wert nicht mit der gewünschten Textur überein, können Parameter wie Rezeptur, Trocknungszeit, Backvorgang, Herstellungstemperatur etc. so lange variiert werden, bis der gewünschte Wert erreicht ist.

Machines used to check textures. The consistency of bread or chocolate is tested by pulling the dough until it tears – or loading the chocolate at certain points until it breaks – and measuring the amount of energy needed to do this. If this value does not coincide with the desired texture, parameters such as recipe, drying time, baking procedure, production temperature, etc. can be varied until the desired value is obtained.

1 | Wir bedanken uns bei Werner Mlodzianowski für diese Information.
2 | Wir bedanken uns bei Werner Mlodzianowski für diese Information.
3 | Wir bedanken uns bei Werner Mlodzianowski für diese Information.
4 | Wir bedanken uns bei Prof. Emmerich Berghofer, bei Werner Mlodzianowski und bei Dipl. Ing. Klaus Dürrschmid für diese Informationen.
5 | Wir bedanken uns bei Prof. Emmerich Berghofer, bei Werner Mlodzianowski und bei Dipl. Ing. Klaus Dürrschmid für diese Informationen.
6 | Wir bedanken uns bei Prof. Emmerich Berghofer, bei Werner Mlodzianowski und bei Dipl. Ing. Klaus Dürrschmid für diese Informationen.
7 | Wir bedanken uns bei Prof. Emmerich Berghofer, bei Werner Mlodzianowski und bei Dipl. Ing. Klaus Dürrschmid für diese Informationen.

1 | We thank Werner Mlodzianowski for this information.
2 | We thank Werner Mlodzianowski for this information.
3 | We thank Werner Mlodzianowski for this information.
4 | We thank Prof. Emmerich Berghofer, Werner Mlodzianowski and Klaus Dürrschmid for this information.
5 | We thank Prof. Emmerich Berghofer, Werner Mlodzianowski and Klaus Dürrschmid for this information.
6 | We thank Prof. Emmerich Berghofer, Werner Mlodzianowski and Klaus Dürrschmid for this information.
7 | We thank Prof. Emmerich Berghofer, Werner Mlodzianowski and Klaus Dürrschmid for this information.

Funktion
function

Form follows function - Die Funktion von Essen

„Design bezeichnete ursprünglich die Verbindung von Ästhetik und Funktionalität. Industriedesigner gestalten Alltagsobjekte und machen sie attraktiver, aber was ist wahre Schönheit? Dinge, die sinnvoll und gebrauchstauglich sind, die ihren Zweck erfüllen. Ich denke, unsere eigentliche Aufgabe, was wir wirklich schaffen sollten, ist Stimmigkeit. Mit anderen Worten, Objekte, die in Aussehen, Geschmack und Funktionsweise ein ausgewogenes Ganzes bilden." Marc Brétillot, Fooddesigner

Warum sind Käselaibe rund und Fischstäbchen eckig? Warum haben Makkaroni, Bagel oder der Gugelhupf ein Loch in der Mitte? Warum werden bei der Kommunion flache, runde Oblaten an Stelle echter Brotstücke verteilt?

Die spezifische Form vieler Esswaren erklärt sich aus ihrer Funktion. Gerichte und Lebensmittel sollen das Auge und den Gaumen erfreuen, wir gestalten sie aber auch nach praktischen Gesichtspunkten und sind sogar bereit, im Gegenzug auf geschmackliche Nuancen zu verzichten. So wie wir manchmal die bequemeren Schuhe anziehen oder das familienfreundlichere Auto kaufen, entscheiden wir uns auch beim Essen für fertige Tiefkühlpizzen oder lagerfähige Dosentomaten.

Essen muss nicht nur gut schmecken und unseren kulturellen Vorstellungen entsprechen, Essen muss auch funktionieren. Unzählige, weltweit erfolgreiche Produkte überzeugen keineswegs aufgrund ihrer ästhetischen oder sinnlichen Qualitäten, sondern punkten mit ganz praktischen Vorteilen. Fischstäbchen zum Beispiel lassen sich industriell herstellen, gut stapeln, lange lagern, einfach zubereiten, gut in den Fingern halten und passen genau in den Mund.

Form Follows Function - The Function of Food

"Design originally referred to aesthetic function. Industrial aesthetics entails making functional objects attractive, but what is real beauty? Things that are meaningful and useful, that are satisfying to use. I think that the task in our projects, what we really should be after, is coherence, in other words producing something that's coherent in terms of its appearance, taste and how it's used." Marc Brétillot, food designer

Why are cheese loaves round and fish fingers rectangular? Why are there holes in the centers of macaroni, bagels or Austrian "Gugelhupf" cakes? Why is the host shared out at Holy Communion a flat round wafer, not a piece of real bread?

The specific shapes of many food products result directly from their function. Food is supposed to be a pleasure to the eyes and palate, but we also design it according to practical criteria and are even willing to forego nuances of taste in return. Just like we sometimes wear comfortable shoes or buy a family car, we also decide in favor of deep-frozen pizza or canned tomatoes with a long shelf life for the sake of convenience.

Food has to be tasty, it has to conform to our cultural expectations and it has to work well, to boot. Numerous products that sell successfully worldwide are not popular because of their esthetic or sensory qualities but because they have very practical benefits. Fish fingers, for example, can be manufactured industrially, they can easily be stacked, have a long life, are simple to cook, can be eaten with our hands and fit exactly into our mouths.

Manche Pastasorten funktionieren wie kleine Löffel, die helfen, die Sauce in den Mund zu transportieren.

There are some types of pasta that function like small spoons helping to transport the sauce to one's mouth.

124 Funktion - function

Wir verarbeiten natürliche Zutaten, um sie unseren Lebens- und Essgewohnheiten anzupassen, um sie billiger herzustellen, müheloser zu transportieren oder bequemer zu konsumieren. Wir mahlen Getreide zu Mehl und backen daraus die unterschiedlichsten Brote, weil die Körner im Ganzen schlecht verdaulich sind. Wir fermentieren Milch zu großen, kleinen, zylindrischen und kugelförmigen Käselaiben, damit sie haltbar wird, und füllen heiße Würstchen in passende, längliche Brötchen, um sie in einer Hand halten zu können, ohne uns zu verbrennen. Essen muss eine große Bandbreite an funktionalen Anforderungen erfüllen, welche die Gestaltung von Nahrungsmitteln entscheidend beeinflussen.

Die Basisfunktion von Essen ist, uns das Überleben zu sichern. Lebensmittel müssen aber noch eine ganze Reihe anderer, weniger offensichtlicher Bedürfnisse befriedigen, um vor dem Konsumenten zu bestehen. Viele Esswaren sind eigens so gestaltet, dass sie möglichst langsam verderben, nicht austrocknen, sich gut tragen lassen oder ohne Besteck problemlos zu essen sind. Wir bevorzugen Nahrungsmittel, von denen wir gut abbeißen können oder die wir, einfach und ohne uns schmutzig zu machen, in gleich große Stücke brechen können. Das Anforderungsspektrum für Nahrung reicht aber noch viel weiter und schließt beispielsweise auch Effizienz bei Anbau, Ernte und Verarbeitung, zweckmäßige Verpackungs- und Verzehrfähigkeit sowie Portionierbarkeit, Take-away-Tauglichkeit, Bekömmlichkeit und Gesundheitsförderung, wie bei Functional Food, mit ein. Das folgende Kapitel widmet sich den funktionalen Aspekten von Food Design und tritt den Beweis an, dass der Grundsatz „form follows function" auch für Esswaren gilt.

We process natural ingredients to adapt them to our lifestyle and eating habits, to make their production cheaper, their transportation less cumbersome or their consumption more convenient. We grind grain and use the flour for baking various types of bread because whole grain is harder to digest. We ferment milk to give it a longer life span in the shape of large or small, cylindrical or spherical loaves of cheese, and we put hot sausages into oblong buns so we can hold them in our hands without getting burnt. Food has to fulfill a wide variety of practical requirements which have a decisive impact on the design of food products.

The basic function of food is to ensure our survival. However, food also has to satisfy a number of other, less obvious needs to be accepted by consumers. Many food products are designed in such a way that they spoil as slowly as possible, do not go stale, can be carried easily or eaten without a knife and fork. We prefer food which we can take a bite of without great effort or which we can break into pieces of equal size without soiling our hands. The range of requirements for food is even wider, including efficient cultivation, harvesting and processing, ease of packaging, consumption and portioning, qualities that make it suited for take-aways, wholesome and conducive to good health (as in functional food). The following chapter is devoted to the functional aspects of food design, and it will furnish evidence that the principle "form follows function" also applies to foodstuffs.

Konservierung
preservation

Haltbarkeit | Neben dem Nährwert ist die Haltbarkeit die zweite zentrale Leistungsanforderung an Nahrungsmittel. Die meisten frischen Lebensmittel beginnen schon nach kurzer Zeit zu faulen oder zu schimmeln und verderben in der Regel schnell. Selbst Wasser ist nur begrenzt lagerfähig. In Europa nutzte man die alkoholische Gärung unter anderem dazu, Getränke wie frische Obstsäfte zu konservieren und das ganze Jahr über aufzubewahren.

Vor der Entwicklung der Kühltechnik bildete die Frage der Konservierung einen, wenn nicht den wichtigsten, Gestaltungsfaktor für Esswaren. Besonders in nördlichen Regionen musste man Methoden entwickeln, um Nahrung für die Wintermonate zu lagern. Die Notwendigkeit der Konservierung führte zur Entwicklung spezieller Zubereitungsverfahren und veränderte gleichzeitig den Geschmack und die Gestalt von Nahrungsmitteln. Ziel der unterschiedlichen Konservierungsverfahren ist es, das Wachstum von Mikroorganismen zu hemmen und Zerfallsprozesse, welche die Struktur und/ oder die Substanz des Lebensmittels verändern und es in Folge ungenießbar machen, zu verzögern. Grundsätzlich unterscheidet man zwischen chemischen Konservierungsverfahren wie Einsalzen, Räuchern, Einlegen, Einzuckern, Säuern oder Zugabe von Konservierungsstoffen und physikalischen Methoden wie Kühlen, Pasteurisieren, Sterilisieren, Einkochen, Trocknen et cetera.

Trocknen | Trocknen ist eine der ältesten Konservierungsmethoden. Schon in ältester Zeit wurden Pilze, Kräuter und Früchte getrocknet, um sie haltbar zu machen. Obwohl die Trocknung ein vergleichsweise einfaches Verfahren ist, hängt die Qualität des Endproduktes wesentlich davon ab, wie die frischen Zutaten geformt oder zugeschnitten sind. Ein gutes

Durability | Apart from nutritional value, durability is the second major requirement food products have to meet. Most fresh food starts to go rotten or moldy very quickly and spoils easily. Even water is limited in durability. In Europe, for example, alcoholic fermentation was one of the methods used to preserve beverages such as fresh fruit juices, and to store them all year long.

Before the advent of refrigeration technology, preservation was one of the most important, if not the most important factor in the design of food. In the northern region, special methods were required to store food for the winter. The need for preservation led to the development of special ways of preparation, changing at the same time the taste and the looks of foods. Preservation methods aimed at inhibiting the growth of micro-organisms and delay decay processes that altered the structure and/or substance of the food, making it inedible. Basically, the main distinction drawn here is between chemical preservation methods (curing, smoking, pickling, candying, acidulation or addition of preservatives) and physical preservation (refrigerating, pasteurizing, sterilizing, canning or bottling, drying etc.).

Drying | Drying is one of the oldest preservation methods. Even in the early days, mushrooms, herbs and fruits were dried to make them more durable. Although drying is a relatively simple procedure, the quality of the final product strongly depends on the shape or cut of the fresh ingredients. Cylin-

Links Rand | Stockfischproduktion in Norwegen: eingesalzener Kabeljau wird zum Trocknen aufgehängt.
Links Mitte | Trocknen ist eine der ältesten Konservierungsmethoden. Schon in ältester Zeit wurden Pilze, Kräuter und Früchte getrocknet, um sie haltbar zu machen. Die österreichische Firma Steirerkraft beschäftigt sich sehr erfolgreich mit der Frage, wie man getrocknete Apfelspalten auf natürliche Weise mit einer knusprigen Oberfläche überziehen kann.
Rechts | Postkarte aus Neapel von 1918: Auf Stöcken und Leinen werden in Gragnano unter freiem Himmel Nudeln getrocknet.

left margin | Dried cod fish production in Norway: salted cod is being hung to dry.
left center | Drying is one of the oldest preservation methods. Already in ancient times mushrooms, herbs and fruits were dried to preserve them. The Austrian company Steirerkraft has been successful in dealing with the question of how dried apple slices can be naturally covered with a crusty surface.
right | Postcard from Naples dated 1918: Pasta being dried on poles and lines in the open air in Gragnano.

Beispiel, wie der Konservierungsprozess die Gestaltung von Lebensmitteln beeinflusst, sind röhrenförmige Nudeln. Vermutlich im elften Jahrhundert nach Christus entdeckten die Araber, dass sich die Haltbarkeit von Nudeln durch Trocknung extrem verlängern lässt. Und sie fanden schnell heraus, dass die Abwehr gegen Schimmel besser funktioniert, wenn Pasta fein und dünn ist.

Wie bei allen getrockneten Waren erfolgt die Konservierung von Nudeln durch einen möglichst weitgehenden Entzug von Wasser, wobei die Verdunstung umso besser funktioniert, je größer die Oberfläche des Objektes ist, das getrocknet werden soll. Aufgrund dieser Erkenntnis begann man den dünn ausgewalkten Teig vor dem Trocknen um Stöcke zu wickeln. So entstanden Formen wie etwa die Makkaroni, die innen hohl sind, dadurch schneller und besser trocknen – und länger halten.

Damals wie heute ist eine gleichmäßige Trocknung des Teiges für Qualität und Haltbarkeit der Pasta ausschlaggebend. Aus diesem Grund fasste die typische Nudel aus Hartweizengrieß und Wasser (ohne Ei!) zuerst im wetterbegünstigten Süditalien Fuß. Im Raum Neapel garantierte das Klima jenen vollständigen Entzug von Feuchtigkeit, der notwendig ist, um das Produkt langfristig vor dem Verderben zu bewahren.

drical noodles are a good example of how the preservation process impacts the design of foodstuffs. Presumably in the eleventh century CE the Arabs discovered that the durability of noodles could be extended enormously by drying. They also quickly found that thin and finely cut pasta was less susceptible to mold.

As is the case with all dried goods, noodles are preserved by withdrawal of water. Condensation works better the larger the surface of the object to be dried. Due to this insight, people started to wrap thinly rolled out dough around sticks for drying. This is how shapes like those of macaroni came about: they are hollow inside, dry faster and better – and thus have a longer life.

Some things do not change at all. Homogeneously dried dough is still decisive for the quality and durability of pasta. For this reason, the typical noodles made of durum wheat semolina and water (no eggs!) were first adopted by the people in the hot south of Italy. The climate in the area of Naples guaranteed the complete withdrawal of humidity required to keep the product from spoiling in the long run. Pasta

So begann die erste industrielle Erzeugung von Teigwaren Mitte des 19. Jahrhunderts an den Hängen des Vesuvs. Noch vor hundert Jahren glichen ganze Küstenstriche dieser Region riesigen Trockenkammern, wo auf Stöcken und Leinen die Nudeln im Meereswind wehten. Damit gewann die Pasta auch als Massennahrungsmittel an Bedeutung.[1]

Auch „Mochi", traditionelle, japanische Reiskekse, wie sie früher in jedem Haushalt hergestellt wurden, entstanden ursprünglich, um einen Teil der Reisernte für den Winter zu konservieren. Die fertig geformten Cracker wurden auf Schnüre gefädelt und im Herbst vor dem Haus zwei bis vier Wochen lang zum Trocknen aufgehängt. Die Cracker wurden dann den Winter über verzehrt. Heute werden Mochi-Cracker ganzjährig in Fabriken erzeugt, wobei der Trocknungsprozess nur mehr vier Tage dauert.[2]

was first produced industrially on the slopes of Mount Vesuvius in the middle of the nineteenth century. A bare hundred years ago, entire coastal stretches in the region resembled vast drying chambers where noodles, hanging from sticks and lines, were blowing in the wind that came in from the sea. This way, pasta also gained importance as food for the masses.[1]

Similar preservation techniques can be found in different cultures. "Mochi", traditional Japanese rice cakes, used to be made by every household. Originally, their purpose was to preserve part of the rice harvest for the winter. Once the crackers had been shaped, they were threaded on strings and hung in front of the house for two to four weeks in the fall. Once dried, they were stored and eaten during the winter. Today, mochi cakes are produced in factories all year round and the drying process takes as little as four days.[2]

Salzen | Eine Möglichkeit, den konservierenden Effekt des Trocknens noch zusätzlich zu steigern, ist die Frischware vorher einzusalzen. Pökelfleisch, Salzgurken und Salat (vgl. ital. „insalata" = eingesalzen) zählen zu den Speisen, die ursprünglich aus der Idee der Konservierung durch Salz entstanden. Schon im alten Ägypten und im antiken Griechenland wurden Lebensmittel mit Salz und Kräutern eingerieben und danach an der Luft getrocknet. In Gebieten, wo Salz ausreichend zur Verfügung stand, kam die Technik auch im Mittelalter zum Einsatz. Stockfisch, eingesalzener und luftgetrockneter Kabeljau, war beispielsweise ab dem 10. Jahrhundert eine begehrte Handelsware. Stockfisch enthält weniger als 15 Prozent Feuchtigkeit, ist nahezu unbegrenzt haltbar und war daher vor dem Zeitalter moderner Kühltechnik eines der wenigen Fischprodukte, die auch in küstenferne Gebiete transportiert werden konnten. Die unzähligen katholischen Fasttage, an denen der Verzehr von Fleisch untersagt war, machten das Gericht zu einer gefragten Spezialität in Zentraleuropa. Später fand Stockfisch, der überwiegend in Norwegen produziert wurde, vor allem als Verpflegung auf langen Schiffsreisen Verwendung.[3]

Curing and Brining | One option of adding to the preserving effect of drying is to put salt on the fresh product. Cured meat, gherkins and salad (derived from the Italian word "insalata", which means salted) are among the foods originally preserved by means of salt. As early as in antiquity, food was covered with salt and herbs and dried in the open air in Egypt and Greece. Where salt was in ample supply, this technique was also applied in the Middle Ages. Stockfish – cod which has been salted and dried – was a sought-after commodity from the tenth century onward. Stockfish has a water content of less than 15%; thus, it is almost non-perishable and was one of the few fish products that could be transported to areas far away from coastal regions before modern chilling technology had been invented. It was also well liked in Central Europe due to the many Catholic fast days when eating meat was prohibited. Later on stockfish, which was mainly produced in Norway, was a valuable part of provisions on long sea voyages[3]

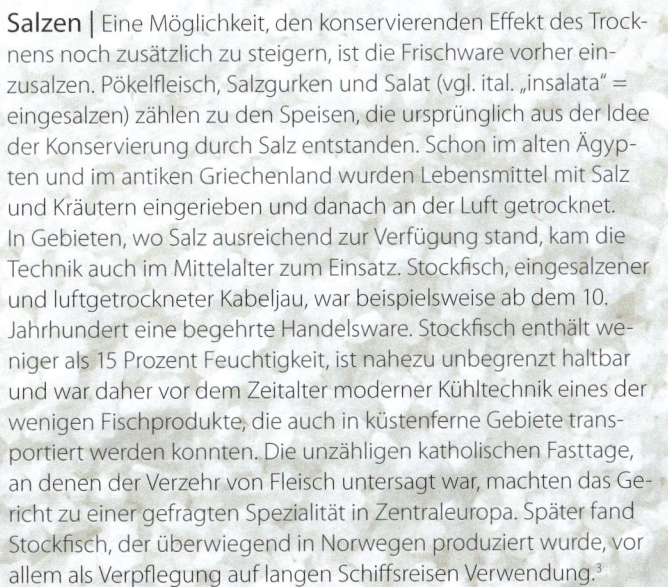

Einsalzen steigert den konservierenden Effekt des Trocknens. Pökelfleisch, Salzgurken und Salate zählen zu jenen Speisen, die aus der Idee der Konservierung durch Salz heraus entstanden sind.

Salting enhances the preserving effect of drying. Salt meat, salted pickles and salads are among the foods that have emerged from the idea of preservation by means of salt.

Auch Fleisch wurde eingesalzen und getrocknet, um es haltbar zu machen. Gegenüber Frischfleisch bedeutete das natürlich enorme Geschmackseinbußen und so galten Rezepte wie italienischer Parmaschinken, Schweizer Bündnerfleisch oder türkischer Pastrami ursprünglich als Armeleuteessen. Heute, da Kühl- und Gefriertechnik Frischfleisch alltäglich machen, sind diese Produkte begehrte Delikatessen.[4]

Zuckern | Das Versetzen von Lebensmitteln mit Zucker erzielt einen ähnlich konservierenden Effekt wie das Einsalzen. Frische Früchte werden mit Zucker vermischt, eingekocht und können danach in luftdicht verschlossenen Gläsern über mehrere Monate hinweg aufbewahrt werden. Das Verfahren stammt vermutlich aus Portugal, wo das Wort „marmelada" eine süße Quittenpaste bezeichnete. Ende des 15. Jahrhunderts gelangte das Rezept nach England, von wo aus es sich in ganz Europa verbreitete.[5]

Räuchern | Die Technik des Räucherns wird nur in Verbindung mit anderen Konservierungsmethoden wie Einsalzen oder Trocknen angewendet. Der heiße Rauch entzieht dem Lebensmittel Wasser und versieht seine Oberfläche mit einer luftdichten, antiseptischen Haut; gleichzeitig dringen Geschmacksstoffe und bakterizide Substanzen ein. Geräuchert werden vor allem Fisch und Fleisch, zu den bekanntesten Rezepten zählen der norwegische Räucherlachs und der alpenländische Räucherspeck. Die konservierende Wirkung des Räucherns ist schon lange bekannt. Bereits 3500 vor Christus räucherten die Sumerer in Mesopotamien Fisch. Auch im antiken Griechenland und später in Rom wurden Käse und Thunfisch über Rauch getrocknet und auf diese Weise länger haltbar gemacht. Im Mittelalter waren geräucherte Heringe für arme Bevölkerungsschichten eine wichtige Fastennahrung.[6]

Links | Konservierung durch Zucker bei der Wiener Delikatessenmanufaktur Staud´s. Das Bild zeigt die Produktion von Marillenmarmelade. Das Rezept für Marmelade stammt vermutlich aus Portugal, wo man schon vor Jahrhunderten Quitten zu „marmelada" verarbeitete.
Rechts | Zur Herstellung von Parmaschinken werden rohe Schweinshaxen mit Salz eingerieben und 12 – 16 Monate lang getrocknet, und zwar erst liegend bei niedriger Temperatur und hoher Luftfeuchtigkeit (-4°, 80%), dann hängend bei zunehmend höherer Temperatur und weniger Luftfeuchte. Heute passiert das in Kühlkammern, früher folgte die Produktion dem Verlauf der Jahreszeiten – im Herbst wurde geschlachtet, im darauf folgenden Sommer war der Schinken fertig.

Meat, too, was salted and dried for the purpose of preservation. Needless to say that, compared to the taste of fresh meat, salting and drying was detrimental, so Parma ham, Swiss bresaola or Turkish pastrami originally belonged to the diet of the poor. Today, fresh meat is nothing special thanks to chilling and refrigeration technology, and these products have become fancy foods.[4]

Sugaring | Adding sugar to food has a similar preserving effect as salting. Fresh fruit are mixed with sugar, heated and filled into airtight jars in which they can be stored for several months. The process is believed to have originated in Portugal, where the word "marmelada" was used for sweet quince paste. At the end of the fifteenth century, the recipe was brought to England, from where it spread all over Europe.[5]

Smoking | Smoking is only used as an adjunct to other preservation methods such as salting or drying. The hot smoke withdraws water from the food and covers its surface with an impervious antiseptic layer, sealing it from the air. At the same time, substances penetrate the food adding to flavor and killing bacteria. Smoking is mostly used for fish and meat, with Norwegian smoked salmon and smoked bacon from the Alps being the most popular products. The preserving effect of smoking has been known for a long time. As early as around 3,500 BCE, the Sumerians in Mesopotamia had smoked fish. In ancient Greece, and later on in Rome cheese and tuna were dried over smoking fires for the purpose of preservation. In the Middle Ages, smoked herrings were an important part of the diet of the poor on fast days.[6]

left | Preservation by means of sugar. The recipe for jam supposedly originates in Portugal where already centuries ago quinces were processed to make "marmelada". The Viennese company Staud's makes marmelade delicatessen products.
right | To produce Parma ham, salt is rubbed into pork knuckles which are then dried for 12 to 16 months – first lying flat at a low temperature and high air humidity (-4°, 80%), then being hung at increasingly higher temperature and lower humidity. Today this happens in cooling rooms, whereas in the past production took place in sync with the seasons – in the fall the pig was slaughtered and by the following summer the cured pork ham was ready.

Einlegen und Fermentieren | In vielen Kulturen werden Nahrungsmittel in Salz- oder Essiglauge eingelegt und damit biochemischen Prozessen ausgesetzt, die konservierend wirken. Schon vor mehr als 2500 Jahren wird in einer chinesischen Gedichtsammlung das Einlegen von Gurken erwähnt. Um Christi Geburt wurden den Koreanern besondere Fähigkeiten bei der Herstellung von fermentierten Speisen wie zum Beispiel „Gimchi" nachgesagt. Dabei wird (hauptsächlich) Chinakohl geschnitten und mit Salz, Gewürzen und Fischbestandteilen (oder Fischsauce) vermengt, fermentiert und dadurch konserviert.

Ganz ähnlich funktioniert die Herstellung von Sauerkraut, das ursprünglich aus China stammen und von den Tartaren nach Deutschland gebracht worden sein soll. Die lange Haltbarkeit und der hohe Vitamingehalt machten Sauerkraut als Verpflegung auf langen Seereisen populär. James Cook beispielsweise führte auf seinen Entdeckungsfahrten stets „sour crout" als Vorbeugung gegen Skorbut mit an Bord.[7]

Auch der berühmte Graved Lachs ist eine Rezeptur, bei der Fisch durch Fermentation haltbar gemacht wird. Er wird traditionell mit Salz und Zucker eingerieben und für mehrere Tage mit Steinen beschwert in der Erde vergraben. Wasserentzug und gleichzeitige Fermentation machen den Fisch haltbar.

Selbst der Inbegriff der Frische, nämlich das japanische Sushi, entstand ursprünglich aus der Idee, rohen Fisch einzulegen und haltbar zu machen. Zu einer Zeit, als die konservierende Wirkung von Salz in Asien noch nicht bekannt war, wickelte man rohen Fisch in gekochten Reis. Dieser gor, säuerte das Fischfleisch und bewahrte es so länger vor dem Verderben. Sushis, wie wir sie heute kennen, nämlich mit fangfrischen Fischstücken und Meeresfrüchten, kamen erst im Laufe des 19. Jahrhunderts in Tokio in Mode.

Die Entdeckung der biochemischen Vorgänge der Fermentation machte Frischkraut zu Sauerkraut, Milch zu haltbarem Käse und Fruchtsaft zu lagerfähigem Most oder Wein. Chemisch betrachtet, ist Fermentation die Umwandlung biologischer Grundprodukte durch bestimmte Enzyme und Bakterien, wie sie zum Beispiel beim Gären, Silieren oder Käsen erfolgt. Fermentieren konserviert aber nicht nur, sondern verändert Aussehen, Textur und Geschmack des Ausgangsmaterials grundlegend. Die typische Farbe und die charakteristischen, meist herben oder bitteren Aromen vieler haltbarer Nahrungsmittel wie Käse, Sauerteig, Sojasauce, Bier, Wein oder Tee entstehen als Folge der Fermentation, die heute ganz gezielt auch bei der Geschmacksentwicklung eingesetzt wird.

Links | Haftbarmachung durch Einlegen oder Fermentation: Sauerkraut, Gewürzgurken und koreanisches „Gimchi" (fermentierter Chinakohl)
Rechts | Käse gehört zu den ältesten Methoden, Milch haltbar zu machen

Pickling and Fermenting | In many civilizations, food is immersed in vinegar or brine and thus exposed to biochemical processes with a preserving effect. 2,500 years ago, a Chinese collection of poems mentioned the pickling of cucumbers. Around the beginning of the Common Era, the Koreans were said to be especially apt in producing fermented dishes such as kimch'i. Its main ingredient is usually cut Chinese cabbage which is mixed with salt, spices and fish (or fish sauce) and preserved by fermentation.

Sauerkraut, which is believed to have originated in China and to have been brought to Germany by the Tartars, is made in a similar way. Its durability and high vitamin content made the pickled cabbage a popular provision of sailors on long journeys. On his voyages of discovery James Cook always had "sour crout" on board to prevent scurvy.[7]

The famous gravad lax is another recipe in which fish is preserved by fermentation. Traditionally, it is cured with salt and sugar and buried in the earth underneath some stones for a few days. Due to the withdrawal of water and simultaneous fermentation, it is made more durable.

The epitome of fish, Japanese sushi, was originally derived from the idea of pickling raw fish to preserve it. At a time when the preserving effect of salt was not yet known in Asia, raw fish was wrapped in cooked rice. The rice fermented, acidifying the fish and keeping it from spoiling. Sushi made of fresh fish and seafood as we know it today only became fashionable in Tokyo in the course of the nineteenth century.

The discovery of biochemical fermentation processes turned fresh cabbage into sauerkraut, milk into durable cheese and fruit juice into must and wine that can be stored. Chemically speaking, fermentation is the transformation of biological products due to certain enzymes and bacteria in processes such as brewing, silaging or cheese-making. Fermentation not only preserves food but also causes fundamental changes in the looks, texture and taste of the basic material. The typical colors and characteristic, mostly tart or bitter aromas of many durable foodstuffs such as cheese, leaven, soy sauce, beer, wine or tea are due to fermentation. Product developers today also deliberately employ fermentation as a means to create certain tastes.

left | Preserving by pickling or fermentation: Sauerkraut, spiced pickles and Korean "gimchi" (fermented Chinese cabbage)
right | Cheese is one of the oldest methods of making milk durable.

Frieren | Dass Lebensmittel länger halten, wenn sie bei niedriger Temperatur gelagert werden, wusste man schon in der Antike und transportierte große Mengen Eis von den Bergen in die Städte, um die Keller mit den Nahrungsvorräten zu kühlen. Auch die Idee des Haltbarmachens durch Frieren existierte bereits vor der Erfindung der Tiefkühltruhe und wurde in entsprechend kalten Regionen unter anderem in Russland, China und von den Inuit angewandt.

Die Geschichte der modernen Kühltechnik begann mit der Präsentation der ersten künstlichen Kühlung durch den schottischen Arzt William Cullen an der Universität Glasgow 1748. Als Begründer des chemischen Kühlschrankes gilt Alexander Twining, der seine durch Luftkompression gekühlten Geräte ab 1834 in den USA kommerziell vermarktete. Zur Standardausrüstung wurden Kühlschränke in den USA in den 1930er Jahren, 1937 hatte bereits jeder zweite US-Haushalt einen Kühlschrank. Der erste europäische Kühlschrank wurde 1929 von den Zschopauer Motorenwerken J.S. Rasmussen in Sachsen gebaut.

Die neue Technik zog eine breite Palette an entsprechenden Produktentwicklungen nach sich, wobei das Potenzial gefrorener Esswaren bis heute sicher noch nicht vollständig ausgeschöpft ist. Eines der ältesten und bis heute populärsten Tiefkühlprodukte ist das Fischstäbchen, dessen rechteckige Form sich aus den verarbeitungstechnischen Bedingungen des damals noch jungen Verfahrens ergab. Der Wissenschaftler und Kochbuchautor Dr. William Kitchiner schlug 1817 in England erstmals vor, Fisch zu frieren, um ihn haltbar zu machen. Rund hundert Jahre später war die Entwicklung der Gefriertechnik so weit fortgeschritten, dass der 1886 in Brooklyn/New York geborene Erfinder Clarence Birdseye die Idee aufgriff und sie schließlich weltweit erfolgreich machte. Um seine Ausbildung zum Biologen am Amherst College in Massachusetts finanzieren zu können, arbeitete Birdseye zwischen 1912 und 1915 in Labrador in Kanada. Dort beobachtete er die einheimischen Inuit, wie sie Fisch bei minus vierzig Grad unter Eis froren. Birdseye faszinierte, dass die

Freezing | As early as in classical antiquity, people knew that low temperatures preserved foodstuffs, and they brought large quantities of ice from the mountains to the towns to cool the cellars where food was stocked. The idea of preserving food by freezing existed well before the deep-freezer was invented, and food was frozen in cold regions such as parts of Russia and China as well as by the Inuit.

The presentation of the first artificial refrigeration system by the Scottish physician William Cullen at the University of Glasgow in 1748 marked the beginning of modern refrigeration technology. Alexander Twining, who sold his appliances using vapor compression commercially in the USA as from 1834, is considered the father of the chemical refrigerator. Refrigerators became standard appliances in the US in the 1930s. In 1937 every second US household had one. Zschopauer Motorenwerke J.S. Rasmussen in Saxony built the first European refrigerator in 1929.

The new technology led to a wide range of product developments, with most of the potential to be found in creating frozen food novelties. One of the oldest types of deep-frozen foods, and still among the most popular ones, is the fish finger. Its rectangular shape was due to the technological conditions of what was a rather new process then. It was in 1817 when Dr. William Kitchiner, a scientist and author of cookery books, first suggested freezing fish for the purpose of preservation in England.

About a hundred years later, the development of refrigeration technology had progressed to the extent that Clarence Birdseye, an inventor born in Brooklyn, New York City, in 1886, returned to the idea and eventually made it a worldwide success story. To be able to finance his training in biology at Amherst College in Massachusetts, Birdseye worked in Labrador, Canada, between 1912 and 1915. There, he watched the indigenous Inuit preserving fish under the frozen surface of waters at minus forty degrees.

Qualität der Tiefkühlfische fast mit jener von frischem Fisch mithalten konnte, und erkannte, dass gefrorener Fisch umso besser schmeckt, je schneller er friert. Heute weiß man, dass sich beim so genannten „fast freezing" kleinere Eiskristalle bilden als bei langsameren Frierverfahren und dadurch die natürliche Struktur des Fischfleisches nahezu vollständig erhalten bleibt.

Fisch als hartgefrorener Quader

1922 begann Birdseye mit künstlichen Kühltechniken und Fisch zu experimentieren, wobei er Fisch mit Hilfe von minus 43 Grad Celsius kalter Luft fror. Das Interesse der Konsumenten blieb jedoch gering und seine erste Firma ging bankrott. Birdseye gab dennoch nicht auf und gründete 1925 die „General Seafood Corporation" in Gloucester/ Massachusetts, die mit Hilfe des so genannten „double belt freezer" gefrorenen Fisch herstellte. Das Gerät, in dem kalte Lauge zwei Nirostastahlbänder kühlt, konnte verpackten Fisch rapide frieren. Um ein gleichmäßiges und konstantes Ergebnis zu erzielen, eigneten sich zur Produktion gleich große, geometrisch geschnittene Fischstücke besser als unregelmäßige Filets und so ließ sich Clarence Birdseye 1927 seine „fish fingers" patentieren.

Praktisch zur selben Zeit, nämlich 1929, entwickelten Nathan Hutching und Petros Persad einen „quick freezer", bei dem zwischen Metallplatten gefrostet wurde. Auch diese Technologie funktionierte besser, wenn die unregelmäßig geformten Fische in schlanke Quader geschnitten wurden, da diese im Vergleich zu kompakten Fischkörpern schneller durchfrosteten. Hutching und Persad brachten ihre gekühlten, panierten Fischstücke um 1930 unter der US-Marke Gorton's auf den Markt.

Am 26. September 1955 wurden Fischstäbchen erstmals auch in Europa, namentlich in England verkauft. Die „europäischen" Fischfinger wurden in der Birdseye-Fabrik in Great Yarmouth, England, hergestellt, von wo aus sie in weiterer Folge ganz Großbritannien und den Rest der Welt eroberten.

Birdseye was fascinated by the fact that the quality of deep frozen fish after thawing was almost the same as that of fresh fish. Moreover, he found that the faster fish was frozen, the better it tasted when thawed. Today, we know that the ice crystals formed by "fast freezing" are smaller than those emerging in slower freezing processes so that the natural structure of fish remains almost completely intact.

Fish as a Hard Frozen Block

In 1922 Birdseye started experimenting with artificial freezing techniques and fish, freezing fish with the help of air cooled down to minus 43 degrees Centigrade. However, consumers were not really interested and his first company went bust. Nevertheless, Birdseye refused to give up, founding "General Seafood Corporation" in Gloucester, Massachusetts, in 1925. The company produced frozen fish using a so-called "double belt freezer". The appliance included two stainless steel belts cooled in brine and could freeze pre-packed fish rapidly. Fish pieces cut to the same geometrical size turned out to be best suited for a homogeneous and consistent result, and in 1927 Clarence Birdseye took out a patent for his "fish fingers".

Around the same time, in 1929, Nathan Hutching and Petros Persad developed a "quick freezer" in which the freezing process took place between two metal plates. Again, the technology turned out to work better on fish cut into slim squares than on irregular natural shapes, which were more compact and took longer to freeze. Hutching and Persad started marketing their breaded and frozen fish under the "Gorton's" brand in the USA in 1930.

On 26 September 1955 the first fish fingers were sold in Europe, or England, to be exact. "European" fish fingers were produced in the Birdseye factory in Great Yarmouth, England, from where they eventually conquered the UK and the rest of world.

Rechts | 1927 von Clarence Birdseye zum Patent eingereichte „Methode zur Zubereitung von Esswaren": der New Yorker gilt als Erfinder der Fischstäbchen.

right | The "method for preparing food" submitted by Clarence Birdseye in 1927 for patenting: the New Yorker is considered to be the inventor of fish sticks.

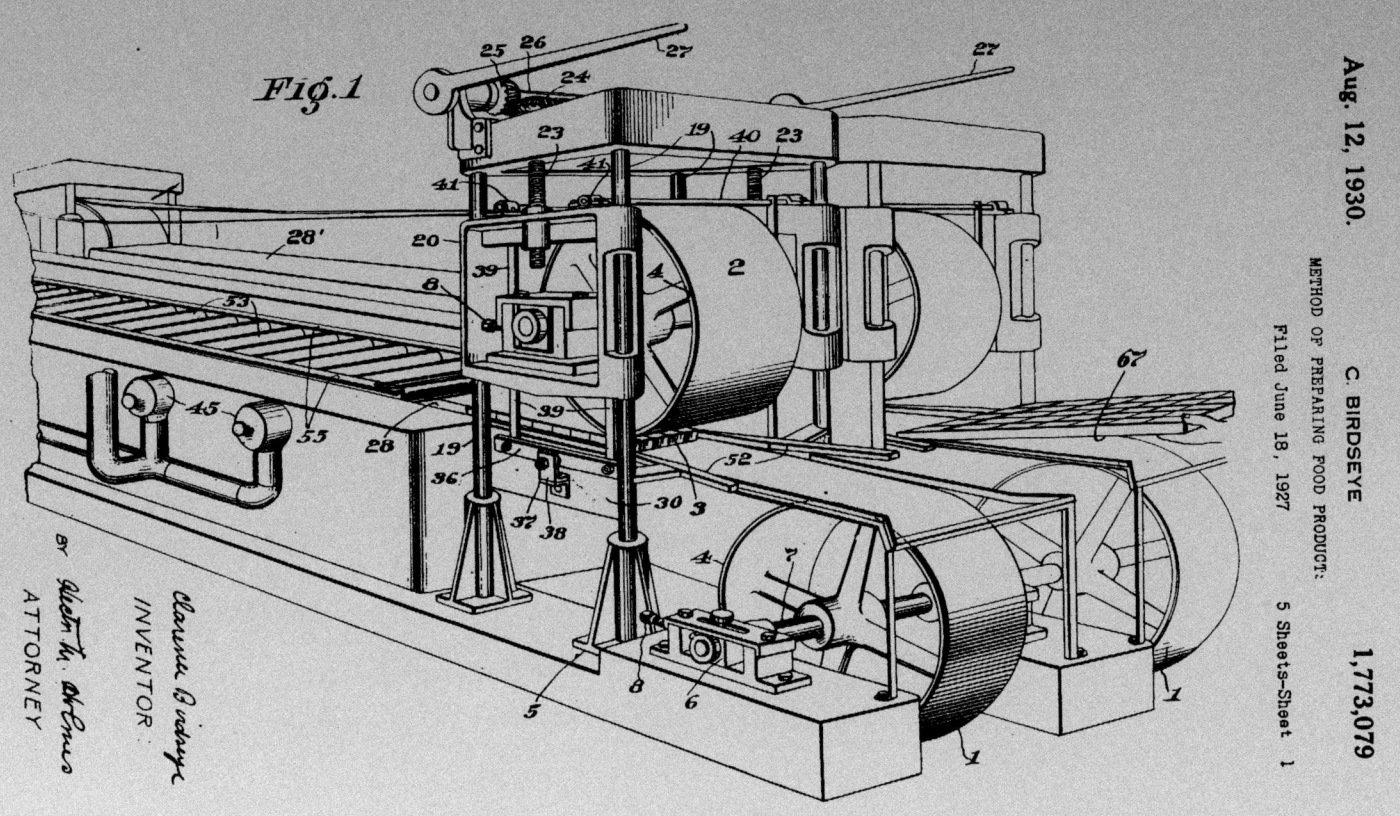

Gefriertrocknen | Nach dem Tiefkühlen ist die Gefriertrocknung die Konservierungsmethode mit dem besten Geschmackserhalt und einem weiteren entscheidenden Vorteil: Gefriergetrocknetes kann Jahrelang bei jeglicher Temperatur gelagert werden. Ein weiterer Pluspunkt ist der enorme Gewichtsverlust im Zuge der Herstellung, der Lagerung und Transport vereinfacht und gefriergetrocknete Speisen als Proviant für Profibergsteiger ebenso interessant macht wie für die Truppenverpflegung in entlegenen Gebieten. Zur Erzeugung eines Kilogramms gefriergetrockneten Spinats benötigt man immerhin ganze 100 Kilogramm Frischware![8]

Die Idee zur Gefriertrocknung lieferte eine Rezeptur aus Nepal, bei der Yak-Milch und Schinken auf Schnee gelegt und im eiskalten Wind gleichzeitig getrocknet, gefroren und dadurch konserviert werden. Zur industriellen Anwendung kam Gefriertrocknung um 1930, als man in Brasilien nach einer Rekordernte eine Technik suchte, um Kaffee zu konservieren. Zu diesem Zweck wandte man sich an die kleine Schweizer Firma Nestlé, die sich mit der Erfindung von Milchpulver einen Namen gemacht hatte. Ein Forschungsteam unter der Leitung von Max Morgenthaler entwickelte in den folgenden Jahren ein Verfahren, das gemahlenen Kaffee aromaschonend haltbar machte. Das Ergebnis, ein Pulver, welches nur mit heißem Wasser aufgegossen werden musste, kam am 1. April 1938 unter dem Namen „Nescafé" in der Schweiz auf den Markt.

Die Herstellung von Instantkaffee erfolgt mittels Gefriertrocknung. Die Zutaten werden zunächst bei -40 bis -35 Grad Celsius schockgefrostet, dann unter Vakuum auf -5 Grad erwärmt und danach 20 bis 35 Stunden getrocknet. Bei der Gefriertrocknung entweichen rund 97 Prozent des enthaltenen Wassers, was zu einem enormen Gewichtsverlust einerseits und zu einer stark verlängerten Haltbarkeit andererseits führt.

Freeze-Drying | Freeze-drying is the preservation method second only to deep-freezing when it comes to maintaining flavor as much as possible, and it has a further benefit: freeze-dried products can be stored for years at any temperature. Another advantage is the enormous reduction of weight during production, which makes storage and transportation easier. For this reason, freeze-dried food is equally interesting as provisions for professional mountaineers as they are for the military stationed in remote areas. It takes no less than 50 pounds of fresh spinach to produce one pound of freeze-dried product![8]

Freeze-drying is a technique that historically was applied in Nepal, where yak milk and ham are placed on snow to be at the time dried and frozen by the ice-cold wind for preservation. Freeze-drying was first used industrially in 1930 when people in Brazil looked for a technique to preserve coffee after a record harvest. For this purpose, they sought help from a small Swiss company named Nestlé, renowned for inventing powdered milk. In the following years, a team of researchers led by Max Morgenthaler developed a process to preserve ground coffee. The result was powdered coffee, and all it required was hot water to pour on. On 1 April 1938 the product was first marketed under the name "Nescafé" in Switzerland.

Instant coffee is produced by means of freeze-drying. The ingredients are first shock frosted at -40 to -35 degrees Centigrade, then heated to -5 degrees in a vacuum and finally dried for 20 to 35 hours. In freeze-drying, roughly 97% of the water is withdrawn, which leads to an enormous reduction in weight and longer shelf life.

Makroaufnahme von gefriegetrocknetem Bergsteigeressen. Ursprung der Gefriertrocknung ist ein Rezept aus Nepal, bei dem Yak-Milch und Schinken auf Schnee gelegt und im eiskalten Wind gleichzeitig getrocknet und gefroren werden. Zur industriellen Anwendung kam Gefriertrocknung um 1930, als man in Brasilien nach einer Rekordernte eine Technik suchte, um Kaffee zu konservieren. Das Ergebnis, ein Pulver, welches nur mit heißem Wasser aufgegossen werden musste, kam am 1. April 1938 unter dem Namen „Nescafé" auf den Markt.

Macro-shot of freeze-dried mountaineer's food. Freeze-drying is based on a recipe from Nepal in which the Yak milk and ham were laid on snow and both dried and frozen in the ice-cold wind. This freeze-drying method was industrially used starting from around 1930 when a technique to preserve coffee was sought in Brazil following a record harvest. The result was a powder which only had to be dissolved in hot water. It was launched on 1 April 1938 under the name "Nescafé".

Ummanteln | Ähnlich wie beim konstruktiven Holzschutz, wo die Lebensdauer des Holzes durch eine entsprechende Gestaltung der Bauteile verlängert wird, kann auch das Design von Esswaren dazu dienen, empfindliche Zutaten vor Austrocknung, Zerquetschen oder Verderben zu schützen. Bei vielen Produkten fungieren essbare Verpackungen oder Schutzhüllen aus anderen Zutaten als eine Art kulinarische Rüstung. Genauso wie Kleidung den menschlichen Körper vor Kälte, Nässe oder Hitze bewahrt, isolieren, stabilisieren und konservieren auch Glasuren, Häute und Rinden ihren labilen oder leicht verderblichen Inhalt. Von der Schokobanane über verschiedene Rohmilchkäse bis zum Frankfurter Würstchen könnten die Füllungen unzähliger Produkte ohne ihre essbaren Ummantelungen nicht lange überleben.

Viele traditionelle Rezepte vom Strudel über Pasteten bis zu Knödeln nutzen den „konstruktiven" Schutz vor Verderben. Zum Beispiel werden in Vorderasien und im Gebiet des südlichen Balkans Reis, gehacktes Fleisch und andere Zutaten in vorgekochte Weinblätter gewickelt, um vor dem Austrocknen bewahrt zu werden. Auch bei Pralinen und getunkten Keksen dient die Schokohülle als Schutz vor äußeren Einflüssen. Marzipan und Nougat trocknen an der Luft rasch ein, werden steinhart und somit zum kulinarischen Albtraum. Die harte und zugleich dichte Schale aus Schokolade hält den Inhalt frisch und in Form.

Kaugummis werden mit einer wasserundurchlässigen Dragee-Kruste überzogen, damit sie während längerer Lagerzeiten nicht austrocknen. Diese Schutzschicht aus Zucker, Farb- und Aromastoffen verleiht den Kauprodukten Formstabilität und eine lang anhaltende, elastische Beschaffenheit. Eine Schutzfunktion der besonderen Art erfüllt die hauchdünne Hülle des so genannten Streuzuckers, einer Staubzuckersorte, die vor allem für die Dekoration von Süßspeisen gedacht ist. Eine kaum sichtbare Ummantelung aus Kokosfett isoliert die Zuckerkristalle gegen Hitze, so dass sie auf warmen Mehlspeisen nicht zerschmelzen.

Statik und Form | Gestalterisches Vorbild vieler Hüllen ist das Ei, das alle Vorzüge einer guten Ummantelung in sich vereint. Die Schale schützt das heranwachsende Küken vor äußeren Einflüssen und bewahrt den Inhalt unbefruchteter Eier längerfristig vor dem Verderben. Seine Stabilität verdankt das Ei unter anderem seiner Form. Die zweifach gekrümmte Schale leitet äußere Kräfte ab, ohne gleich zu zerplatzen. Ein rechteckiges Ei wäre zwar komfortabler zu stapeln, doch ebene Seitenflächen wären wesentlich instabiler und ließen sich leichter eindrücken.

Das Vorbild aus der Natur findet seine Nachahmer beispielsweise in der hauchdünnen Schale der „Schwedenbombe" („Dickmann´s"). Die kuppelförmige und somit stabile Schokohülle nimmt Kräfte, die zum Beispiel durch den Fingerdruck des Essers entstehen, obwohl sie so dünn ist, perfekt auf. Der essbare „Panzer" schützt vor grobem Zupacken und vor dem Raumklima. Die dichte Haut hält die aufgeschlagene Creme in Form, welche ohne Hülle an der Luft rasch in sich zusammenfiele.

Coating | The design of food serving to keep delicate ingredients from drying out, getting crushed or spoiling may resemble the protection of wood in construction where the wooden parts are specifically shaped to increase their life span. In many products, edible packages or protective coatings made of other ingredients act like a culinary armor. Just like clothes keep the human body warm, dry or cool, icings, skins and rinds insulate, stabilize and preserve unstable or perishable content. From mini chocolate bananas to various raw milk cheeses and Wiener sausages, the fillings in numerous products would not survive for a very long time if it were not for their edible coatings and wrappings.

Many traditional recipes, from strudels to terrines and dumplings, use "structural protection" to avoid perishing. In the Middle East and the southern Balkans rice, ground meat and other ingredients are wrapped in precooked wine leaves so they do not dry out easily. In chocolates and chocolate-dipped cookies, the coating serves as protection from outside influences. Marzipan and gianduia dry quickly when exposed to air, they turn hard as rock and become a culinary nightmare. Chocolate is both hard and airtight, thus keeping the contents fresh and in shape.

Chewing gum is covered with a water-tight sugar coat so it does not dry out during longer storage periods. The protective coating made of sugar, coloring and aromas keeps the product in a stable form and makes sure it stays elastic. Castor sugar, a type of superfine powdered sugar mainly intended for the decoration of sweets, forms a thin sugar layer with a special protective function: The sugar crystals are heat-insulated by an invisible coating of coconut oil and therefore do not melt on hot sweets.

Structural Mechanics and Shapes | In many cases, the egg shell is the coating frequently emulated by designers as it combines all the benefits of good protective qualities. It protects the fledgling from outside influences and keeps the content of unfertilized eggs from rotting for some time. Among other factors, the egg shell owes its stability to its shape. The double curvature diverts outside forces without breaking easily. Rectangular eggs would be more convenient when it comes to stacking but flat sides are much less stable and can be crushed more easily.

The model produced by nature is emulated by the super-thin chocolate coating of the "chocolate kiss", sold under the brand name "Dickmann's" in the German-speaking countries, or as "Schwedenbomben" in their Austrian version. It is also known as chocolate teacake, Whippet or Mallomars in the English-speaking countries, and as Krembo in Israel. These treats look very similar but are different in terms of filling: Whippets and Mallomars are in fact marshmallow treats whilst the other types are filled with cream consisting mainly of whipped egg white. Although they all have a very thin outer layer, the dome-shaped and thus stable chocolate coating can stand the force of finger pressure. It works like a shell protecting the inside from rough grabbing and the ambient climate, keeping the creamy interior in shape and from collapsing, which would happen if the cream were exposed to air for too long.

links | Glasuren, Häute und Rinden fungieren als essbare Schutzhüllen. Sie isolieren, stabilisieren und konservieren ihren Inhalt und schützen empfindliche Zutaten vor Austrocknung, Zerquetschen oder Verderben.
Rechts | Die hauchdünne Schale cremegefüllter Schokokuppeln verdankt ihre Stabilität der zweifach gekrümmten Form, die äußere Kräfte ableitet, ohne gleich zu zerbrechen. Die kulinarische „Rüstung" schützt vor grobem Zupacken und hält die aufgeschlagene Creme in Form, welche ohne Hülle an der Luft rasch in sich zusammenfiele.

left | Glazings, skins and crusts function as edible protective shields. They isolate, stabilize and preserve their content and protect sensitive ingredients from drying, squishing or spoiling.
right | The thin shell of cream-filled chocolate domes owes its stability to the two-sided curvature that deflects external forces so that it does not immediately break. The culinary "armor" protects against rough handling and maintains the form of the whipped cream, which would quickly collapse without the shell.

Schutzfilme – Verpackungen – Controlled Atmosphere

Frisches Obst und Gemüse wird an der Luft schnell unansehnlich. Schuld daran sind Stoffwechselreaktionen, welche die Farbe und die Textur von Nahrungsmitteln verändern. Jeder kennt den Effekt, wenn sich an der Oberfläche von Bananen, Äpfeln oder Kirschen braune, weiche Flecken bilden, die stetig wachsen und schließlich die ganze Frucht in Beschlag nehmen, so dass diese verdirbt. Historische Verfahren, die Oxidation an der Oberfläche von Esswaren zu unterbinden, sind unter anderem aus China und dem antiken Griechenland bekannt. Die Chinesen vergruben frische Früchte gemeinsam mit bestimmten Blättern in Erdlöchern, um sie vor dem Verderben zu schützen. In Griechenland tauchte man Früchte in Olivenöl, das einen luftdichten Schutzfilm bildete und so die Reaktionen mit der Umgebungsluft verhinderte.

Moderne Konservierungsverfahren gehen den umgekehrten Weg: Sie schützen nicht das Nahrungsmittel vor der Luft, sondern verändern die umgebende Atmosphäre, unter der es lagert. „Controlled Atmosphere" nennt sich die Idee, Luftfeuchtigkeit, Temperatur sowie die Zusammensetzung der Luft in Lager- oder Transporträumen so zu verändern, dass die Ware langsamer verdirbt oder auch – beispielsweise bei Früchten – beschleunigt nachreift. Stationär wird das Verfahren etwa bei der Lagerung von Äpfeln schon seit Jahrzehnten eingesetzt, relativ neu ist Controlled Atmosphere im mobilen Bereich, beispielsweise in Kühlcontainern. Je nach Produkt werden die Volumsanteile von Sauerstoff, Stickstoff, Kohlendioxid und Ethylen dabei so reguliert, dass der gewünschte Effekt eintritt. Durch die Herabsetzung des Sauerstoffanteils von rund 20 auf nur 5 Prozent beispielsweise wird der Stoffwechsel von verderblichen Gütern reduziert, so dass diese langsamer reifen und verderben. Am Bestimmungsort wird dann durch die Erhöhung des Ethylen-Anteils der Reifungsprozess gezielt

Protective Films – Packaging – Controlled Atmosphere

Fresh fruits and vegetables quickly become unsightly when exposed to air. This is due to metabolic reactions that change the color and texture of food. Everybody is familiar with the brown soggy spots that form on the surfaces of bananas, apples and cherries, growing and finally taking over until the whole fruit has gone bad. From history, we know about processes serving to inhibit oxidization on the surface of food, which were used in China and ancient Greece. The Chinese buried fresh fruits in holes in the ground along with certain leaves to keep them from rotting. In Greece, fruits were dipped in olive oil that formed an airtight protective film and thus prevented reactions with ambient air.

Modern preservation techniques work the other way round. Instead of protecting the food product from contact with ambient air, they change the atmosphere in which the food is stored. "Controlled atmosphere" describes the idea of changing the humidity, temperature and composition of the air in storage and transportation areas in such a way that the goods will spoil less quickly or even after-ripen faster – which is, for instance, the case in fruit. The process has been used for decades in stationary facilities, for example in the storage of apples, whereas controlled atmosphere is relatively new in mobile storage, as in refrigerated containers. Depending on the product, the volume shares of oxygen, nitrogen, carbon dioxide and ethylene are controlled so as to achieve the desired effect. A decrease in the oxygen share from about 20% to only 5% will for example reduce the metabolism of perishable goods so they will ripen and spoil more slowly. By increasing the ethylene share at the point of destination, the ripening process will be restarted in a targeted way until the desired degree of ripeness has

Einwirkende Kräfte:
Transportstöße, Fingerdruck etc.

External forces:
Rough handling in transit, finger pressure, etc.

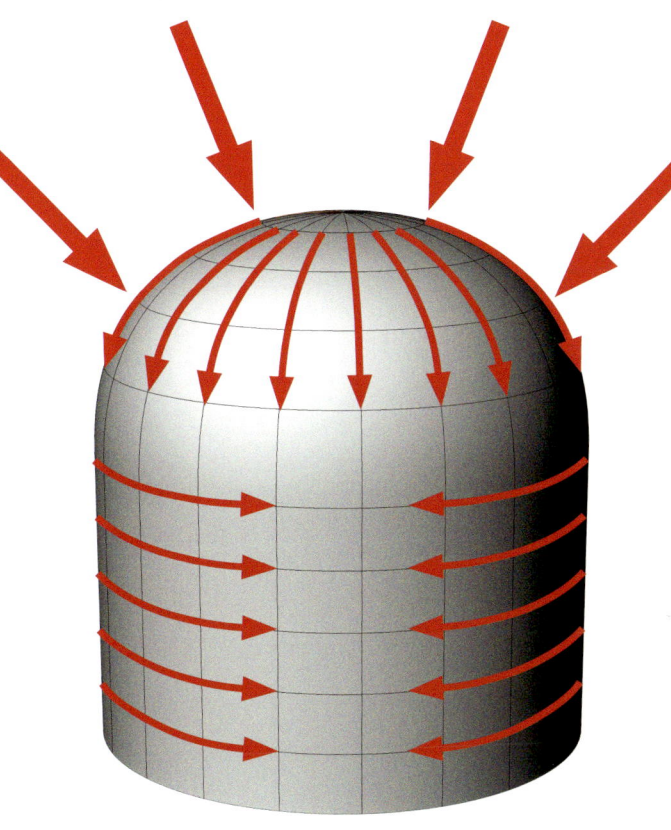

bis zum gewünschten Reifegrad wieder in Gang gesetzt. Die Vorteile sind vor allem wirtschaftlicher Natur, da empfindliche Waren wie Obst, Gemüse, Nüsse, Kaffee oder Gewürze nun nicht mehr per Flugzeug transportiert werden müssen, sondern auch die verlängerte und wesentlich billigere Schiffsreise durchhalten. Für den Transport von Fleisch befindet sich die Anwendung von Controlled Atmosphere noch im Teststadium.

Am Institut für Lebensmitteltechnologie an der Pennsylvania State University in den Vereinigten Staaten beschäftigen sich Professor John D. Floros und sein Team mit einer anderen Art von kontrollierter Atmosphäre: jener in Verpackungen. Noch vor 20 Jahren wäre es undenkbar gewesen, Ananas, Melonen oder Salat gewaschen und vorgeschnitten im Supermarkt zu verkaufen. Durch das Schneiden wird ein Teil der Zellen im Fruchtfleisch aufgebrochen, so dass dieses noch stärker mit der Luft reagiert und noch schneller verdirbt.

John D. Floros ging es darum, die Vorgänge rund um die Oxidation genau zu erforschen und herauszufinden, welche Menge Kohlendioxid wie schnell zugeführt und welche Menge Sauerstoff gleichzeitig abgezogen werden muss, um geschnittene Lebensmittel bestmöglich zu konservieren. Als man die exakten Daten des optimalen Gasaustausches ermittelt hatte, wurde eine entsprechende Kunststoffverpackung entwickelt, welche die erforderliche Diffusion von Sauerstoff und Kohlendioxid bewerkstelligte. Die durchsichtigen Kunststoffbeutel namens „modified atmosphere packages" sind heute mit Salat, Obst oder Gemüse gefüllt in jedem besseren Supermarkt zu finden. Gegenwärtig arbeiten die Forscher rund um Floros daran, eine ähnliche Verpackung auch aus essbarem Material, namentlich aus Stärke, herzustellen.[9]

Hochdruck und Nanotechnologie | Die Suche nach immer neuen, noch besseren Konservierungsmethoden ist trotz Tiefkühltruhe und Vakuumverpackung noch lange nicht abgeschlossen. Eine Idee, die zum Beispiel vom Wissenschaftszentrum Weihenstephan in Deutschland derzeit untersucht wird, ist die Konservierung mittels Hochdrucks. Dabei gelieren die Versuchsobjekte wie Karotten zu einer fruchtgummiartigen Masse, der Nährwert und der Geschmack bleiben aber großteils erhalten. Bislang ungelöst sind allerdings noch die enormen Kosten einer derartigen Hochdruckkonservierung. Neuartige Verfahren in diesem Bereich sind auch Hochdruckpasteurisierung, etwa für Tomatenpüree, Hochdrucksterilisation oder Hochspannungskonservierung.

been reached. The benefits of controlled atmosphere are primarily economic, since delicate goods such as fruit, vegetables, nuts, coffee and spices survive long ship voyages and do not have to be transported by air. The application of controlled atmosphere to meat is still in the testing stage.

At the Department of Food Science of Pennsylvania State University in the United States Professor John D. Floros and his team research another type of controlled atmosphere: packages. Only twenty years ago, it would have been impossible for supermarkets to sell pineapples, melons or salad that have been pre-cut and pre-washed. Cutting breaks part of the pulp cells which react even faster with air and decay more easily.

John D. Floros explored the processes taking place in the context of oxidization and how much carbon dioxide has to be added within what time, and how much oxygen has to be withdrawn at the same time to preserve cut-up food in the best possible way. When the data for optimum gas exchange were available, a plastic wrapping was sought to ensure the required diffusion of oxygen and carbon dioxide. The transparent plastic bag called "modified atmosphere packages" containing salad, fruits or vegetables can today be found in virtually all major supermarkets. At present the researchers around Prof. Floros also work on producing similar packages from edible material, i.e. starch.[9]

High Pressure and Nanotechnology | In spite of deep freezers and vacuum packaging, the search for new and ever improved preservation methods is far from over. One idea currently explored by the Weihenstephan Life Science Center in Germany is preservation by means of high pressure. Test objects such as carrots turn the consistency of fruit jelly while their nutritional value and taste are largely maintained. The issue of cost has not been resolved yet, as high-pressure preservation is still prohibitively expensive. Novel techniques in this field also include high-pressure pasteurization, which is e.g. used in the production of tomato puree, high-pressure sterilization or high voltage preservation.

Auch die Nanotechnologie befasst sich mit dem Haltbarmachen von Lebensmitteln. Der britische Lebensmitteltechnologe David Julian McClements zum Beispiel entwickelt an der University of Massachusetts, USA, essbare Hüllen und Verpackungen aus Nanopartikeln, die das Bakterienwachstum an der Oberfläche von Produkten stoppen und sie dadurch vor dem Verderben schützen sollen. Ein anderes Anwendungsgebiet von Nanometer dünnen Schichten ist, die Wasser-Luft-Diffusion zwischen einzelnen Komponenten der Nahrung – zum Beispiel zwischen dem trockenen Pizzaboden und der darauf liegenden feuchten Tomatensauce – zu unterbinden. Dabei verhindert ein hauchdünner, wasserdichter Überzug, dass der Teig mit der Zeit durchweicht, so dass er länger frisch und knusprig bleibt.

Derartige nanotechnische Ummantelungen können mittels positiv oder negativ geladener Polymer-Lösungen in beliebig vielen, unterschiedlich dicken Lagen nacheinander aufgebracht werden, wobei sich die Funktionsweise jeder einzelnen Schicht kontrollieren lässt. Eine derzeit nur teilweise gelöste Schwierigkeit der Nanohüllen ist allerdings die sensorische Anpassung an die Wünsche und Gewohnheiten der Konsumenten. Damit der Esser den Nanoüberzug im Mund nicht wahrnimmt, müssen nämlich gewisse Parameter erfüllt werden: So muss die oberste Schicht stets negativ geladen sein, die Partikel dürfen eine kritische Größe von 50 Mikrometer nicht überschreiten und müssen so designt sein, dass sie auch optisch nicht auszumachen sind.[10]

Nanotechnology, too, deals with preserving foodstuffs. The British food scientist David Julian McClements, who works at the University of Massachusetts, USA, develops edible films and packages made of nano-laminated layers that are designed to stop bacteria growth on product surfaces to keep them from spoiling. Another potential application of nanometer-thin layers is the prevention of water-air diffusion between food components – e.g. dry pizza dough and humid tomato sauce. A thin layer impervious to water keeps the dough from getting soggy so it continues to be fresh and crispy for a longer period.

Nano-film layers of various compositions, thicknesses and properties can be applied to materials by successive dipping in positively or negatively charged polymer solutions. One issue of nano-coatings which has only partially been resolved to date is their adaptation to the wishes and habits of consumers in terms of sensory qualities. To make sure that eaters do not perceive the nano-coatings, certain parameters have to be fulfilled: The top layer must always be negatively charged, the particles must not exceed a critical size of 50 micrometers, and they must be designed in such a way that they remain undetectable to the eye.[10]

1 | Betriebsbesichtigung Barilla, Parma, im April 2006
2 | Betriebsbesichtigung Firma Kakiyama, Tokio 2008
3 | vgl.: Oxford Companion to Food, S. 688
4 | vgl.: Oxford Companion to Food, S. 689
5 | vgl.: Oxford Companion to Food, S. 483
6 | vgl.: Oxford Companion to Food, S. 727
7 | vgl.: Oxford Companion to Food, S. 696
8 | Informationen von Fa. Simpert-Reiter GmbH, Augsburg
9 | Interview John D. Floros Pennsylvania State University, State College, USA, am 31.3.2009
10 | Interview mit David Julian McClements an der University of Massachusetts, USA, am 26.3.2009

1 | Guided tour of the Barilla factory, Parma, April 2006
2 | Source: Kakiyama co., Tokyo
3 | Cf. Oxford Companion to Food, p. 688
4 | Cf. Oxford Companion to Food, p. 689
5 | Cf. Oxford Companion to Food, p. 483
6 | Cf. Oxford Companion to Food, p. 727
7 | Cf. Oxford Companion to Food, p. 696
8 | Source: Simpert-Reiter GmbH, Augsburg
9 | Interview with John D. Floros, Pennsylvania State University, State College, USA, on 31 March 2009
10 | Interview with David Julian McClements at the University of Massachusetts, USA, on 26 March 2009

Japanische Instantsuppe in einer muschelförmigen, essbaren Verpackung: Mit heißem Wasser übergossen, lösen sich die Zutaten im Inneren auf, die Teighülle wird zur Suppeneinlage.

Japanese instant soup in shell-formed, edible packaging: when hot water is poured over it, the ingredients dissolve on the inside, with the dough shell becoming the soup addition.

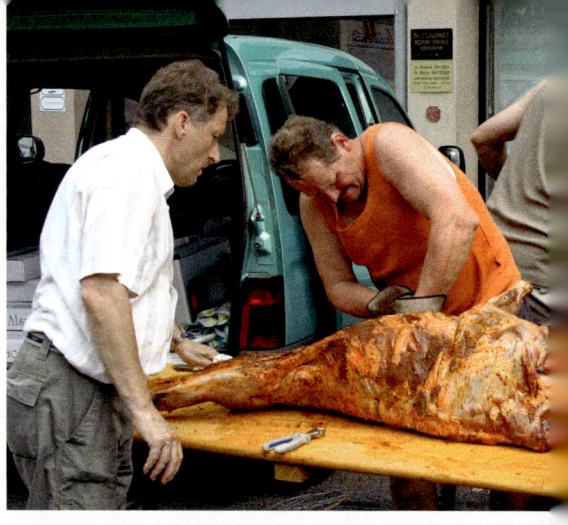

Produktion
production

 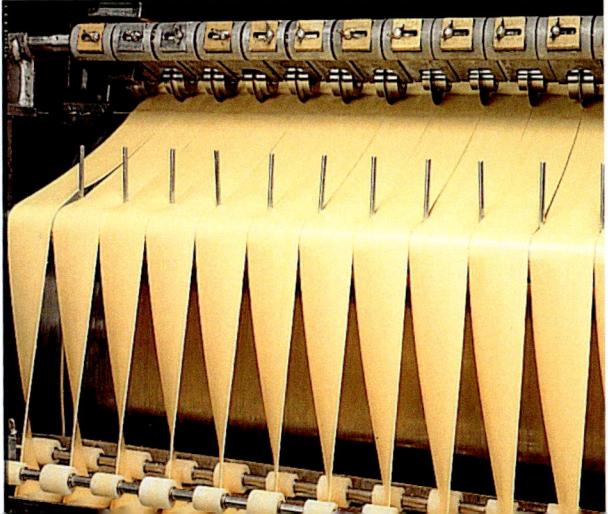 

Essen wird geschnitten, ausgewalkt, gerollt, gegossen, dressiert, mit Hitze und Kälte behandelt, extrudiert, gestanzt, gepresst, getunkt, dragiert und vieles mehr, um es in die gewünschte Form zu bringen. Egal ob Ildefonso, Waffelröllchen, Eislutscher oder Mannerschnitten: Die Art der Erzeugung beeinflusst das spätere Aussehen des Produktes.

Food is cut, rolled out, poured, dressed, treated with heat and cold, extruded, punched, pressed, dipped, coated and much more to bring it into the desired shape. Irrespective of whether it is Ildefonso, waffle roll, ice-cream popsicle or Mannerschnitten wafers: the way food is produced influences its later appearance.

Produktion – und wie sie die Gestaltung beeinflusst | Bandsägen zerteilen gefüllte Waffeln in mundgerechte Stücke, Rundgefrierer gießen Eislutscher. Ungebackene Kekse kleben an Teflonwalzen, Gipsstempel drücken die Form von Gummibärchen in Maisstärke, Zentrifugen dragieren Millionen kleiner Schokolinsen mit bunten Überzügen. Zutaten werden Gieß- und Spritzverfahren unterzogen, mit Nudelwalkern und Fleischklopfern geplättet, durch Laugenbäder und Trockenkammern geführt, in Mixern und Extrudern bearbeitet, unter Dampf erhitzt oder über dem Feuer gebraten.

Vom Grundprodukt bis zum Esstisch durchlaufen Nahrungsmittel verschiedenste Stationen. An erster Stelle steht üblicherweise die Fertigung. Egal ob in der Familienküche, im Handwerksbetrieb oder in der Industriehalle: Wir bevorzugen Gerichte und Produkte, die sich unter möglichst geringem Zeit- und Ressourcenaufwand herstellen lassen. Mit diesem Gedanken im Hinterkopf entscheiden wir, wie wir unsere Nahrung zubereiten und formen: ob und wie wir sie schneiden, rühren, mixen, aufschäumen, kochen, braten oder frieren. Wir wählen das Design einer Speise so, dass sie sich möglichst gut herstellen lässt. Niemand würde auf die Idee kommen, Knödel würfelförmig zu machen, weil sich Kugeln mit den Händen wesentlich leichter formen lassen und außerdem beim Kochen gleichmäßig durchgaren. Italienische Pizzabäcker drehen den Hefeteig um ihre Finger und lassen durch die Fliehkräfte runde Fladen entstehen. Eine Antwort auf die Frage, warum unser Essen genau so und nicht anders aussieht, ist daher schlicht und einfach: Damit es sich gut herstellen lässt.

Vom Werkzeug zur Form | Von der Feuerstelle bis zur Industriehalle steht Köchen und Food Designern ein gewisses Sortiment an Werkzeugen und Verfahren zur Verfügung, welches immer dem technischen Stand der jeweiligen Zeit entspricht. Kochte man einst in einem Erdloch mit heißen Steinen oder über offenem Feuer, so halten heute flüssiger Stickstoff und langkettige Zuckerisotope Einzug in die privaten Haushalte. Die technischen Möglichkeiten der einzelnen Verfahren – vom Küchenmesser bis zum Fließband – beeinflussen dabei die Form des Endprodukts nachhaltig. Werkzeuge und Techniken hinterlassen ihre Spuren, das formale Ergebnis variiert, je nachdem ob man Teige strangpresst, ausnticht oder schneidet, Schokolade gießt, spritzt oder dragiert. Das Verfahren bestimmt die Form.

Im zwölften Jahrhundert beschrieb der arabische Geograph Al-Idrisi gekochte Fäden aus Weizen, die auf Sizilien in der Nähe von Palermo gegessen würden. Das Aussehen dieser Teigschnüre (ital. „spago" = Schnur, daher Spaghetti) oder -würmer („vermicelli" bedeutet kleine Würmer) ist eine direkte Folge ihres Herstellungsprozesses. Der Teig wurde dafür ursprünglich ausgewalkt, geschnitten und zum Trocknen auf Schnüren in die Sonne gehängt. Fadenförmige Pastasorten konnten dank dem geringen Querschnitt besonders schnell und aufgrund ihrer Länge mit geringem Arbeitsaufwand und Schnurverbrauch getrocknet werden. Spaghetti wurden ursprünglich geschnitten und waren deswegen nicht rund, sondern rechteckig.

Production – and its Impact on Design | Band saws divide filled wafers into bite-size pieces, rotary bar freezers cast ice popsicles. Unbaked cookies stick to teflon rollers, plaster stamps print gummi bear shapes into corn starch, centrifuges cover millions of chocolate buttons with colorful coatings. Ingredients are molded or sprayed, flattened out by dough rollers and meat tenderizers, carried through brine and drying chambers, processed in mixers and extruders, steam-heated or roasted over fires.

Food products proceed through various stages from produce to table. The first phase is usually processing. No matter whether it is done in the family kitchen, a small trade operation or a factory, we prefer dishes and products which can be prepared with as little expenditure of time and resources as possible. With this in mind, we decide how we prepare and shape our food – if and how we cut, stir, froth up, cook, roast or freeze it. We choose the design of a dish in such a way as to make sure it will be prepared easily. Nobody would ever consider shaping dumplings like cubes because it is much easier to form spheres with our hands and round shapes cook more homogeneously. Italian pizza bakers turn the yeast dough around their fingers so the centrifugal forces produce the round flat pizza base. One answer to the question why our food looks the way it does is simply: so it can be prepared easily.

From Tool to Shape | From the open fire to the industrial plant, cooks and food designers always use a certain range of tools and processes corresponding to the state of the art at the time. Way back, holes in the ground filled with hot stones or open fires were used for cooking, while today liquid nitrogen and long-chain sugar isotopes are conquering private households. The technical possibilities of the individual processes – from kitchen knife to conveyor belt – have a sustained impact on the shape of the final product. Tools and techniques leave their traces, and the form of the outcome will depend on whether dough is extruded, cut with a cookie cutter or a knife, whether chocolate is poured, applied with a piping bag, or used for coating. Processes determine shapes.

In the twelfth century, the Arab geographer Al-Idrisi described cooked strings of wheat being eaten near Palermo on the island of Sicily. The looks of these strings of dough (Italian "spago" translates as string, the word "spaghetti" being derived from this) or worms ("vermicelli" literally translates as "little worms") result directly from the manufacturing process. Originally, the dough was rolled out, cut and hung out on lines to dry in the sun. Thin threads of pasta had a smaller cross-section, therefore drying faster. Longer pasta needed less space on the lines and less work effort. And moreover, spaghetti were, initially, rectangular, not round, because they were cut.

Links | Dünne Nudeln trocknen besser und sind länger haltbar. Spaghetti haben erst seit der Erfindung entsprechender Pressen einen runden Querschnitt. Ursprünglich wurden sie von Hand geschnitten und waren daher rechteckig.
Unten | Die Herstellungsmethode prägt das Aussehen des fertigen Produktes: Zubereitung von Kürtöskalács, ungarischen Schaumrollen.

left | Noodles which have a thin cross-section dry better and keep longer. Only since the invention of appropriate press forms do spaghetti have a round cross section. Originally they were cut by hand and the cross-section was thus square.
bottom | The production method influences the appearance of the final product: preparation of Kürtöskalács, a Hungarian whipped cream roll.

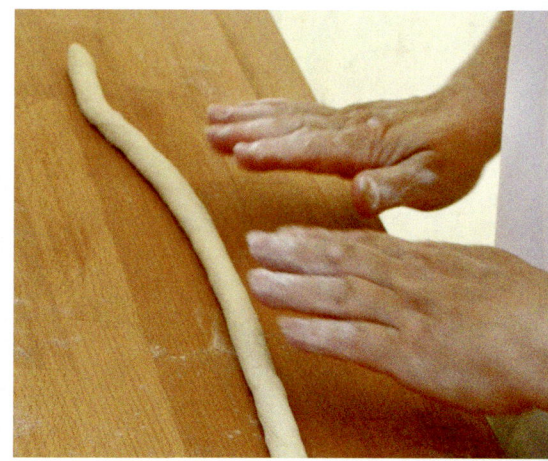

Runde Pasta konnte nur produziert werden, indem man den Teig zwischen den Handflächen wuzelte oder um dünne Stöcke rollte, wie das bei Makkaroni gemacht wurde. Erst die Erfindung von Extrusionspressen ermöglichte die Herstellung von runden Fadennudeln. Im späten achtzehnten Jahrhundert wurde in Neapel eine für damalige Verhältnisse riesige Teigpresse gebaut, mit der bis zu 1,5 Meter lange Spaghetti und – durch austauschbare Pressformen – auch andere Nudelsorten hergestellt werden konnten. Die Spaghetti wurden anschließend in den Straßen von Neapel auf endlosen „Nudelleinen" zum Trocknen aufgehängt, die Hauptstadt Kampaniens in Folge zum Weltzentrum der Pastaproduktion. Auch heute noch werden Spaghetti zum Trocknen aufgehängt, was ihnen eine spezielle Form verleiht: Manche extralange Sorten im Hochpreissegment sind an einer Seite paarweise mit einem 180 Grad gekrümmten Bügel verbunden. Die meisten Hersteller jedoch schneiden dieses herstellungstechnische Merkmal weg, bevor sie die nunmehr einzelnen und geraden Nudeln verpacken.[1]

Auf den Spuren der Fertigung |

Essen wird geschnitten, ausgewalkt, gerollt, gegossen, dressiert, mit Hitze und Kälte behandelt, extrudiert, gestanzt, gepresst, getunkt, dragiert und vieles mehr. Die angewandte Methode beeinflusst das Erscheinungsbild und die sinnliche Wahrnehmung des Zubereiteten. So sind zum Beispiel geschnittene Pralinen wie „Ildefonso" scharfkantig und weisen eine sehr feste Konsistenz auf. Gegossenes Konfekt dagegen hat abgerundete Ecken und kann weiche oder flüssige Füllungen enthalten. Dragiertes wiederum umhüllt das Innenleben, rundet dessen Form ab und punktet mit dem Konsistenzunterschied zwischen weicherem Kern und harter Kruste. Gerollte Produkte wie Hohlhippen oder Croissants sind innen luftig und vermitteln ein anderes Mundgefühl als geschichtete, dichte Lebensmittel à la Lasagne, Mannerschnitten oder Millefeuille. Stanzen lässt flache, gefaltete Pastaformen wie Farfalle entstehen, während Penne oder Hörnchen ebenso wie Zuckerwürfel und Magnum-Eislutscher durch Formstücke gepresst und in die gewünschte Länge geschnitten werden.

Wie vielschichtig die Beweggründe bei der Wahl einer bestimmten Fertigungstechnik sein können, zeigt das Beispiel des vergleichsweise simplen Vorgangs des Rollens. Rollen kann so unterschiedliche Funktionen erfüllen wie die gerechte Verteilung bei Zutaten unterschiedlicher Qualität, etwa beim Rollbraten (niemand bekommt ein Randstück), das „Einrollen" sonst schwer zu bindender Zutaten in Form von Füllen und Cremen, etwa bei Cannelloni, Dosa oder Börek, das Erzielen einer besonders knusprigen, da vergrößerten Oberfläche, etwa bei Salzstangerln oder Croissants, oder die Schaffung optischer Reize, etwa weißes Schlagobers in einer dunklen Schokoroulade.

To produce round pasta, the dough had to be rolled between the palms or wrapped around thin sticks, similar to macaroni. The production of round pasta was only made possible by the invention of the extrusion press. In the late eighteenth century, a press of dimensions that were huge for its day and age was built in Naples. Its screw travel was almost 1.5 meters (62") and it could thus make spaghetti of that length, as well as, with the help of other die shapes, various other types of pasta. The spaghetti were then draped on long racks in the streets of Naples, and the capital of Campania became the world's pasta production center. Even today, spaghetti are hung up to dry, which gives them a special form: some extra long types, actual high-end products, are still connected in pairs when they are sold. However, most manufacturers cut the curved end off before packing straight single noodles[1]

Traces of Production |

Food is cut, rolled out, rolled up, poured, piped on, heat-treated or chilled, extruded, punched out, pressed, dipped, coated with sugar etc. The method used has an impact on the looks and sensory perception of what has been prepared. Cut confectionery such as "Ildefonso" has clearly defined edges and its consistency is solid. By contrast, poured confectionery is rounded and may contain soft or liquid centers. The sugar coating of some products rounds off the shape of the interior it encases and scores because of the different consistencies of soft center and hard crust. Rolled up products such as wafer tubes or croissants are light and airy inside and give us a mouth feel different from that of densely layered food such as lasagna, Austrian "Manner" wafers or millefeuille. Flat folded pasta shapes such as farfalle are created by punching whereas penne or elbow pasta as well as sugar cubes and Magnum ice popsicles are extruded through dies and cut into shape.

The reasons why certain manufacturing techniques are chosen can be quite varied. The seemingly simple process of rolling is a case in point here. Rolling can fulfill a wide range of functions: for example, equitable sharing when parts are of different quality – nobody gets a tapering end piece when rolled roast is distributed – or a way of controlling fillings and creams which are otherwise hard to handle, as in cannelloni, dosa or börek. It can also have to do with getting a bigger, and thus particularly crunchy, surface, as is the case in pretzel rolls or croissants, or with creating optical stimuli, as found in white whipped cream on dark chocolate roll cake.

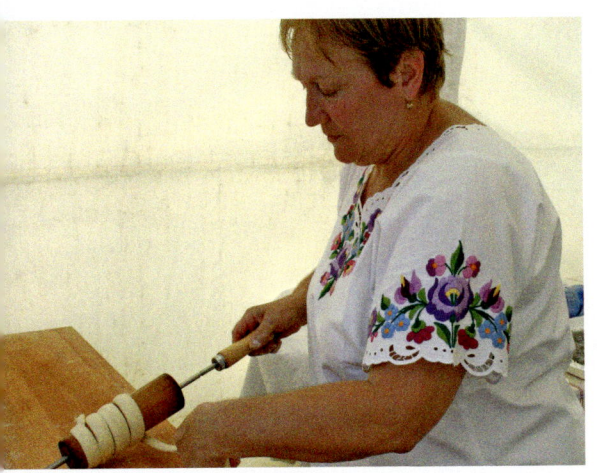
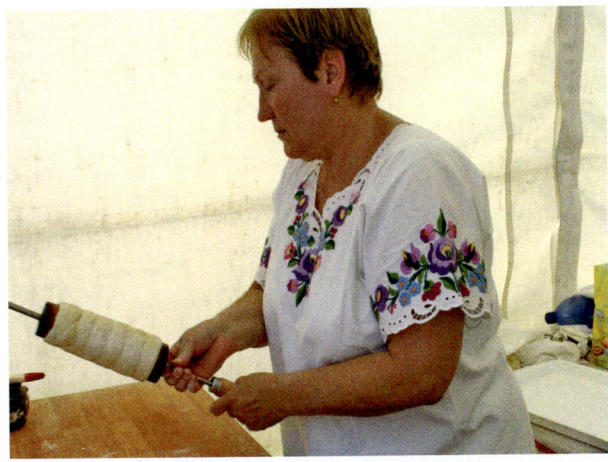

Die Geschichte von Schokolade in fester Form ist vergleichsweise jung. 1847 entwickelte die Firma Fry in Bristol eine Methode, um die davor nur als Getränk bekannte Schokolade auch in festem Zustand herzustellen. Kakaopulver wurde dabei mit Zucker und der vorher abgeschiedenen Kakaobutter vermengt. Der nunmehr zähflüssige Teig konnte in Formen gegossen werden. Unter dem Namen „Chocolat Delicieux à Manger" präsentierte das Unternehmen 1849 die allererste „essbare" Schokolade. Schokolade bietet auf Grund ihrer Konsistenz eine enorme Gestaltungsfreiheit. Allerdings gibt es je nach Herstellungsverfahren produktionsbedingte, formale Einschränkungen. Die meisten Schokoladeprodukte werden gegossen und haben daher leicht schräge Seitenflächen, damit sie anschließend wieder aus der Form fallen. Nur Schokoladebonbons, die geschnitten werden (siehe links), können exakt rechtwinkelig geformt sein.

The history of chocolate in solid form is relatively young. In 1847 the Fry company in Bristol developed a method to create chocolate, which had previously only been known as a beverage, in solid form. Cocoa powder was mixed with sugar and the cocoa butter that had been separated first. The now viscous dough was then poured into molds. The company presented its very first "edible" chocolate under the name "Chocolat Delicieux à Manger"in 1849. Thanks to its consistency, chocolate allows for enormous freedom of design. However, depending on the mode of production, there are formal constraints related to production. Most chocolate products are poured and thus have slightly slanted side surfaces so that they can later fall out of the mold. Only chocolate candies that are cut (see left) can be shaped so that they have a perfect right angle.

Optische Spuren – der Fall aus der Form

Wird ein Produkt geschnitten, dann entstehen rechtwinklige Formen und scharfe Kanten. Beispiele dafür sind das Fischstäbchen oder alle Formen von Schnitten. Werden geometrische Lebensmittel jedoch im Gießverfahren hergestellt, so hinterlässt der Produktionsprozess eindeutige optische Spuren: Zum Beispiel haben Schokoladetafeln oder gegossene Pralinen immer abgeschrägte Seitenflächen und gerundete Kanten. Diese Esswaren werden in Formen gegossen, aus denen sie nach dem Aushärten wieder herausfallen müssen, was (außer bei öffenbaren Formen) nur dann funktioniert, wenn die Seitenflächen leicht konisch auseinander laufen. Scharfe Kanten wiederum haben das Problem, dass sie beim Ausformen leicht hängen bleiben oder abbröckeln.

Selbst die berühmte Sachertorte könnte – wie alle anderen Torten, die nicht in einer Springform hergestellt werden – mit senkrechten Seitenflächen gar nicht produziert werden. Die Konditoren im Wiener Hotel Sacher kaschieren diesen Schönheitsfehler nachträglich mit Hilfe der Glasur, die dann oben einfach etwas dicker ist als unten. Im Falle von industriell erzeugten Pralinen und Schokoladetafeln bleiben die schrägen Flächen für den Konsumenten sichtbar.[2]

Einkerbungen am Rand von Keksen erfüllen eine ähnliche Aufgabe wie die schrägen Seitenwände von Gussformen. So helfen zum Beispiel die berühmten 52 Zähne des Butterkeks, die Teigstücke unbeschadet aus der Form zu lösen. Geradlinige, scharfe Kanten würden beim Ausformen wesentlich leichter brechen. Die vierzehn Zähne pro Länge und zehn pro Breite plus die vier Ecken erfüllen einerseits eine herstellungstechnische Funktion und sind andererseits natürlich auch das unverkennbare Markenzeichen der französischen Kekse „petit LU", die von Jean- Romain Lefèvre und Pauline-Isabelle Utile 1886 in Nantes erfunden wurden, und des deutschen „Leibniz Butterkeks", welches 1891 in Hannover von Hermann Bahlsen auf den Markt gebracht wurde.

Visual Traces – Falling from the Mold

Products which are cut will be rectangular in shape, with clearly defined edges. Fish fingers and all types of wafer blocks are examples of this. If geometrical food shapes are produced by pouring, the manufacturing process will, however, leave visual traces: for instance, chocolate bars or poured confectionery will invariably have beveled sides and rounded edges. The basic chocolate mix is poured into molds from which they have to come out of after hardening. Unless the mold can be opened, this will only happen if the sides are slightly conical. In addition, sharply defined edges cause problems because they easily get caught or break off when the piece of chocolate falls out of the mold.

Like all other cakes which are not baked in a spring form pan, even the famous Sacher cake cannot be produced with perpendicular sides. The pastry cooks of Vienna's Sacher Hotel correct this little flaw by applying more chocolate icing where needed. The beveled sides only remain visible for consumers in industrially produced confectionery and chocolate bars.[2]

Scalloped edges on cookies fulfill a similar task as beveled sides in molds. The famous 52 serrations on the archetypal "petit beurre" butter cookies help remove the dough from the mold without breakage. Straight sharp edges would break more easily. The fourteen serrations on the long side and ten on the short side as well as four corner serrations of the cookie not only serve to keep it from breaking during production but also account for an unmistakable shape that has become the hallmark of "Petit LU", invented by Jean-Romain Lefèvre and Pauline-Isabelle Utile in Nantes in 1886, and their German counterpart "Leibniz Butterkeks", first marketed by Hermann Bahlsen in Hanover in 1891.

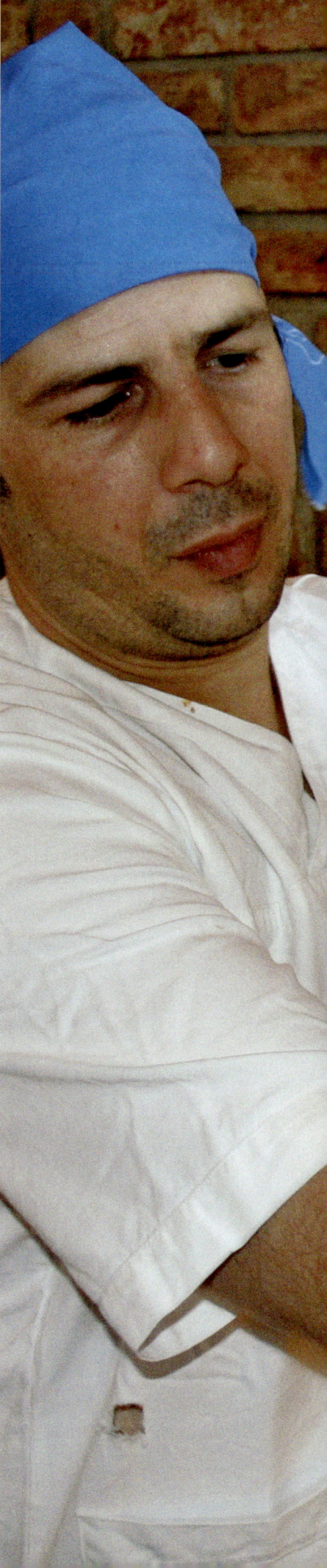

Beide Produkte sind eine Abwandlung der englischen Tradition, kleine Kekse aus Mürbteig (unter anderem „shortbread" oder „butter biscuits") zum Tee zu servieren. In Großbritannien wurde auch das so genannte „docking" erfunden. Dabei werden vor dem Backen kleine Löcher in die Kekse gestochen, damit beim Backen Luft entweichen kann. Diese Technik verhindert, dass sich der Teig im Ofen wölbt. Sie wird heute sowohl bei der Herstellung von Mürbteigtorten verwendet als auch bei den oben genannten Butterkeksen. Die Oberfläche der Leibniz Butterkekse weist zum Beispiel fünfzehn dieser charakteristischen Löcher auf.[3] Die punktförmigen Einkerbungen erinnern an Abnäher von altmodischen Polstermöbeln und verleihen traditionellen Keksen durch diese Assoziation eine gewisse romantische Aura.

Hang zur Geometrie – Mozart ist eine Kugel

Generell ist die industrielle Erzeugung von perfekten geometrischen Körpern eine produktionstechnische Herausforderung, wobei Kugeln besonders problematisch sind. Während sich Knödel in der eigenen Küche relativ einfach und formschön zwischen den Handflächen rollen lassen, ist die maschinelle Herstellung vollendeter Kugeln beinahe unmöglich. Kugelrunde Produkte wie die Mozartkugel weisen – je nach Produktionsverfahren – immer kleine „Schönheitsfehler" auf. So ziert zum Beispiel die Urform der Mozartkugel, wie sie in der Konditorei Paul Fürst in Salzburg seit 1890 händisch erzeugt wird, ein kleiner Extratupfer Bitterschokolade. Bei Fürst wird der Marzipankern auf einen kleinen Stab gespießt, dann mit Nougat umhüllt und schließlich in Schokolade getunkt. Nach Abziehen des Holzstücks wird das Loch mit flüssiger Schokolade verschlossen und dieser Pfropfen bleibt für den Kunden sichtbar. Unter den industriell gefertigten Mozartpralinen konkurrieren zwei Fertigungstechniken, die ebenfalls ihre Spuren hinterlassen: In zwei Hälften gegossen und dann zusammengefügt, verläuft eine feine Naht über das ansonsten makellos runde Produkt der Firma Mirabell. Wird der Marzipankern jedoch mit Schokolade getunkt, entsteht ein gupfartiges Objekt mit Sockel, an dessen Unterseite deutlich die Struktur des Abtropfgitters zu sehen ist. Das Endprodukt ist alles andere als eine Kugel. So geben optische Merkmale Aufschluss über den Hergang der Produktion und verraten einiges über Handarbeit oder Fertigungsstraße.

Herstellungsbedingte Markenzeichen

Technische Notwendigkeiten können durchaus zu ästhetisch reizvollen Ergebnissen führen. Eine Speise mit einer äußerst markanten, herstellungsbedingten „Dekoration" ist der Krapfen, auch Berliner oder Beignet genannt. Zwei goldbraune Rundungen werden von einem waagrecht verlaufenden, hellen Streifen getrennt. Die signifikante Zeichnung entsteht unter anderem durch den Rum, welcher im Teig enthalten ist. Beim Frittieren verdampft der Alkohol und verschafft dem Krapfen mehr Auftrieb, so dass der mittlere Abschnitt das heiße Fett niemals erreicht und daher weiß bleibt.

Both products emulate the English tradition of serving sweet morsels also known as "shortbread" or "butter biscuits" along with tea. Great Britain is also where the technique called "docking" was invented. Prior to baking, little holes are made in the dough so that the air can escape during baking and the cookies do not bulge where air is trapped. This technique is used in shortcrust pastry and butter biscuits alike. For instance, the surface of the "Leibniz Butterkeks" has fifteen of these characteristic holes.[3] The little dots are reminiscent of buttons on old-fashioned upholstered furniture, and by virtue of this association, they lend a certain romantic aura to the traditional cookies.

A Penchant toward Geometry – Mozart and the Spherical Candy

In general, perfect geometrical shapes are a technical challenge for industrial production, and spherical shapes are especially problematic. While dumplings can be shaped nicely and easily by hand in the comfort of our own kitchen, it is almost impossible to produce perfect spherical shapes by means of machines. Depending on the manufacturing process, spherical products like the "Mozartkugel" chocolates always have little flaws. The original shape of the "Mozartkugel", which has been hand-made in the pastry and confectionery shop "Paul Fürst" in Salzburg since 1890, is graced by an extra dab of dark chocolate. According to the Fürst method, the marzipan core is placed on a little stick, then covered with nut paste and finally dipped in chocolate. When the stick is removed, the hole is closed with molten chocolate but the "plug" remains visible. Industrially produced Mozart chocolates are made using two competing methods which both leave their traces on the finished product. The confectionery manufactured by the Mirabell co. consists of two filled halves which are then joined. The product is perfectly round, except for the seam. The other method also has a center of marzipan and nut paste dipped in chocolate but it is dome-shaped rather than round and clearly shows the structure of the grid on which the chocolate is placed during the process of enrobing, so it is anything but spherical. Visual features thus provide insights into the production process as well as the manual work or industrial lines involved.

Hallmarks of Production

Technical necessities may, however, also lead to esthetically appealing results. One product with a characteristic "decoration" due to the process whereby it is made is the "Krapfen", also known under the name "Berliner" or "beignet", a doughnut-type pastry without a hole in the center. Two golden brown halves are separated by a lighter horizontal strip. This feature is due to the rum contained in the dough. The bun is fried in hot oil, and since the alcohol evaporates, the bun is more buoyant so that the middle is never immersed in the frying oil and therefore remains white.

Essbarer Teller: Die Pizza entwickelte sich in der Renaissance in Neapel, wobei die Idee, Teig als Teller für andere Zutaten zu benutzten, durchaus keine Neuerfindung war. Schon vor dem 17. Jahrhundert war der Brauch verbreitet, Bratenstücke auf Brotscheiben zu servieren, die den Saft auffingen. Das Bild zeigt Gigi und Pasquale in der Pizzeria Il Sestante in Wien.

Edible plate: Pizza was developed in Naples during the Renaissance, and the idea of using dough as a receptacle for other ingredients was certainly not a new invention. Already prior to the 17th century the custom of serving pieces of roast on slices of bread that would absorb the sauce was widespread. The illustration shows Gigi and Pasquale at the Pizzeria Il Sestante in Vienna.

Ob der Teig immer schon Alkohol enthielt und der weiße Rumpfstreifen bereits die antiken Hefegebäcke zierte, ist schwer nachzuweisen. Schon im alten Ägypten sollen nämlich krapfen ähnliche Kuchen verzehrt worden sein und auch die Römer kannten eine dem Krapfen verwandte Speise, die sie „globuli", also kleine Kugeln, nannten und während der ausschweifenden Bacchanalien verzehrten. Im zweiten Jahrhundert vor Christus wurde das Rezept für frittierten Hefeteig dann von Cato als römisches Gericht namens „scriblita" beschrieben.[4] Zur Zeit Kaiser Karls des Großen taucht für ein Schmalzgebäck erstmals die Bezeichnung „crapho" auf. Das klösterliche Siedegebäck kam jedoch ursprünglich vermutlich nicht rund, sondern in länglicher Form mit gebogenen, spitzen Enden auf den Tisch. In Wien aß man nachweislich bereits im 15. Jahrhundert ein krapfenartiges Dessert, wie aus einer Kochverordnung von 1486 hervorgeht, welche die Berufsbezeichnung „Krapfenpacherinnen" gebraucht. Der Legende nach soll das heute übliche Rezept eine Kreation der Wiener Hofratsköchin Cäcilie Krapf sein.

Stammt der Krapfen ursprünglich aus der bäuerlichen Küche, so ist die Füllung mit Marmelade eine städtische Erfindung. In Berührung mit Zucker kam der Krapfen, der zuvor ausschließlich

It is hard to prove whether the dough always contained alcohol and whether the stripe also graced ancient yeast pastry. Buns similar to Krapfen are said to have been eaten in ancient Egypt, and the Romans also had something related to the round doughnut-type pastry. They called it "globuli", little globes, and ate it during the licentious bacchanals. In the second century BCE, the recipe for fried yeast-risen dough was described as a Roman specialty called "scriblita" by Cato.[4] During the era of Emperor Charlemagne, an olycook named "crapho" was mentioned for the first time. However, the original shape of the monastic deep-fried pastry seems to have been oblong instead of round, with tapering curved ends. There is evidence that a dessert reminiscent of the Krapfen was eaten in Vienna as early as in the fifteenth century. A decree from the year 1486 refers to the trade of the "Krapfenpacherin" ("Krapfen cook"). Legend has it that today's recipe was created by Cäcilie Krapf, cook to a high-ranking civil servant in Vienna.

While the Krapfen as such originated in rural areas, the idea of filling it with jam was an urban invention. The powdered sugar topping was actually only introduced

Wir wählen das Design einer Speise immer so, dass sie sich möglichst gut herstellen lässt. Niemand würde auf die Idee kommen, Knödel würfelförmig oder Pizzen eckig zu machen, weil sich Rundes mit den Händen wesentlich leichter formen lässt und außerdem beim Kochen gleichmäßiger durchgart.

We always select the design of a food so that it can be produced in the best possible way. No one would ever have the idea of making dumplings cubic or pizzas square, since round shapes can be made much easier by hand and also heat more evenly when cooked.

ungesüßt verspeist wurde, erst im Biedermeier. Von da an war sein Siegeszug nicht mehr zu stoppen. Das Vorbild des heutigen Krapfens könnte somit tatsächlich auf die so genannten Cilly-Kugeln von Frau Krapf zurückgehen. Allein im Kongressjahr 1815 verdrückten die Wiener angeblich an die 10 Millionen Stück von der mit feinster Konfitüre gefüllten Delikatesse.[5]

Von Kuchen und Kapuzen | Die Form des europaweit bekannten Gugelhupfs oder Kugelhopfs mit dem markanten Loch in der Mitte entstand, weil man sich seine Zubereitung optimieren wollte. Der Kuchen ist ungewöhnlich hoch und schmal und wurde in seiner Urform vermutlich in Schüsseln gebacken. Schon die alten Römer aßen Napfkuchen, die rundum mit den charakteristischen, schräg verlaufenden Kerben verziert waren. Das mittige Loch dürfte erst in späterer Zeit aus praktischen Überlegungen hinzugekommen sein: In Schüsseln gebackene Kuchen neigen dazu, im Inneren noch roh und am Rand schon verbrannt zu sein. Die Form des Gugelhupfs spart jenen Teil, der am schlechtesten mit Hitze versorgt ist, einfach aus. Die heiße Luft streicht so auch durch das Loch in der Mitte des Kuchens und erreicht alle Stellen

in the early nineteenth century; before that, it had exclusively been eaten unsweetened. From then on, there was nothing to stop it from becoming ever more popular. In fact, the Krapfen as we know it in Austria today could really be based on the so-called "Cilly-Kugeln" devised by Frau Krapf. In 1815, the year of the Congress of Vienna, the Viennese are said to have eaten no less than 10 million of the tasty pastry filled with fine jam.[5]

Cakes and Hoods | The "Gugelhupf" or "Kugelhopf" which practically all of Europe is familiar with under various names, and which is also known as "Bundt cake" in the US, has a hole in the middle, or is described as ring-shaped; the form ensures optimum baking. The cake is unusually high and narrow, and presumably it was originally baked in a bowl. The ancient Romans are said to have eaten bowl-shaped cakes with the characteristic oblique notches. However, the hole in the center was most likely added later out of practical considerations: cakes baked in a bowl tend to burn on the edge while they are still raw in the center. The ring shape simply leaves out the part that is least well heated. Hot air also passes through the hole in the middle of the cake and

thus reaches all parts evenly. The reason why the shape is also adorned with oblique notches decorating the cake surface – similar to British jelly molds – remains unclear. Apart from the visual effect, they might also help spread the oven heat evenly, or ensure that the cake comes out of the baking mold more easily. Perhaps they are just decorations without specific function. It is e.g. conceivable that they were originally stylized plant-shaped ornaments similar to the fluting on ancient columns.[6]

Characteristically Wavy – Cup Noodles | Just like the hole in Bundt cakes, the design of crimped noodle blocks in the instant soups of the Japanese CupNoodles brand actually aims at optimum preparation. The structure of the noodles is first of all the result of the manufacturing process. The dough is rolled out and cut into thin strips. These fall on a belt which moves much slower than the one transporting the dough towards the cutting blades. This way, the noodles are pushed into the characteristic wavy shape.

Due to the crimped form, they can be compacted in relatively dense and even little loaves which are then fried. As they swim in hot frying oil, the noodles become more compact on top while their structure loosens up below. Different degrees of density in the blocks of noodles contribute to optimum cooking in the cup. When the hot water is added, the temperature in the top part of the cup is higher, and since the blocks are denser on top, this is exactly where there are more noodles to be cooked. The bottom part of the cup is colder but there are fewer noodles so that the noodles cook evenly.[7]

Japanese Fish Sausages | Another example of food whose design is related to the manufacturing process are Japanese fish sausages, so-called "neri-sei-hin". Similar to western sausages, they are a way of using leftovers. Made of fish remnants, they make for a very popular every-day staple. The fish pieces are processed into paste mixed with various flavors and formed in different shapes.

One of the classics are hollow rolls that look like oversized macaroni. The shape of this fish sausage, called "chikuwa", is a result of the historical manufacturing process: the paste was wrapped around a bamboo stick and grilled over an open fire. Another traditional process involved paste spread on a wooden board and steaming it. The final result, the "kamaboko" fish cake, has until today remained oblong and rectangular, with a semi-circular cross-section reminiscent of a barrel vault.[8]

Time Is Money | The meaning of optimum production techniques has not changed over time, it still spells short production or preparation times for dishes and foods. This also seems to be the origin of the Asian tradition of cutting all ingredients into small pieces in the kitchen before cooking. Meat and vegetables were chopped to keep the cooking time short and save on expensive fuel.

In Schüsseln gebackene Kuchen neigen dazu, in der Mitte noch roh und am Rand schon verbrannt zu sein. Die Form des Gugelhupfs spart jenen Teil, der am schlechtesten mit Hitze versorgt ist, einfach aus. Die heiße Luft streicht so auch durch das Loch in der Mitte des Kuchens und erreicht alle Stellen des Teiges relativ gleichmäßig.

Cakes that are baked in bowls tend to be raw in the middle and burned on the edges. The shape of the Bundt cake simply leaves out the part of the cake that is most difficult to heat. The hot air thus also reaches the hole in the middle of the cake, thus heating all parts of the dough relatively evenly.

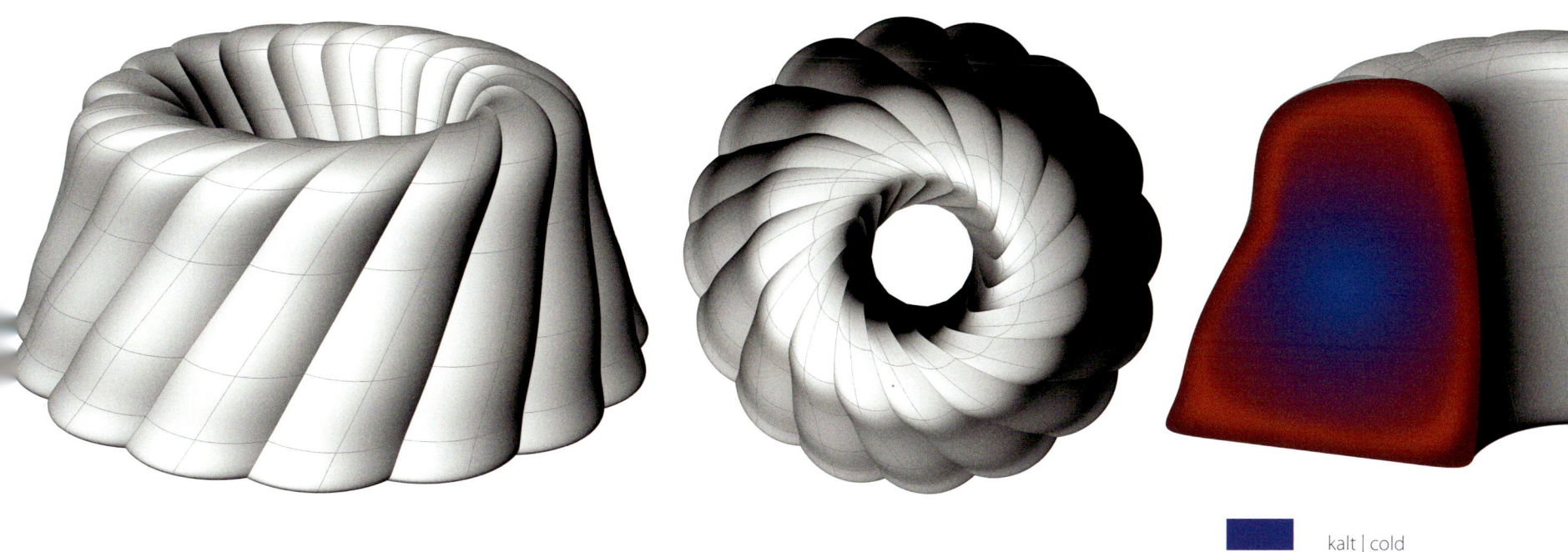

kalt | cold
heiß | hot

Auch oder gerade bei der industriellen Herstellung von Nahrungsmitteln ist Zeit Geld. Man könnte zum Beispiel fragen, warum manche Suppenwürfel Quader sind und keine Würfel. Die Antwort ist einfach: Weil Quader am Fließband effizienter zu produzieren sind als Würfel. Suppenwürfel werden in Formen gespritzt oder gepresst. Um eine tiefe Gussform wie einen Würfel mit der Mischung aus Rindfleischextrakt, getrocknetem Gemüse, Fett, Salz und Gewürzen zu befüllen, benötigt man etwas mehr Zeit als für flache Formen wie Quader. Und da Zeit bekanntlich Geld ist und eine Herstellungsstraße umso profitabler, je schneller sie läuft, sind Suppenwürfel eben keine Würfel, sondern Quader.[9]

Erzeugungsprozesse wirken sich auf die Optik des Endproduktes aus und führen zur Entwicklung entsprechend günstiger Verfahren und Formen. Die Herstellungsdauer von Brot beispielsweise erfuhr mit dem 1961 in England erfundenen „Chorleywood Process", bei dem der Teig mit einem Hochgeschwindigkeitsrührwerk erzeugt wird, eine drastische Verkürzung. Bei dieser Technik wird Mehl mit chemischen Backhilfsmitteln, pflanzlichem Fett, Wasser und Hefe sehr schnell vermischt. Der so entstehende Teig muss nicht mehr rasten, was die Brotproduktion sagenhaft beschleunigt. Der „Chorleywood Process" ermöglicht die komplette Produktion von vorgeschnittenem Toastbrot innerhalb von nur dreieinhalb (!) Stunden, die Konsistenz des fertigen Brotes ist im Vergleich zu herkömmlichem Brot allerdings sehr weich und fluffy, auch der Geschmack leidet.[10]

Inspiration durch Technologie | Neue Technologien liefern stets Inspirationen für neue Designs und erweitern die Möglichkeiten der Formgebung. Die Entwicklung neuer Werkzeuge und Herstellungsverfahren bescherte uns im Laufe der Geschichte immer wieder neue Gerichte, Speisen, Getränke und Lebensmittel. Seit Jahrtausenden experimentiert der Mensch mit unterschiedlichsten Techniken, um lebensnotwendige Güter herzustellen. Einfachste Werkzeuge aus Holz und Stein mutierten im Laufe der Zeit zu Mähdreschern, das Feuer zum Mikrowellenherd. Wir lernten zu schneiden, zu braten und zu kochen. Wir entwickelten Prozesse und Gerätschaften, um Speisen geschmacklich aufzuwerten, sie haltbar und transportfähig zu machen.

Eines der ältesten Beispiele dafür, wie die Entdeckung von Technologien neue Lebensmittel entstehen lässt, ist der Käse. Mit der Domestizierung von Tieren, etwa dem Auerochsen, erkannten die Menschen in der Steinzeit – wahrscheinlich zuerst im mittleren Osten – den Nährwert von Milch. Ungekühlt zerfällt Milch in flüssige Molke und in feste Bestandteile, den Sauermilch- oder Frischkäse. Vermutlich zufällig entstand dann der Hart- oder Labkäse: Um Milch zu transportieren, wurde sie in getrocknete Kälbermägen gefüllt, wo das Lab, ein eiweißspaltendes Enzym, das Kälber zur Verdauung der Muttermilch benötigen, die Milch in harten, haltbaren Käse verwandelte.[11] Die verschiedenen Behälter, in welche die

In the industrial production of food, time is money, too – or rather, this is where time is especially of the essence. We might ask ourselves why some stock cubes are rectangular prisms in spite of their names. The answer is simple: because it is more efficient for a production line to make rectangular prisms than actual cubes. Stock cubes are cast or pressed into molds. It takes more time to fill a deep mold like a cube with a mixture of beef extract, dried vegetables, fat, salt and spices than placing it in a flatter mold, such as a rectangular prism. And since time is money, as we all know, and production lines are the more profitable the faster they run, stock cubes are rectangular prisms, not cubes.[9]

Manufacturing processes have an impact on the looks of the final product and result in the development of ever more efficient techniques and shapes. The production time of bread was for instance dramatically reduced due to the "Chorleywood Process" invented in England in 1961, whereby dough is kneaded at high speed. Flour, baking aids, vegetable fat, water and yeast are mixed extremely quickly so that the dough does not have to rest any longer. The "Chorleywood Process" makes it possible to produce pre-sliced toast in as little as three and a half (!) hours, though the consistency of the bread differs from conventionally produced bread as it is extremely soft and fluffy, and the taste suffers, too.[10]

Inspired by Technology | New technology has always inspired new shapes and extended the possibilities of design. In the course of history, the development of new tools and production processes has again and again given us new dishes, beverages, foodstuffs and products. For millennia, human beings have experimented with a wide variety of techniques to produce vital goods. The simple tools made of wood and stone were turned into combines in the course of time, and open fires became microwave ovens. We learnt how to cut, roast and cook. We developed processes and utensils to improve the taste of food, make them more durable and easy to transport.

Cheese is one of the oldest examples of how the discovery of new technologies leads to new food products. When animals such as the aurochs were domesticated, the people of the Stone Age – most likely the population of the Middle East – discovered the nutritional value of milk. If it is not refrigerated, milk decomposes into liquid whey and solid components, acid curd cheese or fresh cheese. It is believed that hard cheese or rennet cheese developed from this by coincidence: Milk was transported in dried

Esswaren fahren mit Gabelstaplern und Lastkraftwagen, liegen in Einkaufswägen und -taschen und wollen auch im Kühlschrank platzsparend verstaut sein. Ob Tiefkühlgemüse oder Teebutter, Spinat oder Rotkraut, Schokoladetafeln oder Suppenwürfel, die Bewohner der Industrienationen kaufen rechteckige Nahrung in rechteckigen Packungen.

Foods travel with forklifts and trucks, lie in shopping carts and bags and also require being stored in the refrigerator in such a way as to save space. Be it frozen vegetables or butter, spinach or red cabbage, chocolate bars or soup cubes, the inhabitants of the industrial nations buy square food in square packaging.

Milch bei der Käseproduktion in weiterer Folge gefüllt wurde, prägten dann die Formen des Hartkäses: Einfach herzustellende, runde Spanschachteln und flache Fässer ließen runde Laibe in unterschiedlichen Größen entstehen, in Tüchern aufgehängt, erhielt der Käse eine Tropfenform wie etwa der Provolone. Dass sich 80 Kilogramm schwere Emmentaler-, Parmesan- oder Cheddarlaibe aufgrund ihrer zylindrischen Form außerdem rollen und damit auch von einer einzelnen Person transportieren lassen, ist ein weiterer, funktionaler Vorteil, der für das runde Design von Käse spricht.

Gestaltung und Technologie vor 10.000 Jahren – der Sauerteig |
Vor etwa 10.000 Jahren begann der Mensch, systematisch Getreide anzubauen, das zunächst gemahlen und, mit Wasser vermengt, als Brei gegessen wurde. Später erkannte man die konservierenden Fähigkeiten von Hitze. Unsere Vorfahren buken Getreidebrei auf heißen Steinen zu Fladenbroten und schufen damit lagerfähige und transportable Nahrung. Letztlich haben zwei Technologien das Design von Brot nachhaltig beeinflusst. Die eine war der Bau von Backöfen. Auf heißen Steinen lassen sich nur flache Fladen backen, wenn die Hitze jedoch von allen Seiten einwirkt, sind auch gleichmäßig durchgebackene, runde Brotlaibe möglich. Die zweite, wahrscheinlich zufällig entdeckte Technik, welche vollkommen neue Gestaltungsmöglichkeiten eröffnete, war die Sauerteiggärung. Lässt man ungebackenen, dickeren Brotteig stehen, so sorgen Hefebakterien in der Luft für eine Gärung, die den Teig aufgehen lässt und das fertige Brot wesentlich luftiger, poriger und damit besser macht. Im Laufe der Zeit lernte der Mensch, die Sauerteiggärung gezielt zu steuern, indem er von gut gelungenem Teig eine kleine Menge beiseite stellte und diese für das nächste Brot wieder verwendete. Diese 5000 Jahre alte Technologie, die unter anderem im alten Ägypten, im antiken Rom und im Zuge der frühzeitlichen Industrialisierung perfektioniert wurde, wird bis heute eingesetzt.

Durch das Aufgehen des Teiges beim Sauerteigbrot verdoppelte sich die Ausgangsmasse aus Mehl und Wasser und erhöhte somit die Überlebenschancen. Und die

calf stomachs, where rennet, a protein-splitting enzyme which calves need to digest their mothers' milk turned the milk into hard and durable cheese.[1] The containers which the hard cheese is placed in were what then gave it its shape: Round chip boxes, which are easy to make, and flat vats accounted for round loaves of cheese in various sizes. Cheese kept in fabric and hung from the ceiling became drop-shaped, like Provolone. The cylindrical loaves of Emmental, Parmesan and Cheddar cheese, which may weigh up to 160 pounds can also be rolled and thus transported by one person – this turned out to be another functional advantage speaking in favor of round forms of cheese.

Design and Technology 10,000 ago – Sourdough |
About 10,000 ago, humans started to cultivate grain systematically, which was then ground and eaten as a mash mixed with water. Later, our ancestors discovered the preserving effect of heat. They slapped grain mash on hot stones, baking it and turning it into storable and transportable cakes. Eventually, two technologies had a lasting impact on the design of bread. One was the construction of ovens. You can only bake flat cakes on hot stones but if the heat comes from all sides, you will be able to get evenly baked round loaves. The second contribution to the development of bread which opened up entirely new design possibilities was sourdough fermentation, which probably was discovered by coincidence. If you allow the unbaked thick bread dough to rest, airborne yeast bacteria will cause fermentation which makes the dough rise and improves the quality of the baked bread, as it is lighter and more porous. In the course of history, people learnt to control sourdough fermentation; they set aside a small quantity of good dough for use in the next batch. This technique is 5,000 years old and was inter alia used and improved in ancient Egypt, Rome and in the age of early industrialization. It continues to be applied until today.

The rising of the sourdough doubled the original quantity of flour and water, thus increasing chances for survival. The new recipe also had an influence on the shape of

neue Rezeptur beeinflusste die Form des Gebackenen. Nach den flachen Laiben konnte nun eine Unmenge von dreidimensionalen Objekten erzeugt werden. Die unglaubliche Vielfalt an Formen, in denen Brot heute weltweit gebacken wird, zeigt, mit welchem Enthusiasmus die Gestaltungsfreiheit von Sauerteig, der sich nahezu wie Knetmasse formen lässt, bis heute genutzt wird. Die Motive von Gebild- und Festtagsbroten reichen von Tauben, Fischen und Vögeln über Himmelsleitern, Zöpfe und Reiter bis zu Menschen und Heiligenfiguren. Welche geheimnisvollen Vorgänge den Teig beim Backen aufgehen ließen, blieb allerdings über Jahrhunderte hinweg unbekannt. Dem Brot wurden magische Kräfte zugeschrieben, Backöfen als Muttergottheit verehrt und Frauen bis in die jüngste Geschichte nach ihren Kuchen beurteilt. Die perfekt aufgegangene Mehlspeise symbolisierte die Fruchtbarkeit der Köchin und war ihr Garant für eine Eheschließung.

Gestaltung und Technologie heute – Extrusion |
Neue Techniken bringen neue Gestaltungsmöglichkeiten, das zeigt sich im Laufe der Geschichte ebenso wie in aktuellen Entwicklungen. Ein Beispiel aus jüngerer Zeit ist die Extrusion, eine Produktionstechnik, bei der eine schrittweise erhitzte und durch ein Schneckengewinde unter Druck gesetzte Masse durch eine Düse gepresst wird. Die Teigmasse dehnt sich dabei explosionsartig aus, die gleichzeitige Abkühlung an der Umgebungsluft verhindert allerdings, dass die Struktur zerplatzt. Es entstehen „aufgeblasene" Flocken, wie zum Beispiel Erdnussflips. Je nach Ausbildung der Düse erlaubt die Technik unterschiedlichste Formen von Pasta, Knabbergebäck, Süßwaren et cetera.

Extrudierte Produkte zeichnen sich durch runde, oberflächenmaximierte Designs und eine speziell leichte, luftige Konsistenz aus. Bei der Extrusion entstehen Formen, die man mit der Hand so nicht erzeugen kann. Das heißt, durch den Einsatz bestimmter Technologien werden gewisse, neue Formen überhaupt erst möglich. Ein weiterer Vorteil für die Industrie ist, dass extrudiertes Knabbergebäck nach mehr ausschaut, der Rohstoffverbrauch aber durch die eingeschlossene Luft im Inneren vergleichsweise gering ist.[12]

baked goods. Instead of flat loaves, a multiplicity of three-dimensional objects could be baked. The enormous diversity of shapes that bread comes in worldwide testifies to the lasting enthusiasm of humankind for the creative use of sourdough as it can almost be treated like modeling clay. Motifs for festive cakes and bread include, known in the German-speaking countries as "Gebildbrot", for example, include doves and fishes, Jacob's ladder, plaits and horsemen, human shapes and saints' likenesses. For centuries, it was unknown which secret process actually caused the dough to rise in the course of baking. Bread was thought to have magical powers, ovens were adored as mother goddesses, and until the more recent past, women used to be judged by the cakes they baked. If the sweet dough rose perfectly, this was symbolic of the cook's fertility and a guarantee for her to find a husband.

Design and Technology Today – Extrusion |
New technologies come with new design options. This can both be seen in the course of history and in current technological developments. Extrusion is a case in point – this production technique involves dough which is gradually heated and forced through a die by a screw. The dough virtually explodes while expanding but the simultaneous cooling of the ambient air keeps its structure from blowing up completely. This way, "inflated" flakes, such as peanut puffs, are produced. Depending on the die, various shapes of pasta, savory snacks, sweets, etc. can be manufactured this way.

Extruded products are characterized by round designs with a maximum of surface area and specially light and airy consistency. This is to say it is only due to the use of certain technologies that various new shapes can actually be created. Another advantage for industry is that extruded snacks look bigger than they actually are due to the air trapped inside while the quantity of raw material needed is relatively low.[12]

Die Entdeckung der Sauerteiggärung hat das Design von Brot nachhaltig verändert. Die unglaubliche Vielfalt an Formen, in denen Brot heute weltweit gebacken wird, zeigt, mit welchem Enthusiasmus die Gestaltungsfreiheit von Brotteig genutzt wird.

The discovery of sourdough fermentation has had a lasting impact on the design of bread. The unbelievable diversity of shapes in which bread is baked all over the world shows our great enthusiasm for the freedom to design bread dough any desired way.

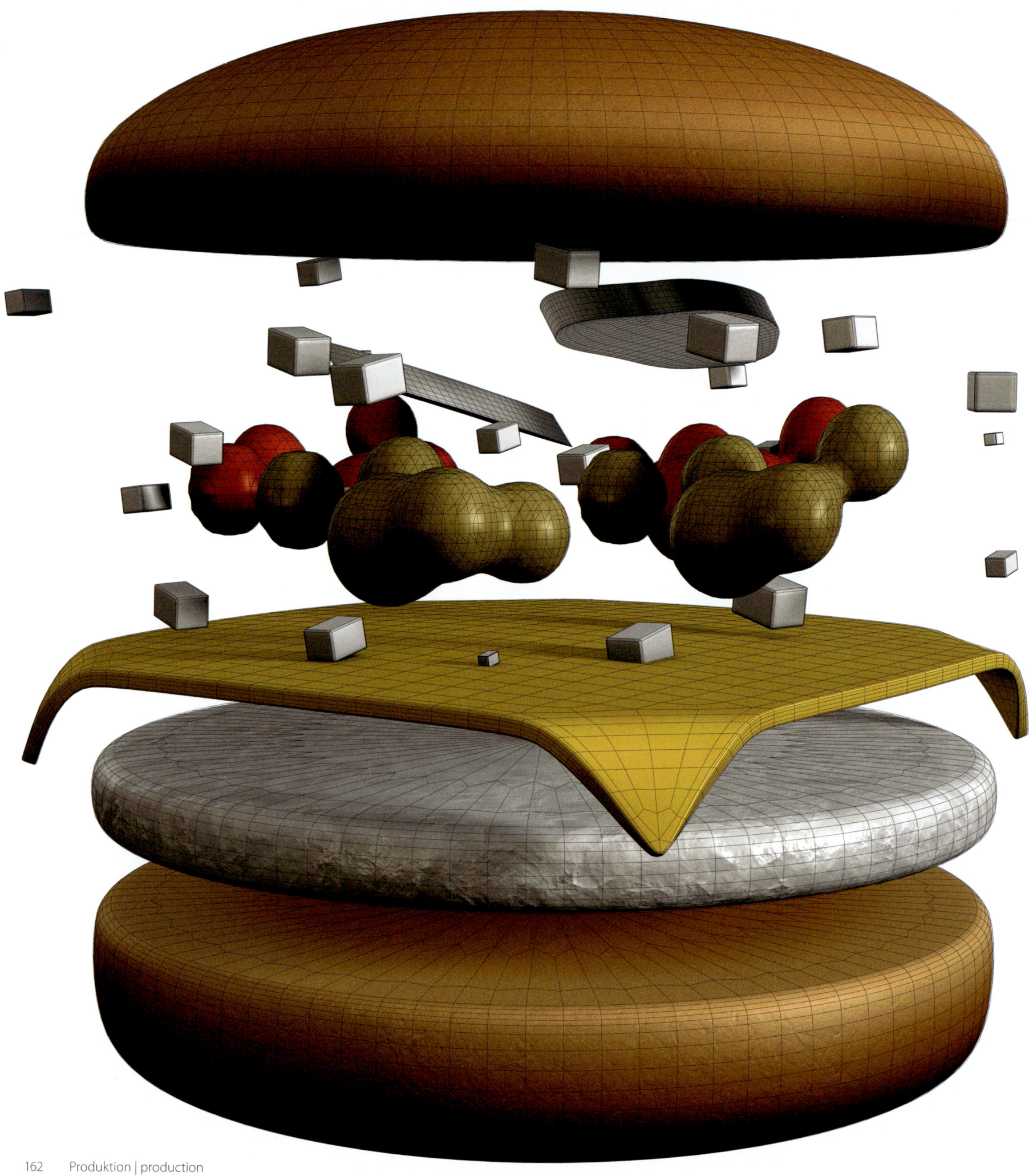

Der Brauch, Reste vom Sonntagsbraten zwischen zwei Brotstücken unter der Woche zu verzehren, wurde im 19. Jahrhundert von deutschen Emigranten nach Amerika gebracht. Vermutlich entstand so aus dem norddeutschen Gericht „Rundstück Warm" der heutige Hamburger. Der Legende nach soll der Bratwurstverkäufer Charlie Nagreen 1892 in Ohio erstmals einen serviert haben. Zur Grundlage der Fast-Food-Industrie wurde der Hamburger, als Roy Croc Mitte der 1950er Jahre die McDonalds-Kette begründete.

The custom of eating the leftovers of a Sunday roast on a week day between two pieces of bread was brought to America by German immigrants in the 19th century. It is said that it developed from the north German dish "Rundstück Warm", today's hamburger. According to legend, the sausage vendor Charlie Nagreen served the first hamburger in Ohio in 1892. The hamburger became the mainstay of the fast food industry when Roy Croc founded the McDonald's chain in the mid-1950's.

McDonald's | Auch beim Fastfood-Konzern McDonald's ist die Gestaltung der Produkte untrennbar mit dem Zubereitungsverfahren verbunden. Der Anspruch von McDonald's ist, weltweit standardisierte, stets gleich schmeckende Speisen anzubieten. Um das zu gewährleisten, müssen die tiefgefrorenen Fleischlaibchen nicht nur nach vordefiniertem Rezept in immer gleicher Größe produziert, sondern auch in allen Filialen von Tokio bis Caracas auf einem standardisierten Grill zubereitet werden. Dieser besteht aus zwei heißen Stahlplatten und lässt sich auf nur zwei verschiedene Temperaturen und zwei unterschiedliche Garzeiten einstellen. Wie bei einem Toaster öffnet sich der Grill von New York bis Mumbai nach der exakt gleichen Zeitspanne automatisch und der Mitarbeiter weiß, dass das Gericht für die Weiterverarbeitung fertig ist. Werden nun neue Burgermodelle entwickelt, dann müssen diese so gestaltet sein, dass sie mit diesem Grill zubereitet werden können. Das betrifft einerseits die Rezeptur und die Zutaten, andererseits natürlich auch die Form: Kommt etwa ein gebackener Fisch neu ins Programm, dann muss die Schnittstärke so gewählt werden, dass er bei der gleichen Temperatur genau gleich lang braucht, um durchzugaren, wie ein klassischer Hamburger. Nach demselben Prinzip funktioniert auch die Zubereitung der frittierten Gerichte wie der Pommes Frites. Die Kartoffelstücke müssen immer gleich groß und gleich temperiert sein, damit sie bei standardisierter Öltemperatur und Backdauer stets gleich schmecken. Außerdem sorgen eigens entwickelte Saucen-Dispenser dafür, dass in jedem Burger eine genau vordefinierte Menge an Ketchup, Mayonnaise oder sonstigen Saucen enthalten ist.[13]

Gelingsicherheit | Eine herstellungsgerechte Gestaltung muss aber nicht unbedingt auf eine industrielle Erzeugung abzielen, sondern soll auch die Zubereitung zu Hause erleichtern. Kann beispielsweise eine entsprechende Gestaltung dazu beitragen, Misserfolge in der Küche zu vermeiden? Eine 1987 vom Pariser Designer Philippe Starck entworfene Nudel kann das angeblich und soll ungeübten Köchen helfen, ein akzeptables Pastagericht zuzubereiten. Die röhrenförmige Nudel hat an der Außenseite zwei dickere Wülste, die im heißen Wasser langsamer durchgaren als der Rest. Die unterschiedlichen Querschnittsstärken sollen trotz zu langer Kochzeit zumindest eine stellenweise bissfeste Pasta garantieren.

McDonald's | The fast food group McDonald's is no exception: here, too, the design of products is inseparably linked with the preparation process. McDonald's claim to fame is that they offer standardized food which tastes the same anywhere in the world. To ensure this, the deep-frozen meat patties have to be produced according to a defined recipe, they have to be the same size worldwide, and in all the outlets from Tokyo to Caracas they have to be cooked on a standardized grill. It consists of two steel hot plates which can only be set to two temperatures and two different cooking times. The two-sided hot plate opens automatically like a toaster after exactly the same period, no matter whether you're in New York or Mumbai, and the employee will know that the patty is ready for further processing. Whenever new burgers are developed, they must be designed so they can be made on the same appliance. This concerns the recipe and ingredients as well as the shape: If fried fish is new on the menu, it has to be cut to a thickness that takes the same time to cook at the same temperature as the other products. The principle also applies to French fries and the like. Potato pieces have to be of the same size and must be stored at the same temperature so they will always taste the same when fried at standard oil temperature. Sauce dispensers are specially designed to make sure each burger gets the predefined quantity of ketchup, mayonnaise or other relish.[13]

Sure to Succeed | While appropriate design can contribute to easier industrial production, it will also be useful at home. Can good design for example help avoid disasters in the kitchen? A noodle shape drafted by the Parisian designer Philippe Starck in 1987 is said to help avoid overcooking, assisting inexperienced cooks in preparing an acceptable pasta dish. The tubular noodle has two bulges on the outside; in hot water, these cook more slowly than the other parts. The different cross-sections are to guarantee pasta cooked al dente, at least in some spots.

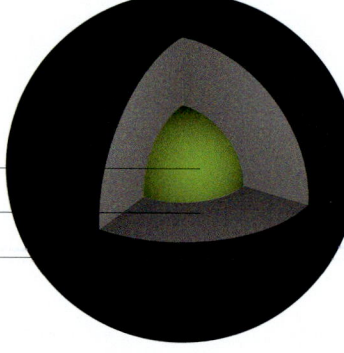

Links | Die Struktur von QimiQ: Träger: Wasser, Lactose, Mineralien

left | The structure of QimiQ: Carrier: water, lactose, minerals

Rechts | QimiQ - Detail
Kern: Milchprotein

right | QimiQ - detail
Core: milk protein

Innenmantel: Milchfett

Interior shell: milk fat

Außenmantel: Gelatine

exterior shell: gelatin

Ein anderes Beispiel für Gelingsicherheit ist QimiQ. Das 1995 von den Österreichern Hans Mandl und Rudolf Haindl erfundene, natürliche Sahneprodukt ist in der Lage, künstliche Emulgatoren und Stabilisatoren zu ersetzen, da es sich dauerhaft mit allen möglichen Flüssigkeiten wie Alkohol oder Essig verbindet und außerdem Emulsionen mit Öl bildet. Bei QimiQ ist die Sahne bereits mit Gelatine vermischt, wodurch man sich bei bestimmten Rezepten das Einweichen und Zugeben von Gelatineblättern erspart. Außerdem lässt sich QimiQ nicht wie herkömmliche Sahne überschlagen, ist hitze-, säure- und alkoholstabil und eröffnet dadurch Anwendungsmöglichkeiten, die für herkömmliche Milchprodukte bislang tabu waren. Es eignet sich sowohl für den Privatgebrauch, die Gastronomie als auch für industrielle Fertigprodukte und wird unter anderem für Terrinen, Dressings, Cremen, Puddings, Saucen, Füllungen, Farcen, Rührei, Brotteig, Buttermischungen und Torten verwendet.

Wie eine entsprechende Gestaltung die Zubereitung erleichtert, zeigt auch die bereits 1958 vom Japaner Momofuku Ando erfundene Instant-Nudelsuppe: eine runde Scheibe aus zusammengepressten, vorgekochten und frittierten Nudeln, die mit Hühneraroma versehen ist und nur noch mit heißem Wasser übergossen werden muss, um eine fertige Hühnersuppe zu zaubern. Später wurde das Design noch optimiert, indem in der Mitte des Nudellaibchens eine Kuhle eingeprägt wurde, in die das für dieses Gericht in Japan übliche Spiegelei treffsicher und gegen Verrutschen gesichert platziert werden konnte. Die erste Instant-Ramen der Geschichte ist in Japan bis heute erhältlich und beliebt – unter anderem als Knabberei direkt aus der Packung zu einem Glas Bier.

Spätes Dogma | Ein gegenläufiges Phänomen zur stetigen Anpassung von Nahrungsmitteln und ihrer Gestaltung an den jeweiligen Stand der Technik ist die bewusste Beibehaltung formaler und optischer Charakteristiken, die während des Herstellungsprozesses entstehen. Oft werden große Anstrengungen unternommen, um das gewohnte Aussehen einer Speise trotz erneuerter Produktionsverfahren nicht zu verändern. Ein Beispiel dafür ist die Wiener Kaisersemmel, die traditionellerweise „geschlagen" wird: Der Bäcker faltet dabei ein flaches, rundes Teigstück von fünfzig Gramm viermal über seinen Daumen zur Mitte hin ein, bevor er den letzten Zipfel in den Hohlraum des ersten steckt und die Semmel somit schließt. Im Zentrum bleiben Hohlräume, die das Gebäckstück innen luftig-flaumig und außen knusprig machen.

Another aid which will make us sure to succeed in the kitchen is QimiQ. The natural cream product invented by Hans Mandl and Rudolf Haindl in 1995 can replace artificial emulsifiers and stabilizers because it blends lastingly with all kinds of liquids such as alcohol or vinegar, and also forms emulsions with oil. In QimiQ, cream is already premixed with gelatin, so that for certain recipes, there is no need to to soak gelatin sheets in liquid and add them. Moreover, QimiQ cannot be overbeaten like conventional whipping cream, it is stable when exposed to heat, or mixed with acids or alcohol, and thus opens up new applications that were hitherto no-go zones for conventional milk products. Suited for households, catering operations and industrial finished products, it is used for preparing terrines, salad dressings, creams, blanc mangers, sauces, fillings, stuffings, scrambled eggs, bread dough, butter mixtures and cakes.

Another example of how design can make preparation easier can be found as early as in 1958. The instant noodle soup devised by the Japanese Momofuku Ando is a disk of pressed, precooked and fried noodles with chicken aroma. You only needed to add hot water to get chicken soup. Later on, the design of the noodle block was optimized – a little recess in the middle of the brick of noodles makes sure that the fried egg usually served with the soup can be placed precisely where it should be and will not slip sideways. The first instant ramen in history continues to be popular in Japan and can still be bought – it is also served uncooked, for snacking straight from the pack with a glass of beer.

Late Dogmatism | The deliberate adherence to formal and visual features due to the manufacturing process is a phenomenon contrary to the continuous adjustment of food products and their design to the state of the art in technology. Often enough, major efforts are made so as not to change the accustomed look of food in spite of new production technologies. The Kaiser roll, also known as a Vienna roll, is a good example. Traditionally, the dough has to be "folded". The baker folds four corners of a flat piece of dough towards the center over his thumb, finally closing the roll by placing the fifth corner in the hollow space underneath the first one. The hollow center makes sure that the dough will always be thoroughly baked.

Da der Formungsprozess der Handsemmel vergleichsweise kompliziert und zeitaufwändig ist, kam bereits in der Zwischenkriegszeit die Idee auf, entsprechende Maschinen zur Arbeitserleichterung zu konstruieren. Ziel war die kostengünstige Produktion einer Imitation der Handsemmel. Der Formungsprozess einer Maschinensemmel unterscheidet sich jedoch wesentlich von jenem der Handsemmel. Bei der maschinellen Erzeugung wird der charakteristische, fünfteilige Stern in einen kompakten Teigklumpen gedrückt. Dabei wird in das fast kugelförmige, nur leicht abgeflachte Teigstück von oben ein Werkzeug mit dem klingenden Namen „Kaisersemmelstempel" hineingedreht.

Nach Auskunft von Herrn Ing. Ferdinand Bodenstorfer, langjähriger Konstruktionsleiter des Brötchenanlagen-Erzeugers Werner & Pfleiderer, stammt die älteste Semmelstanzmaschine aus den Zwanzigerjahren und wurde von der Firma Komenda in Wien entwickelt. Die Maschine besaß zwei nebeneinander liegende Stempel, so dass die Eingabe der Teigstücke mit beiden Händen gleichzeitig erfolgen konnte. In den 1950er Jahren kam dann bei Ankerbrot eine deutlich effizientere Maschine der deutschen Firma Winkler zum Einsatz. Die so genannte „Rekord" war zweireihig und bereits mit dem bis heute üblichen Becherkettentransport ausgestattet.

Die erste vollautomatische Kaisersemmelstanzanlage konstruierte die Firma Werner & Pfleiderer in Wien-Ottakring im Jahr 1959. Diese „Semmelstraße" kam ab 1960 bei Ankerbrot zum Einsatz, wo sie bis Mitte der 70er Jahre ihren Dienst versah. Ab Mitte der 60er Jahre eroberten auch die ersten Erzeugungslinien für Klein- und Mittelbetriebe den Markt. Spätestens seit diesem Zeitpunkt verdrängte die perfekte Regelmäßigkeit der Maschinensemmel sukzessive das asymmetrische Erscheinungsbild der Handsemmel. Mit der Zeit akzeptierte der Konsument auch die veränderte Oberfläche der industriell erzeugten Semmel, die vor dem Stanzvorgang bemehlt werden muss, damit sich der Stempel wieder aus dem Teig ziehen lässt. Die Produktion läuft mittlerweile über vollautomatische Anlagen mit Computersteuerung, die unter anderem auf die Erzeugung halb gebackener und gefrosteter Teilstücke ausgerichtet sind.[14]

As the process of folding the roll is quite complicated and time-consuming, the idea of building a machine to facilitate the work came up in the inter-war years. The goal was to produce an inexpensive imitation of the Kaiser roll. However, the process whereby the machine-made roll is made differs significantly from the manual production method. The characteristic star-shaped pattern is stamped into a compact piece of dough. The dough is almost spherical and only slightly flattened on top. A tool aptly called "Kaiser roll stamp" is used to press it into shape.

According to Mr. Ferdinand Bodenstorfer, who looks back on many years of experience as the head of construction for Werner & Pfleiderer, a manufacturer of bread production lines, the oldest roll stamping machine dates back to the 1920's and was developed by the Komenda co. in Vienna. It was equipped with two stamps next to each other so that the operator was able to place dough pieces underneath them with both hands. In the 1950's Vienna's Ankerbrot co. started using a clearly more efficient machine made by the German Winkler co., the "Rekord" model with two lines and a bucket conveyor system that is still used.

The first fully automatic Kaiser roll stamping line was built by Werner & Pfleiderer in Vienna-Ottakring in 1959. From 1960 and up to the middle of the 1970's the "roll production line" was used by the Ankerbrot co., and in the mid-1960's production lines for small- and medium-sized enterprises were launched. From this moment, the perfectly regular look of the machine-made roll had come to replace the asymmetrical hand-made Kaiser roll. As time went by, consumers got used to the change in the surface of the industrially produced roll as it had to be sprinkled with flour to make sure that the stamp would not be stuck in the dough. Meanwhile production lines are all fully automatic, computer-controlled and geared to the manufacturing of half-baked and chilled pieces.[14]

The five screw thread shaped, thin metal spirals of the Kaiser roll stamp produce an effect similar in appearance to the manually folded look but since there is no folding process, the bakery product is less elastic, i.e. not as crispy as its hand-made counterpart. This is why machine-made rolls often taste soppy and doughy. The original idea of creating the crunchiest bakery product possible has given way to industrialization. The visual hallmark of the Vienna roll – the famous star-shaped pattern on top – was once due to the production process; it has been kept but was adapted to new manufacturing conditions.

Erste Instant-Ramensuppe der Geschichte: Die runde Scheibe aus vorgekochten und frittierten Nudeln mit Hühneraroma kam 1958 in Japan auf den Markt. Später wurde das Design um die Kuhle in der Mitte ergänzt, in die ein Spiegelei platziert werden kann.

First instant ramen soup in history: The round slice of precooked and fried noodles with chicken flavor was launched in Japan in 1958. Later the design was complemented by the hole in the middle where a fried egg can be added.

Die fünf schraubengangförmigen, dünnen Metallspiralen des Kaisersemmelstempels erzeugen einen der Handsemmel ähnlichen optischen Effekt, jedoch entfällt der Faltvorgang, wodurch das fertige Backwerk weniger Spannung – im Fachjargon spricht man von Rösche – besitzt. Das Innenleben von Maschinensemmeln schmeckt daher oft lasch und teigig. Die ursprüngliche Idee, ein möglichst knuspriges Gebäck zu erschaffen, wich der Industrialisierung. Das einst produktionstechnisch bedingte optische Markenzeichen der Semmel – der berühmte Stern – ist zwar geblieben, wurde jedoch den neuen herstellungstechnischen Gegebenheiten angepasst.

Theoretisch könnte man heute Herzen, Sterne oder auch eine Reihe anderer Motive in eine Maschinensemmel stanzen. In den meisten Entwicklungsabteilungen herrscht aber die Meinung vor, dass Konsumenten sehr konservativ sind. Deswegen wagt man sich lieber nicht über neue Formen, sondern ahmt das Aussehen der Handsemmel nach. Designs wie die Wiener Semmel, die aus einer Handwerkstechnik heraus entstanden waren und auch für die geschmackliche Qualität eines Produktes bürgten, werden im Zeitalter der Industrialisierung oft mehr recht als schlecht kopiert. Die Form bedient dann mehr die Erinnerung der Konsumenten als deren qualitative Ansprüche.

Identität durch Produktion | Ein weiteres Beispiel für die sentimentale Beibehaltung herstellungsbedingter Formen ist der Zuckerhut. Zucker in Kegelform, wie er heute gerne für Feuerzangenbowlen oder andere Partygerichte verwendet wird, ist eine Folge der ursprünglichen Art, Zucker zu gewinnen. Im sechsten Jahrhundert nach Christus entwickelte das persische Geschlecht der Sassaniden ein Verfahren, Zucker zu raffinieren, das über 1200 Jahre lang zur Herstellung von weißem Zucker verwendet wurde. Bei dieser so genannten Hutreinigungsmethode wird der Sirup in umgedrehte, kegelförmige Model gefüllt, das Restwasser aus der Melasse rinnt langsam durch eine kleine Öffnung an der Spitze ab. Der Zucker kristallisiert und nimmt dabei die Kegelform an. Anschließend mussten die steinharten, bis zu 50 Kilogramm schweren Zuckerhüte mühsam zerteilt und mit Zuckerzangen zu Brocken zerkleinert werden, welche dann in Mörsern zu Kristall- oder Staubzucker vermahlen werden konnten. Bis in die Mitte des 19. Jahrhunderts war der Zuckerhut die einzige handelsübliche Form von Zucker.

Als sich um 1900 der Zuckerwürfel in Europa immer mehr durchsetzte, läutete er das Ende der Ära des Zuckerhuts ein, dessen Erzeugung allerdings beispielsweise in Österreich erst 1938 endgültig eingestellt wurde. Heute wird der Zuckerhut aus nostalgischen Gründen meist in stark verkleinerter Form gerne wieder gekauft. Ähnlich wie bei der Herstellung von Zuckerwürfeln wird dabei Kristallzucker, der heute mit Hilfe von Zentrifugen direkt aus einer gesättigten Lösung gewonnen wird, in Kegelform gepresst. Mit der Gewinnung von Zucker haben diese Zuckerobjekte jedoch nichts mehr am Hut.[15]

Formal perfektes Kartoffelchip: Die zweifach gegensinnig gekrümmten, ovalen Pringles werden nicht aus ganzen Kartoffeln geschnitten, sondern aus Kartoffelteig geformt.

The perfect potato chip form: Pringles are oval and curved on both sides as they are not cut out of whole potatoes but formed out of potato dough.

Theoretically, we could stamp machine-made rolls with hearts, stars and all kinds of patterns. However, most R&D departments tend to believe that people are rather conservative in their buying decisions. This is why producers do not dare to try out new shapes and prefer imitating the appearance of the hand-made roll. Designs such as the Vienna roll, which we owe to a technique developed by artisans and which also guaranteed high quality in terms of taste, are often rather poorly imitated in the age of industrialization. The shape is then more of an appeal to consumers' memories than to their qualitative demands.

Identity by Production | The sugar loaf is one more example of sentimental reasons causing shapes to survive although this would no longer be required for the purposes of production technology. The sugar cone continues to be used for making brandy punch and other party drinks. Its shape is due to the original process of sugar making. In the sixth century CE, the Sassanid dynasty of Persia developed a process of sugar refining which was used to make white sugar for more than 1,200 years. The method involved conical molds turned upside down. Syrup was poured in and the superfluous water from the molasses slowly poured out again through a little hole at the bottom. As it crystallized, the sugar took on the shape of the mold. The rock-hard sugar loaves weighing up to 100 pounds had to be crushed up in a cumbersome process. Using sugar tongs, they were broken into smaller pieces, which were in turn ground into crystal or powdered sugar in mortars. Up to the mid-nineteenth century, the sugar loaf was the only shape in which sugar was traded.

When the lump sugar in the shape of cubes became ever more popular in Europe around 1900, it ushered in the end of the sugar-loaf era, although in Austria, production was discontinued as late as in 1938, and today, miniature sugar loaves are again sold for nostalgic reasons. However, today's sugar loaves are produced in a way similar to lump sugar – crystal sugar extracted from saturated solution using a centrifuge is pressed into the conical shape. These sugar loaves have nothing to do whatsoever with the ancient way of sugar making.[15]

Another example of using associations with a traditional production process can be found in "Pringles", an oval snack made of potato dough, with a shape reminiscent of potato chips. The typical curving in chips is due to the fact that thin potato slices will curl slightly in hot frying oil. "Pringles" are actually molded out of dough and do not consist of potato slices. They imitate the curved form of chips but come out of the can all identical in size and perfectly shaped. The product was first launched under the name "Pringle's Newfangled Potato Chips" by the US corporation Procter & Gamble in 1968. The "Pringles" part in the name was derived from a street in Cincinnati, the city where Procter & Gamble has its headquarters. "Newfangled" refers to the fact that the chip made from dough was a novelty. Until today, "Pringles" are still the only snack product of its kind with a double curvature.

"Pringles" are in fact made from rolled-out potato dough; first, egg-shaped flat pieces are cut out. These are then placed in a two-part mold made of steel which presses them into shape from the top and the bottom, and carries them through the frying process. In terms of process technology, the most difficult part is to ensure that the dough rises very little, does not form bubbles and stays in the exact curved shape. To achieve this, water content, ingredient mix and frying temperature must be carefully managed and adjusted.[16]

1 | Cf. Oxford Companion to Food, p. 284
2 | Source: Rainer Heilmann, Sacher Hotel, Vienna
3 | Source: Bahlsen co., Hannover
4 | Cf. Oxford Companion to Food, p. 320
5 | Source: Christoph Wagner, Fast schon Food
6 | Source: Hannsferdinand Döbler, Kochkünste und Tafelfreuden
7 | Guided tour through Nissin Foods, Japan, October 2008
8 | Interview with the American anthropologist Elizabeth Andoh, Tokyo, October 2008
9 | Source: Unilever co.
10 | Source: Interview with the American food scientist Kantha Shelke on 4 April 2009
11 | Dr. Ute Paul-Prössler: Käse – Entdecken und Genießen, p. 10, Sigloch Edition
12 | Source: Klaus Dürrschmid, Department of Food Sciences, University of Natural Resources and Applied Life Sciences Vienna
13 | Source: Barbara Taussig, McDonald's/Austria
14 | Sources: Josef Honeder, Honeder's bakery, Weitersfelden; Ferdinand Bodenstorfer, Philip Wagner, Werner & Pfleiderer Food Technology co.
15 | Sources: Martin Doppler, Agrana, A, Christoph Wagner, Süßes Gold
16 | Interview with Artemio Castro, Director Snacks, Procter & Gamble, Cincinnati, USA, on 3 April 2009

Transport
transport

Transport - Rucksack und Container
Neben zeitsparender und ressourcenschonender Fertigung entscheidet aber auch gute Transportfähigkeit über den Erfolg eines Produktes. Heutzutage transportieren Lastkraftwagen Nahrungsmittel quer durch Europa. Der Handel mit Lebensmitteln über weite Distanzen ist jedoch beileibe kein neues Phänomen. Schon die alten Griechen ließen sich beispielsweise Hartkäse aus Marseille liefern. Wenn Reeder heute argentinische Steaks nach Hamburg verschiffen oder Spediteure Tomaten von Anatolien nach Mitteleuropa karren, so tun sie nichts anderes als antike Schiffe, die das alte Rom mit Lebensmitteln aus Ägypten versorgten, oder mittelalterliche Karawanenführer, deren Kamele orientalische Spezialitäten nach Europa trugen. Nahrungsmittel sind Handelsgüter und werden daher seit dem Altertum über weite Strecken von Menschen oder Lasttieren getragen, verschifft, auf Karren oder in Containern verladen.

Natürlich wirkt sich der Wunsch, Essen von A nach B zu befördern, auf das Design der transportierten Ware aus. Welche Formen sich gut befördern lassen, hängt dabei natürlich vom jeweils gewählten Transportmittel ab. Was sich mit bloßen Händen gut tragen lässt, ist für die Fahrt im LKW eventuell ungeeignet. Umgekehrt lassen sich rechteckige Waren zwar stabil und platzsparend in einen Container stapeln, aber schlecht in den Händen halten.

Heute pressen und schneiden Industriebetriebe ihre Produkte meist in stapelbare, platzsparende Quader. Vor dem Zeitalter der Motorisierung erforderte gute Transportierbarkeit aber oft ganz andere Formen: runde Laibe etwa, die sich wie Fässer rollen lassen, oder gelochte Brote, die auf Schnüre oder Stöcke gefädelt werden können.

Transportation - Backpack and Container
Apart from manufacturing that saves time and uses resources efficiently, transportability also has a decisive impact on whether a product becomes a success story or not. Today, trucks take food products all across Europe but food trade over long distances is not a new phenomenon at all. The ancient Greeks had their hard cheese delivered from Marseilles. When ships bring steaks from Argentina to Hamburg or truck drivers take tomatoes from Anatolia to Central Europe, what they do is no different from what people in Antiquity did when they supplied Rome with food from Egypt, or medieval caravans with camels carrying oriental specialties to Europe. Food products are commodities for trading, and they have been transported across continents for ages, by humans or beasts of burden, by ship, by cart or in containers.

Of course, the wish to transport foods from A to B also has a bearing on the design of the goods to be transported. The shape that will be easy to transport will depend on the respective means of transportation. Things that can be carried in our hands might not be perfectly suited for dispatch on a truck. Conversely, rectangular products are stable and can efficiently be stacked in a container but are difficult to carry by hand.

Nowadays, industrial operations press and cut their products into stackable, space-saving rectangular prisms. Before the automotive age, transportability depended on other forms: round loaves which could be rolled like barrels, or bread with holes in the middle that could be hung on strings or sticks.

Links | Brot in Form einer Tragtasche mit Henkel, ein Entwurf des Pariser Industriedesigners Stéphane Bureaux.
Rechts | Nahrung wird transportiert: Bananenblätter und Pfeffer in Indien, vorbestelltes Mittagessen in Tokio

left | Bread designed like a tote bag with a handle – designed by the Parisian industrial designer Stéphane Bureaux.
right | Transporting food: banana leaves and pepper in India, pre-ordered lunch in Tokyo

Bacalhau und Pemmikan | Schon die keltischen Händler der Bronzezeit, die quer durch Europa reisten, waren auf möglichst nahrhaftes, leichtes, unverderbliches und einfach zu verzehrendes Essen angewiesen. Später bildete getrockneter Fisch, auch Stockfisch oder Klippfisch genannt und mittlerweile als „Bacalhau" eine internationale Spezialität, eine wichtige Grundlage für den Aufstieg der kleinen Nation Portugal zur internationalen Seemacht. Weite Seereisen erforderten sehr haltbare Nahrungsmittel, die vor allem auch unempfindlich gegen Spritzwasser waren. Freunden von Wildwestfilmen ist „Pemmikan" ein Begriff, eine Mischung aus gedörrtem, geröstetem und dann zerstoßenem Bisonfleisch, das im Verhältnis drei zu eins mit Talg und Knochenmarksfett verknetet wird. Die Rezeptur wurde von den Indianern Nordamerikas entwickelt und Pemmikan wiegt nur rund ein Fünftel des zur Herstellung verwendeten Frischfleischs. Es kann direkt verzehrt oder zu anderen Gerichten weiterverarbeitet werden. Wegen seines hohen Nährwerts und seiner Haltbarkeit wird Pemmikan mittlerweile sogar als Proviant für Expeditionen und das Militär industriell erzeugt.

Auch das Römische Reich hätte seine Macht ohne ein logistisch ausgeklügeltes Ernährungssystem nicht aufbauen und über Jahrhunderte hinweg halten können. Um den politischen Frieden zu bewahren, musste stets genügend „panis" für die römische Bevölkerung und der Getreidebrei „puls" für die Garnisonen zur Verfügung stehen. Das Getreide dazu wurde aus Ägypten importiert, das Brot in vorindustriellen Bäckereien in Ostia gebacken. Die Marschverpflegung der Legionäre bestand aus Zwieback, Speck, Hartkäse, Knoblauch und Fleisch, bis auf Letzteres alles gut haltbare und transportfähige Lebensmittel.

Bacalhau and Pemmican | As early as in the Bronze Age, the Celtic traders who traveled across Europe depended on durable food that was easy to carry and eat but had high nutritional value. Later on, dried fish, also known as stockfish and meanwhile an international specialty called "bacalhau", contributed greatly to the rise to power of small Portugal as a seafaring nation. Long voyages required highly durable food which was also insensitive to spray water. Friends of movies set in the Wild West will be familiar with "pemmican" – dried, roasted and ground buffalo meat, which is mixed with suet and bone marrow fat in a ratio of three to one and kneaded. The recipe developed by the native North Americans only weighs about one fifth of the fresh meat used to produce it. It can be eaten as it is or processed to prepare dishes. Due to its high nutritional value and long shelf life, today pemmican is even produced industrially as provision for expeditions and the military.

The Roman Empire would not have been able to expand and remain powerful over centuries without a sophisticated system of food logistics. For the sake of political peace, there always had to be enough "panis" (bread) for the population and porridge-like "puls" for the garrisons. The grain needed was imported from Egypt, bread was baked in pre-industrial bakeries in Ostia. Legionaries' provisions consisted of hard biscuits, bacon, hard cheese, garlic and meat, except for the last product all durable and easily transportable food.

Um größere Menschenmengen und Ballungsräume mit Nahrung zu versorgen, müssen Lebensmittel notgedrungen von den Produktionsstätten zu den Verbrauchern transportiert werden. Spätestens mit der Entwicklung der Städte vor rund 9000 Jahren begann die systematische Beförderung von Esswaren aus dem agrarischen Umland in die urbanen Ansiedlungen und machte eine transporttaugliche Gestaltung zu einer zentralen Anforderung an Nahrungsmittel.

Rund und gelocht | So lassen sich zum Beispiel 85 Kilogramm schwere Käselaibe aufgrund der runden Form relativ einfach wie Fässer rollen und sogar von einer einzelnen Person manipulieren. Käse zählt zu den ältesten Lebensmitteln überhaupt und sein Transport, das Käserollen, ist in manchen Gegenden sogar ein kultureller Akt. In den Ursprungsgebieten des Cheddar in der Grafschaft Somerset zum Beispiel rollten die Bauern früher einmal im Jahr die roten Käselaibe die Grundstücksgrenzen entlang, um auf diese Weise ihren Besitz zu markieren. Ungleich skurriler ist das „cheese rolling" – ein Wettbewerb, der alljährlich am Cooper´s Hill im englischen Gloucestershire ausgetragen wird. Auf Kommando wird dabei ein etwa 3,5 Kilogramm schwerer „Double Gloucester" einen stark abschüssigen Hang hinuntergerollt. Sämtliche Teilnehmer verfolgen laufend und stolpernd den Käselaib, derjenige, der den Käse als Erster erwischt, siegt und darf ihn als Gewinn behalten.

Auch große Brotlaibe wurden ursprünglich gerollt. Brot wurde aber auch auf andere Weise transporttauglich gemacht, indem man es beispielsweise zu Bageln, Brezeln oder Simits formte. Mit einer Schnur zusammengebunden oder auf Stöcken aufgefädelt, können ringförmige, gelochte Brote gut

Large groups of people and conurbations can only be supplied with food if it is transported from the production sites to the consumers. The emergence of the first towns about 9,000 years ago also marked the beginning of the systematic transportation of food from the agricultural environs to the urban settlements, making transportability a central requirement in the design of food.

Round and Ring-Shaped | 170-pound cheese loaves can be rolled like barrels and handled by an individual because of their round shape. In fact, cheese is one of the oldest foodstuffs and transporting it by rolling has even become a cultural event. In the county of Somerset, where cheddar cheese originates from, farmers used to roll the red loaves of cheese around the boundaries of their plots to mark them as their property. One of the more bizarre examples of a folk custom is the cheese rolling festival on Cooper's Hill in Gloucestershire, England. A roughly 8-pound loaf of Double Gloucester cheese is sent down a steep hill, followed by contestants who run or slide downhill in pursuit. The person first to reach the bottom or catching the cheese is the winner allowed to take the cheese home.

Big loaves of bread used to be rolled, too. Other ways of making bread easy to transport meant baking it in bagel, pretzel or simit shapes. Ring-shaped bread can simply be bound together with a piece of string for storage or transport from A to B. One person can carry 10, 20, 30 or pretzels or bagels on a string without

Links außen | Snackverkäufer am Bosporus in Istanbul
Unten | Entsprechend geformte Nahrungsmittel können den Transport wesentlich erleichtern. Runde Käselaibe beispielsweise lassen sich wie Fässer rollen.

left margin | Snack vendor on the Bosporus in Istanbul
bottom | Foods with the right shape are easier to transport. Round loaves of cheese can, for instance, be rolled like casks.

Links | Simit-Verkäufer in Istanbul; Mit einer Schnur zusammengebunden oder auf Stöcke gefädelt, können ringförmige Brote gut aufbewahrt und bequem von A nach B gebracht werden.
Rechts | Ähnlich wie bei Stapelstühlen oder stapelbarem Geschirr ist gute Stapelfähigkeit auch bei Nahrungsmitteln ein funktionaler Vorteil.

Left | Simit vendor in Istanbul; Bound together with a string or carried on poles, ring-shaped bread can easily be stored and transported from A to B.
Right | Similar to stackable chairs or stackable dishes, stackability is also a functional asset in food.

aufbewahrt und bequem von A nach B gebracht werden. Ohne weitere Behältnisse wie Säcke oder Tüten kann eine Person 10, 20, 30 oder mehr Brezeln oder Bagel auf Schnüre fädeln und in der Hand tragen. Ringbrote waren auch auf Schiffen lange Zeit eine gefragte Verpflegung: Am höchsten Punkt, also am Mast aufgehängt, war das Brot relativ gut vor Nässe geschützt. Ein modernes Gegenstück zu dieser Idee liefert der französische Designer Stéphane Bureaux mit einem Brotlaib in Form einer Handtasche mit Henkel. Über die Schulter gezogen, lässt er sich einfach und bequem nach Hause tragen, die Hände bleiben frei.

Brot über den Atlantik | Einige Jahrhunderte nach Bagel und Brezel wurde Brot erneut transporttechnisch optimiert, diesmal allerdings mit völlig anderem Ergebnis: Unter anderem weil die Lebensmittelproduktion auf den britischen Inseln von deutschen Bombern während des 2. Weltkriegs stark beeinträchtigt war, mussten die alliierten Soldaten überwiegend von den Vereinigten Staaten aus versorgt werden. Zu diesem Zweck benötigte man Esswaren, die sich möglichst effizient über den Atlantik transportieren ließen. Haltbarkeit war dabei genauso gefragt wie gute Stapelbarkeit, Anforderungen, welche speziell die Broterzeugung vor neue Herausforderungen stellten.

Binnen kürzester Zeit musste ein Brot entwickelt werden, das immer gleich schmeckte, sehr lang haltbar war, nicht austrocknete und möglichst Mann an Mann in die Frachträume der Schiffe geschlichtet werden konnte. Das größte Problem dabei war, dass Brot beim Backen aufgeht. Da sich die Hefe üblicherweise unregelmäßig im Teig verteilt, entstehen beim Backen unterschiedlich große Poren, die wiederum schwer kontrollierbare, unterschiedliche Backhöhen erzeugen.

Die militärischen Brotquader mussten aber immer genau gleich hoch sein, um perfekt gestapelt zu werden. Mit neuen, leistungsfähigeren Teigknetmaschinen gelang es schließlich, die Hefe stärker zu zerkleinern und besser im Teig zu verteilen. Die Zugabe von Stärke und verschiedenen Proteinen verhinderte außerdem, dass die fertigen Brotklötze während

requiring an extra bag. Ring-shaped bread was also popular on board ships: hung on top of the mast, the ship's highest point, it was relatively safe from humidity. The French designer Stéphane Bureaux came up with a modern version: a loaf of bread in the shape of a handbag with handles. Slung over the shoulder, it can easily be carried home hands-free.

Bread across the Atlantic | A few centuries after the invention of bagels and pretzels, bread was again optimized for easier transport, though this time, the result was entirely different. As the food industry in the British Isles had been impaired by German bombs, the allied soldiers mainly procured their supplies from the United States during World War II. For this purpose, food that could be transported across the Atlantic as efficiently as possible was needed, and durability was just as crucial as stackability. This meant a new challenge, in particular for bread production.

Within a very short time, a bread had to be developed which was supposed to taste the same, no matter which batch is was from, was durable, did not go stale and its shape was to be such that it could be closely stacked in the ships' holds. The biggest problem was that bread dough rises during the baking process. As yeast is usually distributed unevenly in the dough, pores can be of different size, leading to different bread sizes hard to control.

The military brick-shaped bread all had to be of the same height for perfect stacking. Eventually, the new high-performance kneading machines reduced the yeast to small particles that spread more in the dough. Added starch and proteins kept the ready-made bread from going stale during the passage. The result, extremely soft white bread with small pores, became the basis of today's pre-cut toast considered to be standard by people in the USA and the

177

der Überfahrt austrockneten. Das Ergebnis, ein extrem weiches, feinporiges Weißbrot, bildete den Grundstein für jenes vorgeschnittene Toastbrot, welches heute in den USA und in Großbritannien Standard ist und von vielen Europäern als zu soft und schwammig empfunden wird. Ron de Santis vom Culinary Institute of America erinnert sich, dass auch den heimkehrenden, amerikanischen Soldaten das neue, weiche und krustenlose Industriebrot anfangs überhaupt nicht schmeckte. Doch das Rad der Zeit ließ sich anscheinend nicht mehr zurückdrehen.[1]

Genmanipulation gegen Quetschen | Von der Herstellung bis zum Teller passieren Lebensmittel viele Stationen. Sie fahren mit Gabelstaplern und Lastkraftwägen, werden in Regale und Tiefkühltruhen geschlichtet und wollen auch in Einkaufswägen, Taschen und Kühlschränken platzsparend verstaut werden. Lebensmittel müssen transportabel sein.

Die Anforderungen an die Beweglichkeit von Nahrung sind mittlerweile enorm und gipfelten 1994 im ersten genmanipulierten Produkt am Lebensmittelmarkt. Die kalifornische Biotechnologiefirma Calgene hatte ein Verfahren entwickelt, bei dem die Erbmasse von Tomaten so verändert wurde, dass die reife Frucht weniger leicht quetscht. Die so genannte „Flavr Savr Tomato" ist widerstandsfähiger gegen Druck und Stöße, zerplatzt in der Kiste nicht so schnell und ist damit transportfähiger als herkömmliche Tomaten.

Nahrung DIN-genormt | Waren Transportbehelfe früher Körbe, Fässer oder Schnüre, so sind die meisten Behältnisse vom Container über Kartons bis zum Tetrapack heute rechteckig. Im industrialisierten Alltag der westlichen Welt ist die so genannte „Europoolpalette", die umgangssprachlich als „Europalette" bezeichnet wird, das Maß aller Dinge, wenn es um eine wirtschaftliche Beförderung geht. Die nach DIN EN 13698 Teil 1 genormte, 1200/ 800/ 144 Millimeter

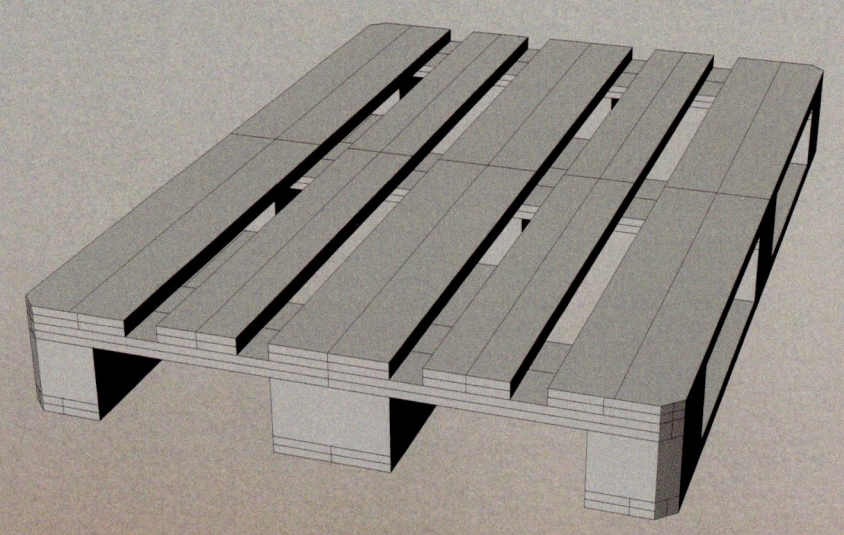

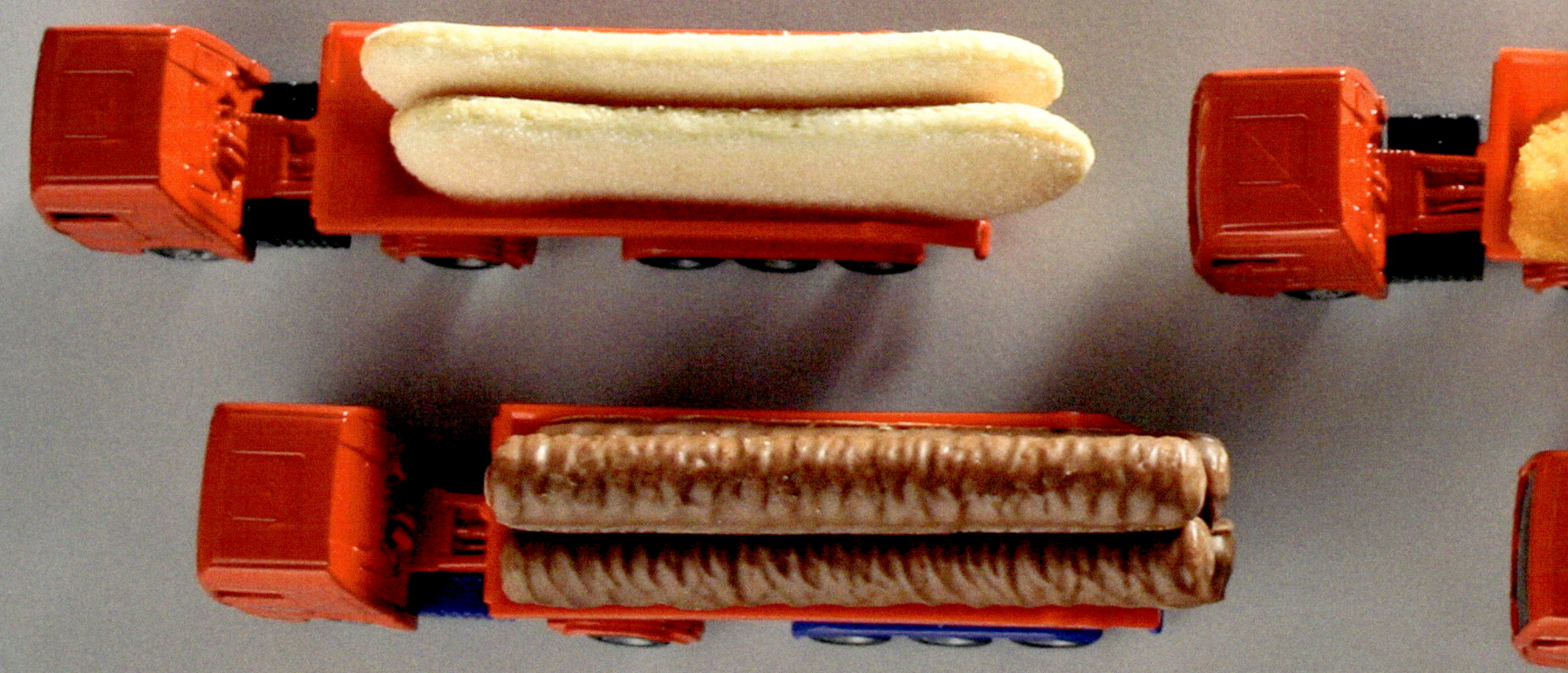

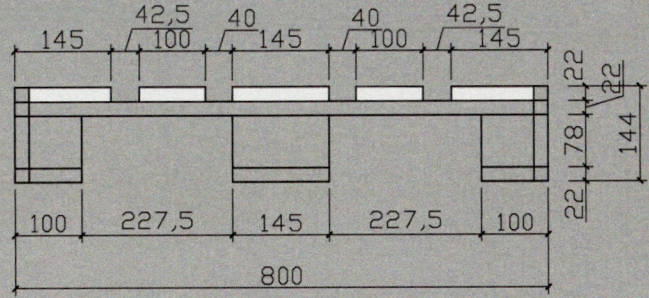

Wenn es um gute Transportfähigkeit geht, ist heutzutage die Palette das Maß aller Dinge. Vor dem Zeitalter der Motorisierung erforderte gute Transportierbarkeit oft ganz andere Formen: 100kg schwere Käselaibe lassen sich auf Grund der runden Form relativ einfach rollen und 10, 20 oder 30 Brezeln können, auf eine Schnur gefädelt, leicht von einer Person getragen werden.

When it comes to good transportability, the pallet is the measure of all things today. Prior to the age of motorization good transportability often called for completely different shapes. Loaves of cheese weighing 100 kilos were relatively easy to roll thanks to their round shape and 10, 20 or 30 pretzels could be easily carried when they hung on a string.

UK, and found too soft and spongy by many Europeans. Ron de Santis of the Culinary Institute of America remembers that even the returning American soldiers did not like the new, soft, crustless industrially made bread in the beginning. However, it seems there was no turning back the hands of time.[1]

Genetic Manipulation for Anti-Mush Food | Foodstuffs pass through many stages on the way from the production site to the table. They ride on forklifts and trucks, are stowed away in shelves and deep freezers, and want to be stacked efficiently in shopping carts, bags and refrigerators.

The demands for improved mobility of food are enormous; in 1994 a culmination points was reached when the first genetically engineered food was brought to the market. The Californian bio-tech company Calgene had developed a process changing the genetic code of tomatoes in such a way that the ripe vegetable did not turn into mush or break easily when squeezed or jostled in a box. This is why the so-called "Flavr Savr" tomato could be transported more easily than other conventional tomato varieties.

Food According to DIN Standards | While transport aids of the past used to include baskets, barrels or strings, most receptacles, from big containers to cardboard boxes and cartons, are rectangular today. The measure of all things in the industrialized everyday life of the western world is best described by the word "euro pallet" when economical transportation is needed. The wooden structure standardized by DIN EN 13698 Part 1 measures 1200/800/ 144 millimeters (42x39x5.6"). It is also used for transporting numerous food products. Therefore, the industry squeezes and cuts many of its products into pal-

Aus rein pragmatischen Gründen werden unregelmäßige Naturprodukte wie Fischfilets zu standardisierten Quadern verarbeitet. Fischstäbchen lassen sich gut industriell herstellen, sind einfach zu transportieren, passen genau in den Mund, und weil sie nicht aussehen wie Fisch, beißen auch Fischmuffel gerne hinein.

For purely pragmatic reasons irregular natural products such as fish fillets are processed into standardized squares. Fish sticks can be easily manufactured industrially and are easy to transport. They fit perfectly in one's mouth and since they do not look like fish, even folks who normally detest fish like them.

große Holzkonstruktion dient unter anderem dem Transport unzähliger Lebensmittel. Daher pressen und schneiden Industriebetriebe viele ihrer Produkte in palettentaugliche, stapelfähige und lagerfreundliche Quader. Ob Tiefkühlgemüse oder Teebutter, Spinat oder Rotkraut, Schokoladetafeln oder Suppenwürfel: Die Bewohner der Industrienationen kaufen rechteckige Nahrung in rechteckigen Packungen.

So orientieren sich beispielsweise die Proportionen des vier Gramm schweren Zuckerwürfels am Schlichtmuster in der Packung, das bei vorgegebenem Gewicht brauchbare Abmessungen ergeben muss. Ein Kilogramm Kristallzucker, also 250 Quader müssen so in eine Schachtel eingeordnet werden, dass alle Lagen vollständig gefüllt sind, kein Luftraum übrig bleibt und letztlich die Packung selbst effizient und bequem auf Europaletten transportiert werden kann.[2]

Transportable Quader | Aussehen und Größe der meisten Waren werden heutzutage von zylindrischen Dosen und rechteckigen Packungen definiert. Selbst Grundnahrungsmittel wie Brot, Milch oder Fisch kommen in Quaderform in die Verkaufsregale. Am augenscheinlichsten ist die formale Adaptation des Grundproduktes vielleicht bei Fischstäbchen. Bereits auf den Fangschiffen werden die Fische ausgenommen, filetiert und zu sortenreinen Blöcken gefroren, deren Abmessungen genormt sind. Die standardisierten Fischblöcke mit festgelegtem Höchstgrätenanteil sind eine internationale Handelsware, welche die einzelnen Fischstäbchenhersteller auf dem Weltmarkt einkaufen. In der Fabrik werden die tiefgefrorenen Fischblöcke dann mit Bandsägen zu länglichen Stäbchen zersägt, paniert und in Becken schwimmend frittiert. Um die Kühlkette nicht zu unterbrechen, bleibt der Fisch während des Backvorgangs tiefgefroren. Temperatur und Frittierdauer sind so gewählt, dass die Panier zwar bräunt, der Fisch im Inneren aber nicht auftaut. Nach einer kurzen Phase der Abkühlung werden die essfertigen Fischstäbchen schließlich in quaderförmige Kartonschachteln verpackt. Die Natur hat eine Vielfalt von unterschiedlich geformten Fischarten geschaffen, der Mensch verarbeitet sie zu standardisierten, transportablen Blöcken. Interessanterweise variieren allerdings die Abmessungen der Fischstäbchen von Hersteller zu Hersteller.[3]

Industriedesigner entwerfen stapelbare Stühle, Kaffeetassen und Biergläser, Food Designer stapelbare Nahrung. Innovativ schafft die Gestaltung der Schweizer „Toblerone" den Spagat zwischen unverwechselbarer Form und brauchbarem Schlichtmuster: Das markante Prisma, dessen Schnittfläche ein gleichseitiges Dreieck ist, lässt sich trotz seiner auffälligen Dreiecksform stabil und effizient stapeln.

Die formale Vereinheitlichung der Verpackung betrifft auch natürliche Produkte wie Obst und Gemüse. So müssen zum Beispiel Ananas enorme Veränderungen über sich ergehen lassen, um in eine Dose zu passen. Obwohl die Frucht keine zylindrische Form aufweist, lagern in der Blechkonserve exakt gleich große, kreisrunde Fruchtscheiben. In ähnlicher Weise träumen Hersteller vermutlich auch von geraden Bananen, eckigen Eiern oder Pfirsichen. Allerdings gingen mit solchen Manipulationen vielleicht auch die Mundgerechtigkeit und die statischen Fähigkeiten mancher Nahrungsmittel verloren. Rundes, Kugelförmiges oder Zylindrisches passt nicht nur perfekt in unseren Mund, sondern leitet auch die auftreffenden Druckkräfte bei Transport und Verzehr wesentlich besser ab als Eckiges.

letable, stackable and easy-to-store rectangular prisms. Deep frozen vegetables or butter, spinach or red cabbage, chocolate bars or stock cubes: the residents of the industrialized world buy rectangular food in rectangular packages.

The portion size of lumps of sugar (four grams) are e.g. in line with the stacking pattern in the box, and the dimensions are simply those that make sense at the given weight. A total of 250 lumps of crystal sugar, weighing one kilogram, or roughly two pounds, have to be put in the box so that there is no unfilled space left and that the packages can be placed and transported efficiently on europallets.[2]

Transportable Bricks | Today the appearance and size of most goods are defined by cylindrical cans and brick-shaped packages. We find that even basic staples such as bread, milk or fish have taken on these shapes when they reach our supermarket shelves and freezers. The most obvious adaptation of a basic product to the needs of transport and stacking is probably the fish finger. Fish are gilled, filleted and frozen in single-variety blocks with standardized dimensions abroad the fishing vessels already. These standardized fish blocks with their defined maximum fishbone content are a commodity traded internationally. The fish finger producers buy them on the world market. In the factory, the deep-frozen fish blocks are then sawed into oblong blocks, breaded and fried in huge vats. The fish has to remain deep-frozen during frying so that the cooling chain is not interrupted. Temperature and frying time are selected in such a way that the breading turns brown but the fish does not thaw. After a brief period of cooling down, the ready-to-cook fish fingers are packed in rectangular boxes. Nature has created a vast diversity of fishes in different shapes and seizes, and we humans process them until they are standardized transportable little blocks, though, interestingly enough, the dimensions of fish fingers vary from producer to producer.[3]

Industrial designers come up with stackable chairs, coffee cups and beer glasses. Food designer create stackable food. One example of innovative design is the Swiss "Toblerone" chocolate, straddling the fence between unmistakable shape and useful stacking features: The distinctive prism with an equilateral triangle for a cross-section can be stacked in a stable and efficient way in spite of its unusual triangular form.

Formal standardization also applies to other natural products such as fruit and vegetables. For example, pineapples have to undergo enormous changes to fit into a can. Although the fruit itself is not cylindrical, cans contain circular slices of exactly the same size. In much the same vein, producers most likely dream of straight bananas, or square eggs and peaches. Manipulating food like that might, however, also have a negative impact on the structural properties and bite-size qualities. Round, spherical or cylindrical foods fit perfectly in our mouths, and they also distribute pressure more evenly during transport and consumption than angular shapes.

Gut geschützt auf lange Wege | Transportfähigkeit ist nicht nur eine Frage der Geometrie, sondern auch der Verpackung bzw. des Schutzes vor äußeren Einflüssen. Die dichte und stabile Teighülle der englischen „pies" zum Beispiel hält klein geschnittenes Fleisch zusammen, schützt es vor dem Austrocknen und macht es damit transportfähig. „Pies" wie der traditionelle „Melton Mowbray pork pie" aus der Grafschaft Leicestershire wurden üblicherweise in ländlichen Gegenden hergestellt und dann als Fastfood in die Städte geliefert. Noch heute produzieren Bäcker und Fleischhauer diese „Fleischeintöpfe" in der essbaren Verpackung und liefern sie aus. Gefüllte Pasta wie Ravioli oder andere Teigtaschen wie die japanischen Gyoza funktionieren nach einem ähnlichen Prinzip. Klein geschnittene, fragile, leicht verderbliche oder zu flüssige Nahrungsmittel werden in essbare Hüllen gewickelt und damit einerseits konserviert und andererseits transport- und verzehrtauglich gemacht.

Im südwestenglischen Cornwall entstanden für die Arbeiter der Zinnminen die so genannten „Cornish Pasties". Die Bergleute mussten ihr Mittagessen unter Tage zu sich nehmen. Dafür füllten ihre Frauen Fleisch (wenn vorhanden), Zwiebel und Kohlrüben in ein zusammengeschlagenes Teigstück aus Mehl, Fett und Wasser. Die Ränder wurden dicht verschlossen und die „Pasty" im Ofen sehr cross gebacken. Die großen Teigtaschen hielten den klein gehackten Inhalt perfekt in Form und waren bequem mit einer Hand zu essen. Der Legende nach hatten die Bergleute durch die staubige Arbeit so schmutzige Hände, dass sie ihre „Cornish Pasties" nur an den eingekerbten Rändern hielten und diese anschließend wegwarfen. Das „Minensandwich" hatte noch einen weiteren Vorteil: Die dichte Hülle hielt den Inhalt bis zu acht Stunden lang auf Temperatur und verschaffte den Arbeitern so eine warme Mittagsmahlzeit. Besonders wohlmeinende Hausfrauen befüllten die halbrunden Mittagskuchen sogar mit verschiedenen Gerichten. Eine Hälfte hatte einen salzigen Inhalt, während die andere Seite süß war. Somit gab es ein zweigängiges Menü in einem einzigen, essbaren Objekt.[4]

Schokoladener Schutzpanzer – die Sachertorte | Auch die 1832 von Franz Sacher in Wien erfundene Sachertorte verdankt ihren weltweiten Erfolg ihrer Transportfähigkeit. Unter den zahlreichen Cremen, Füllungen und Verzierungen anderer Torten sticht der simple Schokoüberzug der Sachertorte durch sein vergleichsweise unauffälliges Aussehen hervor. Eine dünne Schicht Marillenmarmelade ersetzt die übliche Cremefülle, als einziger Schmuck dient die glatte, dunkelbraune Glasur.

Ausgerechnet im so barock verspielten Österreich erlangte damit ein Produkt weltweit Bekanntheit, dessen Form puristischer nicht sein könnte. Als der erst 16-jährige Konditor Franz Sacher (1816–1907) in seinem 2. Lehrjahr am Hof des Fürsten Metternich (1773–1859) die berühmte Schokoladentorte erfand, muss das neue Dessert vor allem durch seine minimalistische Gestaltung fernab aller manieristischer Schnörkel aufgefallen sein. Angeblich kam der Küchenjunge Sacher eher zufällig zu seiner Kreation, als sich eines Abends unerwarteter Besuch im Palais Metternich am Wiener Rennweg ankündigte und der Chef-Patissier krank im Bett lag. Eher notgedrungen

Well Protected on Long Journeys | Transportability is not only a matter of geometry but also of packaging or protection from outside influences. The impermeable and stable crust of English "pies" holds meat cut up into small pieces together, keeps it from going dry and thus makes it transportable. Pies such as the traditional "Melton Mowbray pork pie" from Leicestershire were formerly made in rural areas and taken to the cities as fast food of days gone by. Even today, bakers and butchers still produce and deliver these "meat stews in edible packages". Stuffed pasta, such as ravioli or dumplings of the Japanese gyoza kind, works along similar lines. Minced, fragile, perishable or rather liquid foodstuffs are wrapped in edible "casings" and thus preserved or made better suited for transportation or consumption.

Cornwall in the southwest of England is the home of "Cornish pasties", which were originally developed for the mine workers. The miners had to eat their lunch underground. Their wives filled rolled-out dough made of flour, fat and water with meat (if available), onions and turnips. The pastry was folded and closed tightly on the edges, and the resulting "pasty" was baked until crisp. The dough kept the minced ingredients in perfect shape and could easily be held in one hand when the miners ate them. Legend has it that the miners' hands were so dirty that they only held their "Cornish pasties" by the serrated edges, which they threw away after eating. The "miners' sandwich" came with another benefit: the impermeable crust kept the stuffing warm for up to eight hours so it provided the miners with a hot meal. Especially well-meaning housewives even filled the same half-round pie with different stuffings, e.g. a savory and a sweet one, so they constituted a two-course meal in one crust.[4]

A Chocolate Shell – The Sacher Cake | The Sacher cake, invented by Franz Sacher in Vienna in 1832, also owes its worldwide success to its transportability. In comparison with all the creams, toppings, fillings and decorations there are in the world of cakes, the Sacher cake stands out for its simple chocolate icing and rather low-key look. A thin layer of apricot jam instead of cream filling, and smooth dark-brown chocolate icing – that's all you get.

Ironically, one of the internationally best-known products from Austria, a country notorious for its baroque predilection for adornments, is thus the epitome of purism and simplicity. When pastry cook Franz Sacher (1816–1907) invented the famous chocolate cake during his second year of apprenticeship in the household of Prince Metternich (1773–1859), the new dessert must have attracted attention due to its minimalist look, a far cry from mannerist decoration. It is said that the cake was more of a coincidental discovery when unexpected visitors arrived at Prince Metternich's stately home on Vienna's Rennweg

experimentierte der junge Franz mit Teig und Schokolade, um schließlich vor dem Fürsten zu präsentieren, was später unter dem Titel Sachertorte in die Geschichte einging. Gerade das schlichte Design sicherte jedoch aufgrund seiner funktionalen Vorteile den globalen Erfolg des Schokoladekuchens.

Transportfähigkeit und Haltbarkeit sind neben dem oft als gar nicht so gut beschriebenen Geschmack die eigentlichen Erfolgsfaktoren der berühmten, österreichischen Mehlspeise. Franz Sacher schuf eine Torte, die alle Attribute aufweist, die für transportable Nahrung nötig sind. Ihr Geheimnis liegt in der wasserdichten Glasur, die besonders widerstandsfähig ist und das weiche Innere vor dem Verderben bewahrt. Ähnlich wie die Brotrinde oder die Drageekruste diverser Süßigkeiten schützt die konservierende Schokoladehülle das Innere der Torte vor Austrocknung, wodurch die Sacher vergleichsweise lange – offiziell 21 Tage – frisch und flaumig bleibt. Während die meisten anderen Torten üppige Dekorationen tragen oder leicht verderbliche Zutaten wie Cremen beinhalten, verschwindet bei den 270.000 Sachertorten, die jährlich das gleichnamige Hotel in Wien verlassen, der Teig unter der schützenden Schokoladehaut. Aufwändig dekorierte oder gefüllte Desserts sind meist schon nach wenigen Tagen ungenießbar, die „Sacher" hingegen bleibt tage-, ja wochenlang saftig.

Die besonders widerstandsfähige Glasur schützt das weiche Innere aber nicht nur vor Austrocknung, sondern auch vor Verformung. Ähnlich wie die Karosserie eines Autos bewahrt die stabile Schokohaut vor den Auswirkungen von Erschütterungen oder anderer unliebsamer Behandlung. Um zu jeder Jahreszeit größtmögliche Elastizität zu erzielen, variieren daher auch die Zutaten und das Mischverhältnis des streng gehüteten Rezepts der aus vier verschiedenen Schokoladesorten bestehenden Glasur je nach Monat und Luftfeuchtigkeit. Die harte und dennoch nicht spröde Hülle ermöglicht damit den schadlosen Transport der Fülle und garantiert, dass der Kuchen in Tokio, Los Angeles oder sonst wo auf der Welt unversehrt ankommt.

Vermutlich war Sacher gar nicht der erste Patissier, der eine derartige Schokoladetorte herstellte. Vorgänger der klassischen Sachertorte finden sich, wie Christoph Wagner in seinem „Lexikon der Wiener Küche" schreibt, bereits rund 100 Jahre vor der offiziellen Erfindung, etwa 1718 im „Saltzburger Kochbuch" von Conrad Hagger oder 1749 im „Wienerischen bewährten Kochbuch" von Gartler-Hickmanns. Dass gerade die Rezeptur von Franz Sacher zu internationalem Ruhm gelangte, dürfte wohl an einem überaus populären Kochbuch von Katharina Prato liegen. 1858 veröffentlichte sie in ihrer „Süddeutschen Küche auf ihrem gegenwärtigen Standpunkte" eine „Chocolade-Torte à la Sacher", die jene Sachertorte, die bis heute im gleichnamigen Hotel in Wien in exakt 36 Arbeitsschritten handgefertigt wird, bekannt machte.[5]

street when the chief pastry cook was laid up sick. The cake was born out of necessity when the scullion started experimenting with dough and chocolate, eventually presenting to the Prince the cake later to become famous under Sacher's name. It was precisely the simplicity of design and its functional benefits that made the chocolate cake an international success story.

Apart from the taste (described by some as "not that good at all"), transportability and durability are the secrets to the Austrian cake's success. Franz Sacher created a dessert with all the attributes of transportable food. The icing is impervious to water and especially resilient, and therefore keeps the soft interior safe from spoiling. Similar to bread crusts and sugar coatings around sweets, the preserving chocolate icing keeps the inside of the cake from going dry. This is why Sacher cakes stay fresh, light and airy for a fairly long time – 21 days, officially. Whereas most other cakes are lavishly decorated or contain perishable ingredients such as cream fillings, the dough of the 270,000 Sacher cakes leaving the eponymous hotel in Vienna every year vanishes under the protective chocolate skin. While desserts with sophisticated decorations and fillings usually go bad within a few days, the "Sacher" cake remains juicy for days, or even weeks..

The particularly resilient icing not only protects the soft interior from going dry and stale, it also keeps it from deformation. Similar to the body of a car, the stable chocolate cover guards it from concussions and other types of unwanted handling. To ensure the best degree of elasticity for all seasons, the ingredients and mixture ratios in the secret icing recipe consisting of four different types of chocolate vary according to the month of the year and atmospheric humidity. The icing is hard but not brittle and helps the interior of the cake survive transport without damage. It guarantees that the cake makes it to Tokyo, Los Angeles or elsewhere in the world unharmed.

Presumably, Sacher was not the first pastry cook to make a chocolate cake of this kind. As Christoph Wagner writes in his "Lexikon der Wiener Küche", a veritable encyclopedia of Viennese cooking, precursors of the Sacher cake were mentioned in cookery books such as the "Saltzburger Kochbuch" by Conrad Hagger around 1718 or the "Wienerisches bewährtes Kochbuch" by Gartler-Hickmanns in 1749, about 100 years before the official invention. The international fame of Franz Sacher's recipe is most likely due to an enormously popular cookery book published by Katharina Prato in 1858. In "Die Süddeutsche Küche auf ihrem gegenwärtigen Standpunkte" (On the state of the culinary art in southern Germany), she wrote about a "Chocolade-Torte à la Sacher", thus making the Sacher cake – which keeps being hand-made in the hotel of the same name in Vienna in exactly 36 work stages until this very day – known to the public at large.[5]

Die Sachertorte wurde 1832 von Franz Sacher in Wien offiziell erfunden. Ihre weltweite Berühmtheit verdankt sie weniger dem Geschmack als der Transporttauglichkeit. Während Cremetorten schnell verderben, schützt die dicke Schokoglasur das weiche Innere vor Verformung und Austrocknung und garantiert, dass der Kuchen in Tokio, Los Angeles oder Buenos Aires unversehrt ankommt.

The Sachertorte was officially invented by Franz Sacher in Vienna in 1832. It owes its worldwide fame less to taste than to the fact that it can be transported. While cream cakes spoil quickly, the black chocolate icing protects the soft mass on the inside from deforming and drying, thus ensuring that the cake will arrive undamaged in Tokyo, Los Angeles or Buenos Aires.

Studien der Universität Loughborough, GB, haben gezeigt, dass Kekse in der Packung nicht nur zerbröseln, wenn sie während des Transportes zu grob behandelt werden, sondern auch, wenn sie nach dem Backen einer zu hohen Luftfeuchtigkeit ausgesetzt waren. Dann nämlich zerstören sie sich auf Grund unterschiedlichen Schwund- und Quellverhaltens quasi von selbst.

Studies carried out by the University of Loughborough, UK, have shown that cookies in a package do not just crumble when they are treated roughly during transport but also when they are exposed to overly high humidity after baking. They practically self-destruct as a result of the different shrinking and swelling reactions.

Das Zerbrösel-Problem | Beim Transport werden Lebensmittel oft zerdrückt, gequetscht, zerbrochen oder sonst wie verformt. Besonders bei Knabbergebäck und Keksen sind daher Designs gefragt, die möglichst widerstandsfähig und langlebig sind. Chips zum Beispiel sollen dünn und knusprig sein und in der Packung dennoch nicht zerbröseln. Bei der Entwicklung der geschwungenen Form der „Pringles" wurden daher zunächst eigene Stabilitätssimulationen durchgeführt, Prototypen gebacken und testweise zertrümmert, bevor man sich für die endgültige Formgebung entschied. Ausschlaggebend für den Kräfteverlauf und damit für die Statik des Crackers ist die Art der Krümmung. Für eine möglichst große Stabilität muss der Kräftefluss bei Erschütterungen gleichmäßig über den ganzen Keks verteilt sein, was bei manchen Formen besser funktioniert als bei anderen. Außerdem stützen die Kekse einander in der Packung gegenseitig, wenn sie eng und mit vielen Kontaktpunkten aufeinander gestapelt werden. Sind die einzelnen Cracker dagegen zu stark gekrümmt, liegen sie in der Packung nicht vollflächig aneinander, sondern berühren einander nur an einigen wenigen Stellen und brechen daher wesentlich leichter. Der Vorteil weniger stark gewölbter Cracker ist auch, dass sie sich enger schlichten lassen. Gleich viele Pringles brauchen dann im Supermarkt weniger Regalfläche und werden daher leichter gelistet bzw. können aufgrund der im Vergleich zu herkömmlichen Chips extrem dichten Schlichtung auf derselben Fläche mehr unterschiedliche Sorten angeboten werden.⁶

Mit dem Krümelverhalten von Nahrungsmitteln, im konkreten Fall von Keksen, beschäftigt sich auch eine Studie, welche ein Forschungsteam um Ricky Wildman an der Universität Loughborough in Großbritannien durchgeführt hat. Die Maschinenbauer haben jene Kräfte erforscht, die in Keksen wirken und die Ware zum Leidwesen der Produzenten oftmals frühzeitig in der Packung zerbröseln lassen. Die Schuld daran trägt überraschenderweise nicht etwa eine zu grobe Behandlung während des Transports, sondern der Wassergehalt in den Keksen und die Luftfeuchtigkeit: Wenn die Kekse aus dem Backofen kommen und abkühlen, führt der Eintritt von Feuchtigkeit zu unterschiedlichem Schwund- und Quellverhalten innerhalb der Kekse. Die entgegengesetzt gerichteten Kräfte bewirken, dass sich der Keks quasi aus seinem Inneren heraus von selbst zerstört und zerbricht. Eine Möglichkeit, dieses Phänomen zu vermeiden, ist, die Luftfeuchtigkeit nach dem Backen genau zu kontrollieren und sicherzustellen, dass die Feuchtigkeit im Keks während des Backvorganges vollständig entweicht. Erreichen lässt sich das unter anderem durch eine höhere Backtemperatur, längeres Backen oder die nachträgliche Behandlung mit Mikrowellen.⁷

1 | Informationen: Interview mit der amerikanischen Lebensmitteltechnologin Kantha Shelke am 4.4.2009 & Interview mit Ron de Santis am Culinary Institute of America am 27.3.2009
2 | Interview Martin Doppler, Agrana, Tulln
3 | Betriebsbesichtigung Frozen Fish, Bremerhaven, am 5.2.2007
4 | Interview mit dem britischen Autor Paul Levy am 15.1.2009
5 | Annette Epp, Gerichte und ihre Geschichte, Christoph Wagner, Das Lexikon der Wiener Küche, Direktor Reiner Heilmann, Hotel Sacher, Sacher-Buch, 125 Jahre Hotel Sacher Wien, Gelebte Tradition
6 | Interview mit Artemio Castro, Director Snacks, Procter & Gamble, Cincinnati, USA, am 3.4.2009
7 | Interview mit Dr. Ricky Wildman, Loughborough, GB, am 20.1.2009

The Issue of Crumbling | When foods are transported, they will often be crushed, squashed, broken or deformed in other ways. This is why salty snacks and cookies should be designed in such a way that they are as resilient and firm as possible. For instance, chips should be thin and crunchy without becoming reduced to crumbles in the package. When the curved shape of "Pringles" was developed, special stability simulations were done, and prototypes were fried and crushed in tests before the decision in favor of the final form was made. Curvature is the key when we look at the flow of forces and structural mechanics of a snack. To ensure the highest degree of stability, the flow of forces caused by vibrations has to be distributed evenly across the whole surface, which in some shapes works better than in others. Moreover, snacks support each other in the package if they are stacked closely and with many points of contact. If individual pieces are too strongly curved, they will touch in very few places only instead of having large contact surfaces, and this will cause them to break easily. Another benefit of less strongly curved snacks is that they can be stacked more closely inside the package. The same quantity of "Pringles" will need less shelf space in a supermarket. This way, it will be easier to get on the assortment list, and it will be possible to offer more flavors on the same surface area because "Pringles" are more closely stacked inside their cans than conventional chips.[6]

Experiments conducted by the research team of Ricky Wildman at Loughborough University in the UK dealt with the crumbling behavior of cookies. The mechanical engineers explored the forces at work in cookies that often make the goods crumble while they are still in the packaging, much to the dismay of producers. Surprisingly enough, it is not roughshod treatment during transportation that initiates their fracture, but the interplay between the water in the cookies and the humidity in the air. When cookies come out of the oven and cool down, the redistribution and ingress of humidity leads to differences in shrinkage and expansion within the cookies. These differences give rise to competing forces, which the cookie can only relieve by crumbling. The key to avoiding this phenomenon is careful control of the air humidity and ensuring that the moisture in the cookie is completely removed during baking. Ways in which this can be achieved include elevated baking temperatures, cooking for longer and post-baking treatments such as microwaving.[7]

1 | Sources: Interview with the American food scientist Kantha Shelke on 4 April 2009 & Interview with Ron de Santis at the Culinary Institute of America on 27 March 2009
2 | Interview Martin Doppler, Agrana, Tulln
3 | Guided tour of the Frozen Fish co. factory, Bremerhaven, on 5 February 2007
4 | Interview with the British author Paul Levy on 15 January 2009
5 | Sources: Annette Epp, Gerichte und ihre Geschichte, Christoph Wagner, Das Lexikon der Wiener Küche, Reiner Heilmann, Hotel Sacher, Sacher-Buch: 125 Jahre Hotel Sacher Wien, Gelebte Tradition
6 | Interview with Artemio Castro, Director Snacks, Procter & Gamble, Cincinnati, USA, on 3 April 2009
7 | Interview with Dr. Ricky Wildman, Loughborough, UK, on 20 January 2009

Verzehrssituation
where and how we eat

Verzehrssituation – Couch statt Esstisch | *„Ein Beispiel sind Kinder, die sich beim Essen gerne Finger und Kleidung beschmieren. Wenn wir einen Snack für Kinder entwickeln, versuchen wir das Produkt so zu gestalten, dass es möglichst einfach und ‚sauber' zu essen ist." Christina Jakobsen/ Entwicklungsabteilung Kraft Foods*

„Die Verzehrssituation ist für uns in der Produktentwicklung natürlich eine wichtige Vorgabe, die wir vom Marketing bekommen: Wo soll das Eis verzehrt werden? Wir gucken uns also an, wie das Eis schmilzt – wie schnell es schmilzt, wie lange es am Stil haften bleibt –, um sicherzustellen, dass auch ein langsamer Eisesser das Eis noch am Stil findet und nicht auf dem Boden." Frank Förster/ Entwicklungsabteilung Unilever

Praktische Anforderungen an Esswaren ergeben sich aber nicht nur bei Herstellung und Transport. Denn wir wollen unser Essen nicht nur einfach zubereiten und mühelos transportieren, sondern auch möglichst unkompliziert verzehren. Üblicherweise benützen wir dafür eigens entwickelte Werkzeuge, wie Messer, Gabel, Löffel oder Teller. Doch diese Hilfsmittel stehen nicht immer zur Verfügung, denn oft und gerne essen wir nicht bei Tisch, sondern auf dem Sofa lümmelnd, vor dem Computer tippend, auf einer Picknickdecke liegend oder im Auto lenkend. Zu diesem Zweck haben wir Pizzen als essbare Teller und Pommes, Hohlhippen oder Nachos als essbares Besteck entwickelt.

Esswaren werden in ihrer Form an die Situation, in der sie verzehrt werden, angepasst. Wir gestalten unsere Speisen je nachdem, wann, wo und wie wir sie konsumieren. Wir portionieren Fisch, Fleisch und Gemüse zu kleinen Häppchen, damit sie sich auf einmal in den Mund stecken lassen, und formen Wurst, Müsli und Schokolade zu länglichen Riegeln, damit wir gut von ihnen abbeißen können. Wir füllen Fragiles wie Eiscreme in isolierende, feste Waffelstanitzel, die sich gut in der Hand halten und am Ende aufessen lassen, und füllen heiße, fettige und saucenreiche Zutaten in Brotteig, damit wir sie besser in der Hand halten können. Beispiele dafür finden sich in fast allen Kulturkreisen, von italienischen Tramezzini und französischen Quiches über Hamburger, Hotdogs und Sandwiches bis zu indischen Samosas und mexikanischen Tacos und Tortilla-Wraps.

Die Abstimmung von Lebensmitteln auf die unterschiedlichen Umstände, unter denen Nahrung konsumiert wird, bildet einen wichtigen Eckpfeiler von Fooddesign. Wir gestalten, um bequem zu essen. Je weiter wir uns vom Tisch entfernen, desto komplexer werden die Anforderungen an das Gegessene. Extrembeispiele sind hier sicher Bergsteiger-, Soldaten- oder Astronautennahrung, die speziell gestaltet sein muss, um monatelang haltbar, hitze- und kältebeständig und zudem in der Schwerelosigkeit essbar zu sein. Aber auch auf der Erde können bei der Entscheidung für oder gegen ein

Where and How We Eat – Couch or Dinner Table | *"One example is mothers who give their kids a treat, and the kids come back with all dirty fingers and clothes. And therefore, when we develop children's snacks we need to take into consideration: How can the product be designed to be a clean product." Christina Jakobsen/R&D Department, Kraft Foods*

"For us in product development, the circumstances of consumption are an important piece of information; we get requirements from marketing, so we will know: Where will the ice-cream be eaten? We look at how the ice cream melts. How long does it take, how long does it stay on the lolly stick? We need to make sure that even a slow eater will still have ice cream on the stick and not on the floor before he's done." Frank Förster/R&D Department, Unilever

Practical requirements for food concern production and transport but there are other things to consider as well. After all, we want to prepare our food without much ado, we want it to be easily transportable but we also want to eat it with as few complications as possible. Normally, we use tools developed for that purpose, knives, forks, spoons or plates. However, these aids are not available always and everywhere. Sometimes we do not eat at the dinner table but lounging on the couch, typing away on our computer keyboard, lying on a picnic rug or driving a car. This is why we have pizza for edible plates, and French fries, wafer tubes or nachos for edible cutlery.

Food is also adapted to the situation in which it is eaten. We design what we eat in accordance with when, where and how we eat. We portion our fish, meat and vegetables in bite-size pieces so they fit into our mouths in one go. We produce sausage, muesli and chocolate which is shaped like an oblong bar so we can easily take a bite. Fragile stuff like ice cream is filled into comparatively solid, insulating wafer cones which can be held comfortably in the hand and eaten up as well. Bread dough serves the same purpose for hot, greasy and dripping ingredients. We find examples of this in practically all civilizations, from Italian tramezzini and French quiches to hamburgers, hot dogs and sandwiches, Indian samosas as well as Mexican tacos and tortilla wraps.

Bringing food into line with the circumstances in which it is consumed is yet another important task of food design. We design food so it is convenient to eat. The farther we leave from the table, the more complex are the requirements for the product in question. Food for mountaineers, soldiers or astronauts is definitely an extreme example because it has to keep for months, be resistant to heat and cold, and shaped in such way that it can be eaten in zero gravity. However, we do not have to think of

Eine extreme Verzehrssituation, die extreme Anforderungen an Food Design stellt, ist das All. Space food muss kompakt und matschig sein, damit die einzelnen Zutaten aneinander haften und sich beim Abbeißen keine Brösel lösen, die unkontrolliert herumschweben und Filter verstopfen könnten.

Outer space is an extreme consumption situation that makes enormous demands on food design. Space food has to be compact and soft so that the individual ingredients stick to each other, and when being bitten into no crumbs become loose as they could fly around uncontrolledly and clog filters.

Produkt Fragen ausschlaggebend sein wie: Kann ich mein Essen gut in den Fingern halten? Kann ich es in einer Hand halten und nebenbei noch lenken? Brauche ich ein Messer oder einen Teller und tropfe ich mir Krawatte oder Bluse an, wenn ich hineinbeiße? Konsumentenfreundliches Design kann also durchaus über Erfolg oder Misserfolg eines Nahrungsmittels entscheiden.

Die Anpassung an eine bestimmte Verzehrssituation ist in Zeiten, da immer weniger zuhause bei Tisch gegessen wird, ein ständig wachsender Bereich und führt einmal mehr vor Augen, wie gesellschaftliche Umstrukturierungen auch die Gestaltung von Esswaren beeinflussen. Heutzutage sind praktische Kalorien für alle Lebenslagen eine Selbstverständlichkeit. Ob im Auto, im Flugzeug oder im Schwimmbad: Die Verzehrssituation prägt Aussehen und Beschaffenheit von Lebensmitteln und hält noch viel Entwicklungspotenzial für zukünftiges Nahrungsdesign bereit. Der Trend, immer mehr unterwegs zu essen, wird unser Essen in den kommenden Jahrzehnten stark verändern. In den Vereinigten Staaten gibt es bereits Suppen, die man in der Getränkehalterung im Auto mit dem Zigarettenanzünder warm machen und mit einem speziellen Strohhalm während der Fahrt verspeisen kann. Auch so genanntes Armaturenbrettessen, das speziell für Autofahrer gestaltet wird, ist in gewissen Ländern mittlerweile an Tankstellen erhältlich.[1]

Schinken-Käse-Portsmouth | Konsumenten wollen ihre Mahlzeiten einfach und bequem zu sich nehmen, egal wo sie gerade sind, und das möglichst ohne Zuhilfenahme von Messer, Löffel oder anderem Werkzeug. Um sie einpacken, mitnehmen und unterwegs verzehren zu können, füllen wir ganze Speisen in Brot-, Blätter- oder Waffelteig und kaufen gefüllte Croissants, Taschen und Pasteten, panierte und gebackene Fleischstücke, salzige Wraps und süße Strudel oder Eis zwischen Waffelschichten wie das ovale Häagen-Dazs Cream Crisp.

Den Grundstein zur Idee, Brot und Belag formal aufeinander abzustimmen, legte der Legende nach der Brite John Montagu, der vierte Earl of Sandwich. Um eine Pokerpartie nicht unterbrechen zu müssen, orderte der passionierte Kartenspieler im Jahr 1762 seine Mahlzeit angeblich zwischen zwei Brotstücken. Sein Küchenchef servierte ihm daraufhin gefüllte, zusammengeklappte Weißbrotscheiben und ging als Erfinder des Sandwichs in die Geschichte der Kulinarik ein. Hätte die Familie Montagu – wie im 17. Jahrhundert ursprünglich geplant – den Titel des Earl von Portsmouth und nicht jenen von Sandwich angenommen, würden sich heute Millionen von Hungrigen an Stelle eines Sandwichs vermutlich ein Schinken-Käse-Portsmouth zwischen die Lippen schieben. Wie viel an der Geschichte rund um den spielsüchtigen Grafen Montagu wahr ist, lässt sich schwer sagen. Tatsächlich existierte 1762 ein Gericht namens Sandwich bereits, wie man vom französischen Historiker Pierre-Jean Grosley weiß, der es in diesem Jahr im Magazin von Edward Gibbon als Teil einer „echt englischen" Mahlzeit beschrieb.[2]

space travel – even here on earth, the decision for or against a product may depend on the answer to questions such as: Is this finger food? Can I hold it in one hand while driving a car? Do I need a knife and plate, and does it drip on my shirt or blouse when I take a bite? Consumer-friendly design may thus be decisive for the success or failure of food products.

In our day and age, when ever fewer meals are eaten at the table and in the home, there is growing demand for the adaptation of food to certain situations, a fact that shows how societal changes also have an impact on the design of food. Practical calorie intake in any situation is of course an issue today. No matter if we are in a car, on a plane or by a pool: the situation in which we eat has an effect on the outward appearance and other qualities of food, and the future still holds a lot of development potential in terms of food design in store. The trend of eating while we are out and about will cause major changes in the food sector in the next few decades. For example, in the US there are soups for consumption in the car: they fit into the cup holder, can be heated via the cigarette lighter socket and ingested through a special drinking straw. Gas stations in some countries even offer so-called "dashboard lunches" especially designed for drivers.[1]

Ham and Cheese Portsmouth

People want to consume food easily and conveniently, without using knives, spoons or other tools, no matter where they are . To be able to pack them, take them along and eat then on the road, we stuff bread, millefeuille or waffle dough with whole meals, and we buy filled croissants, pies and pasties, breaded and fried meat, savory wraps and sweet strudels or ice cream between layers of wafer, like the oval Häagen-Dazs Cream Crisp.

Legend has it that the original idea of combining bread and filling in perfect harmony came from John Montagu, the fourth Earl of Sandwich. A passionate card player, the Earl hated to interrupt his game of cribbage, so in 1762 he allegedly ordered his meal to be served between two slices of bread. Consequently, his cook prepared white bread with filling and went down in culinary history as the inventor of the sandwich. Had the Montagu family – as had originally been planned in the seventeenth century – assumed the title of the Earls of Portsmouth, millions of hungry people would be eating ham and cheese portsmouths instead of sandwiches today. It is hard to say if the story about the card-addicted lord is true. In fact, the French historian Pierre-Jean Grosley described the sandwich as part of a "truly English" meal in Edward Gibbon's journal in 1762, so we know that something called "sandwich" existed at that time.[2]

Die Situation, in der wir essen, beeinflusst die Gestaltung dessen, was wir essen. Gehend, stehend, fahrend oder lümmelnd: Handliche Sandwiches, mundgerechte Riegel, Ein-Bissen-Snacks oder sogenanntes Amaturenbrettessen sind in Form und Größe speziell darauf ausgelegt, abseits des Esstischs ohne Messer und Gabel verzehrt zu werden.

A situation in which we eat influences the design of what eat. Walking, standing, traveling or slouching: in terms of form and size, easy-to-eat sandwiches, mouth-sized bars, one-bite snacks or so-called dashboard food are geared to being eaten without a knife and fork – not at the dining table.

Mittlerweile gilt das Sandwich in all seinen Ausführungen und Variationen als Inbegriff globalisierten Fastfoods und selbst der berühmteste Snack der Welt, der Hamburger, wurde schriftlich erstmalig als „Hamburger Sandwich" erwähnt.³ Der Gestaltungsspielraum gefüllter Brote ist groß, vom einfachen Butterbrot bis zu mehrstöckigen Versionen wie dem 1968 erschienenen Big Mac oder dem vornehmeren Club Sandwich. Letzteres wurde 1903 erstmals erwähnt und verdankt seinen Namen angeblich den „Club Cars" der US-Railroads, in denen die drei-lagigen Sandwiches serviert wurden. Unterdessen arbeiten allein in England mehr Menschen an der Herstellung von belegten Brötchen als in der Landwirtschaft. 38 Prozent des gesamten Fastfood-Verzehrs fallen dort auf die gefüllten, zu Dreiecken geschnittenen Weißbrotscheiben.⁴

Essbare Verpackungen | Heute gehört das Sandwich gemeinsam mit Hotdog, Hamburger und Sushi zur Familie des globalisierten Fastfood. Von Singapur über New York bis Buenos Aires verdanken diese Gerichte ihren Erfolg nicht zuletzt dem passgenauen Zusammenspiel von Belag und Trägermaterial. Im Fall von Take-away-Essen werden Bestandteile der Nahrung zu Hilfsmitteln, die einen unkomplizierten Verzehr des Snacks erlauben. Egal, ob Hühner-Nuggets, Milchschnitte oder Prinzenrolle, ohne Trägermaterial wären Inhalt oder Belag nicht einfach auf dem Weg von A nach B zu verzehren. Teigmassen, Panier oder harte Schokolade isolieren gegen Hitze und bewahren die Hände vor Schmutz.

Ähnlich wie der Kunststoff um Stromkabel oder der Holzgriff einer Bratpfanne fungieren diese Hüllen als Isoliermaterial. So schützt die schlechte Wärmeleitfähigkeit eines Hamburger-Buns die Finger vor dem dampfenden Fleischlaibchen. Die Saugfähigkeit des Brotes vermag zudem das Saucenpotpourri einigermaßen im Zaum zu halten. Auf diese Weise wird schmutzigen Blusen und fleckigen Krawatten vorgebeugt. Da der horizontale Schnitt durch das Hamburgerbrötchen allerdings eine weitere potenzielle Gefahrenquelle darstellt, könnten wir uns die zukünftige Entwicklung eines „Ein-Bissen-Burgers" vorstellen. Eine vollständig geschlossene Brotummantelung bei zusätzlicher optimaler Mundgerechtigkeit würde die Fleckengefahr praktisch auf null reduzieren. Ebenso wären Würstchen, die bereits mit Senf oder Ketchup gefüllt in die Supermarktregale wandern, denkbar. Wie wichtig Sauberkeit bei der Konsumation für den Erfolg von Nahrung ist, zeigt ein Hersteller von Mini-Hühnchen Kiew, der nicht umsonst „clean plates guaranteed" – „garantiert saubere Teller" – groß auf die Packung druckt.

Vom Hotdog bis zum Hamburger, vom Tramezzino bis zum Kebab – Die Kombination von Brot und flüssigen, heißen oder fetten Zutaten bildet die Grundlage vieler Fastfood-Gerichte. Die Brothülle fungiert als Isoliermaterial und essbare Serviette in einem, sodass wir heiße Würste oder fettiges Fleisch bequem in der Hand halten können.

From hot dog to hamburger, from tramezzino to kebab – the combination of bread and liquid, hot or fatty ingredients constitutes the basis of many fast-food dishes. The bread shell functions as insulating material and edible napkin at the same time so we can easily hold hot sausages or greasy meat.

In the meantime the sandwich in all its varieties has become the epitome of global fast food, and even the most famous snack in the world. The hamburger was originally called "hamburger sandwich" when it was first mentioned in writing.[3] There are many ways to make a sandwich, it can be anything from a simple combination of bread and butter to multi-layered versions, such as the Big Mac, which was first made in 1968, or the more elegant "Club Sandwich". The latter, which allegedly owes its name to the "club cars" of the US railroads, where the three-layer sandwiches were served, was first documented in 1903. Nowadays, more people in England work in operations preparing sandwiches than in farming. No less than 38% of all fast food eaten there are filled slices of white bread cut into triangles.[4]

Edible Packages | Today, sandwiches, hot dogs, hamburgers and sushi are all members of the globalized fast food family. From Singapore to New York and Buenos Aires, these dishes owe their success to the perfect combination of filling and carrier material. In take-away food, parts of the food become tools enabling the uncomplicated consumption of the snack. No matter whether we are dealing with chicken nuggets, milk slices or chocolate cream sandwich cookies, we would not be able to eat the content or filling on our way from A to B without the carrier material that keeps it together. Dough, breading or chocolate icing insulates the inside against heat and keeps our hands from soiling.

Just like plastic on cables or wooden handles on frying pans, these wrappings serve as insulating material. The thermal transfer properties of hamburger buns are poor, so they keep our fingers safe from the hot meat patty. The bread is also absorbent and encloses the mixture of sauces to a certain extent, thus preventing spots on ties and blouses. However, the horizontal cut through the bun still poses a risk so that in the future, we can well imagine the development of the bite-size burger. A continuous bread wrap in optimum bite size would practically reduce the danger of spots to zero. Sausages filled with mustard or ketchup would also be a potential addition to our supermarket shelves. Clean consumption is important for the success of food: this is most aptly illustrated by the manufacturer of snack-size "Chicken Kiev"; the package reads "Clean plates guaranteed" in big letters.

Kebab-Stand in Istanbul. Ursprünglich bezeichnet Kebab ein orientalisches Rezept, bei dem Fleisch in rechteckige Stücke geschnitten und gegrillt wird. Eine besondere Zubereitungsart ist das türkische Döner-Kebab, bei dem marinierte Fleischstücke kegelförmig auf einen senkrechten Spieß geschlichtet, gebraten, abgeschabt und mit Salat und Reis auf einem Teller serviert werden. Döner-Kebab im Fladenbrot ist eine relativ junge Erscheinung und tauchte vermutlich erstmals in den 1960er Jahren in Istanbul auf.

Kebab stand in Istanbul. Originally Kebab was the name of an Oriental recipe in which the meat was cut into square pieces and grilled. A special mode of preparation is the Turkish döner kebab where the marinated meat pieces are wrapped around a vertical spit to form a sort of cone. Here, the meat is roasted, scraped off and served with salad and rice on a plate. Döner kebab in pita bread is a relatively young phenomenon and is assumed to have appeared for the first time in Istanbul in the 1960's.

Snacken auf japanisch: Reisbälle

Vom Kebab bis zur Frühlingsrolle, von der Apfeltasche bis zum Pain au Chocolat bildet die Kombination eines genießbaren Trägermaterials mit einer „riskanten" Fülle die Grundlage vieler Snacks. In Europa schützen Brot und andere Teige die Finger vor Heißem, Kaltem, Fragilem, Fettreichem oder Flüssigem, in Japan dient dazu das Grundnahrungsmittel Reis. Der beliebteste japanische Imbiss sind so genannte Onigiri oder auch Omusubi, etwa handtellergroße Objekte aus gekochtem Reis, die in jedem Convenience Store und an jeder Tankstelle verkauft werden. Beide Bezeichnungen – Onigiri und Omusubi – beziehen sich auf das Formen der Reisbälle: Musubu bedeutet auf Japanisch „verbinden" oder „zusammenbringen" und nigiru so viel wie „zusammenpressen" oder „kompakt machen". Die typische, dreieckige Form entsteht im Zuge der Herstellung: Der Reis wird dabei auf die flache Hand gelegt und von oben mit der zweiten, abgewinkelten Hand festgedrückt. Reisbälle sind üblicherweise gefüllt und entweder ganz mit Nori (getrocknete Meeresalgen) umwickelt oder nur an der Unterseite mit einer Nori-Schleife versehen. Das Nori-Blatt dient als essbare Verpackung, hält den Reis zusammen und hilft den feuchten, klebrigen Reis in der Hand zu halten. Die Füllungen variieren je nach Region, zu den beliebtesten zählen unter anderem eingelegte Pflaumen, Fischflocken und gegrillter Lachs. Reisbällchen sind in Japan schon seit dem 13. Jahrhundert nachgewiesen, welche Form sie damals hatten, ist allerdings nicht bekannt. Bis heute sind sie jedenfalls das „Sandwich" der Japaner und in ihrem Design ein vorbildliches Take-away-Gericht.⁵

Italienische Reisorangen und Schottische Eier

Auch in Europa kennt man Reisbälle als Snack für unterwegs: Auf Sizilien beispielsweise wird eine fastfood-taugliche Variante von Risotto zubereitet, indem der gekochte Reis zu so genannten Arancini, orangengroßen Bällchen, geformt, mit Fleischragù, Pinienkernen oder anderem gefüllt und anschließend paniert und frittiert wird. Ähnlich verfährt man auf der anderen Seite Europas mit hart gekochten Eiern. Üblicherweise schätzt man aufgeschnittene Eier als Einlage von Sandwiches, Tramezzini oder Wurstbroten. Anders in Schottland, wo ganze, hart gekochte Eier mit Wurstbrät ummantelt, paniert und in heißem Fett frittiert werden. Diese Scotch Eggs sind eine Vorform heutigen Fastfoods, lassen sich gut aufbewahren, mitnehmen und unterwegs mit den Fingern essen. Der traditionelle Imbiss, der kalorienmäßig imstande ist, eine ganze Mahlzeit zu ersetzen, besitzt durchaus das Potenzial, zukünftig als „Edinburger" ähnlich erfolgreich zu werden wie der „Hamburger".

Japanese-Style Snacks: Rice Balls

From kebab to spring roll, from apple pie to pain au chocolat, the combination of edible carrier material and "risky" filling forms the basis of many snacks. In Europe, bread and other products made of dough protect our fingers from hot, cold, fragile, greasy or liquid interiors, in Japan it is the local staple, rice. So-called onigiri or omusubi, palm-sized snacks made of cooked rice sold in convenience stores and at gas stations, are the most popular munchies. Both words – onigiri and omusubi – relate to the shapes of the rice balls: the Japanese word "musubu" translates as "combining" or "bringing together", whilst "nigiru" means "press" or "compact". The typical triangular shape results from the way in which the snack is made. Rice is placed on the palm of one hand and pressed lightly with the cupped palm and fingers of the other hand. Rice balls are usually filled and wrapped in nori (seaweed sheets), either completely or in just a strip on the side facing down. The purpose of the nori strip is to serve as an edible package, to hold the rice together and help keep one's fingers clean, as the rice is sticky and moist. Fillings vary according to the region, with pickled plums, fish flakes and grilled salmon being most popular. There is evidence that rice balls already existed in the thirteenth century in Japan but nothing is known about their exact shape. Until today, they have remained the equivalent of the sandwich to the Japanese and their design makes them ideal take-away food.⁵

Italian Rice Oranges and Scotch Eggs

Rice balls as snacks to have on the road are not entirely unknown in Europe: In Sicily, for example, there is a street food version of risotto where cooked rice is pressed into so-called "arancini", orange-sized balls filled with meat sauce, pine nuts or other ingredients, which are then breaded and fried. A similar procedure is applied to hard-boiled eggs on the other end of Europe: Normally, we would slice the eggs and use them for sandwiches or tramezzini. In Scotland, the eggs are left whole, covered with sausage meat, breaded and fried in hot fat. Scotch eggs, as they are called, are precursors of today's fast food, they can easily be stored, taken along and eaten as finger food while out and about. In terms of calories, the traditional snack can replace an entire meal, and it has potential of becoming as popular as the hamburger – under the name "edinburgher", perhaps.

Das Sandwich der Japaner: Onigiri, sauer gefüllte, dreieckige Reisbälle, eingeschlagen in eine essbare „Serviette" aus Nori

The sandwich of the Japanese: Onigiri, triangular rice balls with sour stuffing, wrapped in an edible napkin made of nori

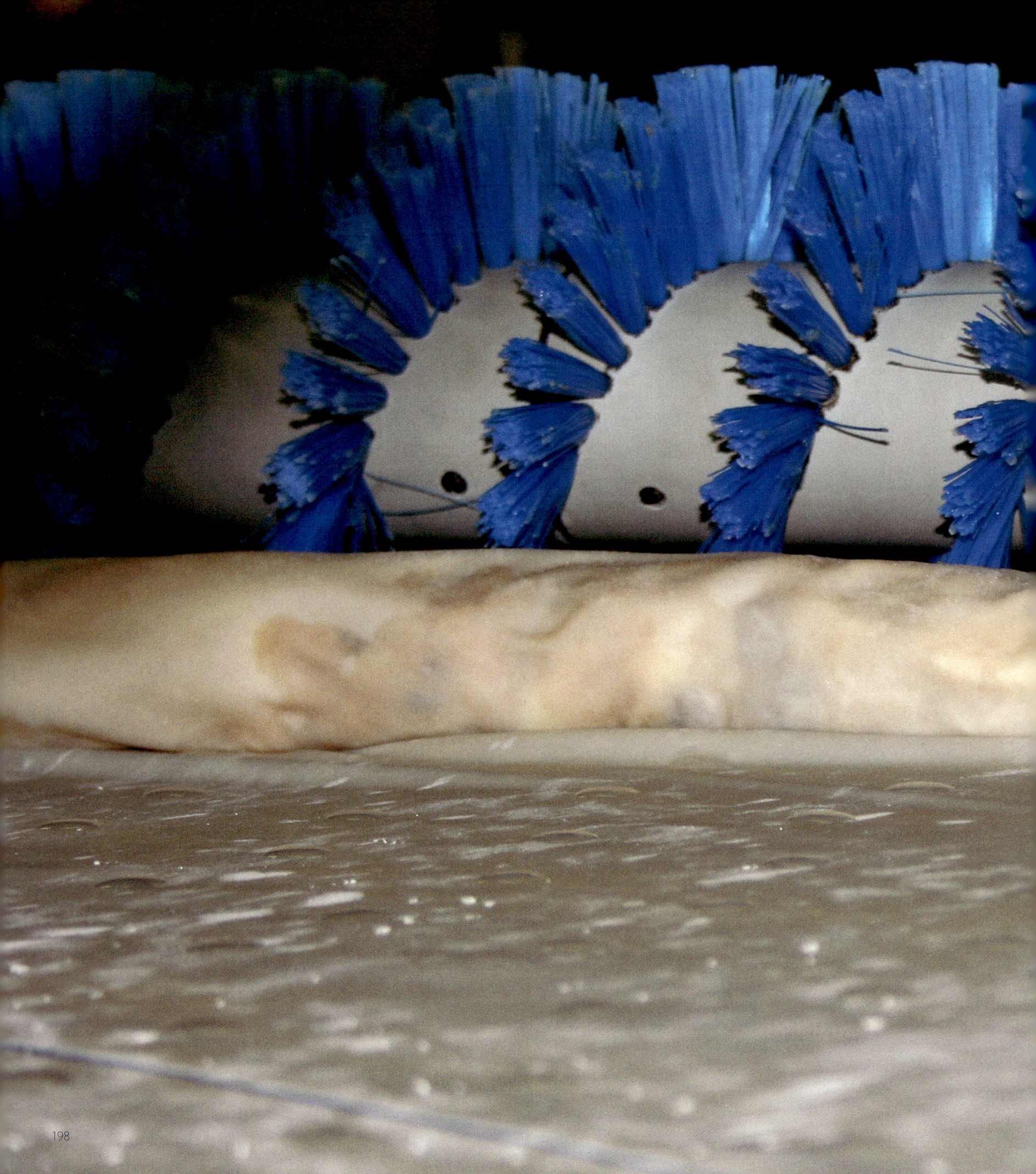

Apfelstrudelerzeugung bei der Firma Frisch & Frost in Österreich. Zu den Vorbildern des Strudels zählt vermutlich das türkische Baklava, ein mit Sirup getränktes Dessert aus Blätterteig, Honig, gehackten Nüssen und Gewürzen. Über Ungarn gelangte das Rezept wahrscheinlich im 18. Jahrhundert nach Mitteleuropa, wo es zu salzigen Suppeneinlagen und süßen Desserts weiterentwickelt wurde. Traditionell muss ein Strudelteig so dünn „ausgezogen" werden, dass man eine darunter liegende Zeitung lesen kann.

Apple strudel production at the Frisch & Frost company in Austria. The strudel is assumed to have been modeled after the Turkish baklava, a dessert made of puff pastry, honey, chopped nuts and spices soaked in syrup. The recipe probably reached Central Europe via Hungary in the 18th century; there, it was further developed and used as salty soup addition and sweet dessert. Traditionally, a strudel dough has to be "stretched" to the extent that it is possible to read a newspaper lying underneath it.

Beilagen statt Geschirr | Die Idee, essbare Ummantelungen als Ersatz für Verpackung, Geschirr und Serviette zu benutzen, ist nicht neu. In Nordafrika und Teilen Kleinasiens existierte bereits in antiker Zeit ein gefülltes Fladenbrot namens Pita, das die Römer nach Italien importierten und das der späteren Pizza ihren Namen gab. Das italienische Nationalgericht, wie wir es heute kennen, entstand in der Renaissance und ist bis heute der Inbegriff erfolgreichen Fingerfoods.

Der Pizzaboden funktioniert wie ein Teller, auf dem sich alle möglichen Zutaten anordnen lassen und der nach dem Essen weder gespült noch weggeworfen werden muss. Wenn heute Pizza auf einem Porzellanteller serviert wird, so entspricht diese Anordnung eigentlich einem Teller auf einem Teller. Designtechnisch eignet sich die Pizza wesentlich besser für Take-away als fürs Restaurant und kann es daher im Zeitalter von Fastfood durchaus mit Konkurrenten wie Hamburger oder Hotdog aufnehmen. Auch der heutige Brauch, Pizza bei Profiköchen zu bestellen, anstatt sie selbst zu backen, ist nicht neu. Pizzen müssen bei sehr hohen Temperaturen zwischen 400 und 500 Grad Celsius gebacken werden, um richtig gut zu schmecken. Im Neapel des 18. und 19. Jahrhunderts, als sich die Pizza als Folge der zunehmenden Popularität der Tomate zum Alltagsgericht entwickelte, verfügte kein Haushalt über einen derart leistungsfähigen Ofen. Deswegen wurde der Hefeteig zuhause vorbereitet, belegt und danach zum Bäcker gebracht, um ihn backen zu lassen.

Die Idee, festen, dichten Teig als Teller für andere Zutaten zu benutzen, ist durchaus keine Erfindung der neapolitanischen Pizzaköche. Schon vor dem 17. Jahrhundert war der Brauch verbreitet, Bratenstücke auf Brotscheiben zu servieren, die den Saft auffingen. Die voll gesogenen „Teller" wurden anschließend gegessen oder an Bedürftige verteilt. Auch das kostbare Salz kam nicht im Fässchen oder Streuer auf den Tisch, sondern wurde den Gästen in ausgehöhlten Brotstücken zugeteilt. Ganz ähnlich arbeiten süße Rezepte wie Teigschälchen, die mit Creme gefüllt und frischen Fruchtstückchen belegt werden und heute auch schon vorgefertigt im Supermarkt erhältlich sind. Aber auch viele salzige Snacks funktionieren nach demselben Prinzip: Fertige Tacos können schnell gefüllt und bequem mit der Hand gegessen werden. Oder „ready to fill biscuit cups" und „pastry cups": Die kleinen Teigschüsselchen lassen sich für Appetithäppchen rasend schnell mit Mayonnaisesalat, Schrimps oder Toamtenstückchen befüllen und im Stehen mit den Fingern verspeisen.[6]

Waffelkegel als essbares Geschirr | Essbare Ummantelungen schützen vor Hitze und fettigen Fingern. Ein wahrer Klassiker des Food Design dagegen schützt vor Kälte: Eiscreme schmilzt bei Berührung mit der Wärme des menschlichen Körpers bekanntermaßen bereits nach wenigen Sekunden zu einem unansehnlichen Saft und kann deswegen unmöglich mit den Fingern verzehrt werden.

Auch die Eiswaffel verdankt ihre Entstehung der Idee des essbaren Geschirrs. Jahrhundertelang war Speiseeis aufgrund der komplizierten Herstellung ein teures Luxusprodukt. Bereits vor über 5000 Jahren holte man in China Schnee von nahe gelegenen Berggipfeln und verfeinerte ihn mit Honig, Früchten und anderen Gewürzen. Alexander der Große motivierte

Side dishes instead of tableware | The idea of using edible wrappings to replace packaging, dishware and napkins is not new. In Northern Africa and parts of Minor Asia there was already in ancient times a flatbread known as pita. The Romans imported this bread to Italy and later named it pizza. Italy's national dish as we know it today was created in the Renaissance age and to this very day it remains the quintessential successful finger food.

The pizza bottom serves as a kind of plate on which all sorts of ingredients can be arranged. After the meal it is neither necessary to rinse it or throw the bottom side away. When pizza is served on a china plate today this arrangement is tantamount to serving a plate on a plate. In terms of design the pizza is much better suited as a takeaway than to be served in a restaurant. In the age of fast food it can thus compete far better with the hamburger or hotdog. Even today's custom of ordering pizza from professional caterers instead of cooking it at home, is not new. Pizzas have to be baked at very high temperatures between 400 and 500 degrees for them to taste good. In Naples of the 18th and 19th century when pizza emerged as a consequence of the growing popularity of tomatoes as an everyday dish there were no households that had such a high-performance oven. For this reason the yeast dough was prepared at home and then taken to a baker who baked it.

The idea of using thick dough as a dish for other ingredients is certainly not an invention of the Neapolitan pizza cooks. Already before the 17th century it was common practice to serve pieces of roast on slices of bread, which would absorb the juice. The fully soaked "plates" were then either eaten or distributed to the needy. Even precious salt was not placed on tables in little bowls or casters but was served to the guests in hollowed out pieces of bread. Sweets recipes such as dough bowls filled with cream and covered with pieces of fresh fruit were similar and are available freshly made at supermarkets today. But there are also many savory snacks that function according to the same principle. Finished tacos can be quickly filled and are easy to eat by hands. Or "ready to fill biscuit cups" and "pastry cups". The small dough bowls can be filled in a flash with mayonnaise salad, shrimps or small pieces of tomatoes and eaten as finger food while standing.[6]

Wafer Cones as Edible Dishes | Edible wrappings protect food from heat and greasy fingers. A veritable classic of food design, by contrast, protects from cold. Ice cream melts when it comes into contact with the warmth of the human body, in just a few seconds becoming an unappealing juice so that it is impossible to eat it with our fingers.

Even ice cream wafers were originally designed to serve as edible dishes. For centuries ice cream was an expensive luxury good because it was so difficult to manufacture. 5,000 years ago snow was collected from nearby mountain peaks and refined with honey, fruit and other herbs. Alexander the Great allegedly spurred on

Zukunftspotenzial für Food Design: Teller, Messer und Gabel aus verzehrtauglichem Material

Future potential for food design: plates, knifes and forks made of material suitable for consumption

seine Truppen angeblich mit gefrorenen Desserts und der römische Kaiser Nero sandte seine Sklaven regelmäßig in die Alpen, um Schnee und Eis für Süßspeisen zu holen. Die persische Kunst der Sherbet-Herstellung traf später in Italien mit einem Rezept zusammen, welches Marco Polo um 1295 nach Norditalien brachte und bei dem Yak-Milch mit Schnee vermischt wurde. Durch Katharina von Medici wurde die Idee, Milch und Früchte zu frieren, am französischen Hof populär, von wo aus Eiscreme in weiterer Folge die Herrschaftshäuser ganz Europas eroberte.

Erst als Jacob Fussel 1851 in Baltimore die erste Eisfabrik der Welt gründete, begann sich Eis vom Luxusgut zur Massenware zu entwickeln. Während Adelige das Dessert jahrhundertelang aus gläsernen Schalen gelöffelt hatten, musste man das Gelato beim Verkauf über die Gasse allerdings in Wegwerfbehälter füllen, die damals teuer und schwer erhältlich waren. Italo Marchiony, ein fahrender Eisverkäufer aus New York, bot mit seinem „icecone", den er 1903 patentieren ließ, die perfekte Lösung für dieses Problem. Praktisch zeitgleich, nämlich bei der Weltausstellung in St. Louis 1904, präsentierte auch E.A. Hamwi die Idee, essbare, kegelförmige Eisschüsseln aus Waffelteig herzustellen.

Schokoguss, Holzstiel und Eiswaffel | Durch die moderne Tiefkühltechnik entwickelte sich Speiseeis zu einem leistbaren Alltagsprodukt. In Wahrheit machten jedoch erst drei weitere, amerikanische Erfindungen Eiscreme zu dem, was sie heute ist, nämlich zu einem der populärsten Take-away-Artikel der Lebensmittelindustrie:1922 entwickelte C.K. Nelson einen Schokoladeüberzug für Speiseeis und 1923 konnten die Amerikaner dank Harry Bust aus Ohio erstmals Eis von einem kleinen Holzstiel schlecken. 1941 schließlich wurde das heutige Eisstanitzel in den Vereinigten Staaten patentiert und Eis in weiterer Konsequenz ein Fastfood-Produkt in ganz großem Stil.[7]

his troops by serving them frozen desserts and the Roman emperor regularly sent his slaves to the Alps to fetch snow and ice to make sweets. The Persian art of producing sherbet later converged in Italy with a recipe that Marco Polo brought to Northern Italy in 1295 – here Yak milk was mixed with snow. At the French court Katharina de' Medici popularized the idea of freezing milk and fruit and from here ice cream subsequently conquered all of Europe's imperial houses.

It was not until Jacob Fussel founded the world's first ice cream factory that ice cream began to develop from a luxury good into a mass commodity. While aristocrats had for centuries scooped the dessert from glass bowls, gelato had to be filled in throwaway receptacles to be sold on the street. These receptacles were still expensive and hardly available at that time. Italo Marchiony, a traveling ice-cream seller from New York, offered the perfect solution to this problem with his "ice cream cone" which he had patented in 1903. At virtually the same time, that is, at the world exposition in St. Louis in 1904, E.A. Hamwi presented the idea of manufacturing edible, cone-shaped ice bowls out of wafer dough.

Chocolate Icing, Wooden Stick and Ice-Cream Wafers | Thanks to modern deep-freeze technology, ice cream soon became an affordable everyday product. In reality, however, it was only three further American inventions that transformed ice cream into what it is today, namely one of the most popular take-away products of the foodstuff industry. In 1922 C.K. Nelson developed a chocolate icing for ice cream and in 1923 Harry Bust from Ohio made it possible for Americans for the first time to eat ice cream from a wooden stick. In 1941 today's ice cream cone was patented in the United States in 1941; as a consequence, ice cream became a fast food product on a very large scale.[7]

Links | Die 1941 in den USA patentierte Eiswaffel fungiert als essbarer Becher. Die schlechte Wärmeleitfähigkeit des Waffelteiges schützt das leicht schmelzende Speiseeis außerdem vor der menschlichen Körperwärme.

Rechts | Ersetzen Brotscheiben und Teigfladen Teller und Schüsseln, so zielt die Gestaltung anderer Lebensmittel darauf ab, für Messer und Gabel einzuspringen. Da sich das Besteck erst nach der Renaissance schrittweise durchzusetzen begann, bestand jahrhundertelang die Notwendigkeit, Speisen so zu gestalten, dass sie problemlos mit den Fingern gegessen werden konnten. Und auch heute tauchen wir Pommes gerne ins Ketchup, Soletti in den Liptauer, Hohlhippen ins Eis und löffeln mit Biskotten (Löffelbiskuit) Apfelmus.

left | The ice cream wafer that was patented in the USA in 1941 serves as an edible cup. The poor heat conductivity of wafer dough also protects the ice cream which melts easily when exposed to human body warmth.

right | If bread slices and pita breads replace plates and bowls, the design of other foods aims at substituting forks and knives. Since cutlery only became gradually accepted after the Renaissance, it was necessary for centuries to design foods so that they could be easily eaten using one's fingers. Even today, we like to dip French fries in ketchup, Soletti cracker sticks in cream cheese, or wafer rolls in ice cream, and to use lady fingers as spoons for apple puree.

So alltäglich uns heute die kleinen Kegel aus Waffelteig auch erscheinen mögen, so genial ist doch genau genommen ihre Erfindung. Einfach und stabil in ihrer Form, haptisch und dekorativ durch die strukturierte Oberfläche, knusprig im Geschmack: So etwa könnte man die wichtigsten Vorteile der goldgelben Tüten zusammenfassen. Von der Wurstsemmel bis zum Schokocroissant bildet die Kombination eines genießbaren Trägermaterials mit einer „riskanten" Fülle die Grundlage vieler Snacks. Und so bedeutete die Kombination von Waffelkegel und Eiskugel eigentlich nur eine Fortführung althergebrachter Imbisse wie jener des Wurstbrotes oder der Apfeltasche.

Essbares Besteck | Ersetzen Brotscheiben und Teigfladen Teller und Schüsseln, so zielt die Gestaltung anderer Lebensmittel darauf ab, für Messer und Gabel einzuspringen. Da sich das Besteck erst nach der Renaissance schrittweise durchzusetzen begann, bestand jahrhundertelang die Notwendigkeit, Speisen so zu gestalten, dass sie problemlos mit den Fingern gegessen werden konnten. Die Römer aßen ihre Gerichte vorgeschnitten und benutzten Brotstücke, um Saucen und Ähnliches zum Mund zu führen. Sowohl Pizza als auch Pasta wurden ursprünglich mit den Fingern verzehrt und dabei als Trägermaterial für andere Zutaten verwendet. Selbst der Sonnenkönig Ludwig XIV. speiste noch mit den Fingern, wohingegen Queen Victoria bereits mit edlem Silberbesteck hantierte. Vermutlich verwenden bis heute weltweit etwa genauso viele Menschen Brot zum Essen wie Metallbesteck oder Holzstäbchen.

Brot statt Gabel | Brot eignet sich aufgrund seiner Konsistenz hervorragend als essbares Besteck. Im mediterranen Raum wird Muschel- und Fischsud mit Weißbrot aufgetunkt, in Mitteleuropa liebt man mit Gulasch- oder Bratensaft voll gesogenes Schwarzbrot. Auch in England verzichtet keiner auf sein Brötchen zur „soup of the day" und in Frankreich werden Croissants gerne in Milchkaffee getunkt und dort bis zur Unkenntlichkeit aufgeweicht. Gebäck und Teige aller Art dienen als Trägermaterial für mehr oder weniger flüssige Zutaten und Speisen. In Indien und Afrika funktioniert man Teile ganzer Gerichte in Besteck beziehungsweise Servietten um. In Äthiopien zum Beispiel werden die gekochten Fleisch- und Gemüsestücke auf großen, weichen,

As ordinary as the small cones made of wafer dough strike us today, their invention was truly a stroke of genius. Simple and stable in form, haptic and decorative thanks to the structured surface, their crispy taste – here's how the main virtues of the golden-yellow cones could be summed up. From the apple pie to the chocolate croissant the combination of an edible wrapping with a "daring" filling constitutes the basis of many snacks. In this sense, the combination of wafer cones and scoops of ice cream simply amounted to the continuation of traditional snacks such as the sausage roll or the pork pie.

Edible Cutlery | With slices of bread and flat cakes replacing plates and bowls, the design of other foods was aimed at substituting knives and forks. Since cutlery only gradually began to win ground after the Renaissance, it was necessary for centuries to design food so that it could be easily eaten with our fingers. The Romans ate their meals pre-cut and used pieces of bread to suck sauces and the like. Both pizza and pasta were originally eaten with fingers and also used as a support for other ingredients. Even the Sun King Louis XIV still ate with his fingers, while Queen Victoria already handled silver cutlery. It can be assumed that just as many people all over the world still use bread to eat as metal cutlery or wooden chopsticks.

Bread Instead of Forks | Thanks to its consistency bread is well suited as edible cutlery. In the Mediterranean region mussel and fish stock is scooped up with white bread, and in Central Europe people love dipping dark bread into goulash sauce or roast gravy. Even in England no one wants to miss out the piece of bread on the 'soup of the day' and in France everyone likes dipping their croissant in milk coffee until it almost dissolves. All manner of bread rolls and bakery products serve as the basis of more or less liquid ingredients and dishes. In India and Africa parts of entire meals are transformed to serve as cutlery

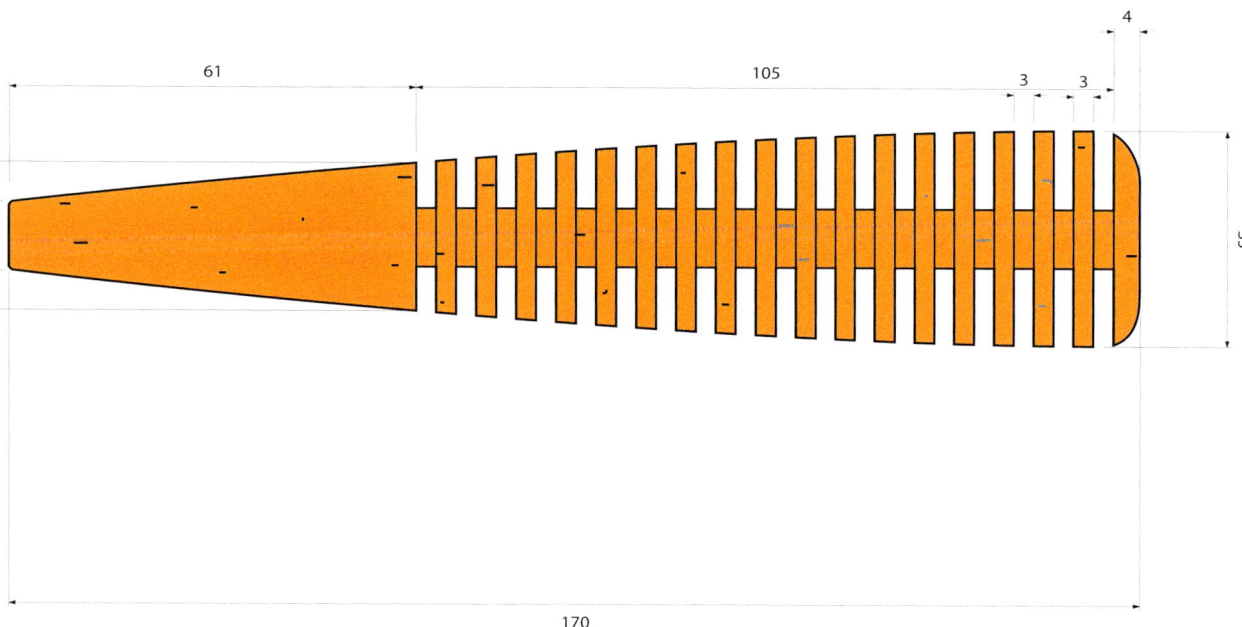

Omelettes ähnlichen Fladen serviert, die man in kleine Stücke reißt, um die Zutaten faltet und die Pakete bequem und sauber in den Mund steckt. Ähnlich läuft die indische Methode ab, Fladenbrot wie „Naan" als eine Art essbare Serviette um „Chicken Tika Massala" oder „Paneer Khofta" zu wickeln und derart mit sauberen Fingern zu essen.

Essen wird bewusst so gestaltet, dass es sich gut mit den Fingern verzehren lässt. Während Brot erst beim Essen in eine passende Form gebrochen wird, sieht die Biskotte – auch bekannt als Löffelbiskuit – von vornherein wie ein Löffel aus. Der kross gebackene Besteckersatz ist mundgerecht und saugfähig, dient meist als Babynahrung und war in früheren Zeiten sogar als Partyfood en vogue. In den zwanziger Jahren tauchte, wer hipp sein wollte, seine Biskotte in Champagner statt in Apfelmus.

Würstchen & Pommes als Prothesen | Auch heutzutage profitiert Fingerfood vom so genannten Fun-Faktor. Gerade weil der Großteil der Nahrung auf die Gabel gespießt wird, bereitet es Spaß, Pommes frites mit den Fingern ins Ketchup zu tauchen, mit Nachos scharfe Saucen zu löffeln oder mit Waffelröllchen Eiscreme zu naschen. Essbares Besteck kehrt als Funfood an den Esstisch des 21. Jahrhunderts zurück. Designtechnisch erscheinen Produkte zum Dippen nicht selten in Gestalt menschlicher Finger: Pommes, Würstchen, Crackersticks oder „Soletti" fungieren als Werkzeuge, die wie unser verlängerter Zeigefinger geformt sind.

Ein Extrembeispiel, wie Esswaren als Besteck zur Aufnahme anderer Zutaten dienen können, schuf der französische Industrie- und Food Designer Stéphane Bureaux mit seinem Projekt „carotte rappeuse". Bureaux drechselt dafür auf einer herkömmlichen Drehbank mehrere ringförmige Einkerbungen in eine ganze, geschälte Karotte, bis diese aussieht wie ein Honiglöffel – und auch so funktioniert. In Gemüsegelee oder Kräuterdip getaucht und gegessen, präsentiert sich die „carotte rappeuse" als ebenso ästhetisches wie funktionales Gericht.⁸

or as napkins. In Ethiopia, for instance, pieces of cooked meat and vegetables are served on large, soft flatbread resembling omelettes, which are torn into small pieces, wrapped around the ingredients so that the resulting "packages" can be easily and neatly put in one's mouth. The Indian method is similar, with pitta bread such as "naan" being wrapped around "chicken tika masala" or "paneer khofta" like a sort of edible napkin.

Food is deliberately designed so that we can easily eat it with our fingers. While bread is usually broken only while eating, the biscuit also known as ladyfinger (in German: "Löffelbiskuit", "spoon biscuit") resembles a spoon from the very outset. The cutlery replacement, baked to a crisp, is mouth-sized, can absorb liquids and generally serves as baby food. In earlier times lady fingers were even en vogue as party food. Whoever wanted to be hip in the twenties, dipped his or her biscuit in champagne instead of applesauce.

Sausages and French Fries as Prosthetics | Even today, finger food benefits from the so-called fun factor. Precisely because most food is skewered on a fork, we find it fun to dip French fries into ketchup to scoop hot sauces with nachos or to spoon up ice cream with a wafer tube. As fun food, edible cutlery returns to the dining room table of the 21st century. In design terms, products for dipping often appear in the guise of human fingers: French fries, sausages, cracker sticks or "Soletti" function as tools that are shaped like an extended index finger.

With his project "carotte rappeuse", the French Stéphane Bureaux created an extreme example of cutlery used for lifting other ingredients. Bureaux used an ordinary lathe to turn several circular indentations into a complete, peeled carrot until it resembled a honey spoon – and also functioned as such. Dipped in vegetable aspic or herbal sauce, the "carotte rappeuse" presented itself as both an esthetic and a functional dish.⁸

Der Pariser Designer Stéphane Bureaux bearbeitet Gemüse mit industriellen Werkzeugen. Für die „Carotte rappeuse" fräst er auf der Drehbank ringförmige Einkerbungen in eine Karotte, bis sie aussieht wie ein Honiglöffel und auch so funktioniert: Tunkt man sie in eine Sauce, bleibt besonders viel davon haften.

Parisian designer Stéphane Bureaux processes vegetables with industrial tools. For "carotte rappeuse" he uses a lathe to make ring-shaped dents on a carrot until it looks like a honey spoon and also functions as such. If you dip it in a sauce, a lot of it will stick to it.

Werkzeug definiert Form | Aber auch wenn wir Messer, Gabel oder Stäbchen benutzen, färbt die Funktionsweise dieser Werkzeuge auf das Design der Nahrungsmittel, die wir mit ihnen essen, ab. Die Art des Verzehrs definiert die Weise, wie die Zutaten in der Küche vorbereitet werden. Während in Europa und in Amerika heute jedem bei Tisch ein Messer zur Verfügung steht, werden in Asien alle Zutaten bereits in der Küche in mundgerechte Stücke geschnitten, damit sie sich mühelos mit Stäbchen essen lassen.

Die Abstimmung des Designs auf das Werkzeug, mit dem es gegessen wird, erfolgt aber auch bei modernen Produktentwicklungen, zum Beispiel bei Instantsuppen. In Indien, wo Cup-Noodles mit der Hand gegessen werden, sind die Becher weniger tief, damit man mit den Fingern leicht hineingreifen kann. Auch die Nudellänge variiert von Land zu Land und passt sich dem jeweils verwendeten Esswerkzeug an. In Japan, wo Instantsuppen mit Stäbchen gegessen werden, sind die Nudeln 30 Zentimeter lang. In Europa und in den USA dagegen, wo die meisten Menschen die Gabel verwenden, um Instantnudeln zu essen, messen sie nur 20 Zentimeter; in Mexiko, wo man Nudelsuppen mit dem Löffel isst, gar nur 10 Zentimeter. In Japan gibt es zudem Sorten mit extra langen Buchweizennudeln. Japaner lieben nämlich das Gefühl in der Kehle, wenn sie Nudeln unzerkaut herunterschlürfen, und um dieses Erlebnis zu intensivieren, bieten Hersteller wie Nissin Instantsuppen mit 60 Zentimeter langer Pastaeinlage an.[9]

Space Food | Wann und wo gegessen wird, beeinflusst naturgemäß das Design der Speisen. Eine extreme – wenn nicht die extremste – Verzehrssituation ist das Weltall. In der Internationalen Raumstation ISS essen die Astronauten schwebend. Space Food muss also so gestaltet sein, dass es in der Schwerelosigkeit mühelos und sicher verzehrt werden kann, ohne dass sich bei hastigen Bewegungen einzelne Bestandteile lösen und unkontrolliert herumdriften. Im Gegensatz zur Erde, wo Krümel auf den Boden fallen, können Essensreste im All in Computern oder anderen lebenswichtigen elektronischen Teilen Kurzschlüsse auslösen. Das Einatmen eines herumschwebenden Tropfens Orangensaft kann einen Astronauten unter Umständen sogar töten. Dünnflüssige Suppen gehören daher nicht zur Astronautenverpflegung, Getränke werden mit Schläuchen aus Beuteln getrunken.

Keine Gase im All | Ein essentieller Gestaltungsparameter von Space Food ist der Flüssigkeitsgehalt. Dieser sorgt für den Zusammenhalt der einzelnen Komponenten, den ohne Hilfe der Erdanziehungskraft eine entsprechende Oberflächenspannung leisten muss. Will ein Astronaut zum Beispiel einen Apfel auf den Tisch „legen", so benetzt er die Unterseite mit etwas Wasser. Die Oberflächenspannung dieses Tropfens hält den Apfel dann an der Tischoberfläche bzw. an seiner Unterseite fest: Da es im

Form-Defining Tools | Even if we use a knife, a fork or chopsticks, the functioning of these tools colors off on the design of the food in question. The tool we use define the way in which the ingredients are prepared in the kitchen. Whereas in Europe and America today everyone has a knife on the table, in Asia all ingredients are already cut into bite-sized pieces in the kitchen so that they can be easily handled with chopsticks.

The coordination of design and eating tools also takes place with modern product developments, for instance with instant soups. In India where people eat cup noodles with their hands, the cups are less deep so that one can easily reach in. Even the length of noodles varies from country to country and is adapted to the eating instrument used. In Japan, instant soups are eaten with chopsticks, so noodles are 30 centimeters (11 inches) long. In Europe and the USA, by contrast, where most people use a fork to eat instant noodles, they measure only 22 centimeters (8 inches) long. In Mexico, people eat instant soups with a spoon, so the noodles are chopped to 10 centimeters (4 inches) in length. In Japan there are even brands with extra long buckwheat noodles. The Japanese love the sensation in their throat when they slurp the noodles without chewing them. In order to enhance this experience, manufacturers such as Nissin offer instant soups with pasta measuring 60 centimeters (22 inches).[9]

Space Food | The design of food is influenced by when and where one eats. An extreme – if not the most extreme – consumption situation is outer space. In the international space station ISS the astronauts eat in zero gravity. Space food must thus be designed so that in a gravity-free environment it can be eaten easily and safely without individual ingredients becoming loose and drifting about uncontrolled. As opposed to the earth where crumbs fall to the ground, remains of food in space can cause short-circuits in computers or other vital electronic pieces of equipment. Inhaling a drop of orange juice flying about could even kill an astronaut in the worst case. Thin soups thus should not be part of astronaut food. Beverages are drunk with straws from bags.

No Gases in Space | One essential design parameter of space food is the liquid content. It makes for the cohesion of the individual components; without the help of gravitation a certain surface tension must be

All keine Gravitation gibt, funktioniert dieses System nach allen Richtungen gleichmäßig. Ein Weltraum-Burger beispielsweise müsste so mit Ketchup und Saucen gefüllt sein, dass die einzelnen Schichten durch die Kohäsion fest miteinander „verkleben". Andernfalls lösen sich die Brothälften von Fleisch, Käse und Tomaten ab und alle Zutaten driften getrennt voneinander durch die Gegend. Hamburger werden deshalb wohl nie zur Standardverpflegung im All zählen.

Hohe Feuchtigkeit sorgt auch bei dem weichen in einzelne bonbongroße Stücke vorportionierten Weltraumbrot dafür, dass es nicht bröselt, wenn man es aus der Packung nimmt oder zerteilt. Allerdings darf das Brot nicht zu feucht sein, damit es bei den Astronauten keine Blähungen auslöst. Da Verdauungsgase den Darm normalerweise mit Hilfe der Gravitation verlassen, können sie im All nicht auf natürliche Weise aus dem Körper entweichen. Für die Bewohner der Raumstation ist jede Art von Gasentwicklung sehr unangenehm und muss extra herausmassiert werden. Deshalb werden im Weltraum auch keine Gerichte mit Zwiebeln oder Knoblauch serviert.

Außerdem müssen Space-Food-Entwickler die äußerst eingeschränkten Möglichkeiten, Essen in der Raumstation zu erwärmen, berücksichtigen. Auf der Erde steigt Hitze durch die Schwerkraft auf und führt zur Konvektion im Topf, dessen Inhalt sich in Folge mischt und relativ gleichmäßig erwärmt. In der Schwerelosigkeit kommt es zu keiner Konvektion. Das Essen brennt unten an, während es oben noch kalt ist. Aus diesem Grund funktioniert ein herkömmlicher Herd im Weltall nicht. Die Mahlzeiten der Astronauten müssen von allen Seiten gleichmäßig erhitzt werden, was auf der ISS mit einer Art von Waffeleisen, in das die Konservendosen eingespannt werden, passiert.[10]

Militärische Kalorien | Nicht ganz so hoch, aber dennoch speziell sind auch die Anforderungen an die Funktionalität von Soldatennahrung: Seit der Erfindung des Krieges beschäftigen sich Militärstrategen nicht nur mit Truppenbewegungen und Lufteinsätzen, sondern auch mit der Beschaffung von ultraleichter, extrem haltbarer und besonders nahrhafter Marschverpflegung. Heutzutage ernähren sich Soldaten auf Spezialeinsätzen von so genannten Combat Rations, in Beutel eingeschweißten, gefriergetrockneten Mahlzeiten, die sicher im Rucksack transportiert und ohne Geschirr zubereitet werden können.

Diese konservierten Fertiggerichte sind 30 Monate bei jeglicher Temperatur haltbar, absolut wasserdicht, können nicht frieren, da sie kein Wasser enthalten, und schwimmen, da die Beutel mit Stickstoff gefüllt sind. Die Augsburger Firma Simpert Reiter ist auf die Erzeugung so genannter Travel-Lunch-Beutel spezialisiert, die zur Verpflegung von Soldaten in Krisengebieten, aber auch

created. If, for instance, an astronaut wants to "lay" an apple on the table, he sprinkles water on the bottom side. The surface tension of these drops keeps the apple on the surface of the table. Since no gravity exists in outer space, this system functions equally in all directions. A space burger, for example, would have to be filled with ketchup and sauce so that the individual layers "stick" to each other through the force of cohesion. Otherwise, the slices of bread would become detached from the meat, cheese and tomatoes, and all the ingredients would go flying through the air. Hamburgers will thus never become standard astronaut nourishment.

A high level of humidity also guarantees that soft bread that has been cut into candy-sized pieces for consumption in outer space does not crumble when it is taken out of the package or divided into pieces. However, the bread should not be too damp so that it does not make the astronaut suffer from flatulence. The digestive gases which normally leave the body through the effect of gravity cannot escape in the same way in outer space. For inhabitants of the space station any kind of gas development can be extremely unpleasant and has to be massaged away. For this reason no dishes with onions or garlic can be served in outer space.

Furthermore, the developers of space food have to take into account the extremely limited possibilities of heating food in the space station. On the ground heat rises through gravity, resulting in convection in the pot whose content subsequently mixes and heats relatively evenly. Without gravity there simply is no convection. Food burns at the bottom, while it stays cold on top. For this reason, an ordinary stove does not function in space. The astronauts' meals have to be heated on all sides, which on the ISS is actually done with a sort of waffle iron which the cans are inserted in.[10]

Military Calories | Not quite as high but still special: these are the demands made on the functionality of soldiers' food. Since war was invented, military strategists have not only been interested in the movement of troops and sorties but also in the procurement of ultra-light, extremely durable and especially nutritious food for marches. Today soldiers on special missions are nourished with so-called combat rations, freeze-dried meals that have been sealed in bags, which can be transported safely in backpacks and prepared without plates.

These preserved ready-to-serve meals can be kept at any temperature for thirty months. They are absolutely watertight and cannot freeze, since they contain no water and can float because the bags are filled with nitrogen. Simpert Reiter is a German company in Augsburg that has specialized in the production of so-called travel-lunch bags geared to feeding soldiers in crisis areas and mountaineers at 8,000 meters above sea level. Simpert Reiter offers about 10 different breakfasts, 30-40 main courses and 10-12 desserts which

Weltraum-Essen der Nasa: Rindfleisch, Studentenfutter, Cremespinat, Cracker, Cashewnüsse, Beef Steak, Käseaufstrich, kandierte Erdnüsse und Orangensaft

Space food from NASA: Beef, nut and raisin mix, cream spinach, crackers, cashew nuts, beef steak, cheese spread, candied peanuts and orange juice.

von Extrembergsteigern gedacht sind. Simpert Reiter hat rund 10 Frühstücksvarianten, 30–40 Hauptmahlzeiten und 10–12 Desserts im Programm, mit denen die Firma unter anderem die deutsche Armee beliefert. Von der Paella bis zur Mousse au Chocolat sind die Gerichte für die Sahara ebenso geeignet wie für den Mount Everest. Verschiedene Getränke in Pulverform runden das Sortiment ab, sogar Cocktails wie Pina Colada sind erhältlich. Derzeit bearbeitet Simpert Reiter die Anfrage einer Baufirma, die im arabischen Raum tätig ist: Sie hätte für ihre Mitarbeiter gerne gefriergetrocknetes Bier. Die Herstellung ist grundsätzlich möglich, Alkohol und eigene Schaumbildner werden nach der Gefriertrocknung extra beigefügt. Hauptproblem des Projektes ist allerdings noch der Preis. 17 Euro je Liter.[11]

Ready to heat, ready to eat: Fertignahrung |

Neben der Biowelle geht die Tendenz bei der Nahrungsaufnahme in der westlichen Hemisphäre in Richtung Fertigprodukt – ein Trend, der bereits vor über 100 Jahren seinen Anfang nahm. 1812 eröffnete die erste Konservenfabrik, 1886 gelangte Maggis flüssige Speisewürze erstmals zum Verkauf. 1929 erschienen Spaghetti in Dosen der Firma Heinz in den Supermarktregalen, 1972 verkaufte Findus eine Tiefkühllasagne. Mittlerweile reicht die Palette vorgefertigter, bequemer Nahrung vom Junghirschbraten bis zur Omelette. So genannte ready-to-(h)eat-Komponenten ersetzen zeitaufwändiges Kochen und befriedigen den Hunger gestresster Europäer, Amerikaner und Japaner. Die Hersteller von Tiefkühlpizza und Fertiggulasch kreieren sowohl im optischen als auch im aromatischen Sinn Einheitsgeschmäcker für einen stetig wachsenden Kundenkreis. Je schneller die Auswahl an Fertiggerichten wächst, umso größer ist auch der Bedarf an Food Designern, die diese gestalten. Und umso entscheidender wird auch der Einfluss dieser bislang eher unbeachteten Design-Sparte als Spiegel von Zeitgeist und kulturellen Strömungen.

Heutzutage sind Fertiggerichte in den westlichen Industrienationen ein fixer Bestandteil des Speiseplans und ein ständig wachsendes Marktsegment. Die funktionalen Vorteile von essfertigen Produkten liegen auf der Hand: Sie sind in einzelnen Portionen abgepackt, lange haltbar, schnell und mit wenig Arbeitsaufwand zuzubereiten. Bei der Entscheidung für Fertigessen tritt die geschmackliche Qualität zugunsten praktischer Überlegung in den Hintergrund: Fertignahrung ist Funktionsessen.

Historisch gesehen, entstanden Fertiggerichte zunächst zu militärischen Zwecken, als Soldatenproviant. Schon für Alexander den Großen, Hannibal oder Julius Cäsar war auf Feldzügen die Versorgung mit geeigneten Nahrungsmitteln entscheidend, ja unter Umständen sogar eine Frage auf Leben und Tod. Die Soldatenkost musste nahrhaft, monatelang haltbar und unempfindlich gegen Hitze, Kälte, Nässe, Zerbröseln und Zerquetschen sein.

the company also supplies the German army with. From paella to mousse au chocolat, the meals are just as suited for the Sahara as they are for Mount Everest. Various drinks in powder form round off the range – there are even cocktails such as pina colada. At present, Simpert Reiter are processing the order of a construction company working in the Arab region. They would like to be able to offer freeze-dried beer, too. This is in principle possible, alcohol and special foam-producing agents would have to be added after freeze-drying. The main problem of the project remains the price: 17 euros per liter.[11]

Ready to Heat, Ready to Eat: Instant Meals |

The organic food wave aside, food consumption in the western hemisphere is moving in the direction of ready-to-serve products – a trend that already began more than a hundred years ago. In 1822 the first canned food factory was opened. In 1886 sales of Maggi's seasoning liquid were launched. In 1929 spaghetti first appeared on supermarket shelves in cans of the Heinz company. In 1972 Findus sold frozen lasagna. In the meantime the gamut of ready-to-serve food ranges from roast venison to omelettes. So-called ready-to-(h)eat components make time-intensive cooking superfluous and satisfy the hunger of stressed Europeans, Americans and Japanese. The manufacturers of frozen pizza and ready-to-eat goulash create standard tastes for a growing clientèle in both an optical and aromatic sense. The faster the selection of ready-to-eat meals grows, the greater is the need for food designers who compose them. And the more important becomes the influence of this design sector, which had to date gone more or less unnoticed, as the reflection of zeitgeist and cultural trends.

Today ready-to-eat meals have become an integral part of the menu in the western industrial nations and a constantly growing market segment. The functional advantage of ready-to-eat products is obvious – they have been packed in individual portions, they keep longer, and can be prepared quickly and with little effort. In opting for ready-to-eat meals, taste qualities yield to practical considerations – ready-to-serve food is functional food.

Historically speaking, ready-to-eat food first appeared to serve military purposes as soldiers' provisions. Already for Alexander the Great, Hannibal or Julius Caesar it was important to supply soldiers on campaigns with proper food and in some instances it was even a matter of life and death. Soldier's food had to be nutritious, keep for months and be insensitive to heat, cold, moisture, crumbling or crushing.

Der Vorteil gefriergetrockneter Fertiggerichte im Vergleich zu Konserven ist das geringe Gewicht. Zur Erzeugung eines Kilogramms gefriergetrockneten Spinats benötigt man ganze 100 Kilogramm Frischware. Gefriergetrocknetes wird daher von Profibergsteigern oder für Soldaten auf Wüsteneinsatz verwendet.

The advantage of freeze-dried ready-to-eat meals over canned meals is the fact that they weigh less. To produce a kilo of frozen spinach a total of 100 kilos of fresh spinach is needed. Freeze-dried products are thus used by professional mountaineers or for soldiers on a mission in the desert.

Fertiggericht und die dafür nötigen Zutaten: Das erste Fertiggericht der Geschichte wurde von Gerry Thomas entwickelt und kam 1954 in den USA auf den Markt. In einer dreiteiligen Packung aus Aluminium musste das so genannte TV-Dinner nur noch im Backofen erwärmt werden. Das TV-Dinner war bereits im ersten Jahr der Renner: Angeblich wurden 10 Millionen Portionen verkauft.

Ready-to-eat meals and the required ingredients: the first ready-to-eat meal in history was developed by Gerry Thomas and was launched in the USA in 1954. In a three-piece aluminum package, the so-called TV dinner had to be heated in the oven. The TV dinner became a huge success the first year already. It is claimed that 10 million portions were sold

Mit der Nudel zum Sieg | Diesen Anforderungen entsprach zum Beispiel die Trockennudel, eine Vorläuferin unserer heutigen Pasta. Sie wurde von den Arabern entwickelt, die aus der chinesischen Nudel, deren Rezept auf dem Landweg nach Vorderasien gelangt war, das vermutlich erste Convenience-Produkt der Geschichte machten. Die heißen Küstenwinde Siziliens sollen die arabischen Besatzer im neunten oder zehnten Jahrhundert nach Christus inspiriert haben, Teigwaren aus Hartweizengrieß und Wasser in der Sonne zu trocknen.

Die neue Speise eignete sich hervorragend als Proviant und konnte problemlos auf Kriegszügen mitgeführt werden. Geringes Gewicht und schnelle Zubereitung sprachen ebenso für die getrockneten Teigwaren wie der Umstand, dass man nur kochendes Wasser benötigte, um sie in eine nahrhafte Mahlzeit zu verwandeln. Noch heute ist die Instantnudel eine beliebte und kalorienreiche Notration, die nahezu endlos im Vorratsschrank ausharrt, ohne zu verderben, und in wenigen Minuten in ein fertiges Gericht verwandelt werden kann.[12]

Prêt à manger: Fertigprodukte | Von der Schokolade bis zum Instantkaffee verdanken viele haltbare Fertigprodukte ihren Erfolg der Suche nach kriegstauglichem Proviant. So führte 1794 eine entsprechend hohe Belohnung von Kaiser

Toward Victory with Noodles | These demands were, for instance, met by dried noodles – a precursor of today's pasta. It was developed by the Arabs who had transformed the Chinese noodle into the first convenience product in history after the recipe had reached the Middle East. Sicily's hot coastal winds are said to have inspired the Arab occupying force in the ninth or tenth century CE to dry noodles made of semolina and water in the sun.

The new food was exceptionally suited as provisions and could be easily carried on raids. Little weight and quick preparation spoke in favor of the dried noodles, and so did the fact that only boiling water was needed to turn them into a nutritious meal. Even to this very day instant noodles are a popular and high-calorie emergency ration that can be stored almost indefinitely in a pantry without spoiling, to be transformed into a ready-to-eat meal within few minutes.[12]

Pret à Manger: Ready-to-Eat Products | From chocolate to instant coffee, many non-perishable ready-to-eat products draw their success from the search for provisions suitable for war. In 1794, an adequately big reward offered by the emperor Napoleon prompted to

Nahrung als Militärstrategie: 1867 erfand der Berliner Konservenfabrikant Johann Grüneberg eine Fertigsuppe aus Erbsenmehl, Speck und Zwiebeln. Die haltbare, leichte „Erbswurst" passte in jeden Soldatenrucksack, war einfach zuzubereiten, sehr nahrhaft und entschied angeblich sogar den französisch-deutschen Krieg von 1870. Die wurstförmig abgepackte Instantmahlzeit erlaubte den Preußen nämlich eine flexiblere Truppenführung ohne Versorgungsprobleme.

Nutrition as military strategy: In 1867 the Berlin canned food manufacturer Johann Grüneberg developed a ready-to-eat soup made of pea flour, bacon and onions. The non-perishable, light "pea sausage" fit into every soldier's backpack, was simple to prepare and is said to have even decided the outcome of the French-German war of 1870. The sausage-formed packed instant meal enabled the Prussians to deploy troops more flexibly without supply problems.

develop canned food. The stable tin cylinders met all the demands made of non-perishable convenience food. The steel shell is able to withstand jolts as well as humidity and temperature fluctuations – the can thus having rightly earned the designation "iron reserve". The cylindrical can subsequently influenced the size and shape of many ready-to-eat foods with which it was filled. Pineapples, for instance, underwent a radical adaptation to its packaging – going from the oval fruit to the standard, always equally big, circular slices of canned pineapples.

In war campaigns far away from home the question as to supplies could be decisive for victory or defeat. The French-German war of 1870 ended – supposedly because of pioneering culinary innovation – in favor of the Prussian troops. In 1867 the Berlin cook and can manufacturer Johann Heinrich Grüneberg invented the so-called "Erbswurst" – a ready-to-eat soup made of pea flour, bacon and onions and sold the recipe to the Prussian state. The practical, light "Erbswurst" (literally: pea sausage) fit in every soldier's backpack, was easy to prepare and nutritious. When the French-German war broke out a separate factory was built where 1,700 workers produced up to 65 tons of Erbswurst daily to meet the demand. The sausage-shaped instant meal subsequently became an important element of modern warfare, since a preservable and easy-to-transport ready-to-eat meal facilitated more flexible management of troops without supply problems. Since 1889 the Erbswurst has been produced by the brothers Carl Heinrich Eduard and Alfred Knorr and can still be found in the product range of the Knorr company to this day.[13]

Industrial Revolution | A number of ready-to-eat and semi-ready-to-eat foods emerged during or after the Industrial Revolution in the middle of the nineteenth century. Industrialization ushered in the end of agrarian society and resulted in a rapid increase in urban population. These societal changes also turned eating habits upside down since the new mass of workers suddenly had to be supplied with endless amounts of cheap food. As many women were working in the new factories and they thus had no time for cooking, the demand for nutritious fast food suddenly grew. The industrial production of steel, fabrics and other goods spurred the industrialization of foodstuffs. "Whoever works faster also has to eat faster," is an allusive quote of Julius Maggi.

Food design is not an invention of the food industry. For thousands of years we have been cutting, steaming and shaping our food. In essence, we all act as 'food designers' on an everyday basis. With the industrialization of food production, however, food design took on a new dimension, since food suddenly became a mass product manufactured under the same premises, in quantities totaling millions. Thus a number of food creations and techniques were born in the nineteenth and early twentieth century. The following may be cited here: soluble roller-dried milk powder (1855, John A. Just), condensed milk (1856, Gail Borden), Liebig's meat extract (1862, Justus von Liebig), margarine (1869, Hippolyte Mège-Mouriès), vanillin (1874, Wilhelm Haarmann), milk chocolate (1876, Daniel Peter, Henri Nestlé), cornflakes (1876, John Harvey Kellog; 1906, Will Keith Kellog) or dried soups (1886, Julius Maggi and Carl Knorr).[14]

About 150 years ago the issue of functionality assumed an entirely new dimension in food. Pioneering inventions such as the steam machine, the weaving loom or the conveyor belt revolutionized our society. Industry conquered Europe and the USA, changing everyday life in a lasting way, and thus also the way people ate. Countless farmers became workers, housewives became workers, and there was hardly any time left for cooking. People ate at work – in the factory. Supplying the explosively growing numbers of workers with inexpensive and nutritious calories resulted in the industrial manufacturing of foods. Food had to be quick and cheap – not tasty and appetizing. The new growingly machine-based production

Schnell und billig musste das Essen sein, nicht schmackhaft und appetitanregend. Die neue, zunehmend maschinelle Erzeugung von Lebensmitteln veränderte die Zubereitung und die Rezepturen der Gerichte grundlegend. Die Industrielle Revolution transformierte dabei nicht nur, was auf den Tisch kam, sondern auch wie es dorthin kam: Flüssige Rindsuppe wurde zu eckigen Suppenwürfeln, frische Milch zu haltbarem Milchpulver, frische Erbsen zu trockenem Erbsenmehl in Wurstform.

Die Entwicklung von Fertiggerichten wurde vorerst vor allem als soziale Notwendigkeit erachtet. Während Thomas Alva Edison die Glühbirne erfand oder Henry Ford das Automobil weiterentwickelte, arbeiteten Wissenschaftler und Unternehmer wie Justus von Liebig, Julius Maggi oder Henri Nestlé fieberhaft an Methoden und Verfahren, um billige Kalorien für die breite Masse herzustellen. Fleischextrakte, Instantsuppen und Milchpulver wurden neben der Profitabilität, die man sich von ihnen erwartete, zunächst auch als soziale Errungenschaften gesehen.

Essen am laufenden Band | Als besonders nahrhaftes und stärkendes Gericht galt zum Beispiel Rindsuppe. Nachdem der deutsche Wissenschaftler Justus von Liebig 1847 erstmals erfolgreich Rindsuppe extrahiert hatte, war die Substanz allerdings so teuer, dass sie nur in Apotheken als Arznei gegen Schwindsucht und andere Krankheiten angeboten wurde. Erst als Liebig entdeckte, dass Rindfleisch in Uruguay unvergleichlich billiger zu produzieren war, wurde „Liebig's extract of Meat" auch für die breite Masse erschwinglich und fand den Weg vom Krankenbett in den Kochtopf. Die neue, reichhaltige Verpflegung auf Fleischbasis ernährte in weiterer Folge viele Arbeiterfamilien – und legte den Grundstein zur Entwicklung des heutigen Suppenwürfels.

Wie wirksam sich sozial schwache Schichten mit den neuartigen Speisen versorgen konnten und mussten, zeigt ein Preisvergleich: Um 1910 kostete ein Maggi-Rindsuppenwürfel fünf Heller, während man für ein Kilogramm Suppenfleisch eine Krone und fünfzig Heller bezahlen musste – also fast dreißigmal so viel. So machten die sozialen Veränderungen infolge der industriellen Revolution breite Teile der Bevölkerung von industriell vorgefertigter Massennahrung abhängig, die ihrerseits wiederum ein komplett verändertes Essverhalten und damit weitgreifende gesellschaftliche Umbrüche bewirkte.

Schnell und nahrhaft | Dass der Nährstoffgehalt für das Design von Instantsuppen noch im 20. Jahrhundert entscheidend war, zeigt das Beispiel der Knorr Goldaugensuppe. In der Zeit des Wiederaufbaus nach dem Zweiten Weltkrieg wünschten sich Europas Konsumenten von ihrem Essen vor allem eines: Möglichst nahrhaft und fettreich sollte es sein. Die Entwicklungsabteilung der Firma Knorr reagierte auf diesen Wunsch, indem sie eine Instant-Rindsuppe mit auffallend vielen Fettaugen schuf, deren Durchmesser mit 5–10 Millimeter normiert wurde, was eine möglichst große, optische Ähnlichkeit mit der echten Rindsuppe bringen sollte. Und weil das Wort „Fettauge" nicht besonders appetitlich klingt, wurde es auf der Packung kurzerhand durch Goldaugen ersetzt.[15]

of food radically changed the preparation and composition of meals. The Industrial Revolution not only transformed what was served at the table, but also what it looked like. Liquid beef stock became transformed into solid soup cubes, fresh milk turned into milk powder, fresh peas into dried pea flour shaped like sausages.

The development of convenience meals was initially considered a social necessity. While Thomas Alva Edison invented the light bulb or Henry Ford refined the automobile, scientists and entrepreneurs such as Justus von Liebig, Julius Maggi or Henri Nestlé were feverishly working on methods and techniques for producing cheap calories for the masses. Meat extracts, instant soups and milk powder were first seen as social innovations in addition to the profits that they were expected to yield.

Food on the Conveyor Belt | Beef stock, for instance, was regarded as a particularly nutritious and invigorating meal. After German scientist Justus von Liebig had successfully extracted beef stock for the first time in 1847, the substance was so expensive that it was only offered in pharmacies as a medicine for dizziness and other illnesses. It was only when Liebig discovered that beef could be produced considerably cheaper in Uruguay that 'Liebig's extract of meat' also became affordable for the masses and found its way from the sickbed to the cooking pot. The new rich food based on meat subsequently nourished many workers' families – and laid the foundation for the development of today's soup cube.

A comparison of prices illustrates how effectively socially weaker strata could feed themselves with the new meals. Around 1910 a maggi beef stock cube cost five hellers, whereas a kilo of soup meat cost one crown and fifty hellers, i.e., almost thirty times as much. The societal changes resulting from the Industrial Revolution made large parts of the population dependent on industrially produced mass food which, in turn, completely changed eating habits and thus revolutionized society.

Quick and Nutritious | That the nutritious value was still decisive for the design of instant soups in the 20th century can be illustrated by the Knorr "Goldaugen" soup brand. In the reconstruction period following World War II, European consumers primarily expected their food to be nutritious and rich in calories. The development department of the German Knorr company responded to this wish by creating instant beef stock with a strikingly large number of drops of fat ("Fettaugen") which had a standard dimension of 5-10 millimeters (0.2-0.4 inches). This was to create the largest possible visual resemblance with real beef stock. And since the German reference to "Fettaugen" ("eyes of fat") does not sound especially appetizing, it was simply replaced by the word "Goldaugen". i.e., "eyes of gold".[15]

Goldaugen Rindsuppenwürfel von Knorr. 1847 entwickelt der deutsche Wissenschaftler Justus von Liebig den Fleischextrakt. Die Flüssigkeit wird zunächst in kleinen Flaschen als Arzneimittel gegen Schwindsucht und andere Krankheiten verkauft. Als Liebig die Billigproduktion von Rindfleisch in Urugay entdeckt, wird „Liebigs extract of Meat" erschwinglich. Die Suche nach geeigneter Truppenverpflegung sowie soziale Gedanken treiben die Entwicklung von Suppenkonzentraten weiter voran. Bereits um 1900 kann der Suppen-Konsument zwischen 40 verschiedenen Geschmacksrichtungen wählen.

Goldaugen ("Golden Eye") soup cubes from the Knorr co. The German scientist Justus von Liebig developed the meat extract in 1847. The liquid was first sold in small bottles as medicine again dizziness and other illnesses. Once Liebig discovered the inexpensive production of beef in Uruguay, "Liebig's extract of meat" became affordable. The search for food suitable for the troops as well as social ideas gave the development of soup in concentrated form new impetus. Already around 1900 the soup consumer was able to select from 40 different flavors.

Der Hauptvorteil von Fertiggerichten liegt heute sicher nicht mehr in der Nahrhaftigkeit, sondern in der schnellen Zubereitung. Der allgemein bejammerte Zeitmangel in westlichen Industrienationen führt nicht nur zu Fastfood und Fertiggerichten, sondern auch zu kuriosen Designvorschlägen wie den Turbospaghetti „Trifogli". Diese haben einen kleeblattförmigen Querschnitt, damit eine wesentlich größere Oberfläche und sind nach nur drei Minuten Kochzeit weich. Die im Vergleich zu herkömmlicher Pasta aufgequollene und äußerst matschige Konsistenz der Schnellnudel kann man allerdings kritisieren.

Mundgerechtigkeit – Essen in aller Munde |
Ein wichtiger Parameter, der bei der Gestaltung von Nahrung berücksichtigt werden muss, ist die Mundgerechtigkeit. Proportionen und Bewegungsabläufe des menschlichen Körpers liefern die Vorgaben für das Design gut sitzender Kleidung und bequemer Möbel, aber auch wenn es um die Formgebung angenehm zu verzehrender Esswaren geht. Kein Kind wäre mit einem Bonbon von zehn Zentimeter Durchmesser oder mit einem scharfkantigen Lollipop glücklich. Pralinen, Kaugummis oder Kornriegel müssen so geformt sein, dass sie unterwegs bequem und appetitlich verspeist werden können, ohne dass sie zuvor zurechtgeschnitten oder zerteilt werden müssen. Die Frage, warum unser Essen so aussieht, wie es aussieht, lässt sich unter anderem mit der menschlichen Physiognomie beantworten.

Sowohl die Größe des menschlichen Mundes als auch seine Form beeinflussen die Gestaltung von Würsten und Brötchen, von Käsesticks, Mozzarellabällchen, Crackern und Schnitten, von Eislutschern und Schokoriegeln. Am bequemsten für den Konsumenten sind vorportionierte Häppchen, die ohne Abbeißen auf ein Mal im Mund verschwinden. Um sich die

The main advantage of instant meals today no longer lies in the nutritious value but in the quick preparation. The generally lamented lack of time in the western industrial nations has brought forth not only fast food and instant meals but also strange design propositions such as the "Trifogli" turbo-spaghetti. This pasta has a clover-leaf-shaped cross-section, i.e., a significantly larger surface, and is soft after only three minutes of cooking. However, the extremely mushy consistency and the swelling of this type of pasta has come under fire.

Bite-Sized Portions – Food for Every Mouth |
An important variable that must be taken into account in the design of food is that it should be bite-sized or easily fit into our mouths. Proportions and movements of the human body are the parameters for designing well-fitting clothing and comfortable furniture. But they are also important when it comes to giving shape to pleasantly edible food. No child would he happy with candy that has a diameter of four inches or sharp-edged lollipops. Chocolate candies, chewing gum or muesli bars have to be shaped so that they can be eaten on the road comfortably and appetizingly without having to be cut into pieces. The question why our food looks the way it does can also be answered in terms of human physiognomy.

Both the size of the human mouth and its shape influence the design of sausages and sandwiches, of cheese sticks, mozzarella balls, crackers and cream slices, popsicles and chocolate bars. For the consumers, preportioned canapés are most convenient, since they vanish in the mouth at once. To avoid getting everything on our fingers, petit fours and chocolate candies that are always eaten without cutlery make for a taste with a

Finger nicht zu bekleckern, sind Petits fours und Pralinés, die immer ohne Besteck verspeist werden, der Genuss eines einzigen Bissens. Wobei Pralinen nicht notwendigerweise aus Schokolade sein müssen: In Japan vertreibt der Molkereikonzern Snowbrand runde Bonbons aus Käse als kleinen Snack für zwischendurch. Die Kugeln aus Streichkäse tragen den Namen „cheese catch", sind in den Geschmacksrichtungen Erdbeere und Milchkaffee erhältlich und schmecken angeblich besonders gefroren gut. Die Käsebonbons sind einzeln verpackt und können bequem mit zwei Fingern in den Mund gesteckt und gekaut werden.[16]

Soll der Verzehr etwas länger dauern als bloß einen Bissen lang, dominieren schlanke Formen zum Abbeißen. Nach dem formalen Vorbild der Banane erlauben Snackwürstchen, Strip Cheese, Shortbread Finger Schokosticks, längliche Eclairs, Schokobananen, Biskotten oder Katzenzungen einen zwar mobilen, aber dennoch appetitlichen Verzehr ohne Messer und Gabel. In England werden rechteckige Sandwiches aus Gründen der Mundgerechtigkeit mit einem Schnitt in zwei Dreiecke zerteilt, ähnlich wie italienische Tramezzini, die auch dreieckig über die Theke gereicht werden. Eine mundgerechte Nobelvariante des herkömmlichen Sandwichs wird zum englischen Afternoon Tea serviert: Aus der elitären Tradition, Weißbrot zu entrinden, gingen die so genannten Fingersandwiches hervor, für die nur das weiche Innere der Toastbrotscheiben in längliche Rechtecke geschnitten wird, deren Form entfernt an Finger erinnert und optimal in den Mund passt.

Auch als die Lebensmittelabteilung der Firma Procter und Gamble begann, Esswaren wie Kuchen, Erdnussbutter und Chips anzubieten, stellte man Überlegungen zum Thema Mundgerechtigkeit an. Da der Markt für konventionelle Chips ziemlich ausgereizt war, wollte man einen Kartoffelchip herausbringen, der sich deutlich von allen Konkurrenzprodukten unterschied. Er sollte immer gleich groß und gleich dick sein

single bite. Praline-type candy does not necessarily have to be made of chocolate. In Japan the Snowbrand dairy corporation sells round candies made of cheese as a snack in-between meals. The balls made of cream cheese, named "cheese catch", are available with strawberry and milk coffee flavor and supposedly taste particularly good when they are frozen. The cheese candies are packed individually so that they can be put in the mouth with two fingers and chewed.[16]

If the consumption should take a bit longer than a bite, narrower shapes are preferred for biting off. Following the formal model of the banana, snack sausages, strip cheese, shortbread fingers, chocolate sticks, longish éclairs, chocolate bananas, biscuits or the popular Austrian "Katzenzungen" ("cat tongues") chocolates permit mobile but appetizing consumption without knife and fork. In England square sandwiches can be divided into two triangles with one cut to make them mouth-sized, similar to Italian Tramezzini that are also served over the counter as triangles. A mouth-sized more sophisticated variant of the ordinary sandwich is served with afternoon tea. Out of the elitist tradition of removing the crust from white bread, so-called finger sandwiches emerged, for which only the soft inside of toast slices is used. These sandwiches are longish squares which vaguely recall fingers and ideally fit in the mouth.

Even when the Food and Beverage Division of the Procter and Gamble Company began manufacturing foods such as cakes, peanut butter, people there began to reflect on offering bite-sized morsels. Since the market for conventional chips had been fairly saturated, they wanted to launch a potato crisp that differed radically from all competing products. The

Aufschnitt- versus Snackwurst: Mundgerechtigkeit ist ein wesentlicher Faktor von Food Design.

Cold cuts vs. snack sausage: making mouth-sized portions was a crucial factor of food design.

und aussehen wie der perfekte Chip. Nachdem Tests ergeben hatten, dass die Mehrzahl der Esser immer einen ganzen Chip mit der Krümmung nach oben in den Mund steckt und nur einige wenige den Cracker umdrehen, so dass die gebogene Seite nach unten weist, wurden die zweiseitige Krümmung und die Größe von Pringles so gewählt, dass sie gut zwischen Gaumen und Zunge liegen und angenehm zu zerbeißen sind..17

Vermutlich wird man in puncto Mundgerechtigkeit zukünftig auch noch mehr auf die Unterschiede zwischen weiblicher und männlicher Physiognomie eingehen. Eine echte Marktlücke sind zum Beispiel schmale, schminkfreundliche Sandwiches, Burger oder Brötchen in Riegelform, von denen auch ein kleiner Mund abbeißen kann, ohne dass der Lippenstift verschmiert. Vermutlich liegt auch ein Gutteil des Erfolges des Sushi darin, dass man respektive frau von den kleinen, schlanken Stücken besser und vor allem eleganter abbeißen kann als von einem Big Mac. (Obwohl man in Japan traditionellerweise nie von einem Sushi abbeißt!)

Portionsgröße | Pralinen, Müsliriegel, Würstchen oder Eislutscher werden als vorverpackte Einheiten mit ähnlichen Größen und Gewichten auf den Markt gebracht. Die Parameter zur Wahl der richtigen Kalorienmenge entsprechen der Verzehrssituation. Handelt es sich um eine Haupt-, eine Zwischenmahlzeit oder um einen Snack? Und welche Zielgruppe soll überhaupt angesprochen werden? Instantsuppen wie Cup-Noodles beispielsweise gibt es in Japan in drei verschiedenen Größen, die den Verzehrportionen von Haupt-, Zwischen- und Snackmahlzeit entsprechen und deren Füllmenge immer geringfügig unter der dabei durchschnittlich verzehrten Kalorienmenge liegt. So hinterlässt das Produkt stets ein positives Gefühl, nämlich jenes, dass man gern mehr davon hätte, und der Konsument öffnet vielleicht eine zweite Packung.18

crisp would be equally large and thick like other potato chips but would be uniform in shape unlike other potato chips. Test panels showed that the majority of the crisp eaters would place one entire crisp in the mouth saddle-side up and that only a few turned it saddle-side down. The double curvature and size of Pringles was specifically designed so that the crisp would fit well between the teeth and the tongue and would be pleasant to bite. 17

As for the right mouth-sized portion, more efforts will be made to focus on the differences between female and male physiognomy in the future. In this context, veritable market niches could open up for thin makeup-friendly sandwiches, burgers or sandwiches in the shape of a bar which can also be bitten by a small mouth, without smearing the lipstick. A large part of the success of sushi is due to the fact that people (especially women) are better able to bite the small, thin pieces and that it is more elegant to eat sushi than a Big Mac. (Even though in Japan no one traditionally takes a bite from a sushi!)

Portion Size | Chocolate candies, muesli bars, sausages or ice-cream lollies are sold on the market pre-packaged, in similar sizes and weights. The parameters for selecting the right amount of calories correspond to consumption situation. Is it a main meal or a snack? And which target group are they geared to? Instant soups such as Cup Noodles, for instance, are available in Japan in three different sizes corresponding to the portions of main and in-between meals as well as snacks, and whose contents always lie slightly below the average number of calories eaten on these occasions. This way the product always leaves a positive feeling, namely that one would like to have more of it, and the consumer will perhaps open a second package.18

Der Entscheidung für oder gegen eine neue Form gehen unter anderem Studien über Ess- und Kauverhalten voraus. Ob und wie oft der Konsument abbeißt, definiert Größe und Geometrie des Produkts. Der menschlichen Physiognomie entsprechend eignen sich runde Formen besser als kantige, um auf ein Mal in den Mund gesteckt zu werden. Rechts außen | Die Kinderwürstchen der steirischen Firma Messner sind speziell auf den Verzehr mit kleinen Händen abgestimmt.

Studies on eating and chewing habits come before a decision is made on one or the other new form. Whether and how often a consumer bites something off determines the size and geometry of the product. In keeping with human physiognomy round forms are better suited for being put in the mouth in one go than angular forms.

Snack-, Süßwaren- und Eiscremehersteller stimmen die Portionsgröße ihrer Produkte ganz gezielt auf die Verzehrssituation, aber auch auf das Alter der jeweiligen Konsumentengruppe ab. Das speziell für Erwachsene entwickelte Magnum passt kaum zwischen die Lippen eines Dreijährigen und wäre mit seinen 86 Gramm ohnehin zu reichhaltig. Jene Eissorten, deren Farb- und Formgebung sie eindeutig als Kinderspeise definieren, wiegen kaum mehr als 50 Gramm und wirken im Munde eines Erwachsenen eher verloren. McDonald's bietet kleine, leichte Ham- und Cheeseburger für Kinder und größere, schwerere Big Macs oder Royalburger für Erwachsene an und Burgerking eigene Minihamburger.

Proportion und Zeitgeist | Die Größe der angebotenen Portionen verändert sich aber auch mit den jeweiligen Vorstellungen von Lebensstil und Zeitgeist und wird von Herstellern immer wieder adaptiert. So brachte etwa der Zuckerwürfel ursprünglich 6 statt wie heute 4 Gramm auf die Waage. Auch Pralinen verändern ihre Größe im Sinne des Zeitgeists. Das italienische Schokoladebonbon „Bacio" von Perugina büßte ebenso wie der Gummibär seit seiner Erfindung im Jahr 1922 fast ein Drittel seines Gewichts ein, auch der Gummibär schrumpfte in diesem Zeitraum. „Klein aber fein" lautet die Devise des 21. Jahrhunderts. Bacetti heißt die 10-Gramm Minivariante, die mittlerweile ebenso wie Schokoriegel, Eislutscher und Hamburger, Würste und Tiefkühlgemüse im Kleinformat angeboten wird. Der Trend korrespondiert mit der Gesundheits- und Schlankheitswelle und steigert die Lust am Erlebnis „Essen". Fabrikanten pressen vermeintlich ungesunde Speisen wie Schokolade in immer kleinere Portionsgrößen und liefern Linienbewussten einen guten Vorwand, zu Naschereien zu greifen. Vom Eislutscher bis zum Fruchtjoghurt findet sich derzeit fast alles im Miniaturformat in den Supermarktregalen.[19]

Snacks, sweets and ice cream manufacturers adapt the portion size of their products specifically to the setting where eating takes place, but also to the age of the given consumer group. The Magnum developed especially for adults hardly fits between the lips of a three-year-old and with its 86 grams, it would be too rich anyway. Those ice cream brands whose color and form clearly identify them as children's food hardly weigh more than 50 grams and look rather out of place in the mouth of an adult. McDonald's offers small, light hamburgers and cheeseburgers for children and larger Big Macs or Royal Burgers for adults, Burger Bing even has its own mini-hamburgers.

Proportion and Zeitgeist | The size of portions on offer also changes with the respective ideas of life style and zeitgeist and is constantly adapted by manufacturers. For example, the sugar cube originally weighed six instead of four grams. Even chocolate candies change their size in keeping with the times. Similar to the gummi bear, which has also shrunken in size, the Italian chocolate candy "Bacio" manufactured by Perugina has lost almost a third of its weight since it was invented in 1922. "Small but nice" is the slogan of the twenty-first century. Bacetti is the name of the 10 gram mini-variant which is meanwhile on the market, just like there are miniature portions of chocolate bars, popsicles and hamburgers, sausages and frozen vegetables. The trend corresponds to the health wave and aspirations to slimness and increases the pleasure in the "eating" experience. Manufacturers offer supposedly unhealthy foods such as chocolate in ever-smaller portions supplying those aware of their waistline a good reason to nevertheless reach for those snacks. From the popsicle to fruit yoghurt, almost everything is available in miniature format on supermarket shelves.[19]

Optimal zum Abbeißen sind längliche Formen mit rundem Querschnitt nach dem natürlichen Vorbild der Banane. Dreiecke sind bei Nahrungsmitteln Sonderformen, da sie schlecht zu essen sind und die spitzen Winkel psychologisch bedrohlich wirken.

Longish shapes with a round cross-section based on the natural model of the banana are optimally suited for being bitten off. Triangles are exceptional forms in foods since they are difficult to eat and the pointed angles have a threatening effect psychologically.

Auch bei „natürlichen" Esswaren sind Portionierung und Packungsgrößen ein Thema, etwa bei Obst und Gemüse. Große Früchte wie Melonen oder Kohlköpfe laufen im Zeitalter von Kleinfamilie und Singlehaushalten Gefahr, sich nicht mehr zu verkaufen. Die steigende Nachfrage nach immer kleineren Verzehreinheiten führte in Japan sogar zu speziellen Gemüsezüchtungen. Nach Meinung des japanischen Detailhandelsriesen Aeon Co. Ltd sind Singles außerstande, große Gemüsesorten allein zu verspeisen. Der Konzern bietet daher speziell gezüchtetes Zwerggemüse an. Auf die Hälfte des üblichen Formats geschrumpfte Rettiche, Mini-Kürbisse, Mini-Karfiol (Blumenkohl) und andere Gemüsesorten sollen den demographischen Entwicklungen Rechnung tragen.

Even with "natural" food, portions and package sizes are an issue, for instance, when it comes to fruit and vegetables. Large fruits such as melons and cabbage heads are in danger of no longer being sold in the age of the small family and single-households. In Japan the growing demand for ever-smaller portions resulted in special vegetables being cultivated. According to the Japanese retail giant Aeon Co. Ltd., singles are simply not able to consume large vegetables on their own. The corporation thus offers specially cultivated miniature vegetables. Radishes shrunken to half their normal size, mini-pumpkins, mini-cauliflower and other types of vegetables reflect demographic changes.

Resteverwertung – von Wurst bis Knödel

Von französischen Pasteten und englischen Pies bis zu chinesischen Jiaozis, von italienischen Tortellini bis zu indischen Samosa dienen essbare Ummantelungen als Schutzschicht, als Konservierungsmethode oder als simples Verpackungsmaterial. Sie verstecken aber auch die oft nicht sehr ansehnliche Wiederverwertung von Resten. In der Koch- und Küchenkultur finden sich viele Tricks für die Verwandlung von Rückständen in neue, appetitliche Gerichte. Überbleibsel oder sonst wenig beliebte Fleischstücke werden zerkleinert, püriert oder faschiert, neu zusammengefügt und mit einer attraktiven Ummantelung optisch aufpoliert. Teige, Därme oder Glasuren dienen oft als Versteck für minderwertige Rohstoffe. Würste beispielsweise dürfen sich in den seltensten Fällen allzu luxuriöser Zutaten rühmen. Ihre Entstehungsgeschichte liest sich wie die des kulinarischen Recyclings schlechthin.

Recycling Leftovers – from Sausage to Dumplings

From French paté and English pies to Chinese jiaozis, Italian tortellini or Indian samosa, edible wrappings serve as a protective cover, as a preservation method or as simple packaging material. However, they also often conceal the not so appealing way of processing leftovers. In culinary culture, we find a number of tricks that can be used for transforming food remains into new, appetizing dishes. Leftovers or not so popular pieces of meat are cut into small pieces, mashed or turned into meatloaf, reassembled and spruced up with an attractive cover. Dough, guts or icings often serve to hide lower-quality raw materials. Only very rarely can sausages, for instance, boast luxurious ingredients – their production can be seen as the quintessential case of culinary recycling.

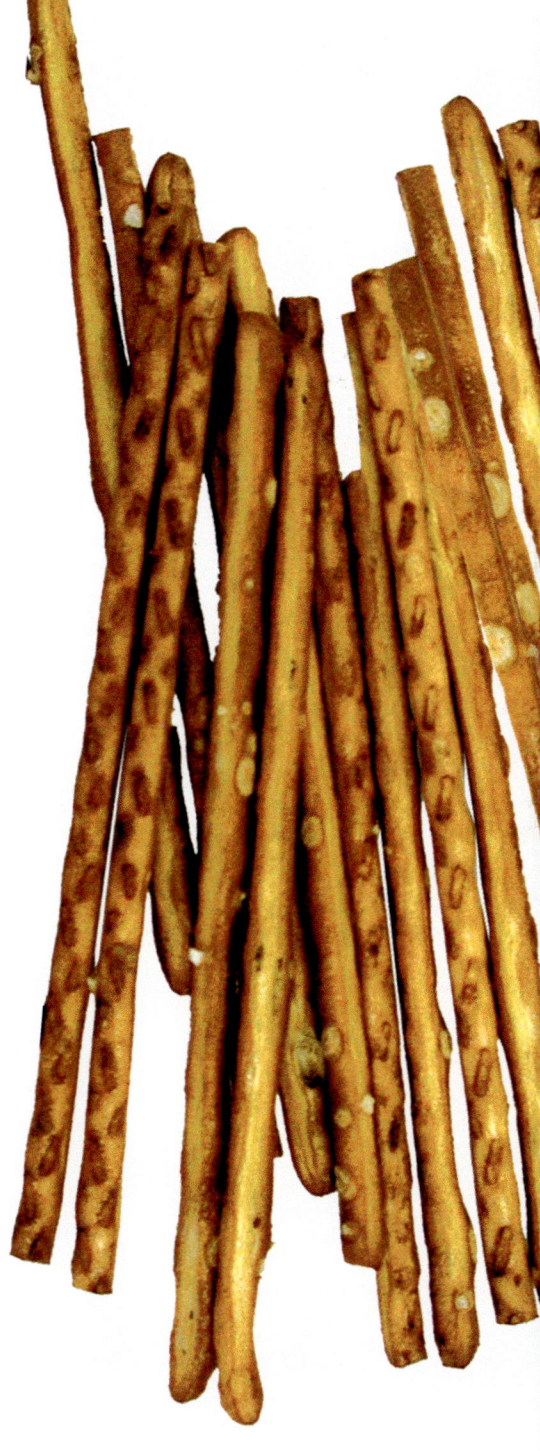

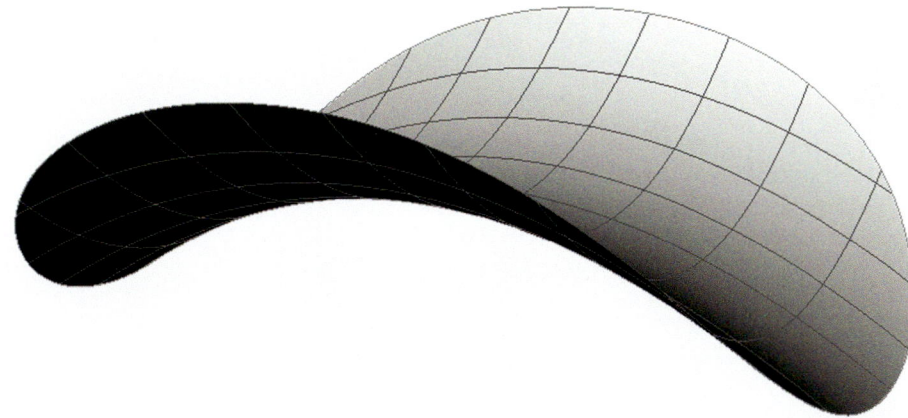

Unten | Die zweiseitige Krümmung von Pringles ist so gewählt, dass sie gut zwischen Gaumen und Zunge liegen und angenehm zu zerbeißen sind. Tests geben dem Design Recht: Die Mehrzahl der Esser steckt Pringles als Ganzes in den Mund, nur einige wenige drehen den Cracker um und beißen ab.
Rechts | Die Firma Kellys produziert Salzstangen, die so dünn sind, dass die meisten Konsumenten gleich mehrere auf einmal in den Mund stecken.

left | The two-sided curvature of Pringles has been selected because it lies well between gum and tongue and is pleasant to bite. Tests prove the design to be just right: the majority of eaters put the entire Pringles in their mouth, only a few turn the cracker and bite off a piece.
right | The Kelly's company produces cracker sticks so thin that most consumers put several of them in their mouth at the same time.

Schon die alten Griechen kannten das Geheimnis des Verwurstens, wie Homer anlässlich eines wurstigen Festes von Penelopes Werbern in der Odyssee berichtet. In der Antike füllte man Fleischstücke, Blut und Innereien in Tiergedärme und schätzte neben dem Geschmack auch die konservierenden Eigenschaften der Wurst. Die Zugabe von Salz sowie das Trocknen oder Räuchern verlängerte die Haltbarkeit des verwendeten Fleisches. Zu den funktionalen Vorteilen gesellte sich schon damals ein leicht frivoler Beigeschmack beim Verzehr der krumm geformten Speise. Der Konsum von Würsten war aufgrund der formalen Ähnlichkeiten mit den männlichen Geschlechtsteilen während des aufkommenden Christentums zeitweise sogar verboten. Fleischer und Esser ließ das allerdings relativ unbeeindruckt. Wie leidenschaftlich in Europa bis heute verwurstet wird, zeigen nicht zuletzt die unzähligen ortsspezifischen Wurstnamen: Wiener, Krakauer, Nürnberger, Göttinger, Debreziner, Mailänder, Polnische, Braunschweiger, Lyoner und andere bilden eine eigene Geographie der Wurstbezeichnungen.

Wiener Frankfurter | Das Frankfurter Würstchen, überall sonst als „Wiener" bekannt, erblickte 1805 im 7. Wiener Gemeindebezirk das Licht der Welt. Der in Bayern geborene und in Frankfurt ausgebildete Fleischermeister Johann Georg Lahner entwickelte damals eine neuartige Wurst aus Rind- und Schweinefleisch. Das Einzigartige an der Rezeptur war die Kombination zweier verschiedener Fleischsorten, was nach der Frankfurter Fleischerordnung als Sakrileg verabscheut wurde, im offenbar freizügigeren Wien jedoch erlaubt war. Zur Herstellung echter Frankfurter wird das Fleisch zunächst von Sehnen befreit, mit Hacken zerkleinert, hernach mit schweren Holzschlegeln weich geklopft und mit großen Wiegemessern zerschnetzelt. In Holztrögen wird das Wurstbrät in weiterer Folge zusammengemischt und anschließend mit großen Handspritzen in Schafsaitlinge eingedrückt. Die daraus abrollende lange Wurst wird in entsprechend kurze Teile geschnitten und in der Mitte zu einem Paar abgedreht. Auf diese Weise können die Zwillingswürste einfach über eine Stange gehängt und in die Räucherkammer geschoben werden, wo sie schließlich den typischen Geschmack erhalten

Die „Wiener Frankfurter", wie die neuen Würste in der Fleischerei Lahner genannt wurden, sorgten für ein sprachliches Durcheinander. Bis heute existieren in der gleichnamigen Finanzmetropole die originalen „Frankfurter" (bereits 1562 als „Bratwerscht" urkundlich erwähnt), während Lahners Würstchen im restlichen Deutschland „Wiener" heißen. Per definitionem handelt es sich bei den Frankfurter „Frankfurtern" um dünne Brühwürste aus reinem Schweinefleisch, die in einem Naturdarm stecken und ihren Geschmack einem speziellen Räucherverfahren verdanken. In Deutschland dürfen nur im Raum Frankfurt produzierte Würstchen den Namen „Frankfurter" tragen, ansonsten nennen sie sich „nach Frankfurter Art" oder ähnlich. Die deutsche Wurstversion unterscheidet sich von den Wiener Namensvettern allerdings nicht nur durch die fehlende Rindfleischfüllung, sondern auch durch ihr Aussehen. Zwar ist das deutsche Produkt ebenso schlank, doch lagern deutsche Fleischer ihr Würstchen gerne Mann an Mann in Kisten. Die Enge verleiht ihnen einen etwas eckigen Querschnitt, während die Wiener Spezialität perfekt gerundet in den Wurstkessel wandert.

Um die Namensverwirrung perfekt zu machen, verzehren auch die Amerikaner eine Wurst namens „Frankfurter". Inhaltlich ist die amerikanische Frankfurter weder mit der österreichischen noch mit der deutschen verwandt, ihre Rezeptur stammt ursprünglich allerdings sehr wohl aus der hessischen Wurstmetropole Frankfurt: Speziell für die jüdische Kundschaft hatte die Frankfurter Metzgerei Gref-Völsing im Jahr 1894 eine koschere Wurst aus Rindfleisch entwickelt und damit eine Rezeptur geschaffen, die in den USA bis heute unter dem Begriff „All-Beef-Frankfurter" gern gegessen wird.[20]

Wurstverwurstung | Fleischreste wandern in die Wurst – und Wurstreste in den Knödel, die Maultasche oder Pierogi. Der übrig gebliebene Braten vom Sonntag gesellt sich zum Mittwochschinken, um in Kombination mit Abschnitten vom Donnerstag am Freitag auf den Tisch zu kommen. Das verwendete „Verpackungsmaterial" definiert dabei Form und Aussehen des „Wochenrückblicks". Während Mehl-Wasser-Mischungen bei slawischen Pierogi, italienischer Pasta oder chinesischen Jiaozis zu halbrunden oder dreieckigen Taschen oft mit formal stark ausgeprägten Rändern verarbeitet werden, resultieren Hefe- oder Kartoffelteige in handgerollten Knödeln oder Dim Sum mit eher dicker, flaumiger Hülle.

Auch die Pastete gehört zu den Klassikern der kulinarischen Resteverwertung. Der Ausdruck leitet sich vermutlich vom italienischen Verb „pastare" ab, was so viel wie „Teig bearbeiten" bedeutet. Schon auf sumerischen Keilschrifttafeln, die um 1700 vor Christus entstanden, finden sich Rezepte für Geflügelpasteten. In der Antike waren Pasteten bei Veranstaltungen in Stadien und Amphitheatern ein beliebter Snack. In Zentraleuropa wurden Pasteten vermutlich um das Jahr 1000 in Frankreich populär. Den Gipfel ihrer Beliebtheit erlebten sie während der Zeit der Renaissance, als betuchte Gastgeber einander mit besonders gewürzten, aufwändig geformten und exaltiert gefüllten Pasteten gegenseitig übertrumpften. Glaubt man einschlägigen Erzählungen, sollen sogar ganze Taubenschwärme und Musikorchester unter Pastetenteig versteckt gewesen sein.

Rundstück Warm – vom Resteessen zum Blockbuster | Auch der mittlerweile weltberühmte Hamburger startete seine Karriere als Resteessen. Den Vorläufer des Inbegriffs US-amerikanischer Essgewohnheiten bildete das so genannte „Rundstück Warm", ein rundes Brötchen, das in Hamburg nachweislich bereits im 17. Jahrhundert existierte und dort mit den Resten des Sonntagsbratens gefüllt wurde.[21] Der Brauch, Reste vom Sonntagsmahl zwischen zwei Brotstücken unter der Woche zu verzehren, wurde im 19. Jahrhundert von deutschen Emigranten via Hamburger Hafen auf Emigrantenschiffen nach Amerika gebracht. Vermutlich war die norddeutsche Sandwich-Variante an Bord auch einfacher zu essen als Suppen oder andere Gerichte, für die man Besteck benötigte. Und bis heute dienen die Brotdeckel nicht nur als Sättigungsbeilage, sondern auch als Isoliermaterial, das die empfindlichen Finger vor dem heißen, fettigen Fleischlaibchen schützt.

The ancient Greeks already knew the secret of making sausage, as Homer relates in the Odyssey on the occasion of a sausage-rich feast celebrated by Penelope's suitors. In ancient times, animal guts were filled with pieces of meat, blood and offal. In addition to the taste, the preserving properties of sausage were highly valued. By adding salt or drying or smoking, the life of the meat used could be prolonged. Apart from the functional advantages, there was also a slightly frivolous connotation to the curved sausages. The similarity with the male sexual organ meant that the consumption of sausages was sometimes even prohibited in the early period of Christianity. However, this did not make a very big impression on butchers and eaters. The great passion invested in the production of sausages in Europe to this very day is reflected in countless local sausage names: Wiener, Krakauer, Nürnberger, Göttinger. Debreziner, Mailänder, Polnische, Braunschweiger, Lyoner and others constitute an atlas in its own right composed of sausage names.

Wiener Frankfurter | The Frankfurter sausage, known everywhere else as "wiener" was born in Vienna's seventh district in 1805. At that time, butcher Johann Georg Lahner, born in Bavaria and trained in Frankfurt, developed a new type of sausage consisting of beef and pork. What made the recipe unique was the combination of two types of meat, which according to the Frankfurt meat code was sacrilegious whilst in more liberal Vienna it was apparently permitted. To produce the real Frankfurter, the meat is first freed from tendons, chopped into small pieces, beaten with heavy wooden mallets and diced with large rocker knives. The sausage meat is then mixed in wooden vats and finally pressed into sheep's guts with large syringes. The long sausage that results from this is then cut into the proper small sizes, two always being bound in the middle to form a pair. These twin sausages are hung over a pole and pushed into the smoking chamber where they finally take on their typical taste.

The "Wiener Frankfurter", as the new sausages had been dubbed in Lahner's butcher shop, spawned linguistic confusion. To this day, the original "Frankfurter" (mentioned in a document as "Bratwerscht" already in 1562) exist only in the eponymous city, whereas Lahner's sausage is called "Wiener" in the rest of Germany. By definition, the Frankfurt "Frankfurter" were thin sausages made of pure pork, which had been inserted in a natural gut and given its taste by means of a special smoking procedure. In Germany only sausages produced in the Frankfurt area are allowed to be called "Frankfurter" – elsewhere they are known as "in the Frankfurt style" or something similar. The German sausage variety differs from its Viennese namesake not just by virtue of the missing beef filling but also through its appearance. While the German product is also slender, the German butchers like to store the sausages side by side in boxes. Their narrow space lent them a somewhat square cross-section, while the Viennese specialty lands in the sausage pot in a perfectly round shape.

In order to make the name confusion perfect, the Americans also eat a sausage called "frankfurter". The American frankfurter is related neither to the Austrian nor to the German one. However, its recipe originated in the Hessian sausage metropolis of Frankfurt. The Frankfurt butcher's shop Gref-Völsing developed a kosher sausage made of beef for its Jewish clientèle and thus created a recipe that is still popular today in the USA as "all-beef frankfurter".[20]

Processing Sausage | Leftover meat ends up in sausage – and leftover sausage in dumplings, or in the "maultasche" or the "pierogi". The remaining Sunday roast joins the Wednesday ham to be served together with Thursday's cold cuts on Fridays. The "packaging material" used defines the form and appearance of the "week's retrospective". While flour-water mixtures are processed to create semi-round or triangular stuffed pastries in Slavic pierogis, Italian pasta or Chinese jiaozis, which often have formally well defined edges, yeast-based or potato dough result in in hand-rolled dumplings or Dim Sum with a rather thick, fluffy outside.

The paté is one of the classics of the culinary leftover recycling. The word is probably derived from the Italian word "pastare", which means as much as "processing dough". Recipes for poultry paté can even be found on Sumerian hieroglyphic tablets from around 1700 CE. In ancient times patés were a popular snack at events that took place in cities and amphitheaters. In Central Europe, patés probably became widespread in France around 1000. They experienced the culmination of their popularity during the Renaissance period when well-to-do hosts tried to trump each other with specially spiced, elaborately shaped and richly filled patés. If one believes the tales, there were even entire swarms of doves and orchestras hidden below paté dough.

Hot Roll – From a Leftover Dish to a Blockbuster | Even the meanwhile famous hamburger started off as leftover food. The precursor of the quintessence of American eating habits was the so-called "Rundstück Warm" (hot roll) – a round roll whose existence was already documented in seventeenth century Hamburg and was filled with the remains of the Sunday roast.[21] The custom of eating leftovers from the Sunday meal between pieces of bread during the week was brought to America by German emigrants who set sail from the port of Hamburg. It was probably easier to eat the northern German sandwich variant on board than soups or other dishes for which cutlery was needed.

Dr. Franz Radatz bei der Frankfurter-Erzeugung in Wien. 1805 erfindet der aus Frankfurt stammende Fleischer Johann Georg Lahner in Wien das so genannte „Wiener Frankfurter". Es besteht aus Rind- und Schweinefleisch, wird in einen Schafssaitling gefüllt und leicht geräuchert. Als „Wiener Würstchen" erobert die neue Rezeptur die ganze Welt, nur in Wien bleibt der Ausdruck „Frankfurter" erhalten.

Frankfurter, or Wiener sausage production at the Radatz company in Vienna. In 1805 the butcher Johann Georg Lahner from Frankfurt invented the so-called "Viennese Frankfurter" in Vienna. The sausage consists of beef and pork, is stuffed in a skin made of sheep's gut and lightly smoked. The new recipe conquered the world as "Vienna sausage" and it is only in Vienna that the name "Frankfurter" remained.

In den USA entwickelte sich aus dem „Rundstück Warm" das Hackfleischsteak zwischen den charakteristischen „bun"-Hälften. Der Bratwurstverkäufer Charlie Nagreen soll die heutige Form des amerikanischen Nationalgerichts erstmals 1892 in Ohio serviert haben. Das Gericht blieb anfänglich jedoch mäßig erfolgreich. Zur Grundlage der Fastfood-Industrie und damit zum globalen Megaseller wurde der Hamburger erst mit der Gründung der McDonald's-Kette. Am 15. Mai 1950 öffneten die Brüder Richard und Maurice McDonald im kalifornischen San Bernadino ihr erstes Restaurant, das durch die Entwicklung rationalisierter Herstellungsmethoden Hamburger zu einem extrem guten Preis anbieten konnte. Die Effizienzsteigerung bei der Zubereitung, die offene Küche und die Umstellung auf Selbstbedienung legten den Grundstein zum späteren Welterfolg.[22]

Außen rosa, innen braun | Auch süße Gerichte bieten gute Möglichkeiten, „gebrauchte" Zutaten rundum zu erneuern und unter Zuckerglasur oder Schokoüberguss wieder einzusetzen. So wandern zum Beispiel beim Wiener Punschkrapfen unansehnliche Tortenreste, in viel Rum getränkt, unter einen knallrosa Zuckerguss und damit in neuem Glanz zurück in die Kuchenvitrine. Zwar sieht das Originalrezept ausschließlich frische Biskuitabschnitte vor, doch vermag die hübsche, pinke Glasur auch andere Teigreste gut zu verbergen. Nach einem ähnlichen Schema funktioniert der Strudel, der ursprünglich aus Vorderasien stammt und mit salzigem Inhalt konsumiert wurde. Über Ungarn gelangte das Rezept nach Mitteleuropa, wo es etwa seit dem 18. Jahrhundert als Mehlspeise bekannt ist und auch mit wenig ansehnlichen, überreifen Früchten oder Fallobst gefüllt werden kann.

Essbare Hüllen verbergen aber nicht nur Reste, sondern auch unerlaubte Zutaten. Manche Teigummantelungen vollführen religiös-motivierte Täuschungsmanöver mit dem Zweck, Fleischspeisen für Fasttage als vegetarische Gerichte zu tarnen. In Schwaben beispielsweise nennt man Maultaschen heute noch gerne „Herrgottsbscheisserl". Die Fleischfülle wird ähnlich wie bei Ravioli in Teig verpackt und damit für den lieben Gott unsichtbar! Maultaschen dürfen somit auch an Freitagen und während der Fastenzeit mit ruhigem Gewissen verspeist werden.

In the United States, the hot roll ("Rundstück Warm") became the ground meat steak that was served between the typical "bun" halves. The bratwurst seller Charlie Nagreen is said to have served today's form of the American national meal for the first time in Ohio in 1892. Initially, the dish was not terribly successful. The hamburger only became the mainstay of the fast food industry, and thus the global mega-seller, when the McDonald's chain was founded. On 15 May 1950 the brothers Richard and Maurice McDonald opened their first restaurant in San Bernadino, California – where they were able to offer the hamburger at an extremely good price thanks to the development of rationalized production methods. The increased efficiency in the preparation, the open kitchen and the switch to self-service laid the ground for the later world success.[22]

Pink on the Outside, Brown on the Inside | Even sweet dishes offer good opportunities to completely renew "used" ingredients and to put them to good use by adding sugar coating or chocolate icing. For instance, unappealing cake leftovers that have been dipped in a lot of rum, covered with bright pink sugar icing and thus make a new striking appearance, end up back in the kitchen cabinet. While the original recipe calls for nothing but fresh pieces of biscuit, the nice pink icing is also well-suited for hiding other dough leftovers. The strudel functions according to the same principle. It originally hailed from Southwest Asia and was eaten with a salty filling. Via Hungary, the recipe reached Central Europe where it has been known as a sweet pastry since the eighteenth century and could also be filled with less appealing, overly ripe fruits or windfall.

Edible wrappings not only cover leftovers but also non-permitted ingredients. Some dough coverings pull off religiously motivated deceptive maneuvers aimed at concealing meat meals for holidays as vegetarian dishes. In Swabia, for instance. "maultaschen" are still known as "Herrgottsbscheisserl" ("Bamboozle the Lord"). The meat filling is packed in dough, much like ravioli, and thus made invisible to God! Maultaschen can thus also be eaten on Fridays and during Lent with a good conscience.

---

1 | Interview mit Werner Mlodzianowski, Leiter des Technologietransferzentrums Bremerhaven, am 18. u. 19.10.2007
2 | Christoph Wagner, Fast schon Food, S. 131ff.
3 | Oxford Companion to Food, S. 370
4 | Oxford Companion to Food, S. 692
5 | Interview mit der amerikanischen Anthropologin Elizabeth Andoh, Tokio, im Oktober 2008
6 | Christoph Wagner, Fast schon Food, S. 119ff.
7 | Catarina Kossuth-Wolkenstein, Die Marke Eskimo, Signum, Wien, 2000, Christoph Wagner, Fast schon Food, S. 285ff.
8 | Interview mit Stéphane Bureaux, Paris, am 21.10.2007
9 | Besichtigung Nissin Foods, Tokio, im Oktober 2008
10 | Interview mit Dieter Isakeit, Head of Erasmus Centre, European Space Agency, Noordwijk, NL, am 8.2.2007
11 | Besichtigung Fa. Simpert-Reiter GmbH, Augsburg, D
12 | Rotraud Degner, Die Welt der Pasta, S. 9f.
13 | Informationen von Marlies Gebetsberger: Fa. Unilever
14 | Dipl.-Ing. Thomas Birus, (c) Bibliographisches Institut & F. A. Brockhaus AG, 2003
15 | Informationen von Marlies Gebetsberger: Fa. Unilever
16 | Betriebsbesichtigung Snow Brand Milk, Tokio, im Okt. 2008
17 | Interview mit Artemio Castro, Director Snacks, Procter & Gamble, Cincinnati, USA, am 3.4.2009
18 | Betriebsbesichtigung Nissin Foods, Japan, im Oktober 2008
19 | Interview mit Corinna Casale, Nestlé Italiana S.p.A., Perugina Historical Museum, Perugia, im April 2006
20 | vgl.: Christoph Wagner, Fast schon Food
21 | vgl.: Christoph Wagner, Fast schon Food, S. 175/176.
22 | vgl.: Christoph Wagner, Fast schon Food

1 | Source: Interview with Werner Mlodzianowski, Director of the Technologie Transfer Zentrum Bremerhaven, on 18 and 19 October 2007
2 | Source: Christoph Wagner, Fast schon Food, pp. 131ff.
3 | Oxford Companion to Food, p. 370
4 | Oxford Companion to Food, p. 692
5 | Interview with the American anthropologist Elizabeth Andoh, Tokyo, October 2008
6 | Source: Christoph Wagner, Fast schon Food, p. 119 ff
7 | Source: Catarina Kossuth-Wolkenstein, Die Marke Eskimo, Signum, Vienna, 2000, Christoph Wagner, Fast schon Food, pp. 285 ff.
8 | Interview with Stéphane Bureaux, Paris, 21 October 2007
9 | Guided tour of Nissin Foods, Tokyo, in October 2008.
10 | Interview with Dieter Isakeit, Head of Erasmus Centre, European Space Agency, Noordwijk, NL, 8 February 2007
11 | Source: Simper-Reiter GmbH., Augsburg, Germany
12 | Source: Rotraud Degner, Die Welt der Pasta, pp. 9f.
13 | Source: Unilever co.
14 | Source: Thomas Birus, © Bibliographisches Institut & F. A. Brockhaus AG, 2003
15 | Source: Unilever company
16 | Guided tour of Snow Brand Milk, Tokyo, in October 2008
17 | Interview with Artemio Casto, Director Snacks, Procter & Gamble, Cincinnati USA, 3 April 2009
18 | Guided tour of Nissin Foods, Japan, October 2008
19 | Interview with Corrina Casale, Nestlé Italiana S.p.A., Perugina Historical Museum, Perugia, April 2006
20 | Source: Christoph Wagner, Fast schon Food
21 | Source: Christoph Wagner, Fast schon Food, pp. 175-176.
22 | Source: Chistoph Wagner, Fast schon Food

Kulinarisches Recycling: Unter der rosaroten Zuckerglasur des Punschkrapfens verwandeln sich Kuchenreste zu einem appetitlichen, neuen Dessert. Der Punschkrapfen besteht aus (offiziell frischem) Biskuitteig, der mit Marmelade bestrichen, in Rum getränkt und mit einer Fondantmasse glasiert wird. Ursprünglich waren Punschkrapfen rund, da sie mit einer runden Keksform, die auch für Krapfen Verwendung fand (daher der Name), ausgestochen wurden. Und weniger knallig: Vor Erfindung der Lebensmittelfarbe färbte man die Glasur mit Rotwein, Orangen- oder Himbeersaft.

Culinary recycling: Below the pink-red sugar icing of the "Punschkrapfen" pastry, cooking leftovers become transformed into an appetizing new dessert. The "Punschkrapfen" consists of (officially fresh) sponge cake dough that is covered with marmalade, soaked in rum and glazed with a fudge mass. Originally the "Punschkrapfen" were round since they were cut with a round cookie cutter which was also used for 'Krapfen' (hence the name). And they were less bright: Before food dyes were invented, red wine, orange or raspberry juices were used to dye the icing.

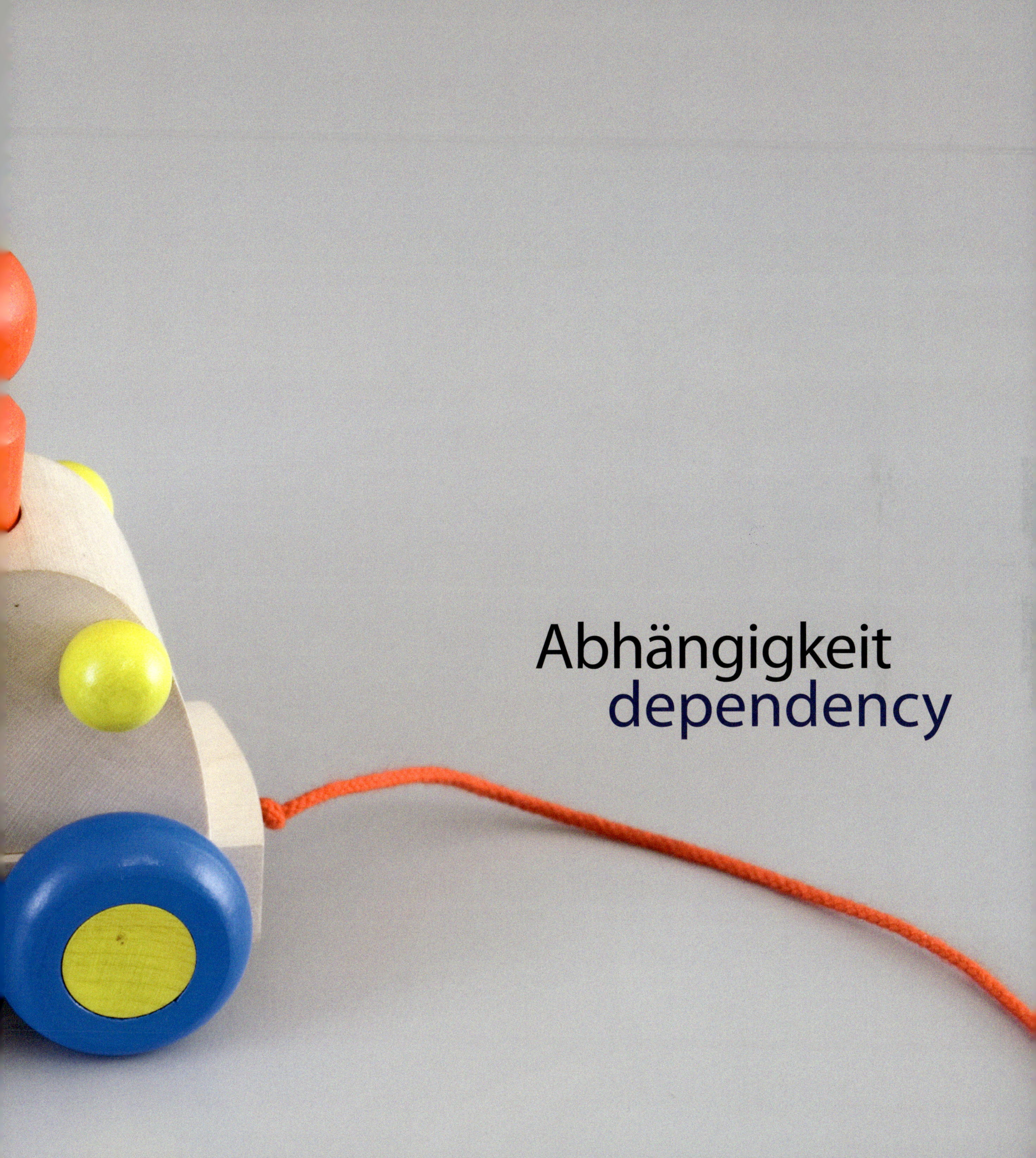

Abhängigkeit
dependency

Kulinarisches Baukastensystem: Wahrscheinlich wäre nie jemand auf die Idee gekommen, Schinken oder Käse in die unnatürliche, quadratische Form zu bringen, hätte es nicht bereits das Toastbrot gegeben. Der Toast geht angeblich auf Katharina von Medici (1519-1589) zurück. Sie erfand den Brauch, in den Weinkelch, der bei Tisch zu Ehren des Königs herumgereicht wurde, ein Stück geröstetes Brot zu legen. Wer den Toast beim Trinken erwischte, hatte einen Wunsch frei. Bis heute versteht man unter einem Toast nicht nur ein knusprig gegrilltes Brot, sondern auch einen Trinkspruch.

Culinary modular system: Probably no one would have ever come up with the idea of styling ham or cheese in an unnatural, square form if it had not been for toast bread. Supposedly, toast goes back to Katharina de Medici (1519-1589). She invented the custom of putting a piece of toasted bread in a wine chalice which was passed around at the table to honor the king. Whoever got the toast while drinking was able to make a wish. To this day, a toast is not just a piece of crunchy grilled bread but also a saying to which one drinks.

Kulinarische Baukästen | Damit Nahrung funktioniert, müssen die einzelnen Komponenten aufeinander abgestimmt sein. Wie ein Schuhlöffel zu Schuhen, ein Rad zum Wagen und ein Knopf zur Hose passt, so werden auch Größe und Gestalt von Lebensmitteln zueinander in Einklang gebracht. Es gäbe keine kegelförmige Eiswaffel ohne Eis, kein Hotdog-Brötchen ohne Würstchen und keinen quadratischen Schinken ohne Toastbrot. Die Form vieler Esswaren ist das Ergebnis ihrer Anpassung an andere Nahrungsmittel.

Die einzelnen Zutaten eines Schinkenkäsetoasts zum Beispiel lassen sich passgenau aufeinander schichten. Sein ältester Bestandteil, auf den die anderen Komponenten formal zugeschnitten sind, ist das rechteckige Weißbrot. Bereits im Mittelalter rösteten die Briten ihr Weizenbrot über offenem Feuer und nannten diese Alltagsmahlzeit „Tost".[1] Zur Zeit der Renaissance aß man in den vornehmen Kreisen Norditaliens teures Weißbrot getoastet, eine Idee, die Katharina von Medici aus Italien an den französischen Hof gebracht und dort zu einem speziellen Ritual weiterentwickelt haben soll. Als Königin erfand sie dort den Brauch, in den Weinkelch, der von der Tafelrunde zu Ehren des Königs geleert wurde, ein Stück geröstetes Brot zu legen. Derjenige, der das Brotstück erwischte, hatte einen Wunsch frei. Bis heute versteht man unter einem Toast daher nicht nur ein knusprig gegrilltes Brot, sondern auch einen Trinkspruch.

Wie der damalige Toast aussah, wissen wir nicht. Spätestens mit Beginn der Industrialisierung jedoch füllte man Brotteig aus Gründen der Produktionseffizienz in Kastenformen und buk auf diese Weise Brotquader anstelle von Laiben.[2] Der rechteckige Querschnitt inspirierte im zwanzigsten Jahrhundert Wurst- und Käsehersteller, die Gestaltung ihrer Produkte dem Brot anzupassen und maßgeschneiderte Beläge für den Toast zu entwickeln. Kein Schwein dieser Welt besitzt rechtwinkelige Oberschenkel – und dennoch erachten Millionen von Kunden die Form des so genannten Toastblocks als völlig selbstverständlich. Auch der 1950 in den USA erfundene Scheiblettenkäse ist von der ursprünglichen Form des Käselaibs stark abstrahiert. So entstand für die quadratischen Weißbrotscheiben ein kulinarisches Baukastensystem, das je nach Geschmack und Hunger in unterschiedlichen Varianten zusammengesetzt werden kann. Aus purer Bequemlichkeit nimmt der Kunde eine gewisse Abwandlung herkömmlicher Produkte in Kauf, was im Fall von eckigem Käse oder Schinken auch adaptierte Herstellungsverfahren und Rezepturen bedeutet. Dabei erscheint das bisher äußerst erfolgreiche Konzept durchaus ausbaufähig. Es bleibt abzuwarten, ob sich im Laufe der Zeit quadratische Tomaten- oder Ananasscheiben dazugesellen werden. Vielleicht füllen wir unsere Toasts auch bald schon mit hart gekochten Straußeneiern im passenden Format.

Culinary Building Blocks | For food to function well the individual components have to be coordinated. Just as a shoehorn fits shoes, wheels fit carts and buttons fit pants, the size and design of foods have to be harmonized. There would not be a cone-shaped ice cream wafer without ice cream, no hotdog bun without a hotdog and no square ham without toast bread. The shape of many foods is the result of their adaptation to other foods.

The individual ingredients of a ham-and-cheese toast, for instance, can be neatly stacked on top of each other. Its oldest component, which the other components formally relate to, is square white bread. As early as in the Middle Ages the British roasted their wheat bread over an open fire and called this everyday meal "tost".[1] In the Renaissance, expensive white bread was eaten toasted in sophisticated circles in Northern Italy – an idea that Katharina de' Medici had brought from Italy to the French court and developed into a special ritual. When she was queen, she invented the custom of placing a piece of toasted bread in the wine goblet, which was emptied by the persons sitting around the table in honor of the king. Whoever got the piece of bread was allowed to make one wish. To this day, toast thus not only refers to a piece of crispy roasted bread but also means "to drink to someone's health".

We do not know what toast looked like in those days. With the onset of industrialization, at the latest, bread dough was filled into box-shaped forms. This way bread squares were baked instead of loaves.[2] The rectangular cross-section inspired producers of cheese and sausage in the twentieth century to model the design of their products after bread and to develop customized fillings for toast. No pig in the world has square thighs – yet millions of consumers take the block shape of ham completely for granted. Even the sliced cheese invented in the USA in 1950 is an abstraction of the original shape of the cheese wheel. This way a system of culinary building blocks emerged which could be put together in different ways depending on your hunger or taste. Out of pure convenience the customer is willing to accept a certain variation of ordinary products which in the case of square cheese or ham also means adapted production techniques or recipes. The strategy that has proven extremely successful to date seems to be expandable. It remains to be seen whether square slices of tomatoes or pineapple will become part of this system. Perhaps our toasts will also be filled with hard-boiled ostrich eggs in the right format.

Das Aussehen vieler Esswaren ist das Ergebnis ihrer Anpassung an andere Nahrungsmittel. Der Durchmesser von deutschen Aufschnittwürsten ist auf 90mm genormt, damit sie gut in 50g schwere, runde Brötchen wie Semmeln passen. Bei Salami und Mortadella funktioniert das bekanntlich nicht so gut.

The appearance of many foods is the result of its adaptation to other foods. The standard diameter of the German cold cut is 90mm so that it fits well in round rolls such as the "Semmel" that weighs 50g. As we know this does not work so well with salami and mortadella.

Jeder Wurst ihr Brot | Nicht nur rechteckiger Schinken, auch runde Würste sind formal auf das Brot, in das sie gefüllt werden, abgestimmt. Ist Ihnen schon einmal aufgefallen, dass die mitteleuropäischen Aufschnittwürste perfekt in eine Semmel oder andere runde, 50 Gramm schwere Brötchen passen? Das war nicht immer so, denn Jahrhunderte lang war Gebäck – sei es die Schrippe in Berlin, das Panino in Italien oder eben die Kaisersemmel in Wien – mit recht unterschiedlichen Wurstformen und -größen konfrontiert. Der verwendete Darm von Schwein, Rind oder Schaf bestimmte Aussehen und Querschnitt der Wurst und bis heute sind internationale Würste wie Mortadella, Chorizo oder Salami nur mäßig wurstsemmeltauglich.

Die Idee der genormten Wurstdurchmesser stammt – wie könnte es anders sein – aus Deutschland, wo es 1925 dem Chemiker Walter Becker an der technischen Universität Darmstadt erstmals gelang, eine künstliche Wursthülle zu erzeugen. Es folgte die Gründung der Firma Becker, Schulze und Co in Hamburg, ab 1933 wurde unter dem Namen „Naturin" produziert, der heute in Fachkreisen vor allem mit der Erfindung der Collagen-Wursthülle 1962 in Verbindung gebracht wird. Eine Weiterentwicklung der damals revolutionären Erfindung führte zu den Collagenfolien, welche 1983 auf den Markt kamen und noch heute Verwendung finden.

Mit der Einführ künstlicher Wursthäute waren die Durchmesser, im Fachjargon als Kaliber bezeichnet, nicht mehr von natürlichen Därmen abhängig, sondern konnten innerhalb herstellungstechnischer Grenzen frei gewählt werden. In der Folge wurde das Aussehen der Würste vereinheitlicht und die Querschnitte wurden normiert. Auch wenn der Kunde bis heute eine Fülle von Rezepturen, Geschmacksrichtungen und Bezeichnungen im Feinkostregal vorfindet, sind die Durchmesser aller klassischen Aufschnittwürste gleich.

Der heutige Querschnitt bringt nicht nur für zehn Scheiben Aufschnitt à 1 Millimeter ziemlich genau 100 Gramm auf die Waage, sondern bildet auch die Grundlage für die Wurstsemmel: Die 90 mm der Extrawurst oder Fleischwurst passen haarscharf in die 50 Gramm schwere Semmel oder jedes andere runde 50-Gramm-Brötchen, während Salami und Mortadella ein eher unbefriedigendes Wurstsemmelerlebnis ergeben. In ähnlicher Weise korrespondieren auch Würste mit Querschnitt in „Brotform", in den USA unter dem Begriff D-Shape bekannt, mit der Schwarzbrotscheibe. Warum allerdings bis heute weder Wurst noch Schinken in der Form von Croissants, Hörnchen, Salzstangen oder anderen Gebäcksorten auf dem Markt ist, bleibt rätselhaft.[3]

Heiße Hunde und essbare Dackel | Während perfekt belegte Wurstbrote und -semmeln relativ jungen Datums sind, hat ein anderes Produkt, das Brot und Belag konsumentenfreundlich kombiniert, eine mehr als 100-jährige Geschichte. Bis heute ist ungeklärt, wer wann, wo und warum den Hotdog erfunden hat, dennoch ranken sich rund um die Entstehung der „heißen Hunde" zahlreiche Erzählungen und Legenden. Eine davon berichtet von einem deutschen Fleischer namens Charles Feltman, der um 1880 auf Coney Island heiße Würstchen zwischen zwei Sandwichhälften verkaufte und so eine Vorform des heutigen Hotdog kreierte. Der Feltman-Imbiss war zwar noch mit Sauerkraut anstelle von Ketchup gefüllt, wies aber schon einen entscheidenden Vorteil des späteren Hotdog auf: Man konnte ihn leicht in der Hand halten und daher bequem im Stehen oder im Gehen verzehren. Zwischen den Jahrmarktbuden und Hochschaubahnen der Vergnügungsviertel entwickelte sich das Würstchen – mit oder ohne Brothülle – rasch zu einem beliebten Gericht, das schließlich auch andere Stadtteile und Lebensbereiche infiltrierte.

Der passende Belag für Salzstangen und Croissants muss noch erfunden werden. Cracker mit formal abgestimmtem Schinken und Käse gibt es bereits.

The right topping for pretzel rolls and croissants is yet to be invented. Crackers which ham and cheese have been adapted to already exist.

To Each Sausage Its Bread | Not just square ham but also round sausages are formally adapted to the bread in which they are filled. Have you ever noticed that Central European cold cuts fit perfectly in a roll or other round breads that weigh 50 grams? That was not always the case. For centuries, rolls – be it the "schrippe" in Berlin, the "panino" in Italy or the "Kaisersemmel" in Vienna – were confronted with rather different forms and sizes of sausage. The guts of pork, beef or sheep used defined the appearance and cross-section of the sausage, and to this day international sausages such as mortadella, chorizo or salami are only moderately suited for sausage rolls.

The idea of the standard sausage diameter came – not so unexpectedly – from Germany where in 1925 the chemist Walter Becker from the Technical University in Darmstadt succeeded for the first time in producing artificial sausage skin. Becker, Schulze und Co was subsequently founded in Hamburg and from 1933 sausageskin was produced under the name "Naturin", which today is known in professional circles mainly for the invention of the collagen sausage skin in 1962. A further development resulted in the collagen sheets which were launched in 1983 and are still in use today.

With the introduction of artificial sausage skins the diameter, known in technical parlance as caliber, was no longer dependent on guts but could be freely selected within manufacturing limits. Subsequently, the appearance of sausage became more uniform and the cross-sections were standardized. Even if the customer still finds a wealth of recipes, tastes and names in the delicatessen section today, the diameter of all classical cold cut sausages is the same.

With today's cross-section, the cold cut does not just weigh exactly 100 grams for ten slices of one millimeter, it also forms the basis of the sausage roll. The 90 mm of Bologna or other meat sausage fit precisely in the 50 gram "Semmel" or any other 50 gram roll, whereas salami and mortadella make for a rather unsatisfactory sausage roll experience. Similarly, sausages with a "bread-shaped" cross-section, known in the US as D-shape, relate to the slice of black bread. Why, however, to this day, neither sausage nor ham is commercially available in the shape of croissants, pretzel stick or other types of bread, remains a mystery.[3]

Hot Dogs and Edible Dachshounds | Whereas sausage sandwiches and rolls with a perfect topping are relatively new phenomena, another product combining bread and topping in a consumer-friendly way has a more than 100-year-old history. To this day, it remains unclear who invented the hot dog when, were and why. The emergence of the "hot" dogs remains shrouded in legends and myths. There is one story about the German butcher named Charles Feltman who sold hot sausages between two sandwich halves on Coney Island around 1880 and thus created a precursor of today's hot dog. The Feltman snack was still filled with sauerkraut instead of ketchup but it already had a real advantage over the later hot dog. You could hold it in your hand and easily eat it standing or walking. Between the fairground booths and the stands in the entertainment districts, the sausage – with and without a bread wrapping – quickly became a popular dish which finally infiltrated other parts of the city and realms of life.

Um die Jahrhundertwende erfreuten sich Würste in den USA großer Beliebtheit als kleine Mahlzeit für zwischendurch, die man an einem Kiosk oder einer Wursttheke im Stehen genoss. Die Verkäufer wickelten die heiße Ware in dickes Papier, damit die Kunden sie in der Hand halten konnten. Der Legende nach erregte der Bayer Antoine Feuchtwanger einiges Aufsehen, als er auf einer Messe in St. Louis 1904 weiße Handschuhe anstelle des Papiers an die Wurstesser verteilte, damit sich diese die Hände nicht beschmutzten. Da ein Großteil der beigestellten Schutzkleidung jedoch in Manteltaschen oder sonst wohin verschwand, sah sich Feuchtwanger gezwungen, eine andere Verpackung für den fettig-heißen Snack zu finden. Sein Schwager, ein gelernter Bäcker, reagierte prompt und entwickelte ein längliches Brot, das in Form und Größe perfekt über das Würstchen passte – und damit war der Hotdog geboren.

Andere Quellen sehen den Hotdog keineswegs als amerikanische Entwicklung, sondern behaupten, der Hotdog gehe auf die um 1906 von Leopold Lahner, dem Enkel des Erfinders der Frankfurter bzw. Wiener Würstchen, kreierten „Würstel im Schlafrock" zurück. Wie dem auch sei, die seltsame Bezeichnung der „heißen Hunde" stammt tatsächlich aus der neuen Welt, auch wenn Siedewürste aufgrund ihrer gekrümmten Form in Frankfurt bereits um 1900 als „Dackelwurst" bezeichnet wurden. Die Bezeichnung „Hot Dog" dagegen soll der Feder des Sportcartoonisten T.A. Dorgan entstammen, der nach einem Spiel auf den New Yorker Polo Grounds die auf den Tribünen verkauften Sandwichs karikierte. Beeindruckt von den Rufen der Verkäufer, die ihre „red hot dachshound sausages" anpriesen, zeichnete er als Füllung an Stelle der „dachshound sausages" einen richtigen Dackel und untertitelte das Bild kurz und bündig mit „Hot Dog".

Mittlerweile ist das Gericht zum Inbegriff amerikanischer Esskultur avanciert. Im Durchschnitt verzehrt jeder Amerikaner 60 Stück im Jahr. Roy Lichtenstein setzte der Ikone Hot Dog 1963 in mehreren Gemälden ein Denkmal und in Washington residiert ein eigenes Hot Dog Information Bureau, das laut dem österreichischen Gourmetjournalisten Christoph Wagner unter Box 3556, Washington, D.C. 20007 kontaktiert werden kann.[4]

Das Ei wird zum Rohr | Eine formale Anpassung ganz anderer Art erfährt das Hühnerei. Um Sandwiches und Brötchen optimal mit formschönen, hart gekochten Eischeiben zu belegen, ersann man das so genannte Stangenei. Dabei handelt es sich um rund 1 Meter lange Würste aus gestocktem Eiweiß und Dotter, wobei letzterer immer perfekt in der Mitte sitzt, so dass sich über die ganze Länge Eischeiben mit stets gleichem Dotter- und Eiweißanteil ohne lästige Randstücke schneiden lassen. Erzeugt wird das Stangenei, indem versprudelte Dotter in gefetteten Rohren mit zirka drei Zentimeter Durchmesser vorgegart, entnommen, in größere, zylindrische Formen mit flüssigem Eiweiß umgegossen und fertig gekocht werden. Verwendung finden die Stangeneier vor allem in Gastronomie- und Lebensmittelbetrieben.

1 | vgl.: Oxford Companion to Food, S. 799
2 | Interview mit Paul Levy in London am 15.1.2009
3 | Informationen von Armin Fiebig, Fa. Naturin, Weinheim, D
4 | vgl.: Christoph Wagner, Fast schon Food, S. 161ff.

Around the turn of the century, sausages became extremely popular in the US as a small meal in between which people could enjoy at a stand or sausage booth. The salespersons wrapped the hot food in thick paper so that it could be held in one's hands. According to legend, the Bavarian Antoine Feuchtwanger caused quite a stir when he distributed white gloves instead of paper to the sausage eaters at a fair in St. Louis in 1904 so that they would not get their hands dirty. Since most of the protective clothing disappeared into coat pockets or elsewhere, Feuchtwanger felt compelled to find a different wrapping for the greasy, hot snack. His brother-in-law, a baker by training, reacted promptly and developed a longish bread with the perfect shape and size for the sausage – thus the hot dog was born.

Other sources do not see the hot dog as an American invention. They claim that the hot dog goes back to the "sausage in a pastry" created in 1906 by Leopold Lahner, the grandson of the inventor of the Frankfurter or Wiener sausage. Whatever the case may be, the strange name "hot dog" actually originates in the new world, even if boiled sausages were already called "dachshound sausage" in Frankfurt around 1900 because of their curved shape. The name "hot dog" by contrast goes back to the sports cartoonist T.A. Dorgan, who after a game at the New York Polo Grounds caricatured the sandwiches sold on the bleachers. Impressed by the calls made by the salesmen who touted their "dachshound sausages", he drew a real dachshound instead of a "dachshound sausage" as the filling and added "Hot Dog" as the drawing's caption.

In the meantime hot dogs have become the quintessentially American meal. Each American eats about an average of 60 hot dogs per year. Roy Liechtenstein created a monument to the hot dog icon in 1963 and in Washington there is a separate Hot Dog Information Bureau that according to the Austrian gourmet journalist Christoph Wagner can be contacted at: Box 3556, Washington D.C. 20007.[4]

Customizing Eggs | The chicken egg underwent a formal adaptation of a completely different type. In order to cover sandwiches and rolls with pleasing shapes of hard-boiled egg slices, the so-called long egg was invented. This is an approximately 1 meter long sausage consisting of solidified egg white and yolk, with the latter always sitting perfectly in the center so that the egg slices can be cut from the entire length; the amount of yolk and egg white is always the same, and there are not unpleasant end pieces. The long egg is produced by cooking mixed yolks in greased pipes with a diameter of approximately three centimeters (about 1 inch), putting them into larger cylindrical forms then and pouring liquid egg whites over before cooking this mixture again. The long eggs are primarily used in the catering industry and in food production.

1 | Oxford Companion to Food, p. 799
2 | Source: Interview with Paul Levy, 15 January 2009
3 | Source: Armin Fiebig, Naturin co., Weinheim, Germany
4 | Source: Christoph Wagner, Fast schon Food, pp. 161 ff.

Die Idee, Brot und Belag handlich zu kombinieren, soll – einer Legende nach – John Montagu, der vierte Earl of Sandwich, gehabt haben. Um eine Partie nicht unterbrechen zu müssen, orderte der passionierte Kartenspieler 1762 seine Mahlzeit angeblich zwischen zwei Brotstücken und ging damit als Erfinder des Sandwichs in die Geschichte ein.

Legend has it that the idea of combining bread and topping in an easy-to-eat way goes back to John Montagu, the fourth Earl of Sandwich. To avoid interrupting a game, the passionate card player allegedly ordered his meal to be served between two pieces of bread in 1762. He thus went down in history as the inventor of the sandwich.

Oberflächendesign
surface design

Haftung und Reibung – Oberflächendesign | Hübsch gestaltete Oberflächen erfreuen das Auge und regen die Geschmacksnerven an. Das Design der Oberfläche von Nahrungsmitteln erfüllt aber auch zahlreiche Funktionen. So stabilisieren geriffelte Strukturen dünne Objekte wie Waffeln und gekörnte fungieren als Abstandhalter zwischen Tiefkühlwaren: Stark strukturierte Materialien frieren nicht so leicht aneinander fest wie glatte. Panierte Fertigprodukte, vom Backfisch bis zu Hühnernuggets, fallen trotz tiefer Temperaturen einzeln aus der Packung, da die extreme Körnung der Panade die Stücke zueinander auf Distanz hält.

Saucengrip – der Winterreifen am Teller | An rauen Oberflächen wiederum bleiben andere Komponenten der Nahrung besser haften. So wie man auf Schneefahrbahn Reifen mit starkem Profil benutzt, um eine möglichst gute Haftung zu erzielen, isst man auch zu Gerichten mit Saucen, Dips oder Säften vorzugsweise Beilagen, die einen guten Grip haben und sich möglichst gut mit der Speise verbinden. Während in vielen Ländern die saugfähige Konsistenz des Brotes zur Aufnahme von Bratensaft, dickflüssigen Suppen oder Curries dient, leistet in Italien die Oberfläche der Pasta diese Aufgabe.

In Italien steht die Frage nach dem optimalen Saucenverhalten der unterschiedlichen Pastasorten im Zentrum vieler Diskussionen. Die Art und Weise, wie sich die geformten Teigstücke mit dem Sugo vermischen, hat eine spürbare Auswirkung auf das Essvergnügen. Glatte Oberflächen wie jene von Spaghetti eignen sich zum Beispiel für intensive, ölhaltige Knoblauch- oder Kräutersaucen, die an den glitschigen Nudeln abperlen und dafür sorgen, dass der Geschmack nicht zu intensiv wird.

Glatt oder geriffelt | Zähflüssige, mildere Sughi dagegen verlangen nach strukturierten Oberflächen, etwa gerillten, die mit dem Attribut „rigato" verkauft werden, bekannt vor allem durch die Penne rigate. Manche Pastahersteller steigern den Effekt noch zusätzlich, indem sie Penne anbieten, deren Oberfläche sowohl an der Innen- als auch an der Außenseite geriffelt ist.

In Italien arbeitet man zur Optimierung des Klebeverhaltens von Pasta aber nicht nur mit geeigneten Formen und strukturierten Oberflächen, sondern auch mit speziellen Herstellungstechnologien. Grundsätzlich gilt: Je rauer und klebriger die Oberfläche, desto besser verbindet sie sich mit dem Sugo. Aus diesem Grund setzt die süditalienische Firma Voiello bei der Erzeugung ihrer Teigwaren Pressköpfe aus Bronze ein. Im Gegensatz zu den herkömmlichen Teflonprofilen raut die Bronze den Teig stärker auf, so dass die fertige Pasta, wenn sie gekocht ist, besonders griffig ist.[1]

Vergleichsweise schwierig gestaltet sich die Oberflächenoptimierung natürlicher Lebensmittel wie Fisch oder Fleisch. So fällt die Veränderung der Saugfähigkeit eines Steaks, um es „saucenfreundlicher" zu machen, naturgemäß schwer. Zwar greifen Köche tief in die Trickkiste, um griffige Hüllen aus Teig und anderen Zutaten zu zaubern, doch die Schnittfläche von Fleisch, die erst der Esser selbst generiert, können sie beim besten Willen nicht manipulieren. Möglicherweise bedeutet dies aber eine kulinarische Chance für die Entwicklung künftiger Technologien. Vielleicht basteln Biologen irgendwann ja das optimale Rind, dessen Fleischoberfläche den perfekten Saucengrip aufweist. „Filet Michelin" wäre dafür ein passender Name …

1 | Betriebsbesichtigung Barilla, Parma, im April 2006

Saucengrip: Die Oberfläche von Teigwaren wird extra strukturiert oder aufgeraut, damit Saucen besonders gut an ihnen haften. Um das Klebeverhalten zu optimieren, wird teurere Pasta mit Pressköpfen aus Bronze und nicht mit herkömmlichen Teflonprofilen erzeugt.

Sauce grip: The surface of pasta is specially structured or roughened so that sauces stick to them well. In order to optimize the stickiness, expensive pasta is made with bronze press heads instead of the common Teflon profiles.

Adhesion and Friction – Surface Design | Nicely designed surfaces are pleasing to the eye and appetizing. The design of the surface of food also fulfills a number of other functions though. Structured surfaces stabilize thin objects such as wafers, and grainy ones function as spacers between frozen products. Strongly structured materials do not freeze together as easily as smooth ones. Breaded foods, from baked fish to chicken nuggets, fall out of the package separately in spite of low temperatures, since the extreme granularity of the breading keeps some space between the individual pieces.

The Grip of Sauces – the Snow Tire on the Plate | Other food components stick better to rough surfaces. Just as you use tires with a deeper profile on snow-covered roads to get as much adhesion as possible, you also preferably eat side dishes that have a good grip and can merge with the food as well as possible. While in many countries, the absorbable consistency of bread is used to collect gravy, thick soups or curries, the surface of pasta performs the same task in Italy.

In Italy the perfect response to sauce of various types of pasta is an endless discussion. The way that shaped pieces of pasta merge with the sauce has a tangible effect on the eating experience. Smooth surfaces like those of spaghetti are well suited for intensive, oily garlic or herb sauces that roll off the slippery noodles and ensure that the taste does not become too strong.

Smooth or Structured | Viscous, milder sughi by contrast require structured surfaces, like the grooved one that is sold with the attribute "rigato" – known primarily as penne rigate. Some pasta manufacturers enhance the effect by offering penne whose surface is structured on both the inside and outside.

In Italy special forms and structured surfaces as well as special manufacturing technologies are used to make the pasta stickier. Basically, the following principle applies: the coarser and stickier the surface, the better it merges with the sugo. For this reason, the southern Italian company Voiello uses hydraulic jacks made of bronze to produce their pasta. As opposed to the ordinary teflon profiles, bronze is able to roughen the surface of the pasta so that the finished pasta has the right texture when it is cooked.[1]

Optimizing the surface of natural foods such as fish or meat is comparatively difficult. Altering the absorbability of a steak to make it more "sauce-amenable" seems (yet) impossible. Cooks dig deep into their box of tricks to conjure non-slippery top layers of dough and other ingredients, but they are simply not able to manipulate the cut surface of meat that is only created when eating. This might possibly mean a culinary opportunity for the development of future technologies. Perhaps some day biologists will come up with the perfect beef whose surface shows the perfect sauce grip. "Filet Michelin" would be an ideal name…

1 | Source: Guided tour of the Barilla plant, Parma, April 2006

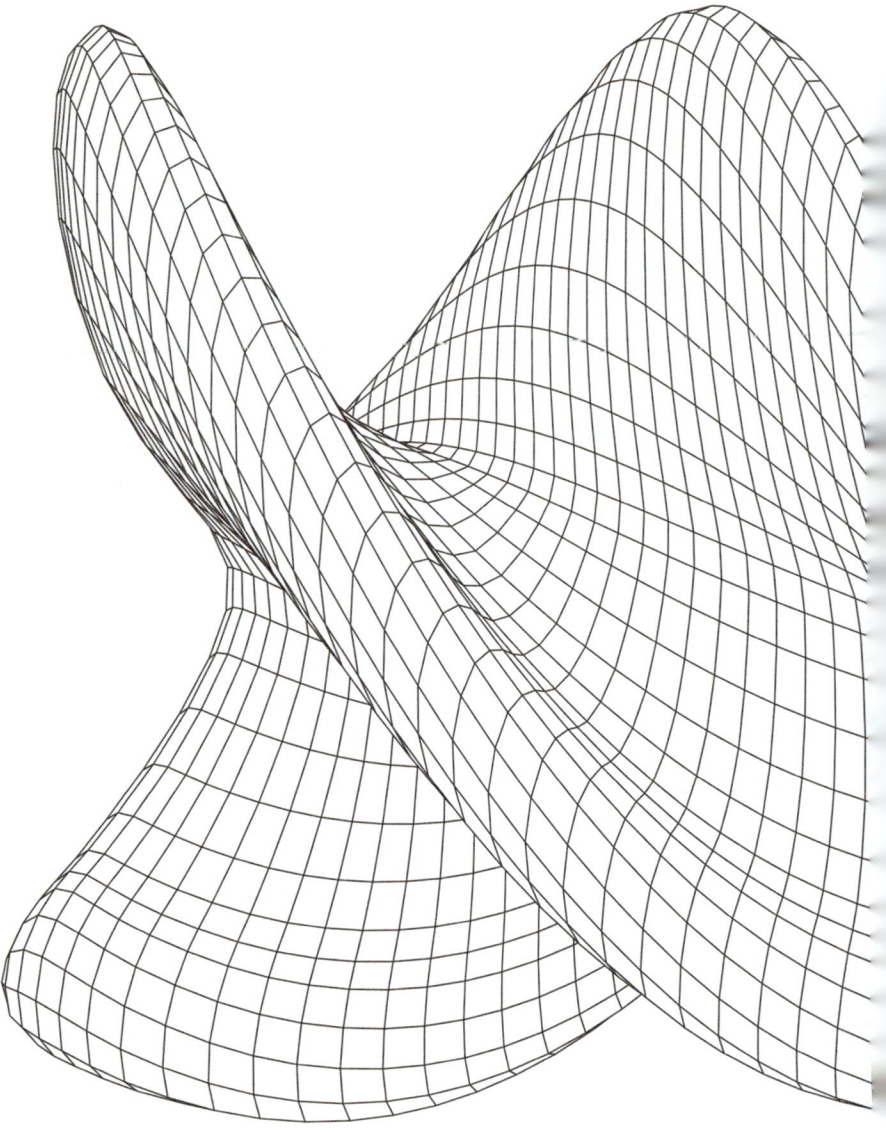

Dass die Form tatsächlich eine Auswirkung auf den Geschmack hat, zeigen die rund 600 unterschiedlichen Sorten italienischer Pasta. Die immer gleichen Zutaten, Wasser und Hartweizengrieß, verursachen je nachdem, wie sich die geformten Teigstücke mit der Sauce vermischen, ein immer anderes Geschmackserlebnis.

There are about 600 different types of Italian pasta, which goes to show that the form actually has an effect on the taste. The same ingredients – water and semolina – always lead to a different taste experience depending on how the shaped pieces of dough blend with the sauce.

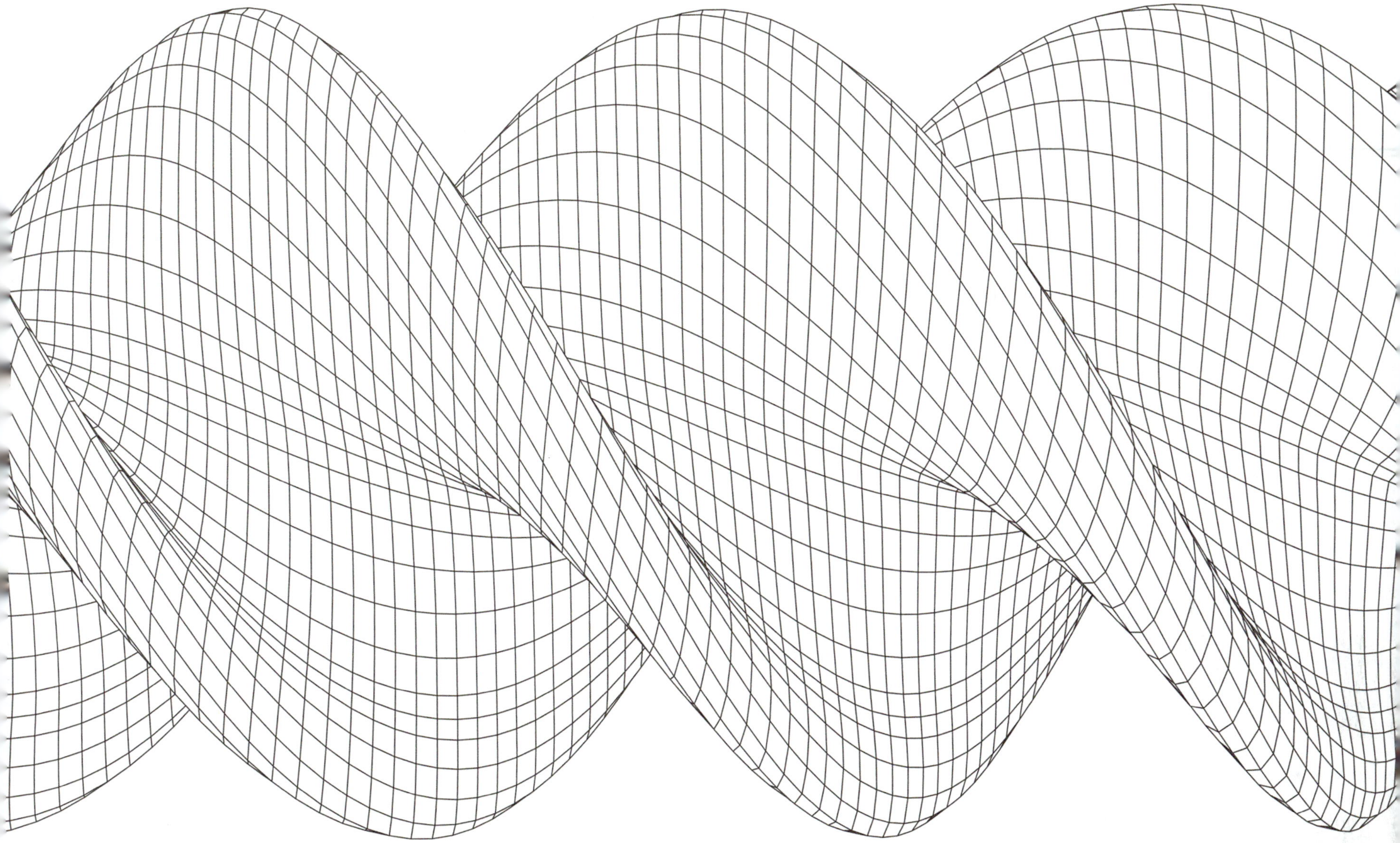

Teilen
sharing

Die Lust am Teilen - Vom Brechen, Schneiden und Portionieren | Teilen ist eine unserer elementaren Handlungsformen. Vom Zerteilen über das Ein- und Austeilen bis zum Umverteilen betrifft es praktisch alle Lebensbereiche. Gerade beim Essen spielt das Teilen eine übergeordnete Rolle, und von wem, wann und wo geteilt wird, beeinflusst die Gestaltung jener Waren, die wir in den Mund nehmen. Wie ein Ganzes in einzelne Stücke zerlegt werden kann, beschäftigt Köche bei der Gestaltung ihrer Speisen, genauso wie Hersteller beim Entwurf ihrer Produkte. Umgekehrt beeinflusst die Art des Zerteilens den Geschmack und die Form, in der ein Gericht anschließend auf den Teller kommt, und schlägt sich daher oft in dessen Bezeichnung nieder. Ob Cremeschnitte, Aufschnitt, Hackfleisch oder Geschnetzeltes: Die Namen vieler Speisen beziehen sich darauf, wie sie im Zuge der Zubereitung zerteilt wurden. Mozzarella geht auf „mozzare", abschneiden, zurück; die Tagliatelle, wörtlich übersetzt die „Abgeschnittenen", verdanken ihren Namen dem italienischen Wort „tagliare" (dt.: ab- oder zuschneiden). Auch das englische Cutlet (Schnitzel) trägt den Vorgang des Schneidens („to cut") in sich und die Bezeichnung für Fleischtranchen kommt vom französischen „trancher", was so viel wie ab- oder zerschneiden bedeutet.

Das Recht zu schneiden | Das Verteilen der Speisen bei Tisch gilt seit Jahrhunderten als verantwortungsvolle Aufgabe, welche die hierarchische Stellung innerhalb einer Gemeinschaft ausdrückt. Das Recht, ein Messer – also eine

The Joy of Sharing - Breaking, Cutting and Portioning | Sharing is one of our most elementary forms of action. Splitting and distributing or redistributing something affects practically all realms of life. In eating in particular dividing plays a superior role and the question of who distributes what, when and where influences the design of those products that we put in our mouths. How something whole is portioned into individual pieces interests cooks in designing their dishes just as it does manufacturers designing their products. By the same token, the way something is divided influences the taste and the form in which something ends up on a plate and is thus often reflected in its name. Be it cream slices, cold cuts or diced meat – the names of many dishes refer to how they were divided while being prepared. Mozzarella is derived from "mozzare", cutting off. Tagliatelle comes from "tagliare" (which can be literally translated as cutting or trimming). Even the English word "cutlet" reflects the processing of cutting and "tranches", the French word for meat slices refers to "trancher", which means cutting up.

The Right to Cut | Passing out the food has been seen as a task involving responsibility for centuries and as one that reflects the hierarchical position within a community. The right to carry a knife, i.e., a weapon, and to divide the meat at the table was

Das Zerlegen von Esswaren in einzelne Teile ist ein Ritual, das seine Spuren auch in der Gestaltung hinterlässt. Egal ob wir schneiden, brechen oder abbeißen, bevorzugen wir Lebensmittel, die sich einfach, schnell und appetitlich teilen lassen.

Dividing foods into individual pieces is a ritual that has also left its traces on design. Irrespective of whether we are cutting, breaking or biting off something, we prefer foods that allow us to do this in a simple, quick and appetizing way.

Waffe – zu tragen und das Fleisch bei Tisch zu zerlegen, war ursprünglich dem Familienoberhaupt vorbehalten. Die übrigen Esser benutzten lediglich Löffel oder die Finger. Auch wenn wir heute gewohnt sind, dass alle Beteiligten mit ihrem eigenen Messer essen, und uns derartige Rituale antiquiert vorkommen, wird die zeremonielle Aufteilung bei gewissen Anlässen bis heute sehr ernsthaft praktiziert. Wie strikt das Vorrecht auf das Zerteilen von Nahrung gehandhabt wird, zeigen zum Beispiel Geburtstags- und Hochzeitstorten. Das Ritual, bei dem das Geburtstagskind – und nur dieses – die Torte anschneidet, zählt zu den feierlichen Höhepunkten jeder Party. Und auf Hochzeiten wäre es ein Sakrileg, würden die Eltern oder die Trauzeugen anstelle des Brautpaares den Kuchen zerteilen!

Gerechtigkeit mit dem Messer | Nicht nur die Frage, wer zerteilt, ist von Bedeutung, sondern auch wie er es tut. Gerechtigkeit bei Tisch ist in allen demokratischen Gesellschaften ein zentrales Thema. Jeder Teilnehmer der Tischgemeinschaft soll ein möglichst gleich großes, gleichwertiges Stück erhalten. Ein funktionaler Anspruch, der sich auch im Food Design niederschlägt. Manche Speisen entstanden aus der Idee heraus, alle Anwesenden mit derselben Menge zu bedienen, und sind in ihrer Gestaltung auf eine einfache Portionierbarkeit ausgelegt. Die Form des Rollbratens zum Beispiel behebt das Problem herkömmlicher Braten mit kleineren und größeren, saftigeren und trockeneren, mageren und fettigeren Partien entlang des Knochens. Beim Rollbraten wird

originally reserved to the head of the family. The other eaters only used a spoon or their fingers. Even if we have become accustomed today to everyone eating with their own knives and such rituals appear obsolete to us, the ceremonial distribution of food is still practised in a serious way on certain occasions. How strictly the privilege of distributing food is handled, can be seen for instance when wedding or birthday cakes are cut. The ritual that the birthday child – and only the birthday child – can cut the first piece of cake is one of the ceremonial highlights at each party. And at weddings it would be sacrilegious if the parents or the bridesmaids would cut the cake instead of the couple!

Justice by Knife | The crucial issue is not just who cuts the food but how this person does it. Justice at the table is central to all democratic societies. Everyone who partakes of a communal table should receive as equal and large a piece as possible. This functional prerequisite is also reflected in the design of food. Some dishes resulted from the idea that everyone present must be served the same amount. Thus they are designed so that they can be easily portioned. The form of a roasted pork shoulder, for instance, eliminates the problem of ordinary roasts by offering smaller and larger, juicy and dry, lean and fattier parts along the bone. In rolled pork the meat is freed from its natural structure, that is bones, tendons, cartilage, etc., made qualitatively the same and then rolled up with further ingredients. The resulting roast thus guarantees equal slices along the entire length without any unpleasant end pieces.

das Fleisch von seiner natürlichen Struktur, also Knochen, Sehnen, Knorpel etc. befreit, qualitativ vereinheitlicht und anschließend mit weiteren Zutaten zusammengerollt. Das neue Gefüge garantiert annähernd gleichwertige Tranchen entlang der gesamten Länge, ohne lästige Randstücke.

Ein anderes Beispiel für gute Teilbarkeit ist die Torte, deren Name an die Form erinnert: Das spätlateinische Wort „torta" bedeutet „rundes Stück Teig". Die kreisrunde Form ist bis heute das Markenzeichen jeder Torte und erleichtert die gerechte Verteilung der Festtagsspeise bei Tisch. Rundes zu teilen bedarf keines außergewöhnlich geschulten Augenmaßes. Halbieren, vierteln, achteln etc. geschieht durch einfache, gerade Schnitte, die immerfort denselben Mittelpunkt passieren und lauter gleich große, formschöne, dreieckige Tortenstücke erzeugen. Zudem werden Tortenfüllungen horizontal aufgebracht und Kirschen sowie anderes Dekor gleichmäßig am äußeren Rand verteilt, damit jedes Stück gleich viel abbekommt.

Zur einfachen Teilbarkeit gesellt sich der starke Symbolgehalt von Form und Vorgang: Die kreisrunde Torte steht für Einheit und moralische Werte der Gemeinschaft. Die keilförmigen Tortenstücke symbolisieren den Einzelnen innerhalb einer Gruppe und werden auch aufgrund ihrer Form stets als Teil eines größeren Ganzen verstanden. Gerade wegen des hohen Symbolwertes empfinden wir es als unverzeihlichen Fauxpas, ja als kulinarische Vergewaltigung, wenn jemand eine „Sacher" oder „Schwarzwälder" in Scheiben oder Würfel schneidet. Das Teilen von Nahrung unterliegt einem strikten, kulturellen Kodex, so dass ein Zuwiderhandeln als unmoralischer, fast anarchistischer Akt gewertet wird.

Beim Teilen runder Festtagsspeisen wird symbolisch die gesellschaftliche oder familiäre Hierarchie aufgehoben, indem zum Beispiel das erst dreijährige Geburtstagskind den Kuchen aufschneidet. Dieser Vorgang findet nicht in jeder Kultur statt. In Japan kommen rituelle Süßspeisen etwa bei der Teezeremonie vorportioniert und zumeist eckig auf den Tisch. Perfekt geschnittene Kanten von Sushi, Sashimi oder traditionellen, streng quaderförmigen Wagashi (Teegebäck) gelten als formales Ideal. Zwar bekommt auch hier jeder ein gleich großes Stück, doch der Akt des Teilens fällt weg, die herkömmliche Rollenverteilung bleibt aufrecht.[1]

Anders als bei der Torte schneidet (traditionsgemäß) die Frau den Brotlaib – auch wenn dieser rund ist – stets in Scheiben. Diese Beobachtung wirft die Frage auf, warum Brotlaib und Torte, obwohl in Form und Größe ähnlich, grundsätzlich anders zerteilt werden. Da Brot nur dann geschnitten wird, wenn wir es belegen oder bestreichen wollen, erscheint die Form der Scheibe aus praktischen Überlegungen günstiger. Keilförmige Stücke sind wesentlich instabiler und lassen sich schlecht belegen. Auch die unterschiedliche Bestimmung der beiden Gerichte spielt eine Rolle. Während das Brot ein alltägliches Nahrungsmittel ist, gilt die Torte als Festtagsessen. Brot wird sukzessive je nach Bedarf aufgeschnitten und danach für weitere Mahlzeiten aufbewahrt. Eine möglichst kleine Schnittfläche schützt da am besten vor dem Austrocknen. Die Torte hingegen kommt nur zu besonderen Anlässen auf den Tisch und wird – zumindest traditionell – sofort zur Gänze aufgeschnitten und verspeist.

Klein, handlich und haltbar | Teilungsrituale bestimmen die Form und die Identität von Nahrung. Während der Form von Brotscheiben keine besondere Bedeutung beigemessen wird, sind manche Formen wie das dreieckige Tortenstück kulturell positiv aufgeladen. Ein Umstand, den sich in weiterer Folge auch andere Lebensmittel zunutze machen, dreieckige Schnitten (Shortbread Triangles, Oakbread) oder Käseecken zum Beispiel. Tatsächlich besteht Schmelzkäse, der 1900 in Thun in der Schweiz erfunden wurde, aus eingeschmolzenem Hartkäse und kann

Dreidimensionale Weiterentwicklung der Schokoladetafel: Terry's Chocolate Orange besteht aus 24 „Orangenspalten", die nur an den Polen aneinander haften und bei leichtem Draufschlagen auseinander fallen. Terry's brachte in den 1920er Jahren zunächst den „Chocolate Apple" auf den Markt, der später von der Orange verdrängt wurde.

The development of the chocolate bar in three dimensions: Terry's Chocolate Orange consists of 24 "orange slices" that only stick together on their ends and break away when lightly tapped. In the 1920's Terry's launched "Chocolate Apple" which later yielded to the Orange.

A further example for food that can be easily divided is the cake. The late Latin word "torta" referred to a "round piece of dough". The circular shape has remained the hallmark of every cake to this day, facilitating the just distribution of the festive food at the table. No specially trained eye is required to divide a cake. Halving, quartering, dividing into eight pieces, etc. is possible by making simple, straight cuts that always begin from the same middle point and produce equally big, harmonious, triangular pieces of cake. The cake fillings are added horizontally and cherries and other ornaments are equally distributed on the outer edge so that each piece gets just as many.

In addition to easy sharing, there is also the strong symbolic content of form and process. The circular cake stands for unity and moral codes of community. The wedge-shaped pieces of cake symbolize the individual within a group and are understood as part of a larger whole due to their form. Particularly because of the high symbolic value we view it as being an unforgivable faux-pas, even as culinary rape, if someone cuts a piece of cake into slices or cubes. The distribution of food is subject to a strict, cultural code so that acting counter to it is seen as an immoral, almost anarchistic act.

In cutting up round festive food, the social or family hierarchy is symbolically suspended when, for instance, the barely three-year-old birthday child cuts the first piece of cake. This act does not take place in every civilization. In Japanese tea ceremonies the ritual sweets are pre-portioned and usually square. Perfectly cut edges of sushi, sashimi or traditional square wagashi (tea pastry) are seen as a formal ideal. Here, too, everyone gets an equally sized piece, but the act of cutting is absent and the conventional distribution of roles remains intact.[1]

As opposed to the cake, it is (traditionally) the woman who cuts the loaf of bread into slices – even if this loaf is round. This observation raises the question why the bread loaf and the cake, though similar in shape and size, are cut in different ways. Since bread is only cut when we want to put a topping on it, the form of the slice appears more favorable from a practical perspective. Wedge-shaped pieces are much more unstable and difficult to cover. Even the different definition of both meals plays a role. While bread is an everyday food, cake is seen as a festive one. Bread is cut successively when needed and then stored for later meals. Keeping the cut surface as small as possible protects it best from drying. The cake is only served on special occasions and is – at least traditionally – immediately sliced and eaten.

Small, Handy and Non-Perishable | Cutting rituals define the shape and identity of food. While the form of bread slices is not attributed great meaning, some forms like the triangular piece of cake is given a positive cultural connotation. This is a factor that was subsequently used by other foods, for instance, triangular waffles, shortbread triangles, oatbread or cheese triangles. Indeed, soft cheese invented in Thun, Switzerland

Links | Die dreieckige Form vorverpackter Käseecken spielt bewusst mit der traditionellen Art, runde Speisen wie Torten oder Käselaibe zu zerteilen. Schmelzkäse besteht aus eingeschmolzenem Hartkäse und kann daher grundsätzlich in verschiedenste Formen gebracht werden. Die Tortenstückform verleiht ihm jedoch ein heimeliges, traditionelles Image.
Unten | Klein, handlich, haltbar: Ähnlich wie bei Waschmittel-Tabs oder Tabletten ist Vorportionierung auch bei Nahrungsmitteln ein gestalterisches Plus. Der Zuckerwürfel wurde 1843 vom mährischen Zuckerfabrikanten Jakob Christoph Rad erfunden. Um 1880 kam dann mit dem Brühwürfel der Firma LEMCO (Liebig´s Extract of Meat Company) auch die Suppe in vorportionierter, eckiger Form auf den Markt.

left | The shape of pre-packaged cheese triangles plays deliberately with the traditional way of dividing round foods such as cakes or loaves of cheese. Processed cheese consists of melted hard cheese and can thus basically be shaped in a number of different ways. However, the shape of a piece of cake gives it a traditional, congenial look.
bottom | Small, easy to handle, non-perishable: Similar to detergent tabs or tablets, the preportioning of food can also be used to enhance the design. Sugar cubes were invented by the Moravian sugar manufacturer Jakob Christoph Rad in 1843. Around 1889 preportioned soup in square form was launched when the stock cube produced by LEMCO (Liebig's Extract of Meat Company) conquered the market.

daher grundsätzlich in verschiedenste Formen gebracht werden. Die dreieckige Form des so genannten „Käseecks" ist also technisch nicht zwingend, sondern frei gewählt, um den positiven Effekt der Tortenstückform zu nutzen.

Die kleinen foliierten Dreiecke aus Streichkäse suggerieren, dass sie sowohl das Ergebnis eines traditionellen Herstellungsprozesses als auch Teil eines großen Käselaibs sind. Sie täuschen die Zugehörigkeit zu einem größeren Ganzen, ja sogar die Existenz einer imaginären Tischgemeinschaft vor. So streichen Hungrige im Singlehaushalt ihre Schmelzkäsebrote gewissermaßen als Teil einer landesweiten kulinarischen Familie. Der Hersteller agiert als sinnbildliches Oberhaupt und verteilt vorsorglich den eingeschmolzenen Riesenlaib. Die Packungsgröße gibt obendrein vor, über das Ausmaß des Hungers Bescheid zu wissen: Lebensmittelkonzerne übernehmen die Portionierung und entheben den Einzelnen der Entscheidung, wie viel Käse er auf sein Brot streicht. Solcherart vordefinierte Stücke benötigen weder Augenmaß noch Schneidemesser und bedienen dadurch die Bequemlichkeit des Konsumenten. Ihre Handhabung ist bequem, zeitsparend, sauber – und müllintensiv.[2]

Vorportionierung | Vorportionierung ist eine funktionale Anforderung an Nahrung, der bei der Gestaltung von Produkten entsprechend Rechnung getragen wird. Sie ist damit ein wesentlicher Designfaktor und verleiht dem entsprechenden Produkt auch einen Mehrwert: Sechs einzeln hübsch in glitzerndem Stanniolpapier abgepackte Teilchen verleihen optisch das Gefühl, mehr gekauft zu haben. Vorportionierung erhöht damit das Preisleistungsverhältnis und ist ein Beispiel, wie entsprechendes Design den Wert des Ausgangsproduktes steigert.

Ähnlich funktional wie das Käseeck begleitet uns ein mittlerweile vollkommen selbstverständlich gewordenes, vorportioniertes Nahrungsmittel durch den Alltag: der Zuckerwürfel. Geometrisch, praktisch, süß: So ließen sich die wichtigsten Eigenschaften des Zuckerwürfels kurz umschreiben. Der Zuckerwürfel zählt mit mehr als 160 Jahren zu den ältesten vorportionierten Produkten auf dem Lebensmittelmarkt. Noch bis zur Mitte des neunzehnten Jahrhunderts waren große Zuckerhüte die übliche Handelsform von Zucker. Das Zerteilen der bis zu fünfzig Kilogramm schweren Kegel erfolgte zuhause und erforderte einigen Kraftaufwand und scharfes Werkzeug. Auch wenn Zuckerstücke wie „Perled Su-

in 1900 consists of melted hard cheese and can basically assume different forms. The triangular shape of the so-called "cheese triangle" is thus not technically mandatory but freely selected to take advantage of the positive effect of the shape of a piece of cake.

The small triangles of cream cheese covered by plastic sheet suggest that they are both the result of a traditional manufacturing process and part of a large cheese wheel. They give consumers the impression of belonging to a nationwide culinary family. The manufacturer acts as a quintessential head of the virtual table community and distributes the huge melted cheese wheel. The size of the package also feigns knowing about the extent of hunger. Food producing companies assume portioning and free the individual from having to decide how much cheese he/she is going to spread on a piece of bread. Thus predefined pieces require neither visual judgment nor a cutting knife and thus make use of the consumer's sense of practicality. Their handling is comfortable, saves time, is clean – and produces a lot of garbage.[2]

Pre-portioning | Pre-portioning is another functional requirement of food that is taken into account in designing products. It is thus a crucial design factor and one that lends the product added value. Six individual pieces nicely wrapped in tin foil paper give us the visual impression of having bought more. Pre-portioning thus enhances the price-performance ratio and illustrates how the right design can increase the value of the original product.

Similarly functional as the cheese triangle, there is a pre-portioned food item that has meanwhile become completely ordinary and accompanies us through everyday life – namely the sugar cube. Geometric, practical, sweet – these are the most important attributes of the sugar cube. With more than 160 years of age the sugar cube is one of the oldest pre-portioned products of the food market. Up to the middle of the nineteenth century, large sugar loaves were the most common commercial forms of sugar. The cones weighing up to 50 kilos (100 pounds) were divided at home, requiring quite a bit of strength and sharp tools. Even if sugar products as "perled sugar" or "almond sugar" were already common from the fifteenth century on, the sugar cube in today's sense was invented by the Moravian sugar industrialist Jakob Christoph Rad in Dačice in 1840. According to legend, his wife Juliane injured herself when attempting to chop up a sugar loaf with

gar" oder „Almond Sugar" bereits ab dem 15. Jahrhundert gebräuchlich waren, wurde die Idee des Zuckerwürfels im heutigen Sinn 1840 vom mährischen Zuckerfabrikanten Jakob Christoph Rad im heutigen Dačice geboren. Der Legende nach verletzte sich seine Frau Juliane beim Zerkleinern eines Zuckerhutes mit der Zuckerzange und bekleckerte den teuren Zucker mit ihrem Blut. Der Vorfall inspirierte Rad zur Entwicklung einer Maschine, die angefeuchtete Zuckerkristalle in kleine, handliche Stücke presste, welche er im Jahr 1843 unter dem Namen „Theewürfel" oder „Wiener Würfel" auf den Markt brachte.

Rads neue Maschine bestand aus einer Schraubtischpresse und einer Stanzplatte mit 400 quadratischen Öffnungen. Der Zucker wurde in nicht ganz trockenem Zustand von der Presse auf die Hälfte seines Volumens reduziert und dabei in Form gebracht. 1843 erhielt Rad ein fünfjähriges Privileg auf die Erfindung, Rohzucker in Würfelform herzustellen. Mit 6 Maschinen, 11 Frauen, 12 Mädchen und 36 Kindern produzierte Rad nunmehr 200 Zentner (10.000 kg) Zuckerwürfel am Tag, die in 1-Pfund-Kisten (0,5 kg) abgepackt und in ganz Österreich-Ungarn verschickt wurden.

Geometrisch, praktisch, süß | In Wien erlangte der neue Würfel unter dem Namen „Wiener Theezucker" schnell Bekanntheit. Die Zubereitung der beliebten Kaffeejause war nun um einen anstrengenden, zeitraubenden Arbeitsschritt, jenen des Zuckerbrechens, verkürzt, die Zuckerdose am Tisch gewann eine neue Bedeutung. Die kleinen Quader erleichterten nicht nur die Dosierung, sondern auch die Kontrolle der Dienstboten, die angeblich gerne vom teuren Zucker naschten: Hatte man ihn zuvor in versperrbaren Schatullen aufbewahrt, so musste man jetzt nur die Stücke zählen, um sicherzugehen, dass nichts fehlte.

Der neue Würfelzucker war zwar äußerst praktisch, jedoch so sündhaft teuer, dass er nach anfänglicher Euphorie bei der Wiener Bevölkerung alsbald wieder in Vergessenheit geriet. Anders in Amerika, wo Rads Maschine in einer New Yorker Raffinerie weiterentwickelt wurde und der Zuckerwürfel später als „Tate Cube" den europäischen Markt quasi ein zweites Mal eroberte.

Vom ursprünglichen Rad'schen Würfel blieb nur der Name. „Echte", würfelförmige Zuckerwürfel waren bis vor wenigen Jahren noch in Tschechien erhältlich, in den meisten Ländern weist der Würfel aus herstellungstechnischen Gründen allerdings nur noch jeweils vier gleich lange Kanten auf und sollte präziser Zuckerquader heißen. Außerdem variieren Abmessungen und Gewicht der Zuckerstücke von Hersteller zu Hersteller. Ein Zuckerwürfel kann zwischen 8 Gramm und 2 Gramm wiegen, wobei der Trend eindeutig in Richtung geringerer Größe geht. In Österreich wiegt ein Zuckerwürfel heute 4 Gramm, was in etwa einem gehäuften Kaffeelöffel entspricht. Auch auf der Packung steht nicht immer das Gleiche: Während die Deutschen und die Tschechen den Ausdruck „Würfel" benutzen, legen sich die Franzosen und die Spanier geometrisch nicht so genau fest und süßen einfach mit „Zuckerstücken". In England wiederum werden Tee und Kaffee mit, wörtlich übersetzt, „Zuckerklumpen" gesüßt.

Würfelzucker wird heute aus angefeuchtetem Feinkristallzucker erzeugt. Die rechteckige Form ist herstellungstechnisch zwar effizienter, aber nicht zwingend. Je nach Art der Stanzform können Zuckerwürfel auch in unterschiedlichsten zweidimensionalen Motiven erzeugt werden, eine Gestaltungsfreiheit, die in den letzten Jahren immer mehr genutzt wird. Zuckerwürfel werden mittlerweile als so genannter Bridgezucker in Form der vier Spielkartensymbole Pik, Herz, Karo und Kreuz oder mit bunten Herzen und Firmenlogos wie etwa jenem des englischen Traditionskaufhauses Fortnum & Mason verziert angeboten. Die französische Marke Canàsuc bietet Zuckerstücke in unzähligen Motiven an, unter anderem in Form fragiler Schmetterlinge, als Tannenzapfen, Edelweiß oder Puzzlesteine.[3]

Eckige Suppe | Vorportionierung ist ein Erfolgsfaktor, der die Bequemlichkeit des Kunden bedient. Schnell und sauber, ohne zusätzliches Werkzeug soll die Nahrung zu konsumieren oder weiterzuverarbeiten sein – Anforderungen, die besonders im Bereich von Snacks und Fertig- oder Halbfertigprodukten gefragt sind. Teebeutel zählen ebenso zu den klassischen, vorportionierten Waren wie getrocknete, zu Quadern verpresste Rindsuppe. Ursprünglich in Fläschchen, später in Pulver-, Tafel- und Wurstform, fand der Rindsuppenextrakt seine gestalterisch populärste Form in vorportionierten

sugar tongs and spotted the expensive sugar with her blood. The incident prompted Rad to develop a machine that would press dampened sugar crystals into small, practical pieces. He launched them on the market in 1843 under the name "Theewürfel" ("tea cubes") or "Wiener Würfel" ("Viennese cubes").

Rad's new machine consisted of a press and a cutting die with 400 square openings. The press was used to reduce the sugar, which was not entirely dry, to half of its volume and form it in the process. In 1843 Rad obtained a five-year privilege for manufacturing unprocessed sugar in cubes. With 6 machines, 11 women, 12 girls and 36 children Rad then produced 200 hundredweights (10,000 kilos) of sugar cubes a day. These sugar cubes were packed in 1-pound boxes (0.5 kilos) and sent out throughout all of Austria-Hungary.

Geometrical, Practical, Sweet | In Vienna the new cube quickly gained renown under the name "Wiener Theezucker" ("Viennese tea sugar"). One arduous, time-intensive step in the preparation of the popular coffee snack had been eliminated, with the sugar bowl on the table gaining new significance. The small squares not only made it easier to dose the sugar, but also to keep tabs on the servants who supposedly liked to nibble from the expensive sugar. Once it had been locked up in caskets, now it was only necessary to count the pieces to make sure that there was nothing missing.

As practical as the new sugar cube was, it was also sinfully expensive so that following initial euphoria among the Viennese population it soon fell into oblivion. The situation in America was very different. There, Rad's machine was further developed in a New York refinery and the sugar cube later conquered Europe a second time, as it were, as the "Tate cube".

Formally only the name of the original Rad cube has remained. "Real" sugar cubes were available in the Czech Republic until a few years ago. In most countries the cube no longer has equally long angles for manufacturing reasons and it would be accurate to refer to it as sugar ashlar or rectangular prism. Moreover, the dimensions and weight vary from manufacturer to manufacturer. A lump of sugar defined as a sugar cube can weigh between 8 grams and 2 grams, with the trend clearly being towards smaller size. In Austria a sugar cube today weighs 4 grams which amounts to about a heaped coffee spoon of sugar. Also the names differ. Whereas Germans and Czechs use the term "cube", the French and the Spanish do not specify so precisely and simply use "pieces of sugar" to sweeten their drinks. In England, "sugar chunks" are used for sweetening tea and coffee.

Today cube sugar is made of dampened fine crystal sugar. The square form is more efficient in terms of production but it is not mandatory. Depending on the die, sugar cubes can also be created with a great variety of two-dimensional motives, a freedom of design that has been exploited ever more in recent years. Sugar cubes are meanwhile available as so-called "bridge sugar" in the form of the four playing card symbols - spade, heart, diamond and club - or with bright hearts and company logos like that of the traditional English department store Fortnum & Mason. The French brand Canàsuc offers sugar pieces in countless motives, including fragile butterflies, pinecones, edelweiss or puzzle pieces.[3]

Square Soup | Pre-portioning makes for success, serving the customer's desire for convenience. The consumption of food should be quick and clean, and not require additional tools for eating or processing – demands that concern particularly snacks and finished or semi-finished products. Tee bags are among the classical, pre-portioned products just as the dried beef stock pressed into cubes. Beef stock extract, originally available in small bottles, later in the form of powder, bar and sausage, later found

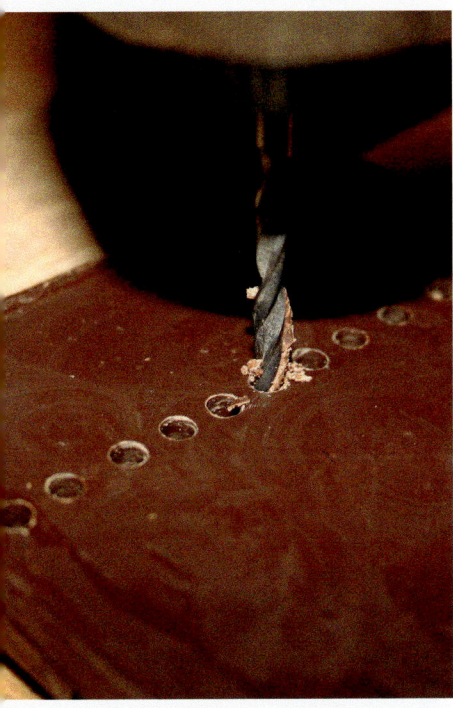

Würfeln und Quadern. In den 80er Jahren des 19. Jahrhunderts erfand die Firma LEMCO (Liebig´s Extract of Meat Company) den Brühwürfel, der aus Fleischextrakt, Rindertalg und Würzstoffen bestand. Im Jahr 1908 brachte dann der aus der Lombardei stammende Ernährungswissenschaftler Julius Maggi in der Schweiz zu Würfeln gepressten Rindsuppenextrakt und damit den Vorläufer des heutigen Suppenwürfels auf den Markt. Heute werden Brühwürfel auf zwei verschiedene Arten erzeugt: Für Österreich und die Schweiz wird dafür Pulver zusammengepresst, wodurch sich der Würfel beim Kochen einfach zwischen den Fingern zerbröseln lässt. Der Rest der Welt verwendet gegossene und damit zähere Suppenwürfel, die sich erst im Wasser auflösen.

Mittlerweile überschwemmen zahlreiche Nachahmer von Zucker- und Suppenwürfel den Markt. Hersteller bieten eine ganze Reihe derartiger Miniportionen an, deren Grundprodukte von einfachen Gemüsesorten bis zu Pastasaucen reichen. Essbare Produkte wie Spinat-Minis (in Österreich als „Zwutschgerln" bekannt), Gemüse- und Würfelsuppen zeugen genauso von der Beliebtheit vorportionierter Waren, wie unessbare Spülmaschinentabs. Das Konzept ist nicht neu, die Annehmlichkeiten portionierter Lebensmittel wussten schon unsere Großeltern zu schätzen: Die Maßeinheiten alter Kuchenrezepte zum Beispiel schreiben den Schokoladeanteil nicht in Gramm, sondern mit Hilfe von Rippen vor. Da diese ohnehin fixen Gewichtseinheiten entsprechen, ersparen sie den Griff zur Waage und sind wesentlich praktischer als heutige Zutatenlisten mit Gewichtsangaben. Ein Konzept, das vielleicht bald wiederentdeckt wird und sich durchaus auch auf andere Nahrungsmittel anwenden ließe. Vielleicht dürfen wir uns zukünftig über Mehl-, Zucker- oder Buttertabs zu 20 Gramm, 50 Gramm oder 100 Gramm freuen. (Manche Butterpackungen verfügen bereits über linealartige Aufdrucke, die das Gewicht anzeigen.)

Kaffeewürfel | Auch die Idee vorportionierter Kaffeetabs ist bereits fast 80 Jahre alt. In den dreißiger Jahren des vergangenen Jahrhunderts gab es Bemühungen, Kaffeepulver nach dem Vorbild des Zucker- oder Suppenwürfels in Würfelform zu pressen. Inspiration für die gestalterischen Ambitionen war eine Rekordernte in Brasilien. „Was tun mit den überschüssigen Tonnen?", fragte sich das brasilianische Kaffeeinstitut und trat an Nestlé mit der Bitte heran, aus den überschüssigen Bohnen Kaffeewürfel zu produzieren. Diese sollten in Wasser löslich sein und möglichst authentisch nach Kaffee schmecken. Nach sieben Jahren Forschung brachte Nestlé am 1. April 1938 den Nescafé heraus – allerdings nicht wie geplant in Würfel-, sondern wieder in Pulverform. Bis heute kommen Kaffeetabs nicht ohne zusätzliche Verpackung, die das Pulver in Form hält, aus.

Vorportioniertes Brot | Die Idee der Vorportionierung führte auch zu Produkten wie vorgeschnittenem Toastbrot, das allerdings einige Probleme hinsichtlich Konsistenz und Geschmack aufwarf. 1912 baute Otto Frederick Rohwedder in Davenport im US-Bundesstaat Iowa die weltweit erste Brotschneidemaschine, die ab dem Jahr 1928 in Chillicothe, Missouri, kommerziell genutzt wurde, um geschnittenes Brot herzustellen. Um 1930 begann dann die Firma „Wonder Bread", „sliced bread" in den gesamten Vereinigten Staaten zu vermarkten. Da die einzelnen Scheiben jedoch wesentlich schneller vertrocknen als ein Laib im Ganzen, muss die Rezeptur derartiger vorgeschnittener Brote entsprechend adaptiert werden. Dem Brotteig werden daher Proteine und Stärke beigemengt, um das kastenförmige Brot auch dann sehr lange weich und „frisch" zu halten, wenn es in einzelne Scheiben vorgeschnitten ist.[4]

its most popular form in pre-portioned cubes. In the 1880s Julius Maggi, a nutritionist from Lombardy, launched beef stock cubes in Switzerland – these cubes being the precursors of today's soup cube. Today stock cubes are manufactured in two different ways. For Austria and Switzerland powder is pressed, so that the cubes simply crumble between one's fingers while cooking. The rest of the world uses cast and thus tougher soup cubes that only dissolve in boiling water.

In the meantime the market is swamped with countless imitations of sugar and soup cubes. Manufacturers offer any number of such miniature portions whose basic products range from simple vegetable brands to pasta sauces. Edible products such as spinach-minis, vegetable and cube soups testify to the popularity of pre-portioned products similar to the inedible detergent tabs. The idea is not new – even our grandparents appreciated the convenience of portioned food. The measures of old cooking recipes, for instance, specified the amount of chocolate not in grams but with the help of rows. Since these rows corresponded to fixed weights anyway, they eliminated the need to use a scale and were much more practical than today's lists of ingredients with weights – an idea that might be rediscovered soon and applied to other types of food. Perhaps we will someday be regaled with flour tabs, sugar tabs or butter tabs weighing 20 grams, 50 grams or 100 grams. (Some butter packages already have ruler-like prints indicating the weight of pieces.)

Coffee Cubes | Even the idea of pre-portioned coffee tabs is almost eighty years old. In the 1930's there were attempts to press coffee into cubes modeled after sugar or stock cubes. These design aspirations were inspired by a record harvest in Brazil. "What to do with the superfluous tons?" the Brazilian Coffee Institute asked, turning to the Nestlé co. to produce coffee cubes. These were supposed to be water-soluble and have an authentic coffee taste. After seven years of research, Nestlé launched Nescafé on 1 April 1933 – however, not cube-shaped as planned, but in powder form. To this day coffee tabs need additional packaging to keep the powder in shape.

Pre-portioned Bread | The idea of pre-portioning also resulted in products such as pre-cut toast, which was accompanied by several problems regarding consistency and taste. In 1912 Otto Frederick Rohwedder in Davenport, Iowa, built the first breadcutting machine that was used commercially from 1928 onward in Chillicothe, Missouri, to produce pre-cut bread. Around 1930 the "Wonder Bread" company began marketing "sliced bread" all over the United States. Since the individual slices dry out much faster than a whole loaf, the recipe of such pre-cut bread had to be adapted accordingly. Proteins and starch are thus added to the bread dough to keep the box-shaped bread soft and "fresh" for a very long time, even if it is cut in single slices.[4]

Ähnlich wie Perforierungen bei Briefmarken oder Küchenrollen helfen Sollbruchstellen auch bei Nahrungsmitteln, ein großes Ganzes in einzelne Stücke zu zerteilen.

Similar to the perforations on stamps or kitchen towel rolls, predefined breaking points also help when large pieces of food need to be divided into individual servings.

Teilen in der Schreibtischlade
Komfort in der Küche und Flexibilität bei Verbrauch und Aufbewahrung gelten heute als unverzichtbarer Maßstab im Food Design. Ob wir schneiden, brechen, abbeißen oder einfach nur auslöffeln: bei Snacks und Fertiggerichten ist eine schnelle, möglichst appetitliche und einfache Teilbarkeit entscheidend.

Viele einzelne Häppchen, die in einer größeren Packung versammelt sind, geben dem Konsumenten die Freiheit, einen individuellen Zeitplan für den Verzehr des Inhalts zu erstellen. Als Josef Manner im Jahre 1898 eine neuartige Waffel namens „Neapolitaner Schnitte Nro. 239" auf den Markt brachte, ging diese zunächst ebenso wie der Gummibär und viele andere Süßigkeiten einzeln über den Ladentisch. Erst seit 1924 liegen zehn Waffeln vorgeschnitten zusammen in der rosaroten Packung. Die praktischen Schnittstellen entstehen im Zuge der Herstellung: Wie der Name schon sagt, wird die „Mannerschnitte" bei ihrer Herstellung aus großen Waffeltafeln herausgeschnitten. Danach werden die einzelnen Lagen mit Nougatcreme bestrichen, zu vieren übereinander gestapelt und mit einer trockenen Waffel abgedeckt. Schließlich werden die „Riesenschnitten" erst mit Bandsägen längs und dann mit Kreissägen quer geschnitten bis schließlich lauter exakt 47 x 17 x 17 mm große, längliche Schnitten entstehen.[5]

Die Sollbruchstelle als Designfaktor
Es bereitet Vergnügen, aus einem Produkt einfach mehrere zu fabrizieren, die entweder klammheimlich verspeist oder großzügig verteilt werden können. Sollbruchstellen ermöglichen ein schnelles und müheloses Auseinanderbrechen und erleichtern die Aufbewahrung. Während halb verzehrten Snacks wie einem angebissenen Sandwich etwas Unappetitliches anhaftet, wartet die angebrochene Packung hygienisch und sauber in der Schreibtischlade. Es bleibt dem Esser überlassen, wie viel er sich wann und wo gönnt. So bietet auch die mehr als hundertjährige „Mannerschnitte" ganz im Sinne des heutigen Zeitgeistes größtmögliche Flexibilität bei Aufbewahrung und Verzehr.

Der Prozess, Lebensmittel zu vermehren, indem ein großes Ganzes in eine Vielzahl von kleinen Stücken gebrochen wird, lässt den Konsumenten überdies im Glauben, mehr gekauft zu haben. Produktentwickler spielen mit diesem psychologischen Effekt und erheben die Teilbarkeit zum wesentlichen Bestandteil ihres Marketingkonzeptes. Eines der diesbezüglich erfolgreichsten Produkte ist das 1968 eingeführte „Twinni" von Langnese (in A: Eskimo, in GB: Wall's), ein Eislutscher auf zwei Stielen. Der Trick liegt in der Sollbruchstelle und der wundersamen Vermehrung von einem in zwei Eislutscher. Twinni kann man also nicht bloß konsumieren, man kann es auch zerteilen, für Kinder ein stets ebenso aufregendes wie schwieriges Unterfangen. Twinni schlecken bedeutet mehr, als bloß Eis essen, Twinni offeriert ein „Zusatzfeature" und wird damit zum Spielzeug.[6]

Ist eine Essware weder klein portioniert noch vorgeschnitten, sollte sie möglichst leicht, bröselfrei und ohne Messer zerteilbar sein. Ähnlich wie Perforierungen bei Briefmarken oder Küchenrollen (Küchenkrepp) helfen Materialschwachstellen wie Einkerbungen auch bei Nahrungsmitteln, eine große Einheit in einzelne Stücke zu zerteilen. Vordefinierte Bruchstellen verringern die Anzahl der Krümel und garantieren optimale Stückgröße.

Division in the Desk Drawer
Comfort in the kitchen and flexibility in using and storing things is today seen as an indispensable standard in food design. Whether we are cutting, breaking, biting off or simply spooning something – in snacks and instant meals it is decisive that something can be divided as quickly, simply and as appetizingly as possible.

A larger package containing a big number of nibbles gives the consumer the freedom to individually time when the content is eaten. When, in 1898, Josef Manner launched a new type of wafer called "Neapolitaner Schnitte Nro. 239", it was first sold individually, just like gummi bears and many other sweets. Only beginning in 1924 pink packages containing ten wafers each were pre-packed and sold. The practical cuts are created during production. As the name already says, the "Mannerschnitten" (Manner slices) are cut out of large wafers. Then the individual layers are covered with nougat cream, four of them stacked and covered with a dry wafer. Finally the "giant slices" are cut lengthwise by means of a band saw and transversally by circular saws until there are long slices each measuring exactly 47 x 17 x 17 mm.[5]

The Predetermined Breaking Point as a Design Factor
It is fun to produce several products out of one that can either be secretly eaten or generously distributed. Predetermined breaking points allow for quick and easy breaking and make it easier to store. Whereas half-eaten snacks like a sandwich that has been bitten off has something unappetizing about it, the opened package of presliced snacks remains hygienic and clean in the desk drawer. It is up the eater to decide on when and where to indulge. The "Mannerschnitte" is more than 100 years old but offers the greatest flexibility in storing and eating – fully in keeping with today's zeitgeist.

The process of increasing the amount of food by breaking a large piece into a number of smaller ones also leaves the consumer believing that he/she has bought more. Product developers play with this psychological effect and make the divisibility of a product an important part of their marketing strategy. In this respect, one of the most successful products of all times is "Twinni" by Langnese (known as Wall's in the UK) that was launched in 1968 – a popsicle on two sticks. Its trick is the predetermined breaking point and the fascinating multiplication in two popsicles. You can eat Twinni, and you can also split it, which for children is a task that is both exciting and challenging. To lick a Twinni means more than just eating ice cream. Twinni offers an "additional feature" and thus becomes a toy.[6]

If food is neither pre-portioned nor pre-cut, it should at least be possible to cut it easily, without crumbling and without having to use a knife. Like the perforations on stamps or kitchen towel rolls, weak material spots such as indentations also help in food to divide

Unzählige Brotsorten weisen ebenso wie Schokoladetafeln, Traubenzucker oder Cracker Sollbruchstellen auf. So brechen die Steirer ihre Langsemmel genauso entlang einer vorgegebenen Einkerbung wie die Deutschen ihre Brotzeile (Ranch Roll), die Engländer die „Swiss Crown" und die Türken ein achtteiliges Gebäckstück namens „Cicek Ekmeği", was übersetzt so viel wie Blumenbrot bedeutet. Eine in der Mitte verlaufende Kerbe unterstützt sogar die christliche Liturgie, indem sie eine formschöne Halbierung der Hostie gewährleistet.

Im Fall der wichtigsten christlichen Kultspeise erfüllt die Sollbruchstelle aber noch eine andere, religiös geprägte Funktion: Jahrhundertelang wurde bei der Kommunion ganz normales Alltagsbrot, das die Gläubigen in die Kirche mitbrachten, gegessen. Da (heute) vor allem bei den Katholiken der Glaube vorherrscht, Jesus Christus sei in jedem noch so kleinen Teil des geweihten Brotes persönlich anwesend, wurde dieses im Mittelalter durch Oblaten ersetzt: Sie sind vorportioniert oder können durch entsprechende Einkerbungen leicht geteilt werden, so dass sich keine Krumen des geheiligten Opferbrotes lösen, die dann verloren gehen oder respektlos behandelt werden könnten.

Ein anderes Extrembeispiel für die Anwendung der Sollbruchstelle ist jenes Brot, das speziell für den Verzehr im Weltall erzeugt wird. In einer Vakuumpackung findet sich eine Miniportion Brot, das sich in 10 bonbongroße Brotklumpen zerteilen lässt. Normales, knuspriges Brot ist in Spaceshuttles und auf Raumstationen tabu, da beim Verzehr unweigerlich Brösel entstehen, die in der Schwerelosigkeit Leitungen und Filter – oder im schlimmsten Fall die Atemwege der Astronauten – verstopfen könnten. Am Moskauer Institut für Biomedizin wurde deshalb ein eigenes „Weltallbrot" entwickelt, dessen Konsistenz durch Weichmacher und Wasser so verändert wurde, dass keine Bröselgefahr besteht.[7]

Kaiserliches Werbegeschenk – die Semmel | Auch die im süddeutschen, österreichischen und slowenischen Raum beliebte Kaisersemmel verfügt über charakteristische Sollbruchstellen. Der Legende nach wurde die Semmel zur Zeit Kaiser Josephs II. zwischen 1780 und 1790 in Wien erfunden. Demnach verdankt der Inbegriff der Wiener Backwerkstradition seinen klingenden Namen der Großmütigkeit seiner Majestät höchstpersönlich. Wiener Bäcker hatten die ungewöhnliche, fünfteilige Form entwickelt und suchten bei Hofe unter dem Namen „gerissene Semmel" um Schutz für das neue Produkt an. Joseph II. zeigte aber besondere Volksnähe und veranlasste, das wohlschmeckende Backwerk „Kaisersemmel" zu benennen. Aus dieser Begebenheit heraus entstand das Präfix „Kaiser-", das der Bezeichnung Semmel ursprünglich voranging.

a larger piece into several individual ones. Pre-defined breaking points reduce the amount of crumbles and guarantee an optimum number of pieces. Countless types of breads feature predetermine breaking points just like chocolate bars, dextrose or crackers. Styrians, for instance, break their long rolls along an indentation just as the Germans have their Brotzeile (ranch roll), the English the "Swiss crown" and the Turks an eight-piece bread known as "cicek ekmegi", which can be translated as flower bread. An indentation in the middle even buttresses the Christian liturgy, ensuring a formally nice halving of the consecrated wafer.

In the case of the leading Christian cult dish, the predetermined breaking point even fulfills another, religiously motivated function. For centuries ordinary bread, which believers brought with them to church, was eaten during communion. Since predominantly Catholics believed (and still believe) that Jesus Christ is personally present in every tiny piece of the blessed bread, this was replaced by wafers during the Middle Ages. They are pre-portioned or can be easily divided thanks to indentations so that the sacrificial bread does not crumble. The danger of losing small pieces, which then could be treated disrespectfully, is thus minimized.

Another extreme example of the use of the predetermined breaking point is space bread which is packed in a vacuum bag and can easily be divided into 10 candy-sized lumps of bread. Normal, crispy bread is taboo in space shuttles and stations since conduits and filters – or in the worst case, the respiratory paths of the astronauts – may become clogged by crumbs uncontrolledly floating around. At the Moscow Institute for Biomedicine a "space bread" was thus developed; its consistency was altered by means of softeners and water so that there is no danger of crumbs.[7]

Imperial Advertising Gimmick – a Special Roll |
Even the "Kaisersemmel", which is also popular in Southern Germany, Austria and Slovenia, features characteristic predetermined breaking points. According to legend, this type of roll was invented in Vienna in the days of Emperor Joseph II between 1780 and 1790. Thus, the allusive name of this embodiment of Viennese baking tradition is owed to the generosity of His Majesty. Viennese bakers had developed the unusual five-piece shape and sought to patent their new product at court under the name "Gerissene Semmel" ("torn roll"). Joseph II, however, proved to be a man of the people and had the tasty roll named "Kaisersemmel" ("Emperor's roll"). This turn of events gave rise to the prefix "Kaiser-", the name originally being hyphenated.

Unzählige Brotsorten, Schokoladetafeln, Traubenzucker und Cracker haben Sollbruchstellen. Vordefinierte Bruchstellen sorgen für gleich geformte, gleich große Stücke und verringern die Anzahl der Krümel.

Innumerable types of bread, chocolate bars, dextrose and crackers have defined breaking points. Predefined breaking points ensure pieces of equal shape and size and reduce the number of crumbs.

255

Andere Quellen sehen die Bezeichnung „Kaiser" hingegen in Zusammenhang mit dem ursprünglich elitären Charakter der Semmel. Im 19. Jahrhundert war die Semmel aus feinem Weizenmehl im Vergleich zu Brot deutlich kostspieliger und daher zunächst vor allem bei der Oberschicht beliebt. Noch bis in die Nachkriegszeit war der Konsum von Kleingebäck für den Normalbürger nicht alltäglich. Semmeln kamen nur zu besonderen Anlässen wie Hochzeiten, Taufen oder an Sonntagen auf den Tisch. In Deutschland und Österreich wurden sie an den Geburtstagen des Kaisers gratis an die Bevölkerung verteilt und daher mit dessen Titel versehen.

Die Semmel als 50 Gramm schweres Weißgebäck existierte schon lange vor der Idee des österreichischen Kaisers, sie als Werbegeschenk zu nutzen. Die sprachliche Wurzel lässt sich bis ins antike Griechenland zurückverfolgen, konkret auf das altgriechische Wort „semídalis", das vermutlich vom assyrischen „samidu" abstammt: Beides bedeutete feines Weizenmehl. Die deutsche Bezeichnung „Semmel" leitet sich vom lateinischen „simila" her, das ebenfalls feinstes Weizenmehl bedeutete. Archäologische Funde bestätigen zudem, dass bereits im alten Pompeji kleine runde Weißgebäcklaibchen unter dem Namen „semlja" verkauft wurden. Weitaus schwieriger ist es, die Entstehungsgeschichte der heutigen Semmelform mit dem charakteristischen, fünfteiligen Stern zurückzuverfolgen. Unter den von Hermann Müller nach Forschungen von Max Währens nachgebackenen antiken Broten in der „Kunstgeschichte des Backwerks" von Hans Jürgen Hansen befinden sich römische Brotlaibchen, die mit den für die Semmel typischen, geschwungenen Einkerbungen verziert sind, allerdings aus acht Segmenten bestehen.[8]

Die Sollbruchstelle als Designfaktor | Warum das kleine Weißgebäck ausgerechnet aus fünf Segmenten besteht und sich damit nicht in zwei gleiche Teile brechen lässt wie die steirische Langsemmel, ist aus heutiger Sicht nicht zu erklären. Manche Quellen sehen in der Semmel ein uraltes Sonnen- oder Blumensymbol, andere Interpretationen führen die fünf Segmente auf die Anzahl der menschlichen Finger zurück. Plausibel erscheint auch eine herstellungstechnische Erklärung: Die fünf Kerben entstehen bei der traditionellen Handsemmel durch das mehrmalige Einschlagen des Teiges über den Daumen des Bäckers. Aufgrund dieser Technik und des gesetzlich vorgeschriebenen Ausbackgewichts von 50 Gramm ergaben sich einfach fünf Segmente. Die menschliche Hand diente demnach nicht nur als Werkzeug, sondern auch als natürliches Maß für die Anzahl der entstehenden Teile.[9]

Wenn man eine Wiener Semmel in zwei gleiche Hälften teilen möchte, braucht man ein Messer, ein Umstand, den man als designtechnisches Manko interpretieren könnte. Absurderweise benötigt man zum Befüllen von kleinem Weißgebäck immer ein Schneidwerkzeug, da es keine einzige Sorte gibt, die sich durch eine entsprechende Sollbruchstelle horizontal teilen lässt. Im Zeitalter des Sandwichs wäre ein Redesign der Semmel mit einer horizontalen Rille anstelle der fünf vertikalen Kerben durchaus eine Überlegung wert.

Other sources, by contrast, see the word "Kaiser" as related to the originally elitist character of the roll. In the nineteenth century rolls made of fine wheat flour were substantially more expensive than bread and thus mainly affordable to the upper class. Even in the post-war era the consumption of small rolls was not an everyday thing for ordinary people. The "semmel" was only served on special occasions such as weddings, baptisms or on Sundays. In Germany and Austria these rolls were distributed free of charge to the population on the Emperor's birthday and were thus known under this name.

The "semmel" as a white roll weighing 50 grams already existed long before the Austrian emperor came up with the idea of using it as an advertising gimmick. The linguistic roots can be traced back to ancient Greece, more specifically to the ancient Greek word "semidalis", which was probably derived from the Assyrian "samidu". Both mean fine wheat flour. The German word "Semmel" comes from the Latin "simila", which was also a word for finest wheat flour. Archaeological findings also confirm that already in ancient Pompeii small round white bread rolls were sold as "semlja". In contrast to the name it is much more difficult to trace the genealogy of today's semmel form with the characteristic five-part star. Among the breads reconstructed and baked by Hermann Müller on the basis of Max Währen's studies in "Kunstgeschichte des Backwerks", an art history of baking, by Hans Jürgen Hansen, we find Roman bread rolls that were adorned with the curved indentations typical for the "semmel", which, however, consisted of eight segments.[8]

The Predetermined Breaking Point as a Design Factor | From today's perspective there is no explanation why the small white roll has to consist of five segments and cannot be broken into two equal pieces like the Berlin roll. Several sources see the "semmel" as being a primeval symbol of the sun or a flower, other interpretations relate the five segments to the number of human fingers. An explanation referring to the manufacturing process also seems plausible. The five dents in the traditional hand-made roll result from the dough being folded several times by the baker's thumb. The five segments were simply a result of this technique and the legally prescribed final baked weight of 50 grams. The human hand thus not only serves as a tool, it is also a natural measure for the number of evolving parts.[9]

If one wants to divide a Viennese "semmel" into two equal pieces, one needs a knife – a factor that could be interpreted as a design flaw. It is absurd that we always need a cutting tool to fill the small white bread roll, since there is no single type that can be divided horizontally in a predetermined breaking point. In the age of the sandwich, it would certainly be worth considering redesigning the "semmel" with a horizontal groove instead of the five vertical dents.

Designte Teilbarkeit | Esswaren mit den Fingern zu zerteilen macht Spaß. Designs, die diesen Faktor positiv nutzen, kommen bei den Konsumenten gut an. Eine der innovativsten Entwicklungen auf diesem Gebiet ist Terry's Schoko-Orange, im Prinzip eine dreidimensionale Weiterentwicklung der Schokoladetafel: einzelne Schokoladestücke, leicht und bröselfrei teilbar in einer gemeinsamen Packung – und das alles in völlig neuer Form. Die „Orange" aus Milchschokolade mit Orangenaroma besteht aus 24 exakt geformten „Orangenspalten", die alle mit einem Oberflächenrelief aus Orangenpapillen versehen sind und nur an zwei Stellen (den Polen) aneinander haften. Bei leichtem Draufschlagen zerfällt die Orange, ohne zu krümeln, in ihre schlanken Einzelteile, die man bequem in den Mund stecken kann. Die 1823 im englischen York gegründete Schokoladenfabrik Terry´s brachte in den 1920er Jahren zunächst den „Terry´s Chocolate Apple" auf den Markt, der später von der „Terry´s Chocolate Orange" verdrängt wurde. Das formschöne Design erfüllt ganz klare Anforderungen – Teilbarkeit sowie Portionierbarkeit – und sprengt dennoch den Rahmen der herkömmlichen Schokoladetafel.

Ein anderes Produkt, dessen Gestaltung den Akt des Zerteilens in den Vordergrund rückt, ist der japanische Faserkäse „Sakeru", der extra so designt ist, dass er – zumindest für Japaner – besonders genussbringend zu zerlegen ist. Ein beliebter japanischer Snack zum Aperitif sind getrocknete Tintenfische. Die Japaner schätzen an ihnen vor allem die zähe Konsistenz, aufgrund derer sich von den einzelnen Stücken mit den Fingern längliche Fasern abspalten und herunterziehen lassen. Vor rund 30 Jahren griff der japanische Lebensmittelhersteller Snow Brand Milk die Vorliebe seiner Landsleute, Esswaren mit den Fingern zu zerfransen, auf, um daraus ein eigenes Produkt zu kreieren. Nach zirka fünf Jahren Entwicklung brachte Snow Brand den „Sakeru-Käse" auf den Markt, wobei Sakeru so viel wie „lässt sich zerfasern" bedeutet. „Sakeru" besteht aus vorportionierten, länglichen Käsestäbchen mit rundem Querschnitt, von denen sich in ähnlicher Weise wie beim Tintenfischfleisch einzelne Fasern herunterlösen lassen.[10]

1 | Interview mit Keiko Nakayama, Toraya Gallery, Toraya Confectionery LTD, Tokio, im Oktober 2008
2 | Informationen: Fa. Rupp
3 | Quellen: Dipl.Ing. Martin Doppler, Agrana, Raffinerie Tulln; Eva Bakos, Gaumenschmaus und Seelenfutter; Hannsferdinand Döbler, Kochkünste und Tafelfreuden; Christoph Wagner, Süßes Gold
4 | Interview mit der amerikanischen Lebensmitteltechnologin Kantha Shelke am 4.4.2009
5 | Betriebsbesichtigung Manner, Wien, im April 2007
6 | vgl.: Catarina Kossuth-Wolkenstein, „Die Marke Eskimo", Signum, Wien, 2000
7 | Interview mit Dieter Isakeit, Head of Erasmus Centre, European Space Agency, Noordwijk, am 8.2.2007
8 | Informationen: Dr. Keller, Institut für Germanistik an der Universität Wien; Etymologisches Wörterbuch von Grimm und Kluge; Hans Jürgen Hansen, Kunstgeschichte des Backwerks, Stalling Verlag, 1968, S. 10
9 | Informationen: Josef Honeder, Bäckerei Honeder, Weitersfelden
10 | Betriebsbesichtigung Snow Brand Milk, Tokio, im Oktober 2008

Designed Divisibility | To divide food by using one's fingers is fun. Designs that make positive use of this factor are popular with consumers. One of the most innovative developments in this sector is Terry's Chocolate Orange, basically a three-dimensional version of the chocolate bar. Individual pieces of chocolate, which can be easily divided without crumbling – now all found in one package and in a new form. The "orange" made of milk chocolate with orange aroma consists of 24 precisely shaped "orange slices" that all have a surface relief of orange papillae and stick to each other in two spots only (the poles). When lightly beaten, the orange falls apart without crumbling, dividing up into individual pieces which can be easily put in one's mouth. Terry's chocolate factory was founded in York in England in 1823; in the 1920s, it first launched "Terry's Chocolate Apple" which later yielded to "Terry's Chocolate Orange". The formally pleasing design met very clear specifications – that it could be divided and portioned – while transgressing the boundaries of the ordinary chocolate bar.

Another product whose design brings the act of dividing front stage, is the Japanese "sakeru" string cheese, which is designed so that, at least for the Japanese, it can be shredded with the greatest possible pleasure. One popular Japanese snack served with drinks is dried squid. The Japanese appreciate their viscid consistency, allowing individual pieces to be plucked off with one's fingers. About thirty years ago the Japanese manufacturer Snow Brand Milk tapped the predilection of his compatriots for tearing food apart with their fingers to create his own product. After about five years of development, Snow Brand launched the "sakeru" cheese brand on the market, with "sakeru" meaning "allows itself to be shredded". "Sakeru" consists of pre-portioned, oblong cheese sticks with a round cross-section from which individual pieces can be pulled as if you were eating squid.[10]

1 | Interview with Keiko Nakayama, Toraya Gallery, Toraya Confectionery LTD, Tokyo, October 2008
2 | Source: Rupp co.
3 | Sources: Martin Doppler; Agrana; Raffinerie Tulln; Eva Bakos; Gaumenschmaus und Seelenfutter; Hannsferdinand Döbler; Kochkünste und Tafelfreuden; Christoph Wagner; Süßes Gold
4 | Source: Interview with the American food scientist Kantha Shelke, 4 April 2009
5 | Guided tour of the Manner co., Vienna, April 2007
6 | Source: Catarina Kossuth-Wolkenstein, "Die Marke Eskimo", Signum Vienna, 2000
7 | Interview with Dieter Isakeit, Head of Erasmus Centre European Space Agency, Noordwijk, 8 February 2007
8 | Source: Dr. Keller, Department of German Language and Literature, University of Vienna; Etymological Dictionary by Grimm and Kluge; Hans Jürgen Hansen, Kunstgeschichte des Backwerks, Stalling Verlag, 1968, p. 10
9 | Source: Josef Honeder, Honeder's Bakery, Weitersfelden
10 | Guided tour of the Snow Brand Milk co., Tokyo, October 2008

Gesundheit als Funktion – Functional Food

Der Kunde wünscht sich nicht nur Esswaren, die bekömmlich und gesund sind, sondern wenn möglich auch noch fit und jung halten, schön machen, vor Krankheiten schützen oder diese im besten Fall sogar heilen. Der Hoffnung auf Nahrung mit Wunderwirkung wird unter anderem Rechnung getragen, indem Produkte nachträglich – also künstlich – mit positiven Stoffen angereichert werden. Speziell präparierte Nahrungsmittel sollen Wohlbefinden und Abwehrkräfte steigern oder gegen Übergewicht ankämpfen helfen. Spätestens seit 1993, als in Japan Nahrungsmittel mit der Bezeichnung „Foshu" („food for specific health use") auf den Markt kamen, ist die künstliche Anreicherung mit Vitaminen oder anderen gesundheitsfördernden Substanzen ein fixer Teilbereich von Food Design. Gesundheit als Funktion von Essen führte in weiterer Folge zu Ideen wie jener, Produkte durch probiotische Bakterien, Ballaststoffe oder Omega-3-Fettsäuren aufzuwerten.

Unter Functional Food oder Performance Nutrition versteht man Lebensmittel, die eine zusätzliche, künstlich erzielte Funktion erfüllen. Diese Zusatzfunktion kann z.B. eine gesundheitsfördernde, cholesterinsenkende, immunstärkende oder verdauungsfördernde Wirkung sein. Im Extremfall soll die Einnahme von Medikamenten ersetzt oder erleichtert werden, der Übergang von Functional Food zu Nahrungsergänzungs- und Arzneimitteln ist daher fließend. Beispiele für Functional Food sind mit Vitaminen angereicherte Fruchtsäfte, mit Spurenelementen und Ballaststoffen versetzte Kornriegel oder Frühstücksflocken, Produkte mit Protein- oder Aminosäurezusätzen wie L-Carnitin, Kreatin oder Taurin. Probiotisches Joghurt und isotonische Sportlerdrinks zählen ebenso zu Functional Food wie jodiertes Speisesalz, Zahnpflege-, Nikotin- oder Reisekaugummis. Neuerdings locken auch so genannte Energy Gums, die durch spezielle Zusatzstoffe aufkommender Müdigkeit entgegenwirken sollen.[1]

Zucker, Cola und Rindsuppe als Arznei

Die Idee, die Nahrungsaufnahme mit der Verfassung des Körpers in Zusammenhang zu bringen, ist nicht neu. In China bildet sie seit Jahrtausenden die Basis traditioneller Heilmethoden. Auch im Westen wurden Nahrungsmitteln im Laufe der Geschichte immer wieder Heilwirkungen nachgesagt. Zucker wurde ursprünglich als Arznei in Apotheken angeboten und wirkte im Krankheitsfall durch seine Nahrhaftigkeit aufbauend und belebend. Auch das von einem Apotheker entwickelte Coca-Cola galt als gesundheitsfördernd, lebensverlängernd und sollte gegen Müdigkeit, Verdauungsstörungen und Nervenkrankheiten helfen. Fleischbrühen und Rindsuppen-Extrakte waren Arzneimittel, die zur Stärkung der Patienten eingesetzt wurden und auch den Begriff des „Restaurants" prägten: Das erste Restaurant der Welt, welches 1760 in Paris eröffnete, war eigentlich eine Suppenküche, die in erster Linie die Gesundheit ihrer Gäste fördern sollte. Serviert wurden neben Früchten und Molkereiprodukten vor allem nahrhafte Bouillons. Letztere verliehen der neu entstandenen Institution schließlich auch ihren Namen: Die stärkenden Suppen, die die körperlichen Kräfte „restaurieren" sollten, trugen nämlich das Attribut „restaurant".

Heute arbeiten Wissenschaftler an Wunderessen, das Plagen wie Übergewicht, Vitaminmangel oder Müdigkeit aus der Welt schaffen soll. Joghurtdrinks versprechen ähnliche Wirkung wie die eines Zaubertranks und erfüllen damit die Sehnsüchte der modernen Leistungsgesellschaft, körperliche und geistige Defizite durch die Einnahme gewisser Substanzen auszugleichen. In fast allen Studien der jüngeren Vergangenheit wird Gesundheit als höchstes Gut bewertet.

Die spezielle Zusammensetzung von Eiern, Kartoffeln oder Brokkoli könnte sogar der Arterienverkalkung oder Krankheiten wie Hepatitis B vorbeugen, so noch die großen Erwartungen vor wenigen Jahren. Mittlerweile ist Ernüchterung eingekehrt, große Lebensmittel- und Pharmakonzerne haben ihre Forschungsprojekte an Functional Food eingestellt oder reduziert. Die Umsetzung der Vision vom heilenden Essen gestaltet sich in vielen Fällen komplizierter, zeitaufwändiger und damit kostspieliger als erhofft.

Functional Food: Gentechniker am John Innes Centre in Norwich, GB, züchten Tomaten, die einen möglichst hohen Gehalt von krebshemmenden Anthocyanen haben. Die „purple tomatos" könnten helfen, die Krebsrate zu senken.

Health as a Function – Functional Food

Customers do not just want food that is appealing and healthy but it should also keep us fit and young, protect us from disease or even heal illnesses in the best instance. The hope for food with a miracle effect is also taken into account when positive substances are added to products later – i.e., artificially. Specially prepared food should enhance wellbeing and resistance or even help combat obesity. At least since 1993, when in Japan food designated as "foshu" ("food for specific health use") was introduced on the market, the artificial enrichment of food with vitamins or other health-boosting substances has become a fixed subsegment of Food Design. Health as a function of food later resulted in ideas such as adding pro-biotic bacteria, ballast substances or omega 3 fatty acids to food.

Functional Food or performance nutrition is defined as food that can fulfill an additional function by artificial means. This additional function can, for instance, be a health-boosting, cholesterol-reducing, immunity-strengthening or digestion-promoting effect. In extreme cases it should be possible to replace or reduce the amount of medication taken. The transition from functional food to nutritional supplements and medication is thus fluid. Examples of functional food are fruit juices to which vitamins have been added, muesli bars or breakfast flakes enriched with trace elements and ballast substances, products with protein or amino acid additions such as L-carnitine, creatine or taurine. Pro-biotic yoghurt and isotonic sports drinks are also included among functional food, as are iodized table salt, and nicotine or travel gums. More recently, so-called energy gums that are supposed to combat fatigue by means of special additives have appeared on the market.[1]

Sugar, Cola and Beef Stock as Medicine

The idea of relating the ingestion of food to the condition of the human body is nothing new. In China traditional healing methods have been based on it for thousands of years. Even in the West food has, in the course of history, always been attributed a curative effect. Sugar was originally offered as a medicine in pharmacies and it was known for its nutritious value in restoring health and its invigorating effect in the case of illness. Even the Coca-Cola, developed by a pharmacist, was seen as promoting health, prolonging life and was said to have helped against fatigue, digestive disorders and nervous illnesses. Meat stocks and beef soup extracts were medicine used for strengthening patients, and they also influenced the idea of "restaurants". The first restaurant in the world, opened in Paris in 1760, was actually a soup kitchen primarily supposed to promote the health of its guests. Apart from fruits and dairy products, nutritious bouillons were mainly served. The latter also gave the newly created institution its name. The attribute "restaurant" had originally been attributed to the invigorating soups which were to "restore" a person's physical forces.

Today scientists are working on miracle foods, which are expected to do away with obesity, vitamin deficits and fatigue. Yoghurt drinks promise a similar effect as a magical potion and thus fulfill the desire of modern competitive society to offset physical and mental deficits by ingesting certain substances. In almost all surveys of the recent past, health is seen as the highest good.

The special composition of eggs, potatoes or broccoli could even prevent the calcification of arteries or illnesses such as hepatits B – we still had great expectations in this respect a few years ago. In the meantime, a certain disillusionment has set in – large food and pharmaceutical corporations have suspended or scaled back their research projects on functional food. In many instances, implementing the vision of healing food is much more complicated, more time-consuming and more expensive than hoped for.

Functional Food: genetic engineers at the John Innes Centre in Norwich, GB, cultivate tomatoes with the highest possible level of anti-cancerogenic anthocyanins. The "purple tomatoes" might help to reduce the incidence of cancer.

Nanotechnologische Hüllen | Intensiv geforscht wird zum Thema gesunde Nahrung weiterhin an Universitäten und staatlich geförderten Forschungseinrichtungen. Der Lebensmitteltechnologe David Julian McClements an der University of Massachusetts in Amherst beschäftigt sich beispielsweise seit rund fünf Jahren mit dem nanostrukturellen Design von Nahrungsmitteln. Ziel seiner Forschungen ist es, Lebensmittel dahingehend zu verändern, dass ihre Inhaltsstoffe gegen Krebs, Herz-, Augen- oder andere Krankheiten wirksam werden. Verschiedene Substanzen wie Omega-3-Fettsäuren, kurzkettige Fettsäuren, Carotinoide oder sekundäre Pflanzenstoffe beeinflussen Leiden wie etwa Darmkrebs zwar positiv, werden aber bereits im Magen verdaut und können ihren gesundheitlichen Effekt daher im Darm nicht entfalten.

McClements und sein Team verkapseln die entsprechenden Komponenten mit Hilfe von Nanotechnologie, damit sie die Magensäure unbeschadet überstehen und erst im Darm aufgespalten werden. Anwendungsgebiete für derartige, nanometrisch dünne Überzüge sind zum Beispiel Milch und Joghurt, die auf diese Weise zur Medikamenteneinnahme herangezogen werden könnten, oder profane Schlankmacher, deren kalorienreiche Bestandteile so ummantelt sind, dass sie unverdaut wieder ausgeschieden werden.

Die Idee, Essen als Transporteur für natürliche Zutaten mit medizinischer Wirkung zu verwenden, wirft jedoch auch gestalterische Probleme auf. Denn die zugesetzten Nanopartikel, welche die gewünschten Stoffe transportieren, dürfen weder den optischen Eindruck noch den Geschmack, die Textur oder die Haltbarkeit der Essware verändern. Beim Test eines Schinkens zum Beispiel, der mit Omega-3-Fettsäuren angereichert worden war, erwiesen sich die nanometrischen Fetttröpfchen, in welche die Fettsäure eingeschlossen war, als problematisch, da sie das Licht zu stark streuten. Sie verliehen dem Schinken eine hellere, milchige Farbe, die von den Konsumenten als ekelhaft beschrieben und abgelehnt wurde. Die Markteinführung von Lebensmitteln, welche die Technologie der Nanobeschichtung nutzen, wird derzeit aktiv verfolgt, ist aber erst in einigen Jahren zu erwarten.[2]

Schöner Brokkoli und violette Tomaten | Am Institute of Food Research im englischen Norwich versuchen Wissenschaftler unter der Leitung von Prof. Richard Mithen, modernen Brokkoli mit gesundheitsfördernden Substanzen von wildem Brokkoli anzureichern. Im Vergleich zu heutigen Züchtungen entsprechen die gesünderen Wildbrokkoli nicht mehr unseren optischen Vorstellungen, wie ein knackiges, frisches und saftig grünes Gemüse auszusehen hat, und sind daher unverkäuflich. Mithilfe von traditionellen Züchtungsmethoden versuchen die Wissenschaftler jene Genome der Wildpflanze, die gesundheitsfördernd wirken, in Zuchtbrokkoli einzusetzen. Das Produkt gelangt nach etwa 18 Jahren Forschungsarbeit demnächst in eine breit angelegte Testphase, in der die gesundheitliche Wirkung der neuen Sorte nachgewiesen werden soll.

Einen Schritt weiter geht der Gentechnologe Cathie Martin am John Innes Centre in Norwich. Die Gentechnik ist ein Teilgebiet der Biotechnologie, bei der die DNA von Lebewesen systematisch modifiziert wird, um Tieren oder Pflanzen gewünschte Eigenschaften zu verpassen. Der Unterschied zur herkömmlichen Züchtung besteht darin, dass Gentechnik auch über Artgrenzen hinweg funktioniert, also gewisse menschliche, tierische und pflanzliche Gene untereinander ausgetauscht werden können und die Verpflanzung einzelner DNA-Abschnitte möglich ist, während bei der klassischen Züchtung immer ganze Genome weitergegeben werden. Ziel der Gentechnik ist die lebensmitteltechnische Verbesserung gewisser Eigenschaften, etwa bessere Lagerfähigkeit, Schädlingsresistenz, Produktionssteigerung oder die kostengünstigere Herstellung von Substanzen. Gentechnisch veränderte Mikroorganismen werden z.B. auch zur Erzeugung verschiedener Zusatz- und Aromastoffe, z.B. Vitamine, Aminosäuren (vor allem Glutamat), Enzyme oder Fette mit speziellen Fettsäuren, eingesetzt.

Nanotechnological Outer Layers | Universities and government-funded research institutions are conducting intensive research on healthy nutrition. Food scientist David Julian McClements at the University of Massachusetts in Amherst, for instance, has been working on the nano-structural design of food for five years. The goal of his research is to alter their contents so that they protect humans against cancer, heart disease, vision disorders or other illnesses. Various substances such as omega-3 fatty acids, short chain fatty acids, carotenoids or phytochemicals may positively influence diseases such as colon cancer but they are digested in the stomach and are thus not able to provide their health-related effects.

McClements and his team encapsulate the appropriate components by means of nanotechnology so that they can withstand the gastric acid and are only decomposed in the large intestine. Applications for such nanometer-thin covers are, for instance, milk and yoghurt which can be used in this way to ingest medicine, or ordinary slimming agents whose high-calorie components are coated so that they are excreted without being digested.

The idea of using foods to transport natural ingredients with potential medical benefits poses some challenges to the food scientists' design skills. The nano-particles that are used to deliver these ingredients should not adversely alter the visual appearance, texture, taste, or shelf-life of the food product. For example, in an experiment with fortification of ham with omega-3 fatty acids entrapped within nano-sized lipid droplets, it was found that the incorporation of the droplets caused a problem because they scattered light too much so that the meat appeared a lighter, milkier color that consumers described as undesirable. The possibility of introducing food products on to the market based on this nano-laminated technology is actively being pursed, and may be viable within a few years.[2]

Nice Broccoli and Violet Tomatoes | At the Institute of Food Research in the English city of Norwich scientists working under the guidance of Prof. Richard Mithen are trying to enrich modern broccoli with healthy substances found in wild broccoli. Compared to broccoli cultivated today, the healthier wild broccoli no longer corresponds to our visual expectations of what crispy, fresh and juicy green vegetable is supposed to look like thus it would be impossible to sell it. Using traditional methods of cultivation scientists try to apply the genomes of the wild plant, which have a beneficial effect on health, to cultivated broccoli. Following some 18 years of research, the product will soon enter an extensive test phase in which the beneficial effect of the new type of broccoli is to be proven.

Genetic engineer Prof. Cathie Martin at the John Innes Centre in Norwich goes one step further. Genetic engineering is a subfield of biotechnology in which DNA taken from living entities is systematically modified. Unlike ordinary cultivation, genetic engineering also functions above and beyond the boundaries of species, meaning that also certain human, animal and plant genes can be exchanged and that it is possible to transplant individual segments of DNA, whereas in classic cultivation, entire genomes have to be passed on. The goal of genetic engineering is to improve specific qualities in processing food, as for instance longer durability, greater pest resistance, increased yield or less expensive production of substances. Microorganisms that have been genetically manipulated can be used, for example, to produce various additives or aromas, vitamins, amino acids (primarily glutamate), enzymes or fats with special fatty acids.

Cathie Martin has implanted the genes of a flower known as snapdragon in tomatoes to produce a fruit containing anthocyanins that are beneficial to health. These bioactive pigments usually occur in cranberries or blueberries but are not eaten often enough or in quantities that are too small to have a relevant effect on our health. Tomatoes, by contrast, are a widely available vegetable and, if they contained a sufficient amount of anthocyanins, could reduce the rate of cancer – at least this is the vision of researchers. Cathie Martin's 'purple tomatoes' containing anthocyanins are almost black and actually do have an anti-cancerogenic effect. However, this has only been tested on cancer in rats.[3]

Living Microorganisms for Our Intestinal Flora

In Europe substances like pro-biotics and pre-biotics aiming at improving intestinal flora as well as general health are among the most important functional substances in foods. Pre-biotics consist of carbohydrates that pass through the stomach in undigested form, settle in the colon where they serve positive, probiotic bacteria as a metabolic basis and inhibit the growth of undesired bacteria. In many years of research work, the Austrian Prolactal company has succeeded in producing a mixture of galacto-oligosaccharides out of animal whey, a byproduct of milk processing. Here, the chain length of molecules as well as the way individual mono-saccharides link up is controlled so that a special pre-biotic substance is created; it is more effective than ordinary pre-biotics, as well as pH-value-stable and heat-stable, and can now be sold to manufacturers of baby food, functional food and drinks.

In the future foods geared to certain target groups can be introduced to the market to cater to the nutrition needs and health conditions of certain population groups. It would, for instance, be conceivable to offer pregnant women food containing all the vitamin components that, at present, are still taken in by means of medication. In Japan the Snow Brand milk corporation already offers small, pre-portioned cheese bars sold under the name "cheese slim pack" to older women. These bars have been enriched with iron and other dietary supplements that are required by menopausal women.[4]

Chewing as a Function – Soft Meals

Even the way in which a product reacts in the mouth, how often and how strongly it has to be chewed can fulfill a special function. In chewing gum, for instance, the added value is found in the fact that it can be chewed forever without its texture changing. The exact opposite effect occurs in food designed so that it can hardly be chewed or not chewed at all. Since 2000 Georg Reier from Homburg an der Saar in Germany has been developing food whose consistency is adapted to persons with chewing and swallowing problems. His "soft meals" are based on the idea that mashed food which does not have to be chewed is designed in such a way that it looks appealing. He processes meat, fish and vegetables into flans and puddings which he then gives various shapes, adding new textures and then putting them together to create combinations of different-sized pieces. At first glance, his soft goulash, for instance, looks no different than regular goulash. Only when using a fork to dig into what looks like diced meat you notice that it is in reality a soft mass consisting of mashed meat, Qimiq, egg white and starch, all tasting like meat but immediately dissolving in the mouth. As we know, foods should also be a feast for the eye. People who for years were only able to feed themselves with mush derive once again more pleasure from eating when they are served these types of dishes.

1 | Source: Thomas Birus, © Bibliographisches Institut & F.A. Brockhaus AG, 2003
2 | Interview with David Julian McClements at the University of Massachusetts, USA, 26 March 2009
3 | Sources: Interview with Richard Mithen and Cathie Martin at Norwich Institute of Food Research, 21 January 2009; information on genetic engineering: Thomas Birus, © Bibliographisches Institut & F. A. Brockhaus AG, 2003
4 | Source: Werner Lorenz; Prolactal co.; guided tour of Snow Brand Milk co., Tokyo, October 2008

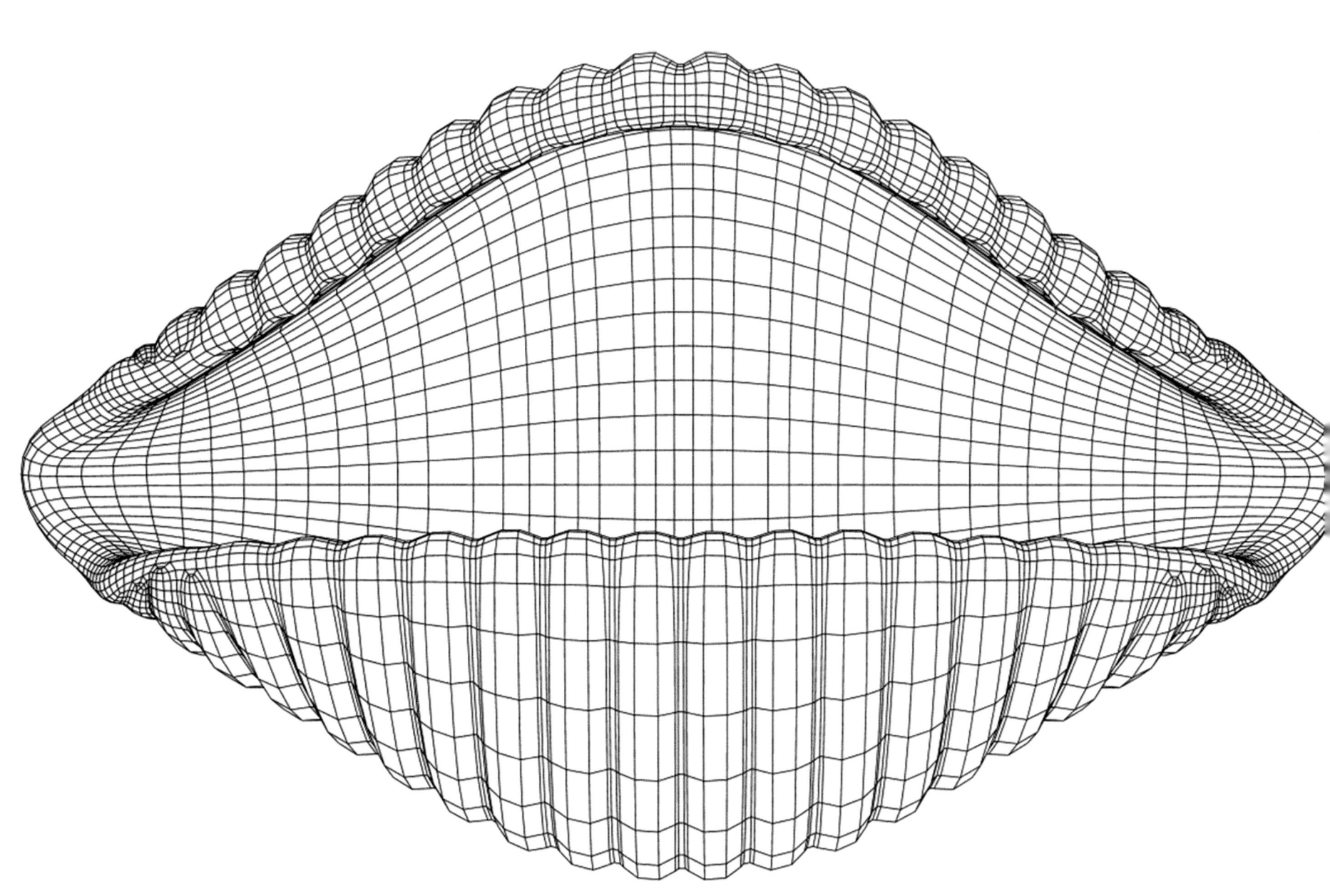

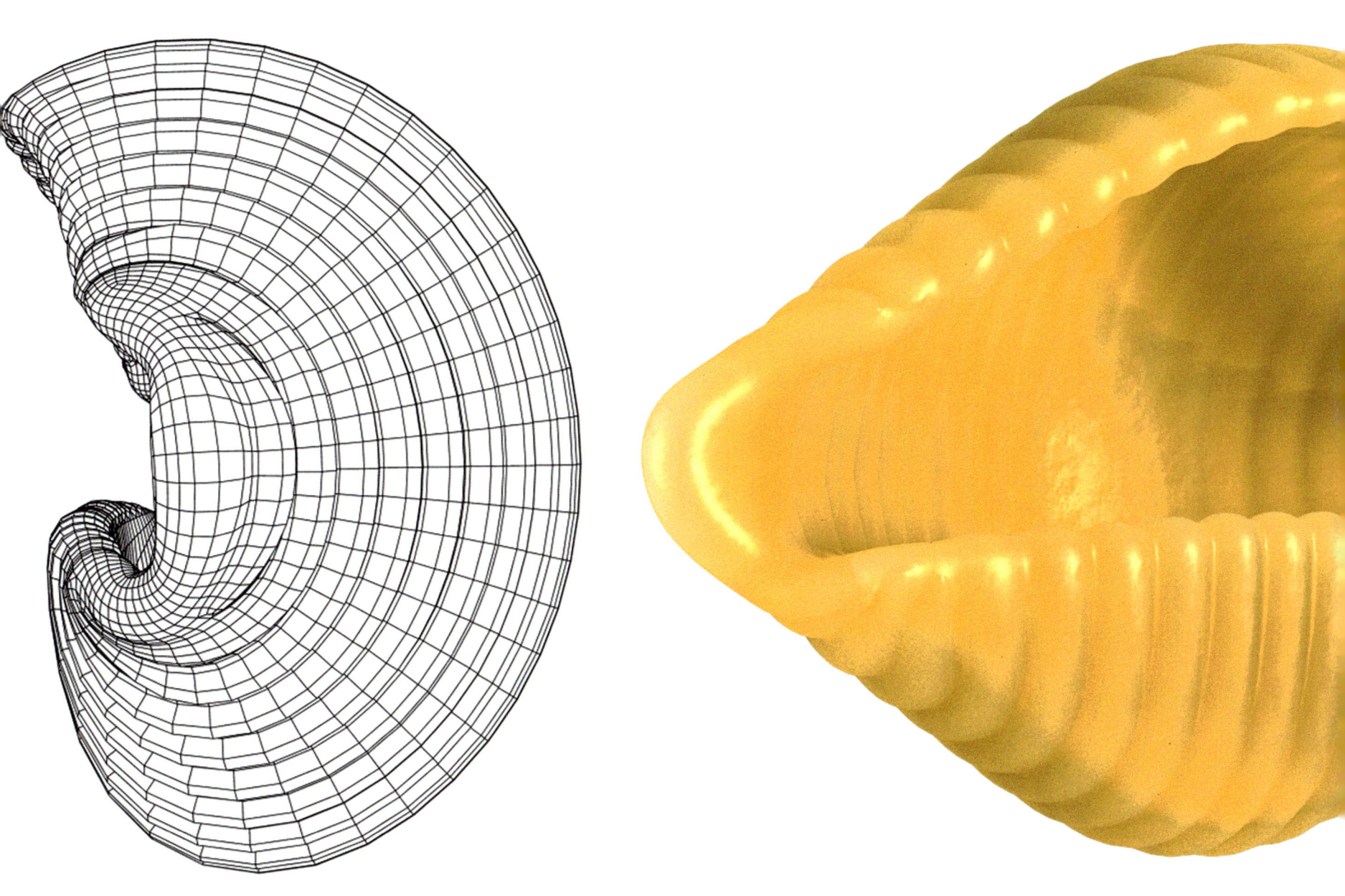

Designprozess
design process

Erfinder und ihre Produkte von links oben nach rechts unten | **Julius Maggi** (1846 - 1912) / Müller / Maggi-Würze 1886; **John Stith Pemberton** (1831 - 1888) / Apotheker / Coca Cola 1886; **Hans Riegel** (1893 - 1945) / Bonbonkocher / Gummibären 1922; **Henri Nestlé** (1814 - 1890) / Apotheker / Milchpulver 1865; **Hermann Bahlsen** (1859 - 1919) / Kaufmann / Leibniz Butter Cakes 1893; **Theodor Tobler** (1876 - 1941) / Konditor / Toblerone 1908; **Justus von Liebig** (1803 - 1873) / Chemiker / Rindfleischextrakt 1852; **Josef Manner** (1865 - 1947) / Kaufmann / Mannerschnitte 1898

Inventors and their products | **Julius Maggi** (1846 - 1912) / miller / Maggi seasoning 1886; **John Stith Pemberton** (1831 - 1888) / apothecary / Coca Cola 1886; **Hans Riegel** (1893 - 1945) / confectioner / gummi bears 1922; **Henri Nestlé** (1814 - 1890) / apothecary / milk powder 1865; **Hemann Bahlsen** (1859 - 1919) / merchant / Leibniz Butter Cakes 1893; **Theodor Tobler** (1876 - 1941) / confectioner / Toblerone 1908; **Justus von Liebig** (1803 - 1873) / chemist / beef extract 1852; **Josef Manner** (1865 - 1947) / merchant / Mannerschnitte 1898

Designprozess – wie entsteht unser Essen?

Die Tatsache, dass in den Vereinigten Staaten in einem Supermarkt rund 100.000 unterschiedliche Produkte angeboten werden, zeigt, dass Food Design heute im wahrsten Sinne des Wortes in aller Munde ist. Auch wenn die bewusste Gestaltung von Nahrung keine Erfindung der Lebensmittelwirtschaft ist, so ist Food Design in seiner heutigen Bandbreite natürlich ein Resultat der modernen Industriegesellschaft. Seit Jahrtausenden schneiden, garen und formen wir unsere Nahrung und agieren im Grunde alle tagtäglich als „Food Designer". Mit der Massenfabrikation hat die Gestaltung von Nahrung allerdings eine neue Dimension gewonnen. Mit einem Mal war es möglich, essbare Produkte nach immer gleichen Prämissen in Millionenstückzahl zu erzeugen, und die Frage des „richtigen" Designs gewann damit schlagartig enorm an Bedeutung.

Die wenigsten Menschen wissen, von wem, wo und wie die Lebensmittel und Gerichte entwickelt werden, in die sie tagtäglich hineinbeißen. Das hat zwei Gründe. Zum einen wird Nahrung zumeist als etwas Ursprüngliches, Natürliches wahrgenommen und ihre Beschaffenheit daher gar nicht erst hinterfragt. So wenig, wie wir die Inhaltsstoffe vieler Produkte kennen, machen wir uns Gedanken über deren Design. Wer kommt schon auf die Idee, dass Hamburger auch eckig, Croissants gerade oder Schokoladetafeln rund sein könnten? Auch wenn Fischstäbchen oder Eislutscher das Resultat menschlicher Kreativität sind, werden sie als genauso gottgegeben wahrgenommen wie eine Kartoffel oder ein Apfel.

Zum anderen findet Food Design hinter verschlossenen Türen statt. Gerade in der Lebensmittelbranche ist die Angst vor Betriebsspionage und Negativpresse extrem hoch (geschätzte 90 Prozent aller Pressemeldungen zum Thema Essen behandeln Skandale). Ideen für Neukreationen, aber auch Rezepturen und Herstellungstricks werden daher wie strenge Geheimnisse gehütet. Als Autor in die Innovationsabteilungen großer Lebensmittelkonzerne vorzudringen, ist äußerst schwierig, oft sogar unmöglich.

Das folgende Kapitel begibt sich dennoch auf Spurensuche zu den Entwicklern von essbaren Blockbustern wie dem Magnum und in die Labors und Studios, in denen Food Design betrieben wird. Wir gehen der Frage nach, wer die Menschen sind, die bestimmen, wie unser Essen aussieht, und begleiten neue Nahrungsmittel von der ersten Idee bis zum fertigen Produkt im Supermarktregal.

Food Designer – wer macht unser Essen?

„Wichtig bei der Entwicklung neuer Produkte ist, dass Off-Flavours, also geschmackliche Störfaktoren, entfernt werden und dass eine hohe sensorische Bewertung langfristig erhalten bleibt. Daher sind Dauerakzeptanz-, Langeweile- und Irritationstest entscheidend, um ein nachhaltig wohlschmeckendes Produkt zu schaffen." Klaus Dürrschmid, Department für Lebensmittelwissenschaften, Universität für Bodenkultur, Wien

Die Urheber und Erfinder traditioneller Nahrungsmittel vom Bagel bis zur Brezel, von den Farfalle bis zum Brotzopf liegen im Dunkeln. Allenfalls Legenden lassen Vermutungen über die Entwicklung und die Bedeutung vieler Brot- und Pastasorten, über die Entstehung von Süßwaren und anderen Gerichten zu. Erst ab dem 18. Jahrhundert kennt man die Namen von Wissenschaftlern und Erfindern, die an Esswaren und deren Herstellung geforscht haben. Im Laufe der Geschichte waren so unterschiedliche Berufsgruppen wie Chemiker, Ärzte, Apotheker, Fabrikanten, Bäcker, Fleischer und Konditoren an der Entwicklung von Lebensmitteln beteiligt. Bis heute ist Food Designer keine eigene Berufsbezeichnung. Vielmehr designen unter anderem Lebensmitteltechnologen, Ernährungswissenschaftler, Köche, Marketingleute, Betriebswirte, Psychologen, Chemiker, Akustiker und Konsumenten unsere Nahrung.

Natürlich ist die Genesis jedes Produktes sehr unterschiedlich. Der Designprozess eines neuen Nahrungsmittels kann sowohl in kurzer Zeit, mit wenig Aufwand und viel menschlicher Intuition erfolgen als auch jahrelange Forschung, unzählige Versuchsreihen und aufwändige Entscheidungsprozesse bedeuten. Nach Auskunft der Lebensmitteltechnologen von Kraft Food dauert es sechs Monate, um einen neuen Geschmack, und zwei Jahre, um eine neue Form zu entwickeln. Bei Uncle Ben's braucht eine durchschnittliche Produktentwicklung 18 bis 24 Monate, es gibt aber auch Ausreißer, deren Entwicklung bis zu 10 Jahre in Anspruch nimmt.

Design Process – How Does Our Food Come About?

The fact that US supermarkets offer 100,000 different products shows that food design is literally on everybody's lips today. Even if deliberate strategies of shaping food are not an invention of the food industry, the broad range of activities included in food design today is causally linked with modern industrial society. We have cut, cooked and shaped our food for thousands of years, and basically, we all act as "food designers" every day. However, with the advent of food mass production, its design took on a new dimension. Suddenly, it was possible to manufacture millions of edible products under the same premises, and the issue of "the right design" gained enormous importance.

Most people do not have any idea who develops the food products they eat every day, or where and how this happens. There are two reasons for this. For one, food is perceived as something primeval, natural, and thus, its qualities are not questioned. We do not know the content of many products we ingest, and by the same token, we do not spend much time thinking about their design. Who would spend any time thinking about whether hamburgers could just as well be square, croissants straight and chocolate bars round? Just like potatoes or apples, most people will take fish fingers or popsicles as given even though they are in fact the result of human creativity.

The second reason is that food design happens behind closed doors. The food industry is extremely wary of industrial espionage and negative headlines in the press (an estimated 90% of all press reports about food deal with scandals). Ideas for novel creations, recipes and manufacturing tricks are thus well-kept secrets. It is very difficult, if not impossible, to get access to the development departments of major food producers.

The following chapter will nevertheless try to track down developments and developers of edible blockbusters such as the Magnum ice-cream popsicle, and look behind the scenes in the labs and studios where food design happens. We will shed light on the question who determines what our food looks like, and we will accompany new foods on the way from the first idea to the finished product on the supermarket shelf.

Food Designers – Who Makes Food?

"The important thing about the development of new products is that off-flavors, i.e. factors which affect taste in a negative way, are removed, and that good sensory ratings are obtained over a long period. Permanent acceptance, dullness and irritation tests are decisive when it comes to creating a tasty product that will remain popular in the long run." Klaus Dürrschmid, Department of Food Sciences, University of Natural Resources and Applied Life Sciences Vienna

The inventors of traditional foods, from bagels to pretzels, from farfalle pasta to plaited bread loaves, are unknown. However, legends may allow us to draw conclusions as to the origin and meaning of many bread and pasta types, the creation of sweets and other foods. Only from the eighteenth century onward do we know the names of scientists and inventors who worked on food and their production. In the course of history, a wide variety of professions and trades were involved: chemists, physicians, pharmacists, industrialists, bakers, butchers and confectioners. To this very day, food designer is not a professional designation. Modern food design is pursued by food scientists, nutritionists, cooks, marketing specialists, business managers, psychologists, chemists, acoustic engineers and consumers.

Of course, the genesis of products varies in duration and effort. The design process required for a new food product may be short and need little expenditure, just a lot of human intuition. Conversely, it may also take years of research, numerous test series and complicated decision-making. According to food scientists working with Kraft Foods, it takes six months to develop a new flavor and two years to come up with a new shape. People at Uncle Ben's say that the average product development period is 18 to 24 months, with some exceptions where it takes up to ten years.

Generell herrscht das Prinzip „speed to market", da sich der Zeitgeist in puncto Lebensmittel rasch wandelt. Was heute gefragt ist, kann morgen schon wieder unpopulär sein. Im Allgemeinen ist die Lebensdauer von Nahrungsmitteln, also die Zeitspanne, in der ein Produkt im Verkauf bleibt, relativ kurz. Nur eines von zwanzig neuen Lebensmittelprodukten hält sich länger als zwei Jahre auf dem Markt. Konsumentenumfragen im Vorfeld einer Neuentwicklung fragen daher nicht nur danach, ob ein Kunde ein Produkt zu einem bestimmten Preis kaufen würde, sondern auch danach, wie oft er es tun würde.[1]

Innovationsdruck | In Europa kommen jährlich rund 10.000 neue Lebensmittel auf den Markt. Der japanische Lebensmittelhersteller Nissin bringt pro Jahr durchschnittlich unglaubliche 300 neue Produkte auf den Markt. Davon überleben nur zwei oder drei länger als einen Winter. Man könnte meinen, dass der Kunde gar nicht nach so vielen neuen Esswaren verlangt, sondern mit der ohnehin großen Auswahl an vorhandenen Produkten durchaus zufrieden ist. Dazu kommt, dass der Durchschnittsesser äußerst konservativ ist. Essgewohnheiten ändern sich mit den Generationen, also sehr langsam.[2]

Dennoch ist der Innovationsdruck in der Lebensmittelbranche sehr hoch. Da der Konsum von Nahrungsmitteln mehr oder weniger konstant bleibt, können Lebensmittelkonzerne nur wachsen, indem sie neue Segmente erobern oder anderen Produzenten Marktanteile abspenstig machen. Der Konkurrenzkampf innerhalb der Sparte ist also sehr stark. Zudem werden börsennotierte Unternehmen unter anderem nach ihrem Innovationspotenzial bewertet. Auch wenn sich der Kunde mit kleinen Neuerungen, die von Zeit zu Zeit auftauchen, zufrieden geben würde, müssen Hersteller die Produktpalette ständig erweitern, um dem Image eines innovativen Unternehmens gerecht zu werden.[3]

Impulsgeber und Research | Inspirationen zur Gestaltung von Nahrungsmitteln kommen von der Wissenschaft bis zum ganz normalen Konsumenten aus den verschiedensten Ecken. Technische Innovationen, neuartige Grundprodukte sowie Forschung und Entwicklung im Bereich von Lebensmitteltechnologie, Herstellung, Transport, aber auch Gesetze, Normen und Verordnungen schaffen Anreize für neue Produkte.

Klassische Impulsgeber sind zunächst einmal Markt- und Trendforschung. Konzerne unterhalten dafür eigene Abteilungen oder beauftragen Agenturen, die Wünsche der Konsumenten aufzuspüren und das Anforderungspotenzial zukünftiger Produkte zu recherchieren. Kundenbefragungen gehören genauso zum Prozedere wie die Analyse von Ernährungsratgebern, Koch- und Gourmetmagazinen, die Beobachtung von Bio- und Delikatessenläden oder die Nutzung des Internets. Der US-Konzern Kraft Foods betreibt unter www.innovatewithkraft.com eine eigene Website, auf der alle Interessierten ihre Ideen für neue Kraft-Produkte einbringen können. Für die Firma Uncle Ben's durchforstet eine Marketingagentur aus Chicago Foren und Blogs nach Trends

In general, the key principle is "speed to market" because the food zeitgeist changes quickly. What is much in demand today may be a shelf warmer tomorrow. The life cycle of foodstuffs, i.e. the period for which a product will be on the market, is generally rather short. Only one out of twenty new food products remains on the market for more than two years. Consumer surveys prior to new developments not only inquire if a customer would buy something at a certain price but also how often he or she would buy it.[1]

Under Pressure to Innovate | In all of Europe, roughly 10,000 new foodstuffs are launched on the market every year. The Japanese food producer Nissin alone markets an incredible average of 300 new products annually. Out of these, only two or three survive for longer than one winter. We would tend to think that consumers do not demand that many new food products and that they are satisfied with what seems to be a broad range to choose from, anyway. Moreover, the average eater is generally very conservative and eating habits change with generations, that is to say, very slowly.[2]

Nevertheless, the food industry is under enormous pressure to innovate. As the consumption of food remains more or less constant, food corporations can only grow if they conquer new segments or take market shares from other producers. Competition in the market is stiff, and listed companies are also rated according to their potential for innovation. Even if customers were satisfied with a minor novelty here and there, producers have to expand their product range continuously to be seen as innovative enterprises.[3]

Impulses and Research | Inspiration for the design of food comes from all directions, from scientists to normal consumers. Technical innovation, novel basic products and R&D in food science, production, transportation, as well as legislation, standards and regulations provide impulses and incentives for new products.

Market research and trend scouting are classic sources of impetus. Corporations have their own departments or commission agencies to identify customer wishes and research the specifications for future products. Customer surveys are part of the procedure, as are the analysis of books and brochures on nutrition as well as cookery and gourmet magazines, the observation of organic food stores and delicatessen, or searches on the Internet. The US corporation Kraft Foods runs a special website at www.innovatewithkraft.com where all interested parties are invited to make proposals for new Kraft products. For Uncle Ben's a marketing agency in Chicago checks Internet fora and blogs for trends and consumers' predilections. If for example websites specializing in cooking and recipes discuss dishes with rice, bell peppers and Indian spices, the product managers at Uncle Ben's start thinking about how to create a new product using these ingredients. By the same token, the sites might identify a regional cuisine that is particularly en vogue at a time, and inspire the company to launch a "Mexican product line".[4]

und Vorlieben der Konsumenten. Wird zum Beispiel auf Koch- und Rezeptseiten viel über Speisen mit Reis, Paprika und indischen Gewürzen diskutiert, so überlegen die Entwickler von Uncle Ben's, wie sich daraus ein neues Produkt kreieren ließe. Oder man erkennt, welche Regionalküche gerade en vogue ist, und bringt daraufhin eine „mexikanische Linie" auf den Markt.[4]

Research in Form von Forschungsreisen betreibt der nordamerikanische Speiseeishersteller Ben & Jerry's. Einmal pro Jahr unternimmt das gesamte „Research & Development"-Team einen Betriebsausflug in einen anderen Teil des Landes, um Köche zu treffen, Restaurants, Bäckereien und Konditoreien zu besuchen und Unmengen an Desserts zu verkosten. Die Ergebnisse werden dann in rund 120 Kurzkonzepte für potenzielle neue Eissorten verwandelt und der Marketingabteilung unterbreitet. Diese wählt daraus je nach Marktlage und in Hinblick auf die Gesamtkollektion etwa 50 Konzepte aus, die dann weiter bearbeitet werden.[5]

Wertvolle Inspirationen kommen aber auch aus den Reihen der eigenen Mitarbeiter, die in vielen Firmen dazu angehalten werden, Verbesserungsvorschläge einzubringen. Auch Lieferanten von Zutaten oder Verarbeitungsgeräten können die Initialzündung für ein neues Produkt geben. Etwa neue Aromen, die sich leichter verarbeiten lassen, besser hitzebeständig oder länger haltbar sind, können eine Produktidee anregen. Impulse für Neuentwicklungen kommen auch aus dem Ausland, von der Konkurrenz oder – im Falle von Geschmacksrichtungen – aus der Getränkeindustrie. Aromen, die sich in einer Saison bei Mineralwasser und Co. bewährt haben, finden sich mit großer Sicherheit in der nächsten Saison auch in Joghurts, Eis und Süßigkeiten wieder.

Entwicklung | *„Bei der Produktentwicklung wird oft so vorgegangen, dass zunächst grobe Parameter wie mechanische Eigenschaften definiert werden und erst dann der Geschmack", Klaus Dürrschmid, Department für Lebensmittelwissenschaften, Universität für Bodenkultur, Wien*[6]

Ist die Idee erst einmal gefunden, werden die Entwicklungsparameter festgelegt und der eigentliche Gestaltungsprozess beginnt. In einem internen Workshop wird das Konzept konkretisiert oder eine externe Agentur wird beauftragt, Vorschläge zu erarbeiten, wie das neue Produkt aussehen und funktionieren könnte. In enger Zusammenarbeit zwischen Firmenleitung, Marketingabteilung, Produktionsleitung und Rezeptentwicklung beginnt dann die tatsächliche Entwicklung des neuen Produkts hinsichtlich Verkaufskonzept, Zielgruppe, Portionsgröße, Verzehrssituation, Geschmack, Farbe, Form, Mundgefühl, Verpackung, Herstellbarkeit, Haltbarkeit etc. .

Das Image Lab | Dabei kommt eine Vielzahl von Methoden zum Einsatz: Zeichnungen und Modelle, Computersimulationen, Fotocollagen und essbare Prototypen. Im Image Lab, dem Herzstück der europäischen Entwicklungsabteilung von Kraft Foods, werden zum Beispiel Fotos von existierenden Produkten am Computer in neue Schokoladen und Pralinen verwandelt. Dazu werden im angeschlossenen Fotostudio

New impulses for the American ice-cream producer Ben & Jerry's, come from research trips. Once a year, the entire Research & Development team goes on a staff outing, traveling to a different part of the country to meet with cooks, visit restaurants, bakeries and pastry shops, and taste an enormous array of desserts. The results are summarized in about 120 brief product descriptions for potential new ice-cream flavors and submitted to the marketing department. Depending on the market situation and considering the company's existing product range, about 50 summaries are then shortlisted for further steps.[5]

Ideas from the shop floor can be very valuable inspirations for new products, and many companies encourage their staff to come forward with suggestions for improvement. Suppliers of ingredients or processing equipment, too, may bring the initial spark to manufacturers and prompt them to develop a novelty. This could be a new aroma which is easier to process, or something making a product more heat-resistant or less perishable. Impulses might also come from abroad, from competitors or – in case of flavors – from the beverage industry. Aromas which were successful sellers in mineral water or soft drinks will surely be back in yoghurts, ice cream and sweets the next season.

Development | *"In product development rough parameters, such as mechanical properties, are often defined first, and only then is taste dealt with." Klaus Dürrschmid, Department of Food Sciences, University of Natural Resources and Applied Life Sciences Vienna*[6]

Once an idea has been born, the development parameters are defined and the actual design process starts. The proposition will either take more concrete shape in an in-house workshop, or an external agency will be commissioned to prepare suggestions for what the product could look like or how it could work. There is close cooperation between management, marketing department, production management and recipe developers when the design work as such starts. A sales concept for the product addressing issues such as target group, portion size, consumption setting, taste, color, shape, mouth feel, package, production conditions, shelf life etc. has to be decided on.

The Image Lab | In this context, a wide range of methods are used: drawings and models, computer simulations, photo collages, edible prototypes. At the Image Lab, the heart piece of the European R&D Department of Kraft Foods, photos of existing products are turned into new chocolates and confectionery on the computer screen. At the in-house photo studio, pictures of inspiring objects are taken for this purpose, and then re-designed into new developments with the help of image pro-

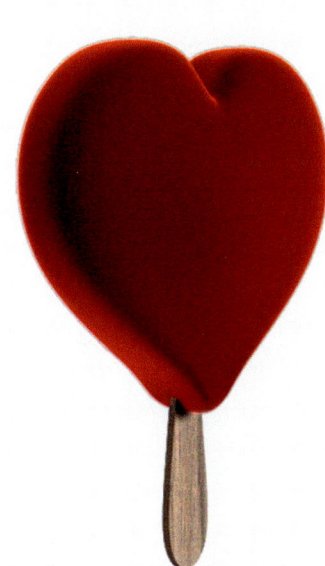

Es dauert 6 Monate, um einen neuen Geschmack, und zwei Jahre, um eine neue Form zu entwickeln. Dennoch hält sich nur eines von 20 neuen Lebensmittelprodukten länger als 2 Jahre auf dem Markt.

It takes six months to develop a new taste and two years to develop a new form. Still only one of twenty new food products remains longer than two years on the market.

Inspirationsobjekte fotografiert und danach mit Hilfe von Bildbearbeitungssoftware zu Neuentwicklungen umgestaltet. Die virtuellen Pralinen, Snackriegel oder Schokoladetafeln werden dann online gestellt und von ausgewählten „Testkonsumenten" nach rein optischen Kriterien bewertet. Ohne den Geschmack zu kennen, entscheiden die Versuchspersonen nur anhand der Abbildungen, ob sie ein Produkt kaufen würden oder nicht.

Auf diese Weise können Produktideen in unzähligen Varianten – unterschiedlichsten Formen, Farben und Größen – relativ schnell und kostengünstig ausgetestet werden, bevor die eigentliche Entwicklung überhaupt beginnt. Empfinden die Konsumenten etwa die Farbe als zu intensiv und künstlich, so wird das Aussehen des Produkts entsprechend adaptiert und erneut getestet. Kommt der Entwurf an, werden bis zu 300 formale Varianten entwickelt, bevor man sich endgültig für eine bestimmte Formgebung entscheidet. Erst wenn das Dreigespann aus Produkt, Konzept und Preis zusammenstimmt, treten die Rezeptentwickler, Produktionstechniker, Aroma- und Marketingexperten auf den Plan, um die Rohfassung in ein konkretes Produkt zu verwandeln.[7]

Rezeptentwicklung | Üblicherweise werden Bilder oder Prototypen von Neuentwicklungen nicht im Internet, sondern zunächst von Firmenmitarbeitern getestet. Form, Farbe, Geschmack et cetera durchlaufen auf diese Weise einen ersten „Marktfähigkeitscheck". Findet eine Idee Anklang, wird diese in einem nächsten Schritt auf Verkaufstauglichkeit, Produktionskosten, Herstellungsmöglichkeiten und ähnliche Parameter hin überprüft und ein erstes Vermarktungskonzept erarbeitet. Parallel dazu erfolgt die Rezeptentwicklung des neuen Lebensmittels. In Versuchsküchen werden erste Rohversionen des neuen Produkts oder seiner Einzelteile hergestellt, die von den Projektbeteiligten laufend getestet und optimiert werden.

Konsulenten: von Psychologie bis Sounddesign | Das firmeneigene Entwicklungsteam lässt sich dabei möglicherweise auch von externen Konsulenten beraten. Psychologen und Motivforscher beispielsweise stehen bei, wenn es um die psychologische Wirkung von Farben und Formen geht, informieren über die kulturellen und rituellen Bedeutungen von Speisen und wissen, welche Assoziationen bestimmte Texturen oder Farbkombinationen in uns auslösen.

Bei der Entwicklung von Produkten mit markanter Geräuschperformance wie Keksen, Crackern oder Bier werden unter Umständen auch eigene Sounddesigner beigezogen. Konsumenten schenken dem Faktor Geräusch beim Essen zwar kaum Aufmerksamkeit, dennoch gibt das Krachen, Schnalzen oder Blubbern unbewusst Auskunft über die Frische und die Qualität einer Speise bzw. eines Getränks und hat dadurch einen wesentlichen Einfluss darauf, ob etwas gut schmeckt oder nicht.

cessing software. The virtual confectionery, muesli bars or chocolates are then uploaded and rated according to purely visual criteria by selected "test consumers" who access the pictures on the Net. Without ever having tasted the product, the testers decide on the basis of the product look whether they would buy it or not.

Many product ideas and innumerable variants – different shapes, colors and sizes – can thus be tested at relatively low cost before the development job in the actual sense of the word has begun. If consumers for instance think that a color is too gaudy and artificial, the look is changed accordingly and the product is re-tested. If a draft is well liked, up to 300 variants of form are developed until the final design is eventually chosen. Only when the troika of product, proposition and price converges, the recipe developers, production engineers, aroma experts and marketing specialist take center stage to transform the blueprint into a concrete products.[7]

Recipe Development | Usually, pictures or prototypes of new developments are not tested on the Internet but by staff members. Shape, color, taste etc. are thus given a first "marketability check". If an idea does well at this stage, the next phase is to look at salability, production costs, manufacturing possibilities to draw up an initial marketing plan. In parallel, the recipe of the new food product is developed. In test kitchens, the first basic versions of the new product or its components are prepared, continuously subjected to testing and optimized by the people working on the project.

Consultants: From Psychology to Sound Design | The company's in-house development team might also seek advice from external consultants. Psychologists and motive researchers, for instance, assist when it comes to the psychological effect of colors and forms, providing information on cultural and ritual meanings of food and which associations certain textures and color combinations trigger in us.

In the development of products with striking sound output such as cookies, crackers or beer, a sound designer might be consulted While consumers do not pay so much attention to the sound factor in eating, a cracking, snapping or bubbling sound might unconsciously provide information on the freshness and quality of food or drink, and thus have substantial influence on whether something tastes good or not.

Das Magnum wurde von Unilever Deutschland in Heppenheim bei Frankfurt entwickelt und kam 1990 auf den Markt. Ausgangspunkt war die Idee, ein Stieleis für Erwachsene zu schaffen, das sich in Form, Farbe und Geschmacksrichtungen von klassischen Kindereislutschern unterscheidet. Ein wichtiger Eckpunkt des Konzepts war die Portionsgröße – daher auch der Name „Magnum" –, die durch die kompakte Form optisch noch hervorgehoben wird.

The Magnum, developed by Unilever Germany in Heppenheim near Frankfurt, was launched in 1990. The original idea was to create an ice-cream popsicle for adults that differed in terms of form, color and taste from classical children's ice-cream products. An important aspect of the concept was the portion size – thus the name "Magnum" which is optically reinforced by the compact form.

Der Akustiker Friedrich Blutner und sein Synotec Psychoinformatik Studio in Geyer nahe Chemnitz haben sich auf das Design von Geräuschen, die beim Öffnen der Packung, beim Portionieren, beim Eingießen, beim Zubeißen oder Kauen von Essen entstehen, spezialisiert. Im Auftrag der Lebensmittelindustrie bringen sie Mineralwasser dazu, wie ein sprudelnder Gebirgsbach zu klingen, komponieren das Zerplatzen von Prosecco-Perlen zu einem Rauschen, das an das Schwanken großer Bäume im Wind erinnert, und versehen Schleckeis mit auffälligen Sound-Substanzen wie Nüssen, Waffeln oder Schokostreusel.[8]

Pilot Plants: Entwicklung des Verfahrens |
Gleichzeitig beschäftigen sich Produktionsleitung und Verfahrenstechniker mit der praktischen Umsetzung des neuen Produktes. Dabei geht es um Fragen wie: Welches Produktionsverfahren eignet sich am besten? Welche Arbeitsschritte sind notwendig, welche Maschinen kommen zum Einsatz? Müssen neue Fertigungsstraßen angeschafft werden? Wie kann das Produkt adaptiert werden, um die Herstellungskosten möglichst gering zu halten? Wie können die Anlagen gereinigt werden? Welche Hygienevorschriften sind an welchem Betriebsstandort zu erwarten? Rezepturen, die sich in kleinen Entwicklungsküchen oder Labors als problemlos erwiesen haben, müssen nun in so genannten Pilot Plants auf ihre Massenherstellungstauglichkeit hin überprüft werden. Pilot Plants sind „verkleinerte" Industrieanlagen, in denen industrieähnliche Bedingungen simuliert und Produktionsverfahren direkt getestet werden können. Nicht zuletzt aufgrund der Mengen schmecken die Ergebnisse dieser Produktionsanlagen oft ganz anders als die im Labor entwickelten Rezepturen. Mit der Hilfe von Verfahrenstechnikern werden die Fertigungsstraßen in den Pilot Plants entsprechend adaptiert und optimiert. Der Geschmack muss letztlich genau derselbe bleiben wie jener der handproduzierten Prototypen aus der Testküche.[9]

Acoustic expert Friedrich Blutner in Geyer, Germany has specialized in designing sounds created when a package is opened, food is portioned, poured, bitten or chewed. He and his Synotec Psychoinformatik Studio have been commissioned to make carbonated water sound like a bubbly mountain brook, to compose the bursting sound of Prosecco bubbles so that it sounds like large trees swaying in the wind, or to add striking sound substances like nuts, wafers or chocolate crumbles to ice cream.[8]

Pilot Plants: Development of a Procedure |
At the same time production management and process engineers are working on the practical implementation of the new product. Here the focus is on questions such as: Which production procedure is best suited? Which working steps are necessary, which machines will be used? Do new assembly lines have to be purchased? How can facilities be cleaned? Which hygiene rules can be expected at which operation site? Recipes that have proven to be unproblematic in small development kitchens or labs now have to be tested in so-called pilot plants to see if they are fit for mass production. Pilot plants are "miniature" facilities in which industrial conditions are simulated and production methods can directly be tested. Given the quantities produced in these facilities, the products often taste completely different than the recipes developed in a lab. With the help of process engineers the assembly lines in the pilot plants will be adapted and optimized accordingly. The taste must ultimately remain the same as in the hand-made prototypes from the test kitchen.[9]

Ein entscheidendes Element im Entwicklungsprozess ist natürlich die Preiskalkulation. Während des gesamten Arbeitsablaufs muss stets der Verkaufspreis im Auge behalten werden. Denn letztendlich entscheidet die Preiskalkulation und somit die Frage: „Kann das neue Produkt zu einem Preis angeboten werden, den der Kunde auch bereit ist, dafür zu bezahlen?", darüber, ob die Neuerscheinung jemals in Produktion geht oder nicht.

Konsumententests | Sind alle internen Tests zur Zufriedenheit verlaufen und stimmen die Zahlen, werden externe Verkostungen mit Konsumenten und / oder geschulten Testpersonen durchgeführt. In speziellen Sensoriklabors werden die unterschiedlichen Wahrnehmungskomponenten wie Geruch, Farbe, Form et cetera dabei durch gezielte Fragestellungen getrennt voneinander abgefragt und mittels Fragebögen ausgewertet. Ergibt der Test zum Beispiel, dass das Produkt als unnatürlich aromatisiert empfunden wird, weil es zu rot ist, wird die Rezeptur entsprechend verändert und erneut getestet.

Schließlich werden Samples hergestellt, die in ausgewählten Geschäften in einem Probelauf zum Verkauf angeboten werden. Bei Nissin beispielsweise beträgt die erste Testphase überhaupt nur 28 Tage. Werden in dieser Zeit nicht mindestens sechs Packungen pro Woche und Geschäft verkauft, wird das Produkt zurückgezogen. Läuft es dagegen erfolgreich an, wird mit der Massenfertigung begonnen und die Verkaufsregale werden gefüllt. Dann entscheidet nur noch der Kunde darüber, ob die Neuentwicklung ein profitabler Erfolg oder ein teurer Flop wird.[10]

Stieleis für Erwachsene | Das Herzstück jeder Neuentwicklung ist das gelungene Zusammenspiel von Konzept, Design und Marketingstrategie. Wie perfekt diese Taktik aufgehen kann, wenn die einzelnen Parameter zusammenstimmen, demonstrieren Blockbuster wie das Magnum. Das Magnum wurde im Entwicklungslabor der Firma Unilever in der zweitgrößten Eisfabrik Europas in Heppenheim entwickelt und kam in Deutschland 1989 auf den Markt.

Die Genesis von Magnum begann mit der Idee der Marketingabteilung, das klassische Kinderprodukt „Eislutscher" auch für Erwachsene interessant zu machen. Ein elegantes Stieleis für Großjährige versprach gutes Verkaufspotenzial. Um diese Kundengruppe gezielt anzusprechen, entschieden sich die Entwickler für eine schlichte Ziegelform mit leicht abgerundeten Ecken. Auch bei der Wahl der Farben und Geschmacksrichtungen übte man entsprechende Zurückhaltung und setzte auf die unspektakuläre Kombination von brauner Schokolade und hellem Vanilleeis. Ein wichtiger Eckpunkt des Konzepts war die Portionsgröße – daher auch der Name „Magnum" –, die durch die kompakte Form optisch noch hervorgehoben wird. Der Konsument sollte gleich sehen, dass es sich um ein „ordentliches" Stück handelt.

Auch das Herstellungsverfahren von Magnum unterscheidet sich von jenem für klassische Eisschlecker. Ein adaptiertes Tiefkühlverfahren lässt extrem kleine Eiskristalle entstehen, wodurch das Magnum eine cremigere Konsistenz erhält als herkömmliches Stieleis. Außerdem wird Magnum nicht gegossen, sondern extrudiert und in echter Schokolade getunkt. Auch die geschwungene Form des Stiels unterscheidet sich klar von den einfachen Stäbchen der Kindereissorten.

In den letzten Jahren wurde in Heppenheim noch ein zweites bekanntes Eisprodukt der Marken Wall's und Langnese entwickelt: die Drops. Die Idee zu den einzelnen kleinen Eiskügelchen, die getrennt voneinander aus der Packung fallen, entstand aus einer praktischen Überlegung heraus: Damit sich Kinder beim Essen nicht so leicht ankleckern, wollte man ein Eis schaffen, das nicht vom Stiel tropft. Entstanden sind daraus die kleinen Drops, die aus der Packung direkt in den Mund geschüttet werden und so weder mit den Händen noch mit der Kleidung in Berührung kommen.

Die Entwicklung der Drops stellte vor allem eine herstellungstechnische Herausforderung dar. Es galt kleine, gleich große, nicht leicht schmelzende Eiskügelchen zu erzeugen, die in der Packung nicht aneinander festfrieren, sondern schön einzeln nacheinander lose herausfallen. Wie das technisch funktioniert, wollte man uns leider nicht verraten …[11]

Of course, the crucial element in the development process is price calculation. Throughout the entire work procedure, the sales price must be kept in mind. Ultimately, the price calculation along with the question "can the new product be sold for a price that the customer is willing to pay?" decides whether a new development will ever be produced or not.

Consumer Tests | If all internal tests have been satisfactorily completed and the figures are right, the external samplings with consumers and/or trained test persons follow. In special sensor labs various perceptual components such as smell, color, shape etc. are tested separately on the basis of targeted questions and evaluated by means of questionnaires. If a test, for example, shows that the product is perceived as having an unnatural aroma, because it is too red, the recipe is altered accordingly and tested anew.

Finally, samples are produced and offered for sale in selected shops in a trial run. For instance, in Nissin, the first trial phase is only 28 days. If the minimum amount of six packages per week and store are not sold within this time, the product will be withdrawn. If, by contrast, it starts off successfully, mass production will be launched and the store shelves will be filled. Then it is only up to the customer to decide whether the new development will be a profitable success or an expensive flop.[10]

Ice-Cream Popsicles for Grown-ups | At the heart of every new development there is the successful interplay of concept, design and marketing strategy. Blockbusters such as Magnum ice cream show how successful this approach can be if individual parameters harmonize. Magnum was developed in the R&D lab of the Unilever company in Europe's second largest ice-cream factory in Heppenheim and introduced on the market in Germany in 1989.

The idea for Magnum came from the marketing department, i.e., to make a classic children's product, the ice-cream popsicle, interesting for adults as well. An elegant ice-cream popsicle for older folks held out good sales potential. In order to target this group of customers, the developers opted for a simple brick form with slightly rounded corners. In the selection of colors and tastes they also showed restraint and banked on a non-spectacular combination of brown chocolate and light vanilla ice cream. A key point of this concept was the size of the portion – hence the name "Magnum" – which was also visually highlighted by the compact form. The consumer is supposed to immediately see that it is a "substantial" piece.

Even the production procedure of Magnum differs from that of conventional ice-cream popsicles. An adapted deep-freeze procedure allows extremely small ice crystals to emerge, giving the Magnum a creamier consistency than ordinary ice-cream. Moreover, Magnum is not frozen in molds, but extruded and dipped in real chocolate. The curved shape of the stick also clearly stands out from the simple sticks of children's ice cream.

In recent years, a second well-known ice cream product of Wall's and Langnese brands has been developed in Heppenheim: the Drops. The idea of creating the individual small ice-cream balls that fall out of the bag separately was the result of practical considerations. To prevent children from sullying themselves while eating, the company wanted to create an ice cream that would not drip from the stick. They eventually came up with the small drops that can be poured directly from the package into the mouth so that they do not come into contact with hands or clothing.

The development of the drops posed a challenge primarily in terms of manufacturing. The idea was to create small ice-cream balls that would not melt easily or freeze into a lump in the package but fall out one after the other. How that is supposed to function in technical terms is unfortunately confidential.[11]

External Research Institutions

Externe Forschungseinrichtungen | Kraft und Unilever zählen zu den Global Players auf dem Lebensmittelmarkt. Sie betreiben eigene Innovationsabteilungen, ziehen aber ebenso wie viele andere Erzeuger für gewisse Teilbereiche – sei es Sensorik, Akustik oder Technologieentwicklung – auch Spezialisten als Konsulenten bei. Fallweise wird sogar die gesamte Erarbeitung neuer Produkte komplett ausgelagert oder die Lösung spezifischer Problemstellungen an externe Forschungseinrichtungen vergeben.

Das Technologietransferzentrum in Bremerhaven beispielsweise ist ein marktorientierter Forschungsbetrieb, der für Konzerne, öffentliche Stellen und internationale Organisationen angewandte Forschung, Entwicklung und Umsetzung in den Bereichen Lebensmitteltechnologie und Bioverfahrenstechnik betreibt. Der Auftraggeber – sagen wir: ein großer Lebensmittelkonzern – möchte zum Beispiel eine Schokolade, die sehr fest ist, laut knackt und trotzdem schnell schmilzt, weil Marktumfragen sagen, das wäre ein kommender Trend. Die beiden Eigenschaften schließen einander aber eigentlich aus. Um dieses Problem technisch zu lösen, beauftragt er dann das TTZ.

Oder ein Erzeuger will einen neuen Schokoriegel herausbringen, der aus weichen, härteren und knusprigen Elementen besteht; vorerst weiß er nicht, ob dieses Produkt dem Konsumenten tatsächlich schmeckt oder ob er nicht einfach ein normales Stück Schokolade vorzieht. Zu diesem Zweck beauftragt er das TTZ, welches Testesser zu sich einlädt, um den Riegel zu verkosten und mit anderen, schon auf dem Markt befindlichen Produkten zu vergleichen. Parallel dazu wird das Produkt chemisch analytisch untersucht, um festzustellen, welche chemischen Inhaltsstoffe für ein bestimmtes Geschmacksmuster ausschlaggebend sind und z.B. einen vollmundigen Wein oder eine gut schmeckende Schokolade ausmachen. Auf Grundlage dieser Daten entscheidet der Auftraggeber dann, ob er die Entwicklung weitertreiben, die Rezeptur adaptieren, das Verfahren, die Form oder die Farbe abändern möchte und so weiter.

Ziel derartiger Projekte ist es, der Industrie Daten an die Hand zu geben, um Flops zu vermeiden. Man kann mit Sicherheit sagen, dass ein Produkt auf dem Markt versagen wird, wenn es bei den Verkostungen im Sensorik-Labor negativ beurteilt wird. Umgekehrt ist es schwieriger. Nur weil die Tests positiv ausfallen, muss das entsprechende Lebensmittel noch nicht unbedingt ein Megaseller werden. Denn bei Erfolg oder Nichterfolg einer Ware kommen auch äußere Einflüsse wie gesellschaftliche Rahmenbedingungen und kulturelle Hintergründe zum Tragen.[12]

Grundlagenforschung | Forschungseinrichtungen und Universitäten, die staatlich finanziert sind und sich unabhängig von Markt und Industrie mit Grundlagenforschung oder ernährungswissenschaftlichen Problemen befassen, leisten einen wesentlichen Beitrag zur Weiterentwicklung von Nahrungsmitteln. Das Institute of Food Research in Norwich, Großbritannien, beispielsweise forscht primär im Auftrag der britischen Regierung,

External Research Institutions | Kraft and Unilever are among the global players on the food market. They run their own innovation departments, but like other manufacturers they also enlist experts for certain special areas – be it sensorics, acoustics or technology development – as consultants. In some cases, the entire development of new products is completely outsourced, or external research institutions are commissioned to solve specific problems.

The Technology Transfer Center in Bremerhaven, for instance, is a market-oriented research center which does applied research, development and implementation in the fields of food and bio-processing technology for corporations, public institutions and organizations. The commissioning party – let's say a large food company – would, for example, like a chocolate that is very solid, makes a loud crunching sound yet still melts quickly since market studies show that this is a coming trend. However, these qualities contradict each other. To solve this problem in technical terms, the corporation will turn to the TTZ for assistance.

Or a manufacturer might want to launch a new chocolate bar which combines soft, harder and crunchy elements, but he does not know whether the consumer likes such a combination or simply prefers a normal piece of chocolate. To this end, it consults the TTZ, which invites test eaters to try the chocolate bar and to compare it with other products on the market. Parallel to this, the product is chemically analyzed to determine which chemical substances are decisive for a certain taste pattern and make a wine full-bodied or chocolate tasty. On the basis of this data, the commissioning company or institution then decides whether it will continue the development, adapt the recipe, the procedure, or alter the form and color, etc.

The goal of such projects is to provide the industry with data so that it can avoid flops. It can be said with certainty that a product will fail on the market if it gets negative reviews in the sensorics lab. The opposite case is more complex. If the test results are positive, this does not necessarily mean that the product will become a mega-seller. External influences such as social conditions and cultural factors come to bear in determining the success or failure of a product.[12]

Basic Research | Other research institutions and universities are government funded and deal with food and nutrition related issues independent of the market and industry. The Institute of Food Research in Norwich, Great Britain, for instance, does research for the British government in various fields, including healthy nutrition. In the context of the "Food Structure & Health

Wer macht Fooddesign? Unter anderen gestalten Lebensmitteltechnologen, Ernährungswissenschafter, Marketingleute, Betriebswirte, Psychologen, Chemiker, Industriedesigner, Akustiker und Konsumenten unsere Nahrung. Fooddesigner lassen sich grob in vier Gruppen einteilen: Produktentwickler, die für Lebensmittelkonzerne arbeiten, Wissenschaftler, die Grundlagen- oder Auftragsforschung betreiben, Experten für gewisse Teilbereiche wie Psychologie oder Akustik sowie freie Designer und Künstler. Von links nach rechts | Christina Jakobsen ist Managerin für Chocolate Innovation Europe beim Lebensmittelkonzern Kraft Foods; Frank Förster war für Unilever in Heppenheim, D, maßgeblich an der Entwicklung des Magnums beteiligt; Kantha Shelke arbeitet als selbstständige Lebensmitteltechnologin in Chicago

Who designs our food? To name a few: Food engineers, nutritionists, sales peolple?, economists, psychologists, chemists, industrial designers, acousticians and consumers. Food designers can be roughly divided into four groups: product developers who work for food companies, scientists who do basic or commissionend research, epxerts for certain special fields such as psychology or acoustics as well as free-lance designers and artists. From left to right. | Christina Jakobsen is manager for Chocolate Innovation Europe at Kraft Foods; Frank Förster contributed decisively to the development of Magnum at Unilever Germany; Kantha Shelke works as free-lance food engineer in Chicago, USA

unter anderem zum Thema gesündere Nahrung. Im Rahmen des „Food Structure & Health Research Programme" erarbeitet das Institut grundlegende, wissenschaftliche Prinzipien zur Reduktion des Fettkonsums. Ziel ist es, gesundes Essen genussvoller und damit für den Konsumenten akzeptabler zu machen. Dr. Peter Wild betreibt Food Design auf der wissenschaftlichen Ebene, sein Spezialgebiet ist die Entwicklung fettarmer Leichtprodukte, etwa Mayonnaise mit nur 15% Fettgehalt.

Warum Leichtprodukte nicht schmecken | Fast alle flüssigen und cremigen Nahrungsmittel sind Emulsionen, also Fett-Wasser-Gemische. Das Fett transportiert dabei nicht nur Geschmack, sondern ist in vielen Fällen hauptverantwortlich für die Textur. Die Art und Weise, wie sich die Fett- und Wassertröpfchen miteinander vermischen, bestimmt die Konsistenz, die wir im Mund wahrnehmen. Das Problem ist, dass Emulsionen äußerst instabil sind. Die Fetttröpfchen müssen eine solide Oberfläche haben, damit sie beim Zusammenprall mit anderen Fettteilchen nicht zerplatzen und sich mit diesen zusammenschließen. Dann entmischt sich die Substanz, was meist als geschmacklich negativ empfunden wird: Jeder kennt das, wenn Wasser aus der Senftube rinnt oder sich Flüssigkeit am Rand von Cremen absetzt.

Je weniger Fett in Produkten wie Joghurt oder Mayonnaise enthalten ist, desto schlechter funktioniert dieses Emulsionsgefüge und als dementsprechend schlecht empfinden wir daher die Textur von Leichtprodukten. Eine herkömmliche Methode, die Viskosität fettarmer Produkte in den Griff zu bekommen, ist die Beimengung von Stärke, was jedoch gesundheitlich nicht viel Vorteil bringt, wenn im Prinzip Fett durch Zucker oder Kohlenhydrate ersetzt wird.

Peter Wild und sein Team suchen nach physikalischen Methoden, um fettarme Produkte mit einer für uns angenehmen, also gewohnten Textur auszustatten. Eine Möglichkeit, Emulsionen auch mit einem niedrigen Fettgehalt geschmeidig zu halten, ist den Kern der einzeln herumschwirrenden Fetttröpfchen mit Wasser zu füllen. Damit bleibt die Anzahl der Fettteilchen und das gewohnte Mundgefühl (fast) erhalten, der Gesamtfettgehalt ist jedoch geringer. So genannte WOW (Water in Oil in Water)-Emulsionen erreichen bei sensorischen Tests fast die Werte von Produkten mit herkömmlichem Fettgehalt und werden etwa doppelt so gut eingestuft wie Produkte mit gleichem Fettgehalt in einer klassischen Emulsion. Hergestellt werden WOW-Emulsionen durch eine spezielle Membran-Emulsifikation, wobei geringe Scherkräfte den Verlust interner Wassertröpfchen reduzieren. Das System wird bereits für fettarme Mayonnaise, für Streichkäse und Dressings getestet. Praktische Probleme treten derzeit noch bei der Herstellbarkeit, bei Lager- und Transportfähigkeit auf.[13]

Freie Designer und Künstler | An der Schnittstelle zwischen Essen und Kunst, Produktdesign und Eat Art beschäftigen sich auch selbstständige Industriedesigner und Künstler mit dem Thema Nahrung. Freiberufliche Food Designer sind in vielen Ländern noch eine Rarität, in Frankreich oder Spanien schon seit einiger Zeit etabliert. Vor mehr als zehn Jahren gründete der französische Designer Marc Brétillot an der École Supérieure d'Art et de Design in Reims eine eigene Klasse für „Design Culinaire", die er bis heute leitet. Der Wiener Künstler Peter Kubelka leitete bis 2000 die Klasse für Film und Kochen als Kunstgattung an der Städelschule in Frankfurt und in Holland wird an der „hogeschool HAS DenBosch" Food Design and Innovation unterrichtet.

Research Programme" the institute is developing basic scientific principles for reducing the consumption of fat. The goal is to make healthy food more enjoyable and thus more acceptable for the consumer. Dr. Peter Wild runs food design on a scientific level. He has specialized in the development of extra low-fat products such as mayonnaise with only 15% fat content.

Why Light Products Do Not Taste Good | Almost all liquid and creamy foods are emulsions, that is to say fat-water mixtures. Fat does not just transport taste, in many instances it primarily accounts for texture. The way in which fat and water drops blend determines the consistency which we feel in our mouths. The problem is that emulsions are extremely unstable. The drops of fat must have a solid surface so that they do not merge with other particles of fat when colliding with them. If the single drops burst, the emulsion will dissolve, which is usually perceived as negative in terms of taste. Everyone has experienced this when water drips out of the mustard tube or liquid forms on the edges of creams.

The less fat products such as yoghurt or mayonnaise contain, the worse this emulsifying structure works and the worse we thus perceive the texture of light products to be. One way to keep emulsions smooth even if they have a low fat content is to fill the individual drops of fat floating around with water. This way, the number of fat particles and the usual sensation in the mouth will remain (almost) the same. However, the total fat content is lower. In sensory tests, the so-called WOW (water in oil in water) emulsions obtain almost the same values as products with ordinary fat content in classic emulsions. WOW emulsions are produced by means of special membrane emulsification with low shear forces reducing the loss of internal water drops. The system is already being tested for cream cheese and dressings. Practical problems still arise with regard to manufacturing, storage and transportation.[13]

Free-Lance Designers and Artists | At the interface between food and art, product design and Eat Art there are also free-lance designers and artists working on the subject of food. In many countries free-lance food designers are still a rarity but they have been able to establish themselves more recently in France or Spain. Already more than ten years ago, French designer Marc Brétillot founded his own class for "Design Culinaire" at the École Supérieure d'Art et de Design in Reims which he is still head of. The Viennese artist Peter Kubelka directed a class for film and cooking as an art genre at the Städelschule in Frankfurt until 2000, and in Holland food design and innovation are taught at the Hogeschool HAS Den Bosch.

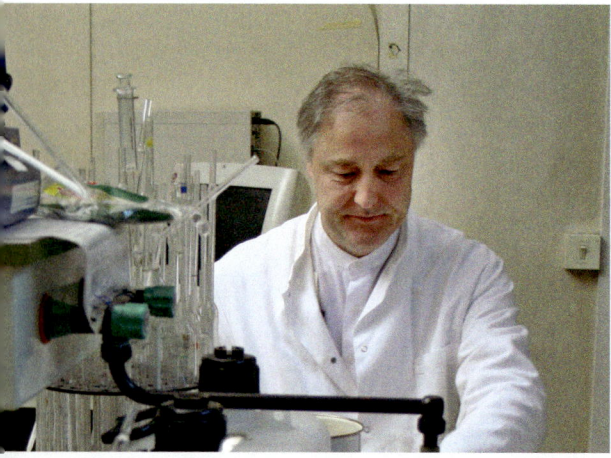

Designer wie die Jelly Mongers in London, Stéphane Bureaux in Paris oder Martí Guixé in Barcelona entwickeln Produkte für den Massenkonsum, arbeiten mit Köchen, Konditoren und Caterern zusammen oder entwerfen Einzelobjekte bis hin zu essbaren Installationen und Kunstprojekten. Gerade diese Gruppe von „Food Designern" liefert wichtige Impulse für die Weiterentwicklung unserer Nahrung, eine Weiterentwicklung nämlich, die sich nicht notgedrungen an den ökonomischen Vorgaben des Marktes, sondern an kulturellen Zusammenhängen orientiert.

Der katalanische Designer Martí Guixé zum Beispiel beschäftigt sich seit Jahren intensiv mit den kulturellen und praktischen Aspekten des Essens. Er entwickelt dreibeinige Lollipops, die man im Gegensatz zu herkömmlichen Schleckern während des Essens jederzeit bequem abstellen kann, Cracker, deren Oberfläche Muster als „Abbeißanleitung" zieren, oder Bohnen mit IBM-Aufdruck und Tortillas mit dem Schriftzug der Firmen CK und Fuji: Sein so genanntes „Sponsored Food" ist mit Logos internationaler Firmen ausgestattet und soll in einer Restaurantkette weltweit gratis angeboten werden. Für den modernen „nomadic worker" entwirft er Nahrungspillen und „Pharma-Food", das aus so kleinen Partikeln besteht, dass es eingeatmet und nicht gegessen wird.[17]

1 | Wir bedanken uns bei Christina Lund Jakobsen, Image Lab Kraft Europe, München, u. Mike Wilson, Uncle Ben's, Los Angeles, für die Informationen
2 | Betriebsbesichtigung Nissin Foods, Tokio, im Oktober 2008
3 | Interview mit Werner Mlodzianowski, Leiter des Technologie Transfer Zentrums Bremerhaven, am 18. u. 19.10.2007
4 | Interview mit Mike Wilson, Uncle Bens, LA, am 25.03.2009
5 | Betriebsbesichtigung Ben & Jerry's, Stowe, Vermont, USA, am 23.03.2009
6 | Interview von Klaus Kamolz in „Forschen und Entdecken" 2009
7 | Wir bedanken uns bei Christina Lund Jakobsen, Image Lab Kraft Europe, München, 2007 für die Informationen
8 | Interviews Prof. Gisla Gniech, Bremen, 11.2007, und Dr. Friedrich Blutner, Geyer, 2007
9 | Betriebsbesichtigung Ben & Jerry's, Stowe, Vermont, USA, am 23.03.2009 und wir danken Dr. Floros für diese Information.
10 | Betriebsbesichtigung Nissin Foods, Tokio, im Oktober 2008
11 | Frank Förster, Unilever, Heppenheim
12 | Interview mit Werner Mlodzianowski, Managing Director des Technologie Transfer Zentrums Bremerhaven, am 19. Okt. 2007
13 | Interview mit Dave Heart, Peter Wild, Institute of Food Research Norwich, am 21. 1.. 2009
14 | 1:1, Martí Guixé, 010 Publishers, Rotterdam 2002

Designers such as Jelly Mongers in London, Stéphane Bureaux in Paris or Marti Guixé in Barcelona develop products for mass consumption, collaborate with cooks, pastry bakers and caterers or design individual objects, edible installations or art projects. This group of "food designers" in particular provides important impulses for the further development of food, a development that is not forced to follow the economic exigencies of the market but orients itself after cultural factors.

Catalan designer Martí Guixé, for instance, has been focusing on the cultural and practical aspects of food for years. His developments include three-legged lollipops that, unlike ordinary ones, are easy to put down at any time while you are eating, crackers whose surface is adorned with patterns serving as "biting directions", beans with an IBM logo, and tortillas with the letters of the CK or Fuji company. His so-called "Sponsored Food" boasts the logos of international corporations and is to be offered free of charge in a worldwide restaurant chain. Guixé also develops nutritional pills and "pharma food" for the modern "nomadic worker". This type of food consists of particles so small that they are inhaled and not ingested.[14]

1 | Informations by: Christina Lund Jakobsen, Image Lab Kraft Europe, Munich, and Mike Wilson, Uncle Ben's, Los Angeles
2 | Guided tour of Nissin Foods, Tokyo, October 2008
3 | Source: Interview with Werner Mlodzianowski, Director of the Technologie Transfer Zentrum Bremerhaven, on 18 and 19 October 2007
4 | Interview with Mike Wilson, Uncle Ben's, LA, 25 March 2009
5 | Guided tour of Ben & Jerry's, Stowe, Vermont, USA, 23 March 2009
6 | interview with Klaus Kamolz in "Forschen und Entdecken" 2009
7 | Christina Lund Jakobsen, Image Lab Kraft Europe, Munich, 2007
8 | Interview Prof. Gisla Gniech, Bremen, November 2007, and Dr. Friedrich Blutner, Geyer, 2007
9 | Guided tour of Ben & Jerry's, Stowe, Vermont, USA, 23 March 2009 and we thank Dr. Floros for this information.
10 | Guided tour of the Nissin Foods co., Tokyo, October 2008
11 | Frank Förster, Unilever, Heppenheim
12 | Source: Interview with Werner Mlodzianowski, Director of the Technologie Transfer Zentrum Bremerhaven, 19 October 2007
13 | Interview with Dave Heart, Peter Wild, Institute of Food Research, Norwich, 21 January 2009
14 | 1:1, Martí Guixé, 010 Publishers, Rotterdam 2002

Food Designer von links nach rechts | Der französische Chemiker Hervé This erforscht als Universitätsprofessor die chemischen Zusammenhänge des Essens; Nanotechnologe David Julien McClements forscht am Institut für Food Science an der University of Massachusets, USA, unter anderem an fettreduziertem Essen; Wahrnehmungspsychologin Gisla Gniech berät Hersteller über die psychologische Wirkung von Nahrung, Der französische Designer Marc Brétillot betreibt ein Atelier in Paris und gründete 1999 die Klasse für „Design Culinaire" an der École Supérieure d'Art et de Design in Reims, die er seither leitet; der Industriedesigner Stéphane Bureaux gründete 1989 sein Studio für Global-Design in Paris, das sich seit 1997 auch zunehmend mit Culinary Design befasst; unter dem Namen Jellymongers betreiben die Briten Sam Bompas und Harry Parr ein Atelier in London, das sich vor allem mit dem Entwurf und der Herstellung traditioneller, englischer Jellies beschäftigt.

food designers from left to right | French chemist and university professor Hervé This studies the chemical background of food. Nanotechnologist David Julien McClements researches at the Institute for Food Science at the University of Massachussets, USA, focusing on fat-reduced food. Perception psychologist Gisla Gniech advises manufacturers about the psychological effect of food. French designer Marc Brétillot runs a studio in Paris and established a class for "Design Culinaire" at the École Supérieure d'Art et de Design in Reims in 1999, which he has directed ever since. In 1989, industrial designer Stéphane Bureaux founded his studio for Global Design in Paris, which since 1997 has dealt increasingly with culinary design. Under the name Jelly Mongers, Brits Sam Bompas and Harry Parr run a studio in London which mainly focuses on the design and manufacture of traditional English jellies.

Kultur
culture

Du bist, was du isst – der kulturelle Faktor von Food Design

| „Jede Speise bildet eine somato-psycho-soziale Einheit. Das heißt, unser Essen besitzt nicht nur einen Nähr-, sondern immer auch einen Genuss- und einen Symbolwert." Gisla Gniech, Wahrnehmungspsychologin

„Geschmack ist eine kulturelle Kategorie, ein Parameter, der den Lebensstil widerspiegelt. Jede Zeit, jede Epoche hat Ihren eigenen Geschmack." Werner Mlodzianowski, Leiter des Technologie Transfer Zentrums Bremerhaven

Die Auswahl der Nahrungsmittel und die Art, wie sie konsumiert werden, definieren den persönlichen Lebensstil und grenzen von anderen Gesellschaftsschichten ab. Vom Shrimps-Cocktail bis zum Döner, vom Ciabatta bis zur Instantsuppe: Was wir essen, signalisiert, wer wir sind. In derselben Weise, wie wir Mode tragen, uns mit Möbeln umgeben oder Autos fahren, um unserem Weltbild Ausdruck zu verleihen, erwarten wir auch von Nahrungsmitteln, dass sie nicht nur unsere Bäuche füllen, sondern auch das gewünschte Lebensgefühl vermitteln.

Mit dem Essen nehmen wir nicht nur Kalorien zu uns, sondern auch Werte. Die Gestaltung von Speisen spielt dabei eine zentrale Rolle, denn sie verwandelt simple Zutaten in National- und Liebessymbole, sexuelle Anspielungen, Opfergaben oder religiöse Gerichte. Denn wir essen nicht nur, was nahrhaft ist, was uns schmeckt, was verfügbar und einfach zu konsumieren ist. Wir essen, was unseren kulturellen Standards entspricht, uns in unserer Identität und unserer Existenzvorstellung bestätigt.

Erfolgreiche Lebensmittel schmecken daher nicht nur gut, sondern wecken Gefühle und erzählen Geschichten. Schon vor tausenden von Jahren transportierte entsprechend zubereitete und geformte Nahrung symbolische Inhalte und religiöse Werte und signalisierte die Zugehörigkeit zu Lebens- und Interessensgemeinschaften. Über Jahrhunderte hinweg prägte die rituelle Komponente des Essens die Gestaltung von Speisen und Getränken. Zu den ältesten bekannten Beispielen der Gestaltung essbarer Produkte zählen symbolisch aufgeladene Opferbrote. Manche dieser antiken Motive existieren sogar noch heute, wie zum Beispiel der Zopf oder das Croissant. Mittlerweile sind die Inhalte und Bedeutungen dieser Formen zwar in Vergessenheit geraten, doch ursprünglich hatten sie den Ausschlag für ihre Gestaltung gegeben.

Auch wenn die Industrialisierung einen quantitativen Sprung bei der Herstellung von Esswaren bewirkt hat, so ist das Prinzip der Nahrungsgestaltung als menschliche Ausdrucksform an sich nicht neu. Die systematische Erzeugung und die bewusste Formgebung von Lebensmitteln entstanden genau genommen bereits mit der Abkehr vom Jäger- und Sammlertum. Mit der Entwicklung sesshafter Gesellschaften mussten Esswaren gezielt erzeugt, den Lebensbedingungen angepasst und somit gestaltet werden. Der Mensch begann die Natur zu kontrollieren und seine Nahrung so zu verändern, wie er sie brauchte, wie sie ihm schmeckte und wie sie seinen Wertvorstellungen am besten entsprach.

Bis heute ist Food Design ein kultureller Akt. Mit den unterschiedlichst geformten und gestalteten Lebensmitteln konsumieren wir nach wie vor auch Werte, Traditionen und Symbole. Reichte man im alten Ägypten den Göttern pyramiden- oder fischförmige Brote als Opfergaben, so wollen wir heute mit Nahrungsmitteln Gesundheit, Schönheit und ewige Jugend inhalieren, das Gefühl haben, im Einklang mit der Natur zu leben, oder moralischen Grundsätzen wie Nachhaltigkeit, Umweltgerechtigkeit oder Fairtrade Ausdruck verleihen.

Japanisches Konfekt in der Form von No-Theater-Masken, ein Symbol für langes Leben. Ein typisches Geschenk für ältere Leute.

Japanese confectionery in the form of No theater masks, a symbol for a long life. A typical gift for older people.

You Are What You Eat – The Cultural Factor in Food Design |
"Every dish is a somatic-psychic-social entity. Thus, food has not only nutritional value, but also pleasure value and symbolic value." Gisla Gniech, perceptual psychologist

"Taste is a cultural category, a parameter which reflects life-style. Each time, each epoch has its own taste." Werner Mlodzianowski, Director of the Technologie Transfer Zentrum Bremerhaven

The food we choose and the way in which we consume it defines our personal life-style and distinguishes us from other social groups. From shrimp cocktail to döner, from ciabatta to instant soup: what we eat reflects who we are. We express our view of the world in the fashion we wear, the furniture we buy and the cars we drive – and we expect food not only to fill our tummies but also to give us the desired feel for life.

Eating not only serves calorie intake, but also conveys values, meanings and emotions. The design of food is crucial in this context as it turns simple ingredients into sexual innuendos, offerings or religious meals. We do not just eat what is nutritious, tasty, available or comfortable. We choose food that conforms to our cultural standards, confirms our identities and our views of existence.

Thus, successful food products do not just taste good, they also stir up emotions and trigger fantasies. Already thousands of years ago, appropriately prepared and shaped food transported symbolic content and religious values, signaling a sense of belonging to communities of life and interests. For centuries, the ritual component of eating left an imprint on the design of food and beverages. One of the oldest examples of edible product design loaded with symbolic meaning is found in various types of sacrificial bread. Some of the ancient motives, such as the plait or the croissant, have survived to this day. The contents and meanings of these shapes may have been forgotten but they originally contributed significantly to their design.

Even if industrialization led to a marked increase in the production of food, the principle of food design as a form of human expression per se is nothing new. To be precise, the systematic production and deliberate shaping of food started at the end of the age of hunters and gatherers. Once human beings became sedentary, societies had to be supplied with food in a targeted way; food had to be adapted to living conditions and thus designed. Humankind started to control nature and change food as needed, as liked and as it conformed best to the value system.

Until this very day, food design is a cultural act. By consuming foods in many varieties of shapes and forms, we also consume values, traditions and symbols. Whereas in ancient Egypt pyramid- or fish-shaped bread was offered to the gods, today we seek to inhale health, beauty and eternal youth with modern food products, or express ethical principles such as sustainability, environmental equity and fair trade.

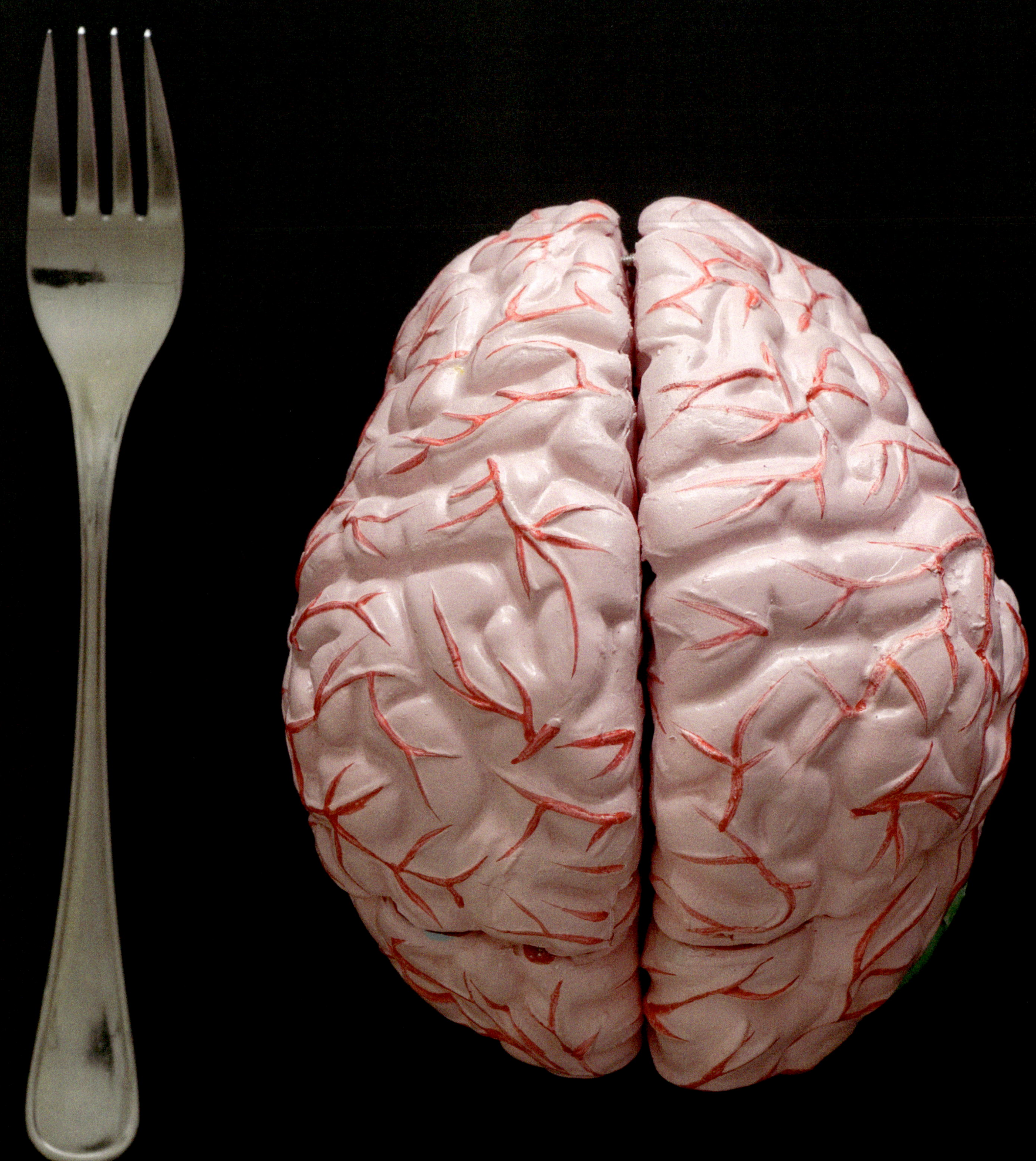

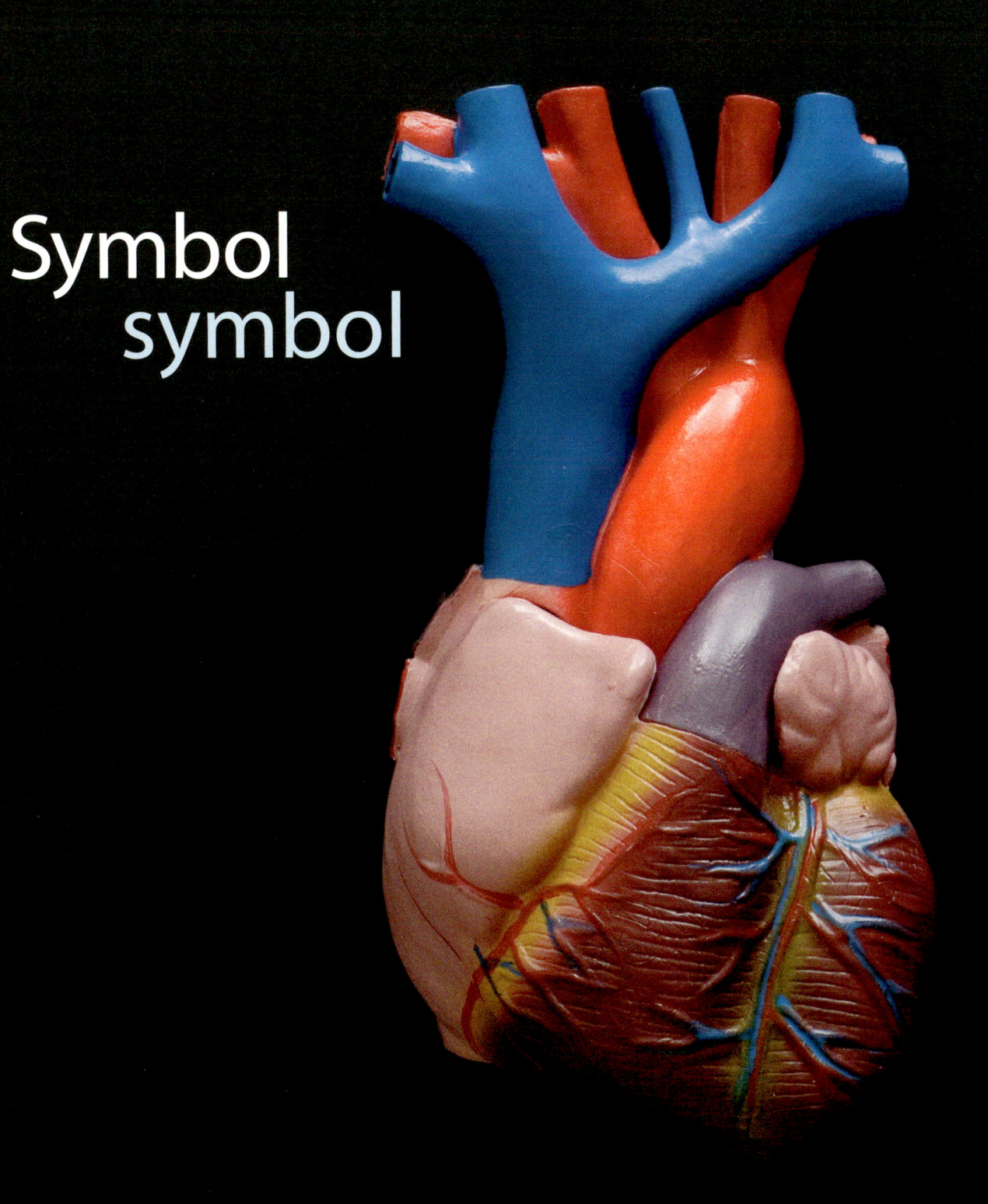

Symbol
symbol

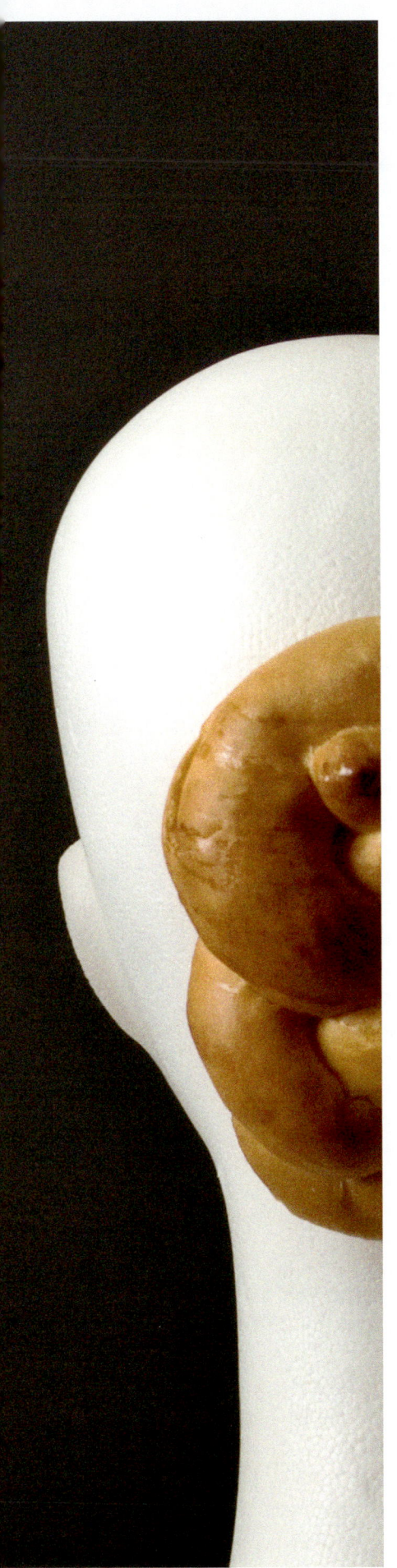

Religiöse Symbole – Kultspeisen und Opferbrote | Zu den ältesten bekannten Beispielen bewusst geformter Nahrungsmittel zählen symbolische Darstellungen aus Brotteig, die als Kultspeisen, Brotgaben oder Opferersatz dienten. Die Strategie, zu religiösen Anlässen anstelle von Tieren entsprechend geformtes Brot zu essen, wirkt bis in die heutige Zeit nach: Zu Ostern schlachten wir kein Lamm, wie es der Tradition des jüdischen Pessach-Festes entspräche, verschenken aber sehr wohl Osterlämmer in Form von Kuchen oder Schokolade.

Der Striezel oder Brotzopf entstand vermutlich infolge des früher üblichen Opfers der eigenen Haare. Aus Trauer oder Ehrerbietung schor man sich die Haare und brachte sie den Göttern dar, ein Brauch, der beispielsweise in Indien noch heute praktiziert wird. Andere Interpretationen sehen Zöpfe als Weiterentwicklung von Broten in Drachen- und Schlangenform, wie sie schon bei den alten Sumerern gebräuchlich waren. Brote in geflochtener Form besaßen ursprünglich jedenfalls kultischen Charakter und noch heute verspeisen wir Striezel, Zopf oder große Brioche zu Allerheiligen, Weihnachten oder Ostern als religiöse Festspeise. Im Laufe der Jahrhunderte wandelte sich die Gabe für die Götter zum Geschenk für Tote und Lebende: Paten überreichen den Striezel traditionell an ihre Täuflinge und junge Mütter erhalten ihn nach der Geburt als Symbol für das Neugeborene, wobei die verschlungenen Teigwürste als Darstellung eines in Windeln gewickelten Kindes interpretiert werden. Geflochtenen Brotsorten schreibt der Volksglaube zudem besondere Kräfte zu: Sie sollen vor Krankheit schützen und Unheil von Mensch und Vieh fern halten.[1]

Religious Symbols – Cultic Food and Sacrificial Bread | Symbols made of bread dough are among the oldest examples of deliberately shaped food; they were used as cultic food, bread offerings or substitutes for sacrifices. The strategy of eating bread in the shape of animals instead of the actual animal meat for religious holidays is echoed in present-day conventions. At Easter time, we do not slaughter lambs, although this would be in keeping with Jewish Passover tradition, but instead we give away cakes or chocolates shaped like Easter lambs.

Plaited buns or bread presumably replaced an earlier practice of sacrificing one's own hair. People cut their hair in mourning or in deference, bringing them to the gods - a rite still practiced in India today. Other interpretations consider plaits to be developments of bread in dragon or snake shapes which were common among the ancient Sumerians. Plaited bread was originally cultic in nature, and even today, we eat plaited buns, bread or brioche pastry as festive food on religiously motivated occasions, such as All Saints' Day, Christmas or Easter. In the course of the centuries, the gift to the gods was transformed into a traditional present to the dead and the living: godparents give it to their godchildren for their baptism, and young mothers receive it after childbirth, as a symbol of the newborn baby, because the plaited dough is thought to depict a child in swaddling clothes. According to popular belief, plaited bread has – the ability to protect from illness and keep misfortune away from humans and animals.[1]

Geflochtene Brote haben sich vermutlich aus dem früher üblichen Opfer der eigenen Haare entwickelt. Auch in der christlichen Kultur hatten Zöpfe ursprünglich rituellen Charakter: als Allerheiligen-, Weihnachts- oder Osterstriezel und – da die geschlungene Form optisch an ein Wickelkind erinnert – als Geschenk für junge Mütter.

Plaited breads were said to have developed from the sacrificial offering of one's own hair, which was once common. Even in Christian culture, plaits originally had a ritual character: as plaited bread made for All Saints' Day, Christmas or Easter – since the wrapped form visually recalled a baby in swaddling clothes – as a gift for young mothers.

Mondsichel und Teufelshorn | Auch die gekrümmte Form des Croissants, Kipferls oder Hörnchens entstand ursprünglich als Merkmal einer religiösen Kultspeise. Der Legende nach wurde sie von Wiener Bäckern anlässlich der zweiten Wiener Türkenbelagerung im Jahr 1683 erfunden. Demnach hörten die nachts arbeitenden Bäcker, wie die osmanischen Angreifer einen Tunnel unter die Stadtmauer treiben wollten, schlugen Alarm und konnten so das Eindringen des feindlichen Heeres verhindern. Als Erinnerung an die Rettung Wiens schufen die Bäckergesellen ein Gebäck in Form des osmanischen Halbmondes: Das Kipferl war geboren. Andere Quellen weisen den Wiener Bäckermeister Peter Wendler als Kipferl-Erfinder aus und setzen das Entstehungsdatum schon mit der ersten Wiener Türkenbelagerung 1529 fest.

In Wahrheit sind sichelförmige Brote jedoch wesentlich älter und waren im Rahmen von Mondkulten vermutlich bereits in der Antike gebräuchlich. Im antiken Griechenland kamen gebogene Gebäckstücke wahrscheinlich als Opfergabe für die Mondgöttin Selene zum Einsatz und symbolisierten schon damals – wie heute auf der türkischen Flagge – die Mondsichel. In späteren Jahrhunderten wandelte sich die Bedeutung des traditionellen Backwerks. Im Zuge der Christianisierung wurde die Mondsichel als gestalterisches Vorbild verdrängt und durch christliche Assoziationen ersetzt. Wie die deutsche Bezeichnung „Hörnchen" noch heute suggeriert, symbolisierte das gebogene Gebäckstück nunmehr die Hörner des Teufels. Ein alter Brauch der Donauschiffer etwa, allesamt Nichtschwimmer, empfiehlt, die sichelförmige Backware nicht selbst zu essen, sondern zur Besänftigung der Wassergeister in die damals noch gefährlichen Fluten zu werfen.[2]

Kampf gegen das Böse | Im Mittelalter wurde das Böse oft mit dem Bösen bekämpft. So hinderten aus Stein gemeißelte Fratzen an den Gesimsen gotischer Kathedralen den Teufel daran, das Gotteshaus zu betreten. Noch heute sind derartige Wächter an so mancher Kirche, etwa an der Fassade der Notre-Dame in Paris, zu sehen. Ob die Wiener die aus Teig gebackenen Teufelshörner 1683 auf ähnliche Weise auch im Kampf gegen die Osmanen einsetzten, fällt in den Bereich der Spekulation. Wir halten es für denkbar, dass das Kipferl während der Belagerung als eine Art politisches Propagandamittel fungierte. Der Verzehr des türkischen Halbmonds könnte einer sinnbildlichen Vernichtung des Feindes gleichgekommen sein.

Der Glaube, das Kipferl habe seinen Ursprung in Wien, ist übrigens nicht nur in Österreich, sondern auch in anderen Ländern verbreitet. So schreibt etwa der belgische Soziologe Leo Moulin in seinem umfassenden Werk „Augenlust & Tafelfreuden", dass das Croissant vom deutschen Hörnchen abstamme und ein Wiener Gebäck bezeichne, das 1689 zur Feier des Abzuges der Türken entstanden sei. Auch in der englischen Literatur finden sich ähnliche Darstellungen, woraus wir schließen, dass das Kipferl zwar nicht seinen Ursprung, wohl aber seine europaweite Verbreitung der zweiten Wiener Türkenbelagerung verdanken könnte. Vermutlich erlangte der essbare Halbmond nach dem Ende der Belagerung derartige Beliebtheit, dass er weit über Wien hinaus bekannt wurde und bald auch an der Seine in Mode kam. Dort mutierte der kulinarische Export aus Österreich dann zum Croissant und somit ausgerechnet zum Inbegriff französischer Lebensart.

In der französischen Bezeichnung „croissant", was so viel wie „zunehmender Mond" oder „Mondsichel" bedeutet, ist überdies der Bezug zum Mond als gestalterischem Vorbild erhalten geblieben. Der Begriff Kipferl oder Kipfel leitet sich hingegen vom mittelhochdeutschen „kipf" und in weiterer Folge vom lateinischen "cippus" (Pfahl) her, wobei sich der Zusammenhang laut Duden aus der länglichen Form der beiden Objekte erklärt.

Mondkulte und mondförmige Opferspeisen finden sich nicht nur in Europa, sondern auch in vielen anderen Kulturen. In Japan zum Beispiel werden zu Neujahr mondsichelförmige Reiskuchen gegessen und in China verschenkt man zum alljährlichen Mondfest im Herbst zylindrische Mondkuchen namens „yuèbǐng". Sie werden sowohl salzig als auch süß gefüllt

Der Legende nach entstand das Croissant, Kipferl oder Hörnchen 1683 in Folge der 2. Wiener Türkenbelagerung. Tatsächlich waren mondsichelförmige Brote schon in der Antike bekannt und dienten vermutlich als Opfergabe für Mondgottheiten wie die griechische Göttin Selene. Im Mittelalter wurde die gebogene Form des heidnischen Gebäcks dann als Hörner des Teufels interpretiert.

According to legend, the croissant or crescent emerged in 1683 in the wake of the 2nd Turkish Siege of Vienna. Crescent-shaped breads were already known in Antiquity and are said to have served as sacrificial offerings for moon deities such as the Greek goddess Selene. In the Middle Ages, the curved form of the heathen bread was then interpreted as the devil's horns.

Crescent and Devil's Horn | Also the curved shape of the croissant, known as "Kipferl" or "Hörnchen" in German, was originally cultic food with a religious background. According to legend, Vienna bakers invented it when the Turks laid siege to the city for the second time in 1683. While working at night, the bakers were said to have heard the Osmanic attackers dig a tunnel underneath the city fortifications; they sounded the alarm and were thus able to prevent the enemy forces from entering the city. In remembrance of the rescue of Vienna, the bakers created pastry in the shape of the Osmanic crescent – the "Kipferl". According to other sources, the Viennese baker Peter Wendler invented the crescent pastry at the time of the first Turkish siege in 1529.

In reality, crescent-shaped bread is much older than that and most likely already part of moon cults in Antiquity. In ancient Greece, curved pastry was probably used as an offering to the moon goddess Selene. Its shape symbolized the crescent moon – as it does on the Turkish flag. In later centuries, the meaning of the traditional bakery product changed with the advent of Christianity. The crescent moon as a model for design was replaced by Christian associations with the horns of the devil (hence the literal translation from German, "little horn" for "Hörnchen"). The boatmen on the river Danube, who were all non-swimmers, practised the tradition of tossing the crescent-shape pastry into the river instead of eating it - to appease the water spirits.[2]

Fighting Evil | In the Middle Ages, evil was often used to drive out evil. Gargoyles chiseled from stone on the cornices of Gothic cathedrals were supposed to keep the devil out of the Lord's house. These guards can still be seen today, for example on the façade of Notre-Dame in Paris. Whether the Viennese actually used the baked devil's horns in their fight against the Turks in 1683 is left to speculation. We think that the crescent-shaped pastry may well have been a means of political propaganda during the siege. The act of eating a crescent could have been tantamount to the symbolic annihilation of the enemy.

The idea that the crescent-shaped pastry originated in Vienna is not only a wide-spread belief in Austria but also in other countries. The Belgian sociologist Leo Moulin published a comprehensive work on the cultural history of eating and drinking under the title "Les liturgies de la table. Une histoire culturelle du manger et du boire". Moulin states that the croissant was derived from the German "Hörnchen", which in turn described Viennese pastry invented in 1689 to celebrate the retreat of the Turks. We have found similar reports in English literature, which has led us to the conclusion that the pastry may not owe its origins but at least its Europe-wide distribution to the Second Turkish Siege of Vienna. Presumably, the "edible crescent" became so popular far beyond Vienna after the end of the siege that it was even adopted by the French. On the banks of the river Seine, it turned into the "croissant" and thus, of all things, became the epitome of French lifestyle.

The crescent moon is actually reflected in the French word "croissant", so that the reference to the moon as a model for the shape of the pastry lives on. The word "Kipferl" or "Kipfel" is derived from the Middle High German "kipf", which in turn is based on the Latin "cippus" (post, stake). According to the Duden dictionary, this points to the oblong shape of the pastry.

und enthalten manchmal ganze, gesalzene Eidotter, die den Vollmond symbolisieren. Verziert werden sie mit Schriftzeichen, mit denen man dem Beschenkten langes Leben oder Harmonie wünscht. Auch in der europäischen Weihnachtsbäckerei ist der Mond ein klassisches Motiv, obwohl er mit dem christlichen Weihnachtsfest eigentlich in keinem Zusammenhang steht.

Gekreuzte Arme | Auch die ungewöhnliche Form der Brezel, einer Backware, die mehr ihrer Form als ihres Geschmacks wegen Berühmtheit erlangte, hat symbolischen Charakter und entstand vermutlich als kultisches Gebäck. Der Legende nach soll das klösterliche Fastengebäck im Jahr 610 von einem Mönch in Norditalien erfunden worden sein. Die charakteristische Form der Doppelschlinge stellt, so die häufig zu lesende Entstehungsgeschichte, zwei vor der Brust gekreuzte Arme dar – eine damals übliche Gebetshaltung, die auch Tote in zahlreichen Darstellungen einnehmen. Der Name „Brezel" leitet sich demnach vom lateinischen Wort „bracchium" her, was so viel wie Arm bedeutet.

Der wahre Ursprung des zweifach geschlungenen Gebäcks könnte allerdings bis zu antiken Ringbroten zurückreichen. Tatsächlich bezeichnete „bracchium" auch ein römisches Ringbrot, das die Christen schon im zweiten Jahrhundert als eucharistisches Gebäck verwendeten. Aus diesem entwickelte sich zunächst ein Abendmahlsbrot in Form der Zahl 6 und schließlich im elften Jahrhundert das heutige Erscheinungsbild der Brezel. Die Beliebtheit des mit Salz bestreuten, mehrfach durchlöcherten Brotes nutzten später die Bäcker, indem sie die Brezel im 12. Jahrhundert zu ihrem Zunftzeichen erhoben, welches bis heute am Portal unzähliger Bäckereien prangt und an die tausendjährige Handwerkskunst des Brezelformens erinnert. Die Herstellung selbst übernimmt mittlerweile freilich oft eine automatische Brezel-Schlingmaschine. Ihr charakteristisches Aroma und die glänzende Kruste verdankt die Brezel einem kurzen Bad in Natronlauge, erst danach wird sie im Ofen knusprig gebacken. Das ungewöhnliche Design ist übrigens kein europäischer Einzelfall: Brot und Backwaren in Form auf unterschiedlichste Weise verschlungener Teigwürste finden sich in vielen Kulturen, so auch in China und in Japan.

Kein Jesus, keine Kreuze | Religiös oder kultisch motivierte symbolische Darstellungen sind der Ursprung für das Aussehen vieler heute üblicher Gerichte und Nahrungsmittel: von Marzipanrosen und Lebkuchenherzen über Friedenstauben, Ostereier oder Osterhasen bis zu Weihnachtsmännern, Weihnachtsbäckerei und Totenköpfen. In Mexiko verschenkt man zu Allerheiligen bunte Schädel aus Zucker, auf deren Stirn der Name des Beschenkten steht und die ihn an die Vergänglichkeit des Lebens erinnern. Im Laufe der Geschichte wurden vielen Speisen magische Kräfte nachgesagt: Die Brezel war ein Liebessymbol, der Lebkuchen sollte Fruchtbarkeit bringen und der Brotzopf vor Bösem schützen.

Die meisten Formen christlicher Festtagsgerichte entstanden aus der Uminterpretation heidnischer Kultspeisen und Symbole. Die Formen der Weihnachtsbäckerei beispielsweise versinnbildlichen nicht – wie man vielleicht annehmen könnte – das Weihnachtsevangelium. Wir essen keinen Zimtchristus, keine heiligen Zuckerkönige oder Vanillejungfrauen. Stattdessen backen wir abstrahierte Blumen, geometrische Formen wie Quadrate, Kreise und Rhomben, Weihnachtsbäume, Kleeblätter, Hasen, Rehe und Hirsche – allesamt Motive, die mit Christi Geburt in keinem Zusammenhang stehen. Der Mangel an essbaren christlichen Symbolen deutet auf den Ursprung heutiger Kultspeisen in vorchristlicher Zeit hin.

Andererseits erscheint uns der Verzehr von Jesusfiguren, Schokokruzifixen, Darstellungen des Heiligen Geistes oder gar Gottes selbst als blasphemisch. Das zentrale Ritual der Liturgie ist zwar die Einverleibung des Göttlichen durch die Gläubigen, wir sprechen auch davon, den Leib Christi zu essen, stecken ihn aber tatsächlich nur in sinnbildlicher Form als Opferlamm oder als völlig abstrakt geformte Hostie, deren Aussehen eher an ein rundes Stück Karton als an Brot erinnert, wirklich in den Mund. In anderen Kulturen werden essbare Abbildungen von Gottheiten sehr wohl verspeist, in Japan etwa isst man Glücksgötter aus Backteig und verkauft Buddha unter anderem in der Form von Lollipops. Auch den alljährlichen, eigentlich kannibalischen Verzehr eines christlichen Schokoladeheiligen am Nikolaustag empfinden wir als ganz normal.

In der Weihnachtszeit beißen wir dem heiligen Nikolaus bzw. Santa Claus gedankenlos den Schokokopf ab, ein essbares Kruzifix empfinden wir dagegen als Blasphemie.

At Christmas time we think nothing of biting off the chocolate head of Saint Nicholas or Santa Claus, while we see an edible crucifix as being blasphemous.

Moon cults and moon-shaped offerings are not restricted to Europe, they also exist in many other civilizations. In Japan, crescent-shaped rice cakes are eaten on New Year's Day, and in China cylindrical moon cakes called "yuébǐng" are given as presents for the Moon Festival in the fall. Moon cakes contain savory or sweet fillings, and sometimes whole salted egg yolks to symbolize the full moon. They are decorated with ideographs wishing the recipients a long life or harmony. The moon is also a classic shape of European Christmas cookies even though there is actually no connection with the Christian nativity celebration.

Crossed Arms | The unusual shape of the soft pretzel, a baked product which is famous for its form rather than its taste, is also symbolic and probably reflects its origin as cultic food. According to legend, the monastic fasting bread was invented by a monk in northern Italy in 610 CE. Most sources say that the characteristic shape in which the strip of dough is folded symbolizes arms crossed in prayer in front of the chest – the usual prayer position at the time, which can also be seen in many depictions of dead bodies. The word "pretzel" (from the German "Brezel") is said to be derived from the Latin "bracchium", which means "arm".

The true origin of the folded strip of dough might even be found in ancient ring-shaped breads. In fact, the Latin word "bracchium" denotes a Roman ring-shaped bread used by Christians in the Eucharist as early as in the second century. The communion bread which developed from the ring later had the shape of the figure 6 which by the eleventh century had turned into the pretzel as we know it today. The bread with several holes, sprinkled with salt, was so popular that bakers started using it as the symbol of their guild as from the twelfth century. It continues to hang over the doors of many bakeries, reminding us of the art of folding pretzels, which looks back on a thousand years of history. Even though today, the actual folding work is often done by a machine. The characteristic taste and shiny crust of the soft pretzel emerges when it is briefly dipped in soda lye before it is baked until crisp. The unusual design is not unique to Europe, though. Bread and bakery products consisting of dough strips folded in various ways exist in many civilizations, including China and Japan.

No Jesus, No Crosses | Many popular dishes have their origins in religious or cultic symbols, from Easter eggs to Easter bunnies, from doves of peace to chocolate Santas, from marzipan roses to gingerbread hearts, Christmas cookies and skulls. People in Mexico give colorful sugar skulls with the recipient's name on the forehead evoking the transitoriness of life. Many foods in history were said to have magical powers – the soft pretzel was a symbol of love, gingerbread was for fertility, and plaited bread should protect from evil.

Most shapes of Christian festive dishes are re-interpretations of pagan cultic food and symbols. The forms of Christmas cookies, for example, do not symbolize the Christmas story according to Luke, as one would expect. We do not eat cinnamon Christs, three Magi made of sugar or vanilla-sprinkled Holy Virgins. Instead, we bake abstractions of flowers, geometrical forms such as squares, circles and diamonds, Christmas trees, trefoils, rabbits, does and stags – all of them motives totally unrelated with the birth of Jesus. The lack of edible Christian symbols points to a pre-Christian origin of what we consider cultic food today.

Eating figures of Jesus, chocolate crosses, symbols of the Holy Spirit or even God the Father is something that seems blasphemous to us, anyway. Although the central liturgical ritual calls for believers to take in what is divine, even calling it "the body of Christ", this involves the sacrificial lamb, or rather the abstract shape of the host which looks more like a round piece of cardboard than bread. By contrast, edible likenesses of gods are eaten in other cultures. Fortune gods made of batter are gobbled up, and buddhas are even sold as lollipops. We also consider it normal to eat a Christian saint made of chocolate on St. Nicholas' Day every year though this could actually be construed as cannibalistic.

Letzte Seite | Mit dem Essen konsumieren wir nicht nur Kalorien, sondern auch Werte. Die Bedeutung vieler Gerichte – von der Hostie über die Torte bis zum Cola – geht weit über die simple Nahrungsaufnahme hinaus.
Diese Seite | Esswaren mit Symbolgehalt: Geburtstagstorte, englische Hostie mit Kruzifixprägung, Buddha-Lollipop als Kindersouvenir eines japanischen Tempelfestes.

previous page | With food we consume not only calories but also values. The meaning of many dishes – from the host via the cake to Cola goes way beyond simple intake of food.
this page | Food with symbolic meaning: birthday cake, English host with imprinted crucifix, Buddha lollipop as a child's souvenir from a Japanese temple celebration.

Eine andere Ausnahme von diesem Tabu bilden „Hot Cross Buns", kleine Brötchen aus Hefeteig mit Rosinen, die an der Oberseite mit einem Kreuz aus Zuckerguss oder Reispapier verziert sind. Cross Buns werden in Großbritannien traditionellerweise am Karfreitag getoastet, mit Butter bestrichen und als Symbol für die Kreuzigung gegessen. Aber auch das Design der Cross Buns, die unter diesem Namen 1733 erstmals erwähnt wurden, dürfte eigentlich heidnischen Ursprungs sein. Der Überlieferung nach handelt es sich um ein altes Gebäck der Sachsen, das der sächsischen Göttin der Morgendämmerung und des Frühlings, der „Eostrae", geweiht war, wobei das Kreuz an der Oberfläche die vier Mondphasen symbolisierte. „Eostrae" beziehungsweise „Eastre" könnte auch die Wortwurzel für die heutigen Bezeichnungen „Easter" beziehungsweise „Ostern" sein.

Food Design – ein kultureller Akt |
Food Design ist ein kultureller Akt. Um die Inhalte und Bedeutungen hinter den vielfältig gestalteten Speisen und Gerichten verstehen zu können, muss man mit der jeweiligen Kultur vertraut sein. Wem Christentum und Kreuzigung nicht geläufig sind, der kann das Cross Bun nicht verstehen, wer das Haaropfer nicht kennt, dem sagt der Striezel nichts, und wer mit Mondkulten nichts anfangen kann, für den ist das Kipferl einfach nur ein krummes Gebäckstück. Ähnlich unbedarft betrachten wir Europäer die unzähligen, aufwändig gestalteten japanischen Nahrungsmittel. Optisch faszinieren vor allem die bunten Süßigkeiten, die traditionellerweise zur Teezeremonie gereicht werden und ein fernöstliches Beispiel für Kultspeisen mit starkem Symbolgehalt sind. Die so genannten Wagashi und Higashi bestehen hauptsächlich aus gesüßtem Bohnenmus und haben ihren Ursprung vermutlich in China, wo historisch ähnliche Häppchen als Opfergabe für die Götter zubereitet wurden.

Das Konzept, das hinter dem enormen Gestaltungsreichtum des Teekonfekts steckt, erscheint ziemlich philosophisch: Aufgrund der Kürze des Lebens soll jedes menschliche Aufeinandertreffen und damit jede Teezeremonie einmalig und perfekt sein. Die detailreiche Gestaltung von Wagashi trägt dazu bei, die Besonderheit des Augenblicks einzufangen. Form und Farbe des Konfekts bilden die Schönheit der Natur und der jeweiligen Saison ab oder korrespondieren mit dem speziellen Anlass, mit Feiertagen und Festen. Ursprünglicher besaßen die kleinen symbolträchtigen Kuchen auch kultischen Charakter und dienten beispielsweise dazu, Böses zu vertreiben, um ein langes, gesundes Leben oder eine gute Ernte zu bitten.

Essbare Poesie – japanisches Teekonfekt |
Die Art der Darstellung erscheint dem westlichen Betrachter äußerst abstrakt: Ein rechteckiges Konfekt aus zweifärbigem Bohnenmus kann schon genügen, um bei Japanern eine ganz bestimmte Assoziation aufzurufen. Eine traditionelle Sorte zum Beispiel besteht aus einem Quader mit gelben und braunen Streifen und symbolisiert eine bestimmte gelbe Blume, die entlang eines Weges oder eines Flusses blüht. Das Motiv, das den Japanern aus traditionellen Gedichten vertraut ist, wurde bereits 1707 erzeugt. Unterstützt wird der Symbolgehalt des Konfekts durch die poetischen Namen, welche die einzelnen Designs tragen: „Im Schatten junger Blätter" ist ein

"Hot cross buns", raisin buns with a cross of sugar icing or rice paper on top, are another exeption from this taboo. In the UK, they are traditionally toasted and eaten with butter on Good Friday in remembrance of the Crucifixion. However, the design of the cross buns, which were first mentioned under this name in 1733, seems to be of pagan origin, too. Tradition has it that the buns were actually devoted to Eostrae, the Saxon goddess of dawn and spring, with the cross symbolizing the four moon phases. "Eostrae" or "Eastre" is also considered be the etymological origin of today's word "Easter".

Food Design – A Cultural Act |
Food design is a cultural act. You have to be familiar with cultures and civilizations to understand the meanings behind the highly diverse designs of foods and dishes. People who are not familiar with Christianity and the Crucifixion, will not see the point in the cross bun. If you know nothing about hair sacrifices, plaited bread will mean nothing to you. Without understanding the cultic significance of the moon, the croissant will be nothing but a curved piece of pastry. This is what we experience in Japan since we have a rather naïve approach to the many sophisticated Japanese foods. We are mainly fascinated by the visual appearance of the colorful confectionery traditionally served during tea ceremonies: however, they are examples of cultic food from the Far East with great symbolic content. So-called wagashi and higashi mainly consist of sweetened bean paste. It is assumed that they originated in China, where similar tidbits used to be prepared as offerings to the gods.

The idea behind the tea confectionery and its enormous diversity seems rather philosophical: As life is short, every occasion when people meet, and thus every tea ceremony, must be unique and perfect. The design of wagashi is rich in detail and helps seize the special character of the moment. Colors and shapes of the confectionery reflect the beauty of nature and the respective season, or fit the occasion. Originally, the symbolic little cakes were cultic food, serving to ward off evil or ask for a long life in good health or a plentiful harvest.

Poetry to Be Eaten – Japanese Tea Confectionery |
To western beholders, the depictions seem rather abstract: a rectangular piece made of two-toned bean paste may be enough to trigger certain associations in the Japanese. One traditional type is a rectangular prism with yellow and brown stripes. It symbolizes a special yellow flower blooming along a path or river. The motive was created in 1707 and the Japanese are familiar with it from traditional poems. The symbolism in the confectionery is enhanced by the poetic names of the individual designs: "In the Shadow of Fresh Leaves" is a transparent cube with little fishes splashing about. It is to remind the Japanese eater of the refreshing effect of cool water in the hot Japanese summers. "Ivy in the Mist" is sprinkled with rice flour to create the impression of foggy air. "Maple over Flowing Waters" is produced of multi-colored paste in a wooden mold, showing maples leaves painted red and yellow above two blue spirals

transparenter Quader, in dem sich kleine Fische tummeln und der auf den Esser in den heißen japanischen Sommern die erfrischende Wirkung von kühlem Wasser ausüben soll: „Efeu im Nebel" ist mit Reismehl bepudert, um den Eindruck von Nebel zu erwecken; „Ahorn über fließendem Wasser" wird aus mehrfärbiger Masse in einem Holzmodel hergestellt und zeigt rot und gelb verfärbte Ahornblätter über zwei blauen Wasserspiralen; „Tautropfen auf grünen Blättern" ahmt mit winzig kleinen, durchsichtigen Geleewürfelchen das Glitzern einzelner Wassertropfen nach und „Dünnes Gras" stellt auf den Boden gefallenes Herbstlaub dar, welches unter dünnem Gras hervorschimmert.

Bachblüten und Kimonostoffe |
Traditionelle Teekonfekt-Formen, die als Opfergabe, Glücksbringer oder Talisman fungieren, sind unter anderem die Schildkröte, der Kranich, die Meerbrasse (Thai, Seabream), das Wildschwein, der Hase, der Hund und der Karpfen. Manche Motive aus der Edo-Zeit (1603 - 1868) sind heute noch gebräuchlich, wie etwa ein Model mit einem Sonnenaufgangsmotiv aus dem Jahr 1796. Inspiration für die Gestaltung von Teekuchen liefern auch die verschiedenfärbigen Lagen von Seidenkimonos, Muster alter Holztempel, traditionelle Keramik, das Design von Lackboxen oder Hofkostüme. Zum Beispiel ist der spezielle Hut kaiserlicher Hofmusikanten in verkleinerter Form als Wagashi erhältlich. Jüngere Designs müssen nicht unbedingt japanischen Traditionen entspringen. Zum Beispiel bestellte 1926 eine reiche Dame Konfekt in Form von Golfbällen beim traditionellen Wagashi-Produzenten Toraya. Dieser „Fernost-Sportgummi" ist mittlerweile ein fixer Bestandteil der Toraya-Kollektion, die nach wie vor regelmäßig erweitert wird: zum Beispiel wenn Kosmetikunternehmen für eine Produktpräsentation Wagashi in Form oder Farbe eines neuen Lippenstiftes bestellen.

Ähnlich wie bei der christlichen Hostie ist der Geschmack von Wagashi relativ zweitrangig. Den mehr als 3000 verschiedenen, teils sehr aufwändigen Designs stehen nur wenige, relativ ähnliche Geschmacksrichtungen gegenüber. Der Verzehr von Teekonfekt wird weniger mit sinnlichem als vielmehr mit geistigem Genuss in Verbindung gebracht. Wagashi essen ist wie ein Gedicht lesen, es versetzt einen für einige wenige Augenblicke in eine andere Welt. Nicht der Geschmack, sondern die mögliche Assoziation bereiten dem Esser Vergnügen.[3]

Regionale Symbole – Essen als Identität |
„Wir essen nicht, was uns schmeckt. Uns schmeckt, was wir essen", Klaus Dürrschmid, Department für Lebensmittelwissenschaften, Universität für Bodenkultur, Wien[4]

Essen verkörpert Zugehörigkeit zu Familie, Kulturkreis und nationaler Einheit. Was den Briten ihre Fish and Chips und den Amerikanern die Hamburger, sind den Japanern Sushis und den Spaniern die Paellas. Regionale Spezialitäten fungieren als kulinarische Wahrzeichen ihres Herkunftslandes. Wir identifizieren Staaten nicht nur mit Sehenswürdigkeiten wie Eiffelturm, Tower Bridge oder Akropolis, sondern auch mit ihren Nationalspeisen. Neben religiösen Symbolen liefern auch regionale Symbole Gestaltungvorlagen für Food Design.

Um den identitätsstiftenden Faktor des Essens zu verstärken, werden entsprechende Speisen kreiert oder in speziell geformte Gerichte und ihre Geschichte nachträglich Bedeutungen

symbolizing water. "Dew Drops on Green Leaves" imitates the gleam of single drops of water in the shape of tiny transparent jelly cubes, and "Thin Blades of Grass" shows fallen leaves under narrow blades of grass.

River Flowers and Kimono Fabrics |
Traditional tea confectionery used as offerings, lucky charms or talismans include turtles, cranes, thais or seabreams, wild boars, hares, dogs and carps. Some motives from the Edo era (1603 – 1868) are still in use today, for example a mold depicting a sunrise that dates from 1796. Tea confectionery is inspired by the many colored layers of silk kimonos, patterns found on ancient wooden temples, traditional pottery, the design of lacquer boxes or courtiers' attires. A miniature copy of the hat worn by court musicians can, for example, be bought as a wagashi. More recent designs do not necessarily originate from Japanese traditions. In 1926 a wealthy lady ordered confectionery shaped like golf balls from Toraya, a producer of traditional wagashi. The out-of-the-ordinary design became a fixture in the Toraya range, which continues to be extended regularly, e.g. when a cosmetics manufacturer orders wagashi in the shape or color of a new lipstick for a product presentation.

The taste of wagashi is more or less secondary, a feature it shares with the Christian host. There are around 3,000 different designs, some of them extremely intricate, but only a few, relatively similar flavors. The consumption of tea confectionery is not so much relished with the senses but is rather an intellectual enjoyment. Eating wagashi is like reading a poem, it takes you to another world for a fleeting moment. It is not the taste but potential associations which make it a pleasant experience for the eater.[3]

Regional Symbols – Food as Identity |
"We don't eat what we like. We like what we eat." Klaus Dürrschmid, Department of Food Sciences, University of Natural Resources and Applied Life Sciences Vienna[4]

Eating means belonging – to a family, a civilization, a nation. What fish and chips are to the Brits and hamburgers to the Americans, sushi is to the Japanese and paella to the Spaniards. Regional specialties are culinary landmarks of their countries of origin. We do not only identify countries by their main sights – the Eiffel Tower, Tower Bridge or the Acropolis – but also by their national dishes. Apart from religious symbols, regional symbols are also models for food design.

The identity-bestowing factor of food is enhanced by creating special dishes or by giving food a particular shape and subsequently interpreting it accordingly. Thus, the triangular shape of "Toblerone" is associated with the Swiss Alps

Mitte von oben nach unten | Japanisches Konfekt namens „Tatta no fuchi" und die dazugehörigen Holzmodel. Tatta ist ein Ort, der für sein verfärbtes Herbstlaub berühmt ist; „Kokage no mizu": traditionelle japanische Süßigkeit für die heiße Jahreszeit, die einen klaren Fluss im Schatten eines Baumes darstellt; „Wakabakage": japanisches Sommerkonfekt, das einen Goldfisch in einem Teich im Schatten überhängender Zweige zeigt.
Rechts | Kulinarisches Wahrzeichen: Die Toblerone wurde 1908 von Theodor Tobler in Bern erfunden und wird heute weltweit mit den Schweizer Bergen in Zusammenhang gebracht.

center from top to bottom | Japanese dry sweet named „Tatta no fuchi" and wooden moulds: Tatta is a place famous for its colored leaves in autumn. Traditional Japanese sweets for the summer season called „Kokage no mizu" evoking the Image of a limpid stream in the shade of a tree; "Wakabakage" Japanese summer confectionery showing a goldfish in a pond under the shade of overhanging branches
right | Culinary landmark: Theodor Tobler of Berne invented Toblerone in 1908. Today it is associated worldwide with the Swiss Alps.

hineininterpretiert. So wird die dreieckige Form der „Toblerone" weltweit mit den Schweizer Bergen in Verbindung gebracht und agiert als kulinarisches Logo für die Alpen. Ob Theodor Tobler davon träumte, ein Symbol für sein Heimatland zu erschaffen, als er 1908 in Bern die „Toblerone" erfand, ist allerdings fraglich. Was Tobler tatsächlich zur außergewöhnlichen Form inspirierte, fällt in den Bereich der Spekulation: Eine Geschichte besagt, dass eine Pyramide aus rot und beige bekleideten Mädchen in einem Pariser Nachtclub den Schweizer Chocolatier zu seiner Kreation anregte. Denkbar ist auch, dass die dreieckige Form auf ein Symbol der Freimaurer zurückgeht, bei denen Tobler Mitglied war. Wie dem auch sei, heute wird der dreieckige Schokoladeriegel vornehmlich mit dem Matterhorn assoziiert und in rund 120 Länder weltweit exportiert, wobei die typische Dreiecksform erst seit 1994 auch gesetzlich geschützt ist.

Essen als Propaganda | Kulinarischer Patriotismus liefert ein erfolgsträchtiges Konzept für die Gestaltung und die Vermarktung von Esswaren. Der neapolitanische Pizzakoch Raffaele Esposito benannte 1889 eines seiner Rezepte nach Regina Margherita, der ersten Königin des geeinten Italiens. Die Farben des Gerichts – der weiße Teig, mit roten Tomaten und grünem Basilikum belegt – repräsentieren die italienische Flagge und appellieren damit an den Nationalstolz der Italiener.

Essen wird als Träger von Ideologien benutzt, ja sogar als politisches Propagandamittel eingesetzt. Gerichte werden mit den Namen von Herrschern, Helden und berühmten Persönlichkeiten versehen oder gezielt mit politischen Ereignissen und traditionellen Bräuchen in Zusammenhang gebracht. Die italienische Pastasorte Tripolini, an den Rändern strukturierte Bandnudeln, erinnert bis heute an die Eroberung Libyens und seiner Hauptstadt Tripolis im Jahr 1912. Das Filet Wellington setzt dem britischen Herzog, der Napoleon 1815 bei Waterloo besiegte, ein kulinarisches Denkmal und das Ragout Marengo erinnert an die Schlacht im piemontesischen Marengo, als Napoleon 1800 gegen Österreich zu Felde zog.

Wir definieren uns über unseren Speiseplan und grenzen uns mit ihm von anderen Gruppen ab. Angelernte Essrituale bedeuten Zusammenhalt – in der Familie, im sozialen Umfeld, in der eigenen Kultur. Gemeinschaften identifizieren sich über ihre Nahrungsgewohnheiten und stellen damit ihre Überlegenheit gegenüber anderen klar. Unser Essverhalten ist der Spiegel unserer kulturellen Standards.

Geschmack ist keine absolute Größe, sondern individuell, situativ und kulturell erlernt. Internationale Lebensmittelkonzerne müssen auf die unterschiedlichen Geschmacksvorstellungen von Europäern, Amerikanern oder Asiaten reagieren und arbeiten ganz bewusst mit den geschmacklichen Vorlieben der verschiedenen Kulturen. Procter & Gamble beispielsweise vertreibt derzeit weltweit rund 100 Sorten Pringles.

worldwide and even functions as a culinary logo of the mountains. However, it is not so sure that Theodor Tobler was really dreaming of a symbol for his home country when he invented "Toblerone" in Berne in 1908. We can only speculate about the true source of inspiration, though according to one story the idea for the unusual shape came to the the Swiss chocolate-maker's mind when he saw a pyramid formed by showgirls clad in red and tan in a Paris nightclub. The triangular shape might also have been a tribute to the Freemasons, as Tobler was a member of a lodge. Be that as it may, the triangular chocolate bar is primarily associated with the Matterhorn today. It is exported to about 120 countries around the world, and the characteristic shape is copyright-protected, albeit only since 1994.

Food as Propaganda | Culinary patriotism is a promising proposition for the design and marketing of food. In 1889 the pizza maker Raffaele Esposito from Naples named one of his creations after Regina Margherita, the first queen of united Italy. The colors – white dough with a topping of red tomatoes and green basil – represent the Italian flag, thus appealing to the national pride of the Italian population.

Food is used as a conveyor of ideologies, and even for political propaganda. Dishes are named after rulers, heroes and famous personalities or connected with political events and traditional practices in a targeted way. The Italian "tripolini" pasta, thick ribbons with structured edges, commemorates the conquest of Libya and her capital Tripoli in 1912. Beef Wellington recalls the British duke who defeated Napoleon in the Battle of Waterloo in 1815. Marengo stew was created in remembrance of the battle in Piemont in which Napoleon confronted the Austrians in 1800.

We define our identity through what we eat, and we deliberately dissociate ourselves from other groups by doing so. Our eating rituals have an aspect of coherence – within the family, in a social environment, in one's own culture. Communities express their identities via their eating habits and seek to show their superiority over others this way. Our eating behavior mirrors our cultural standards.

Taste is nothing absolute...it is individual, situational and culturally acquired. International food corporations should respond to the different ideas of tastes of Europeans, Americans and Asians. They deliberately cater to the predilections of various cultures. At present, Procter & Gamble distributes about 100 types of "Pringles" worldwide. The seasonings and intensity of aromas differ from country to country.

Essen bedeutet Zugehörigkeit zu Familie, Kulturkreis und nationaler Einheit. Was den Italienern die Pasta und den Amerikanern das Barbecue, ist für Japaner gekochter Reis und Thunfisch.

Food signifies allegiance to family, a culture and national entity. What pasta is for the Italians and Barbecue for Americans, cooked rice and tuna fish is for the Japanese.

Die Würzarten und die Stärke der Aromen unterscheiden sich dabei von Land zu Land: So sind „Sour Cream & Onion"-Pringles in den USA cremiger und saurer als in Europa, wo sie dafür stärker nach Zwiebel schmecken. In Japan steht man auf Sorten wie Consommé, Umami, Pilze, Wasabi oder Fisch, alles jedoch weniger geschmacksintensiv als in der westlichen Welt. Für China wiederum gibt es eine eigene Linie namens „natural ocean" mit Aromen wie Seegras, Shrimp oder Krabbe, die für unseren Gaumen vielleicht etwas eigenartig anmuten.[5]

Auch die Fülle von Fischstäbchen wird den kulinarischen Vorlieben in den einzelnen Ländern angepasst. Die fischmuffigen Alpenländer bevorzugen das nahezu geschmacklose Filet vom Polardorsch, während in Küstenregionen der vollmundigere Seehecht oder Seelachs in den Fischstäbchen landet. Auch an der Art der Panier scheiden sich die Geister, so dass sie je nach Region heller oder dunkler, grobkörniger oder glatter zubereitet wird. Geschmack ist eben eine Frage der kulturellen Zugehörigkeit, die das Design der Nahrungsmittel von Kultur zu Kultur variieren lässt.

Ekel und kulturelle Abgrenzung | Die kulturellen Unterschiede beim Essen betreffen aber nicht nur das, was wir essen, sondern vor allem auch, was wir nicht essen. Gerade beim Essen liegen Lust und Frust, Genuss und Ekel sehr nah beisammen. In der Regel ekeln wir uns vor Fäkalien, verfaulten Früchten und verrottetem Fleisch. Auch Bitteres oder Dunkles löst instinktive Ablehnung aus, denn es besteht die Gefahr, dass solche Nahrungsmittel giftig sind. Dennoch ist Ekel nur teilweise genetisch vorprogrammiert: Der Abscheu vor gewissen Speisen ist Ausdruck kultureller Standards und dementsprechend anerzogen. Wir lernen von Kindesbeinen an, uns vor bestimmten Dingen zu grausen und vor anderen nicht. Es gibt zum Beispiel keinen ernährungstechnischen Grund dafür, dass Europäer keine Heuschrecken essen. Wenn uns bei der Vorstellung, dass die Chinesen Hunde und Meerschweinchen verkochen, der kalte Schauer über den Rücken läuft, so reagieren die Japaner völlig entsetzt, wenn sie hören, dass man in Europa süße kleine Kaninchen verspeist.

"Sour Cream & Onion" Pringles are, for example, creamier and more sour in the USA than in Europe. In Europe, the onion flavor is in the foreground. In Japan people like flavors such as consommé, umami, mushrooms, wasabi or fish. All the seasonings are less intense than in the Western world. For China there is a separate product line called "Natural Ocean" with flavors such as seaweed, shrimp and crab which might be a bit strange for our palates..[5]

Fish fingers also differ according to the culinary preferences of various countries. People in the Alpine region are not particularly fond of fish so they like fish fingers made of nearly tasteless arctic cod. In coastal regions, the fish used more frequently in fish finger production is hake or coalfish which has a more full-bodied fish flavor. The breading differs, too – it can be lighter or darker, grainier or smoother, depending on the region. Taste is a matter of cultural belonging, thus the design of food varies from civilization to civilization.

Disgust and Cultural Dissociation | Cultural differences in the choice of food not only concern what we eat but also, or even primarily, what we do not eat. Pleasure and frustration, gusto and disgust are very close to each other. Normally, we find feces, rotting fruit and putrefying meat disgusting. Instinctively, we also reject bitter or dark stuff because there is a risk that it is poisonous. However, disgust is only in part genetically pre-programmed. Aversions against certain things are an expression of cultural standards and we are brought up to reject them. From the time when we are little, we are taught that some things are yucky while others are not. For example, there is no nutritional reason why Europeans do not eat locusts. The idea that the Chinese turn dog and guinea pig meat into meals for human consumption sends a chill down our spines. Conversely, the Japanese are taken aback when they hear about sweet little rabbits being eaten in Europe.

Essen wird zu einem symbolischen Akt, wenn Gruppen klar definieren, was sie sich inkorporieren und was eben nicht. Wenn Christen Schweinefleisch essen, Moslems jedoch nicht und Hindus überhaupt Vegetarier sind, so tragen diese religiös motivierten Ernährungstabus ebenso zur Sinnstiftung und zum kulturellen Selbstverständnis bei wie heute beispielsweise die unterschiedlichen Regionalküchen. Lebensmittel sind ein Teil unserer Identität, ihre Auswahl eine Frage des Lifestyle. Kritik am eigenen Nahrungsverhalten wird als Angriff auf die Persönlichkeit und damit als schwere Beleidigung gewertet. Die Essgewohnheiten anderer Gemeinschaften hingegen werden bestenfalls belächelt, meist verurteilt. Wir können Fremde nicht „riechen" und meinen damit Ausdünstungen infolge des andersartigen Essens. Auch Religionen benutzen Nahrungsgewohnheiten, um „Ungläubige" zu brandmarken.

Beim Essen sind wir nicht nur Gewohnheitstiere, sondern kompromisslos gegenüber Andersartigem. Gesellschaftlich angelernter Ekel wird benutzt, um „Fremde" abzuqualifizieren oder zu beleidigen. Ausdrücke wie „Kraut-", „Spaghettifresser" oder „Kümmeltürke" bezeichnen wenig schmeichelhaft die Angehörigen anderer Kulturen. In den USA nennt man die Franzosen abfällig „frog-eaters". Kultureller Ekel offenbart das rassistische Potenzial unseres Nahrungsverhaltens. Viele Asiaten bezeichnen die Europäer als Butterfresser und behaupten, die Europäer stänken nach Butter.

Eating becomes a symbolic act whenever groups clearly define what they do and do not embrace. When Christians eat pork while Muslims do not, and Hindus are vegetarians, these religiously motivated nutrition-related taboos contribute to cultural self-definition, just like various regional cuisines do today. Foodstuffs are part of our identity, and their selection is a matter of lifestyle. People consider criticism voiced about their eating habits as personal attacks and thus as serious insults. At the same time, however, they sneer at the eating habits of other communities or even condemn them. In German, we express absolute dislike by the idiomatic phrase "I cannot stand your smell", which has to do with the odor of perspiration due to food that is different from our own. Religions also stigmatize "infidels" for different eating habits.

When it comes to eating, we are creatures of habit and we are uncompromising vis-à-vis others who are different. Socially acquired disgust is used to debase or insult "aliens". Expression such as "krauts", "beaners" or "limeys" are rather unflattering references to people from other cultures. In the US, the French are snidely called "frog-eaters". Culturally motivated disgust reveals the racist potential of our eating habits. Many Asians call Europeans "butter eaters" and claim that they reek of butter.

Genuss oder Ekel? Geschmack ist kulturell erlernt. Wir lernen von Kindesbeinen an, uns vor bestimmten Dingen zu ekeln und vor anderen eben nicht.

Relish or disgust? Taste is culturally acquired. From childhood on we learn to feel disgust for certain things but not for others.

Novel Food
Die Europäische Union hat für das alte Phänomen „Was der Bauer nicht kennt, frisst er nicht" einen neuen Terminus gefunden und eine Verordnung dazu erlassen. Unter so genanntem Novel Food versteht man Lebensmittel, die – aus europäischer Sicht – von ihrer Art her neu oder kulturell fremd sind. Novel Food umfasst mithilfe von modernen Verfahren hergestellte Lebensmittel wie z.B. hochdruckpasteurisiertes Tomatenpüree, Lebensmittel aus neuartigen Ausgangsstoffen, etwa Fleischersatz aus Schimmelpilzkulturen, sowie mittels Gentechnik erzeugte Produkte, aber auch exotische Früchte. Die Novel-Food- Verordnung der EU unterscheidet u.a. Lebensmittel mit neuer oder gezielt modifizierter primärer Molekularstruktur, wie z.B. synthetische Fettersatzstoffe, Lebensmittel aus Mikroorganismen oder Algen und Lebensmittel, bei deren Herstellung ein nicht übliches Verfahren angewandt wurde, z.B. Hochdrucksterilisation oder Hochspannungskonservierung. Butter fällt wenig überraschend nicht darunter.[6]

Schlecker als Mutprobe
Der gezielte Einsatz von Fremdartigkeit und Ekel kann ein erfolgreiches Designkonzept liefern, das ganz bewusst mit Abscheu und Grauen spielt: Die britische Firma Edible vertreibt transparente Schlecker mit Schnapsgeschmack, in deren Innerem essbare Skorpione und Ameisen eingegossen sind. Die Tiere sind echt, denn der Esser soll sich beim Zubeißen ruhig ein bisschen ekeln. Nach einem ähnlichen Prinzip funktioniert auch der mexikanische Mescal-Schnaps, der mit einem Wurm genossen wird. Das beliebte Kindergartenspiel „Wer traut sich einen Regenwurm zu essen?" wird damit in die Erwachsenenwelt gebracht und das Abenteuer Ekel zum Designparameter.

Saisonale Symbole – shun und kisetsu-kan
Neben Regionalität und religiösen Symbolen ist auch der saisonale Bezug eine Inspirationsquelle für das Design von Nahrung. In der japanischen Küche werden Gerichte als Abbildung der jeweiligen Jahreszeiten gestaltet, wobei die Japaner dabei zwei Begriffe unterscheiden: „shun" und „kisetsu-kan". „Shun" bedeutet die kulinarische Freude, eine Speise zur genau richtigen Saison zu essen. Eine zum gerade rechten Zeitpunkt gepflückte Frucht zum Beispiel ist „shun". Das können mehrere Monate oder auch nur eine ganz bestimmte Woche des Jahres sein.

„Kisetsu" bedeutet im Gegensatz dazu ein saisonales Gefühl, das durch eine bestimmte Formgebung, eine Farbe oder ein Motiv ausgelöst wird. Gerichte, die kein ausgeprägtes „shun" haben, sondern ganzjährlich erhältlich sind, wie zum Beispiel Reis oder Reiscracker, werden der jeweiligen Saison mithilfe von speziellen Designs angepasst. So wird zum Beispiel gekochter Reis im Frühjahr in kleinen Metallformen zu einer Kirschblüte und im Herbst zu einem Ahornblatt gepresst, bevor er serviert wird. Reiscracker werden mit bunten Sujets bedruckt, welche von der Kirschblüte über verfärbte Herbstblätter bis zu anderen gerade aktuellen Zutaten wie Fisch-, Gemüse- oder Pilzsorten reichen. Motive für die saisonale Gestaltung von Gerichten liefern aber auch traditionelle Erzählungen, Gedichte und Legenden. So wird der Herbst unter anderem durch ein kleines Häschen symbolisiert, das einer japanischen Kindererzählung entstammt. Demnach schlägt im herbstlichen Vollmond ein kleiner Hase Reis (eine traditionelle japanische Art, Reis zuzubereiten).

In Japan wird jeder Monat mit gewissen Fischen, Meeresfrüchten, mit ganz bestimmten Obst- und Gemüsesorten in Zusammenhang gebracht. Nicht nur die Zutaten, auch die Zubereitung, die Gestaltung, das Geschirr und das Tischgespräch werden auf die jeweilige Jahreszeit abgestimmt. Im Frühjahr prägen Motive der Kirschblüte die Gestaltung der Speisen. Verschiedenste Zutaten werden in die Form ganzer Blüten oder einzelner, herzförmiger Blütenblätter gebracht.

In den feucht-heißen Sommern Japans herrschen transparente Speisen vor, die an kaltes Wasser oder Eis erinnern und dem Esser Kühle vermitteln sollen. Metaphorische Abkühlung versprechen zum Beispiel durchsichtige Süßigkeiten aus Agar-Agar, in denen farbige Blüten oder kleine Fische „schwimmen". Ein beliebtes Motiv ist Wasser auch in Form von Wasserstrudeln, Wellen und Muscheln. Im Herbst sind verfärbte Ginko- oder Ahornblätter, Kastanien, Kaki-Früchte, Tannennadeln, Zapfen, Nüsse, Matzutake-Pilze, Chrysanthemenblüten oder -blätter gestalterische Vorbilder für Gemüseschnitzer, Konfekterzeuger und Köche. Im Winter herrschen kleine Schneebälle oder Schneeflocken sowie weiße, kristalline Oberflächen vor, die Schnee ins Gedächtnis rufen. Wintersymbole sind auch die rote Kamelie, die japanische Yuzu-Zitrone und die weißen und violetten Pflaumenblüten.

Motiv, Form, Farbe
Der saisonale Bezug erfolgt in der japanischen Küche auf drei verschiedene Arten: durch ein appliziertes Motiv, durch eine bestimmte Form oder durch die Farbe. Bei Rosa, Pink oder pfirsichfarbenem Orange beispielsweise denken Japaner sofort an die Kirschblüte, das vermutlich wichtigste Naturschauspiel des Jahres. Der Sommer wird mit tiefem Blau und saftigem Grün assoziiert. Leuchtendes Gelb, Orange und Rot stehen für verfärbtes Laub und damit für den Herbst. Weiß erinnert an Schnee und Zwetschgenbäume, die in Japan gegen Ende des Winters blühen. Die vielleicht stärkste Farbsymbolik Japans allerdings ist saisonunabhängig und das ganze Jahr über allgegenwärtig: das so genannte Ko-Hako, die Kombination von Rot und Weiß, bedeutet „Glück". Viele Geschenkboxen und Bonbonnieren enthalten daher ausschließlich rote und weiße Süßigkeiten.[7]

Ekel als Designkonzept: spanischer Schlecker mit Schnapsgeschmack, in dessen Innerem essbare Ameisen eingegossen sind.

Disgust as a design concept: Spanish lollipops with brandy taste, with edible ants sealed inside.

Novel Food | The European Union found a new expression for the old phenomenon that people do not eat what they do not know, and even adopted legislation to regulate it. "Novel food" is defined as foodstuffs which – from a European perspective – are new or culturally alien. Novel food comprises foodstuffs produced by means of modern processes, such as tomato puree manufactured by high pressure pasteurization, or food made of novel ingredients, e.g. substitutes meat produced from fungus cultures, as well as genetically engineered products, but also exotic fruit. The EU Regulation concerning Novel Foods also applies to foods and food ingredients which present a new or genetically modified primary molecular structure, such as synthetic substitutes fats, foods which consist of micro-organisms or algae, and foods manufactured by means of unusual production processes, such as high pressure sterilization or high voltage preservation. It comes as no surprise that butter does not enter the picture.[6]

Lollipops for a Dare | The targeted use of alienness and disgust can be a successful design idea deliberately playing on revulsion and horror. The UK-based company Edible sells transparent lollipops with liquor taste and scorpions or ants inside. The animals are real and the eater is supposed to feel slight disgust when biting into them. Mexican mescal liquor, known for being enjoyed with a worm in it, works along similar lines. The kindergarten dare "Who's afraid of eating an earthworm?" is thus brought to the world of adults, and the adventure of disgust becomes a design parameter.

Seasonal Symbols – Shun and Kisetsu-kan | Alongside regional characteristics and religious symbols, reference to the seasons is another source of inspiration for food design. In Japanese cuisine, dishes reflect the seasons. In this context, the Japanese distinguish between "shun" and "kisetsu-kan". "Shun" is the culinary pleasure of eating something when the time is right. A fruit picked at exactly at the perfect moment when it is in season is considered "shun". This may refer to several months or just a specific week of the year.

By contrast, "kisetsu" stands for the season-related feeling triggered by a certain shape, color or motive. Dishes not marked "shun", which are available all year round, such as rice or rice crackers, are adapted to the respective season with the help of special design. In the spring, cooked rice is shaped like a cherry blossom by means of a metal mold, whereas it is served in the form of a maple leaf in the fall. Rice crackers are imprinted with colorful decorations, from cherry blossoms to colored fall leaves or other seasonal ingredients such as fish, vegetables or mushrooms. Traditional stories, poems and legends also provide themes for the seasonal design of food. The fall, for example, is symbolized by a little hare from a Japanese children's story, so in the fall, a little hare is beating rice (a traditional preparation method) in the full moon.

In Japan every month is associated with certain varieties of fish, seafood, fruit and vegetables. Not only the ingredients but also the preparation, decoration, tableware and conversation are adjusted to the season. In the spring, the cherry blossom is the predominant motive. Ingredients are shaped like entire blossoms or single heart-shaped petals.

In the humid and hot Japanese summers, transparent dishes are all the rage. They remind eaters of cold water or ice, and are supposed to cool "the mind". Welcome, albeit metaphorical chill comes from transparent sweets made of agar-agar enclosing colorful flowers or little fishes. Water in the shape of spiraling eddies, waves and sea shells is also popular. In the fall, vegetable carvers, confectionery makers and cooks are inspired by colored ginkgo and maple leaves, chestnuts, khaki fruits, fir needles, pine cones, nuts, matsutake mushrooms, chrysanthemum flowers or leaves. In the winter, little snowballs or snowflakes and white crystalline surfaces reminiscent of snow predominate. Winter symbols also include the red camellia, the Japanese yuzu lemon and white as well as purple plum blossoms.

Motive, Shape, Color | There are various ways to create references to the seasons in Japanese cooking: one is by adding a motive, and then there are specific shapes and colors. Delicate or hot pink as well as peachy orange will remind the Japanese at once of the cherry blossom, probably the most important natural spectacle of the year. Summer is associated with deep blue and rich green. Bright yellow, orange and red stand for colored leaves, and thus for fall. The color white recalls snow and plum trees which begin to bloom toward the end of winter in Japan. However, the strongest Japanese color symbolism is unrelated to any season and ubiquitous all year round: it is called "ko-hako", the combination of white and red, which stands for "happiness". This is why many gift and confectionery boxes exclusively contain red and white sweets.[7]

Innen & außen – der Mythos von Fruchtbarkeit und neuem Leben | Der uralte Mythos von Metamorphose, Geburt, Vergangenheit und Zukunft ist einer der häufigsten symbolischen Inhalte von Esswaren. Das Öffnen oder Zerbeißen gefüllter Lebensmittel assoziieren wir mit Vorgängen aus der Natur wie dem Schlüpfen eines Schmetterlings aus seinem Kokon oder eines Kükens aus dem Ei. Das Zusammenspiel von Hülle und Fülle steht für das Gegensatzpaar von Alt und Neu, für Fruchtbarkeit und die Entstehung von neuem Leben. Der umhüllte Kern oder die Fülle einer Speise erinnern an das heranwachsende Kind im Bauch der Mutter oder den Samen, der im Inneren einer Frucht reift. Glücks- und Zukunftsbotschaften finden sich daher nicht zufällig im Inneren von Speisen. Auf den britischen Inseln freut man sich alljährlich über Münzen im Christmas Pudding, welcher der Legende nach erstmals 1714 von König George I. verspeist worden sein soll, und in China über kleine Zettelchen in den so genannten Glückskeksen.

Wobei Glückskekse, wie wir sie heute kennen, nicht – wie bei uns fälschlicherweise oft angenommen – bei den chinesischen Neujahrsfeierlichkeiten verteilt werden, sondern eigentlich gar nicht aus China stammen. Der Überlieferung nach wurden zwar zur Zeit der mongolischen Besatzung Chinas im 13. und 14. Jahrhundert aufrührerische Botschaften in die noch heute beliebten Mondkuchen gefüllt. Man sagt, dass der Revolutionär Chu Yuan Chang anti-mongolische Propaganda in der süß-sauren Näscherei versteckte, um gegen den Feind mobil zu machen. Ausgerechnet ein Japaner, namentlich Makato Hagiwara, Betreiber eines Teehauses in San Francisco, soll die Idee viel später aufgegriffen und ab 1909 Glückskekse in der heute bekannten Form an seine Gäste verteilt haben. In Los Angeles existiert eine Konkurrenzgeschichte, welche die Erfindung der Glückskekse David Jung, dem Inhaber der Hongkong Noodle Company, zuschreibt. Belegt ist jedenfalls, dass Glückskekse in China erst in den 1990er Jahren auftauchten – und zwar als US-Import.

Spannung in Hülle und Fülle – das Überraschungsei | Gefüllte Gerichte bedienen den Mythos von Fruchtbarkeit und Neubeginn und erzeugen zudem Spannung, was in ihrem Inneren wohl zu Tage treten wird. In Europa galten Frauen, deren Kuchen beim Backen nicht aufgingen, lange Zeit als unfruchtbar und waren schwer unter die Haube zu bringen. Auch die Form der traditionellen japanischen Hochzeitstorte verleiht dem Wunsch nach Fruchtbarkeit Ausdruck: Die weiße Kuppel erinnert in ihrer Form an den Bauch einer Schwangeren und verbirgt in ihrem Inneren lauter kleine bunte Kugeln aus süßem Bohnenmus. Die „Babies" kommen beim rituellen „Kaiserschnitt" durch die Brautleute zum Vorschein.

Diese Seite | Die Gestaltung von Wagashi, traditionellen japanischen Süßigkeiten, soll die Einmaligkeit des Augenblicks einfangen.
Rechts | Herstellung einer Kastanie aus süßer Bohnenmasse bei Toraya in Tokio, wo man Wagashi in rund 3000 verschiedenen Designs produziert.
Links | Toraya-Katalog von 1707, zu sehen ist unter anderem „Yambuki", ein Quader mit einem gelben und einem braunen Streifen, der einen Weg oder Fluss darstellt, an dessen Seiten eine bestimmte gelbe Blume blüht. Ein Motiv, das den Japanern aus traditionellen Gedichten vertraut ist.
Nächste Seite | Japanisches Konfekt in jahreszeitlichen Motiven: Wasserstrudel, Ginkoblatt, Ahornblatt, Chrysanthemenblüte, Herbstpilz, Kastanie und anderes. Im Vordergrund Schleifen- und Knoten- Motive, die das „Zusammenbringen" symbolisieren und zu freudigen Anlässen verschenkt werden. Die Farbkombination Rot und Weiß bedeutet Glück. Die kleinen Zuckerkügelchen in der Mitte sind so genanntes „Kompeito", welches ursprünglich von den portugiesischen Einwanderern aus Nagasaki stammt. Die spezielle Form entsteht bei der Produktion: die Zuckermasse wird in einer Trommel mit kleinen Löchern dragiert. Die Farben ändern sich je nach Saison.

this page | The design of wagashi, traditional Japanese confectionery is to capture unique moments.
right | Production of a chestnut out of sweet bean mass at Toraya in Tokyo where traditional Japanese confectionery is produced in about 3,000 different designs.
left | Toraya catalog of 1707, including "yambuki", a block with a yellow and brown stripe which symbolizes a path or river along which a certain yellow flower grows. The Japanese are familiar with this motive from traditional poetry.
next page | Traditional Japanese confectionery in seasonal motives: vortex, ginkgo leaf, maple leaf, chrysanthemum blossom, autumn fungus, chestnut, etc. In the foreground, bow and knot motives to symbolize the act of "bringing together", given on joyful occasions. The color combination red and white symbolizes happiness. The small sugar balls in the center are so-called "kompeito" which originated from Portuguese immigrants in Nagasaki. The special form emerged in production: the sweet mass is sugar-coated in a drum with small holes. The colors vary depending on the season.

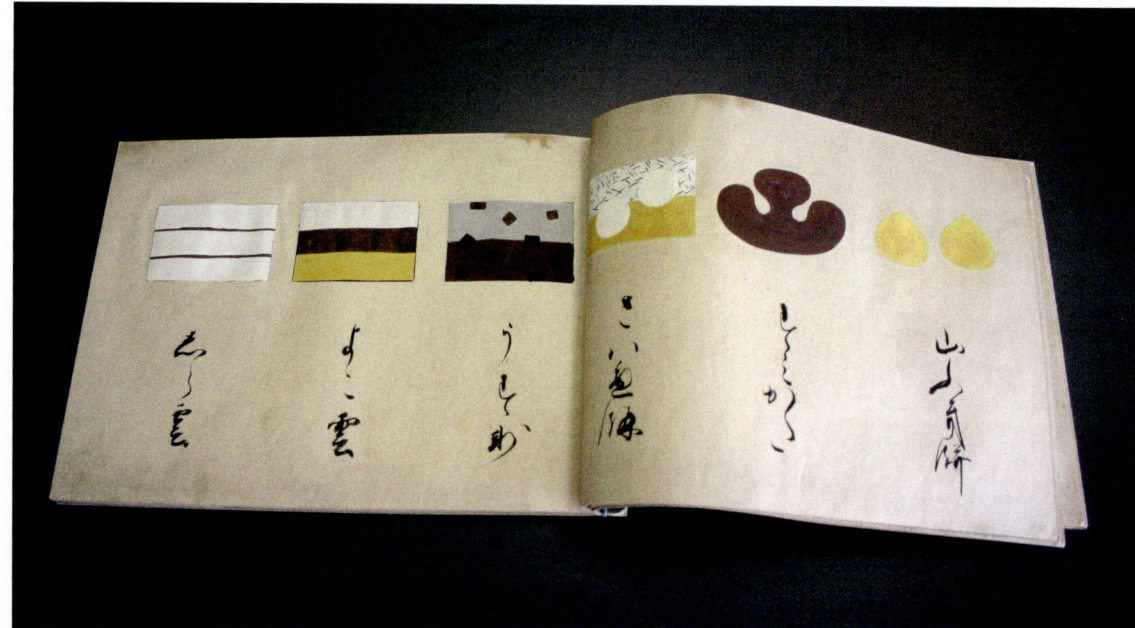

Inside & Outside – The Myth of Fertility and New Life | The age-old myth of metamorphosis, birth, past and future is one of the most common symbolic contents of food. The act of opening or biting into filled or stuffed food is associated with processes in nature, such as a butterfly emerging from a cocoon or a hatchling from a chicken egg. The interaction of shell and interior symbolizes the opposites of old and new, fertility and the beginning of new life. The enclosed center or stuffing of a dish reminds us of the child growing in the mother's belly, or the seed inside a fruit. It is no coincidence that messages for good luck and prophecies of the future are usually hidden in food. On the British Isles, people are delighted when, every year, they find coins in their Christmas pudding, which according to the legend was first eaten by King George I. in 1714. Little pieces of paper in so-called Chinese fortune cookies are equally popular.

In the West fortune cookies are usually associated with the Chinese New Year's celebrations, but in fact, they did not actually originate in China. Chinese hid rebellious messages in moon cakes – which are well-liked until this very day - when they were occupied by the Mongolians in the thirteenth and fourteenth centuries CE. The revolutionary Chu Yuan Chang is said to have hid anti-Mongolian propaganda in the sweet and sour snacks to mobilize the masses. Nevertheless, of all people, it was a Japanese tea-house owner named Makato Hagiwara, who unearthed the idea in San Francisco much later, starting to hand out fortune cookies in the shape we know today to his guests from 1909. In Los Angeles the story is told differently: The invention of the fortune cookie is ascribed to David Jung, the owner of the Hong Kong Noodle Company. One thing is sure: fortune cookies first came to China in the 1990's – imported from the USA.

Suspense Inside and Out – The Surprise Egg | Dishes with fillings perpetuate the myth of fertility and new beginnings, and they keep us in suspense about what we will discover inside. In Europe, women whose cakes failed to rise were long considered infertile and hard to marry off. The shape of the traditional Japanese wedding cake also expresses the wish for fertility: The white dome is reminiscent of a pregnant woman's belly. Inside, there are colored little balls made of sweetened bean paste. These "babies" are delivered by a ritual "cesarian section" when the newlyweds cut the cake.

Der Mythos von Hülle und Fülle: Zu Ostern werden Eier bunt gefärbt oder als gefüllte Schokoeier verschenkt. Der italienische Süßwarenhersteller Michele Ferrero griff diese Tradition auf und vermarktet seit 1974 ganzjährig Schokoladen-Eier mit Spielzeugfüllung. Obwohl die Kinderüberraschung eine alltägliche Nascherei ist und anstelle eines Kükens kleine Spielsachen zum Vorschein kommen, trägt sie immer noch die Symbolik von Fruchtbarkeit und neuem Leben in sich.

The myth of shell and filling. At Easter eggs are dyed in bright colors or given away as chocolate eggs with filling. Italian confectioner Michele Ferrero took up this tradition and has been marketing chocolate eggs with a toy surprise on the inside all year round since 1974. Even though the children's surprise is an ordinary snack and instead of a chick small toys emerge, it still bears the symbolism of fertility and new life.

The Surprise Egg is a textbook example of how food design can use an old fertility myth for modern food. Eggs are a traditional symbol of fertility, they are dyed in many colors or given away as chocolate eggs at Easter time. Especially in Italy, this is a very widespread custom, and huge, if not over-sized chocolate eggs filled with toys are popular gifts. Around Easter times, veritable contests over sizes, pricing and fillings take place among pastry shops and sweets producers. The Italian confectionery manufacturer Michele Ferrero followed the tradition but started marketing chocolate eggs containing small toys all year round in 1974. Similar to the yolk of hard-boiled eggs, the capsule containing the toy is yellow, the inside of the egg is white chocolate. Although Kinder Surprise Eggs are part of everyday life and although they contain toys instead of hatchlings, they are still associated with the myth of fertility and new life. In Austria alone, fourteen million units of the special chocolate eggs are sold every year.

Female Body Parts | The popular Italian chocolate known as "bacio" copies a different model: a human body part associated with fertility. When the confectionery was launched in 1922, the name was quite daring – bacio meaning kiss – but the chocolates also owed their success to their shape: the softly rounded form, topped by a whole hazelnut in chocolate coating recalls the female chest. Thus, the bacio chocolate symbolizes eroticism and pleasure on the one hand, and maternal love, caring and affection on the other.

Another symbol of fertility is the characteristic form of "pane ferrarese". The traditional white bread from Ferrara, also known as "coppia", which translates as "pair", looks like the abstraction of two female abdomens with strongly accentuated genitals.

1 | Hannsferdinand Döbler, Kochkünste und Tafelfreuden, Orbis, 2002
2 | Source: Christoph Wagner
3 | Interview with Keiko Nakayama, Toraya Gallery, Toraya Confectionery Ltd., Tokyo, October 2008
4 | Interview with Klaus Kamolz in "Forschen und Entdecken" 2009
5 | Interview with Artemio Castro, Director Snacks, Procter & Gamble, Cincinnati, USA, 3 April 2009
6 | Thomas Birus, (c) Bibliographisches Institut & F. A. Brockhaus AG, 2003
7 | Interview with the American anthropologist Elizabeth Andoh, Tokyo, October 2008

Als fooddesignerisches Paradebeispiel, wie ein alter Fruchtbarkeitsmythos für ein modernes Lebensmittel genutzt werden kann, gilt die Kinderüberraschung. Eier sind ein traditionelles Fruchtbarkeitssymbol, zu Ostern werden sie bunt gefärbt oder als gefüllte Schokoeier verschenkt. Dieser Brauch stößt speziell in Italien auf großes Echo, wo sich mit Spielwaren gefüllte Schokoladeeier – zumeist in deutlich überdimensionierter Form – als Geschenke großer Beliebtheit erfreuen. Rund um Ostern überbieten Konditoren und Industriebetriebe einander hinsichtlich Größe, Preisklasse und Füllung. Der italienische Süßwarenhersteller Michele Ferrero griff diese Tradition auf und vermarktet seit 1974 ganzjährig Schokoladeneier mit Spielzeugfüllung. In Anlehnung an ein hart gekochtes Hühnerei färbt Ferrero den Kunststoffdotter gelb und verwendet für die Innenseite des Schokoeis weiße Schokolade. Obwohl die Kinderüberraschung eine alltägliche Nascherei ist und anstelle eines Kükens kleine Spielsachen zum Vorschein kommen, trägt sie immer noch den Mythos von Fruchtbarkeit und neuem Leben in sich. Allein in Österreich werden pro Jahr über vierzehn Millionen Stück der gefüllten Schokoeier verkauft.

Weibliche Körperteile | Eine Vorlage ganz anderer Art bildet die populäre, italienische Praline „Bacio" ab: einen mit Fruchtbarkeit in Zusammenhang gebrachten menschlichen Körperteil. Nicht nur der bei ihrer Einführung im Jahr 1922 äußerst gewagte Name – Bacio bedeutet Küsschen – besiegelte den Erfolg der Baci, sondern auch ihre Form: Die sanften Rundungen des Konfekts, gekrönt von einer ganzen Haselnuss im Schokoüberzug, erinnern markant an eine weibliche Brust. Der Bacio symbolisiert damit einerseits Erotik und Genuss, zugleich aber auch Geborgenheit, Zuneigung und mütterliche Liebe.

Auch die markante Form des so genannten Pane Ferrarese kann als Fruchtbarkeitssymbol gedeutet werden. Das traditionelle Weißbrot aus Ferrara, das auch als „coppia" bezeichnet wird, was so viel wie Paar bedeutet, sieht aus wie zwei abstrahierte, weibliche Unterkörper mit stark akzentuiertem Geschlechtsteil.

1 | vgl.: Hannsferdinand Döbler, Kochkünste und Tafelfreuden, Orbis, 2002
2 | Wir danken Christoph Wagner für diese Information
3 | Interview mit Keiko Nakayama, Toraya Gallery, Toraya Confectionery LTD, Tokio, im Oktober 2008
4 | Interview von Klaus Kamolz in „Forschen und Entdecken" 2009
5 | Interview mit Artemio Castro, Director Snacks, Procter & Gamble, Cincinnati, USA, am 3.4.2009
6 | Dipl.-Ing. Thomas Birus, (c) Bibliographisches Institut & F. A. Brockhaus AG, 2003
7 | Interview mit der amerikanischen Anthropologin Elizabeth Andoh, Tokio, im Oktober 2008

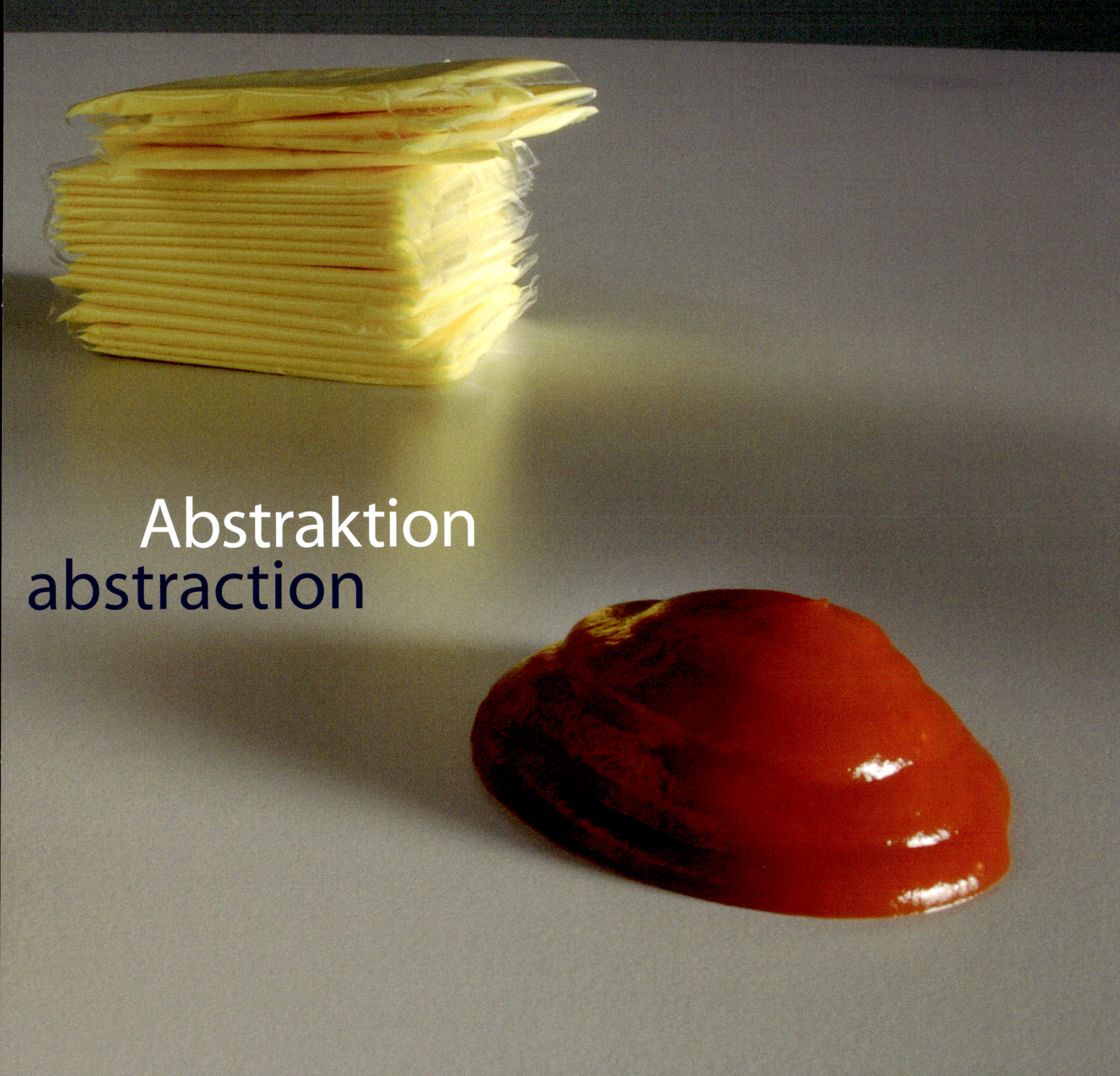

Entfremdung vom Ausgangsprodukt ist ein Erfolgsfaktor für Food Design: Tote Tiere werden zu zartrosa Würsten, Fische zu panierten Quadern, Kartoffel zu Chips abstrahiert.

Alienation of original product is a success factor for food design: dead animals become light-pink sausages, fish is turned into breaded squares, potatoes into chips.

Abstraktion und Entfremdung

„Seit tausenden von Jahren verarbeiten wir Lebensmittel, wir machen sie warm, setzen sie aufs Feuer oder verarbeiten sie biochemisch, das heißt, wir fermentieren sie, etwa bei Käse, Wein oder Brot. Alle diese Vorgänge führen zu einem Ergebnis, das mit dem Ausgangsprodukt, dem Naturprodukt eher wenig zu tun hat. Ich sehe das nicht negativ. Ganz im Gegenteil: Die Verarbeitung natürlicher Nahrung hat uns kulturell und sogar evolutionär sehr viel weitergebracht. Ernährungstechnische Errungenschaften wie die Entdeckung des Sauerteigs haben zur Entwicklung unserer Spezies wesentlich beigetragen." Werner Mlodzianowski

So unterschiedlich die Inhalte und Bedeutungen sind, die Speisen transportieren, so vielfältig sind auch die Hintergründe und Ursachen, aus denen heraus sie entstehen. Ist Symbolik ein Beweggrund zur Gestaltung von Nahrung, so ist das Gegenteil, die bewusste Unterbindung von Assoziationen – also Täuschung und Entfremdung zwischen Esser und Gegessenem – ein anderer. Die Motive dafür können religiös, politisch oder einfach nur modebedingt sein.

Im alten China imitierte man (bei Strafe) verbotene Fleischspeisen durch Gemüsegerichte. Auch im antiken Rom war der Verzehr mancher Ingredienzien zeitweilig untersagt. Die Küchenmeister ahmten Verbotenes und nicht Verfügbares mithilfe anderer Zutaten nach und wurden so zu wahren Meistern der Tarnung und Täuschung. Sie verfremdeten Ausgangsprodukte bis zur Unkenntlichkeit, bastelten Seeigel aus Quitten, Vögel aus Teig und verwandelten Kürbisse mit dem Schnitzmesser in Thunfische, Champignons oder Blutwürste. Schließlich wurde die Irreführung bei Tisch dermaßen geschätzt, dass man Köche, deren Kreationen trotz üppiger Dekorationen die Zutaten erahnen ließen, angeblich auspeitschte.

Bis heute verschleiern kulinarische Täuschungsmanöver Inhaltsstoffe und Zutaten. Nicht selten servieren Ärzte bittere Arzneien oder Impfstoffe auf einem Zuckerwürfel. Zahlreiche Bonbons entstanden ursprünglich als Tarnung einer medizinischen Funktion, um Kindern und genervten Eltern die mühsame Prozedur der Medikamentengabe zu ersparen. „Hustinetten" zum Beispiel helfen gegen Husten – und schmecken dennoch süß.

Meister der Abstraktion – das Fischstäbchen

Die Idee des Täuschens gipfelt in Produkten, die eine formale sowie geschmackliche Abstraktion als Mittel zum Zweck benutzen. Die Zubereitung führt zu einer völligen Entfremdung vom Grundprodukt und einer kompletten Veränderung von Aroma und Aussehen. Das Fischstäbchen etwa begeistert seit knapp 40 Jahren Kinder und Erwachsene ebenso wie erklärte Fischmuffel. Die panierten Quader sind ein konsumentenfreundlich vorportioniertes Fertiggericht, das einfach zuzubereiten ist und aufgrund seiner abstrakten Form selbst eingeschworene Fischhasser überzeugt.

Viele Menschen mögen keinen Fisch – aufgrund der Gräten und des Aussehens, aber auch wegen des intensiven Geschmacks. Wird der Fisch jedoch in ein lebensfernes Rechteck geschnitten und mit viel Panier umhüllt, wirkt er gar nicht mehr wie Fisch. Die streng geometrische Form entfremdet das Grundprodukt hervorragend und täuscht über die tatsächliche Zusammensetzung hinweg.

Abstraction and Alienation

"For thousands of years, we have processed foodstuffs, we have heated them, placed them on the hearth or subjected them to biochemical processing, that is to say, we have fermented them, as is the case in cheese, wine or bread. All these processes result in a product that has little to do with the natural material we started out from. To me, this is not a negative development. On the contrary: the processing of natural foodstuffs meant great steps forward in cultural and evolutionary terms. Nutritional feats like the discovery of sourdough have greatly contributed to the development of our species." Werner Mlodzianowski

Food conveys a wide variety of contents and meanings, and the background from which they emerge are equally diverse. Symbolism is one motivation to design food, and the deliberate prevention of associations – by deception and the alienation of the eaters from their food – is another. Motivation can be religious, political or simple a matter of fashion.

In ancient China meat dishes (consumption of which was punishable by law) were imitated by vegetables. In ancient Rome, too, some ingredients were prohibited at certain times. The cooks copied what was forbidden or unavailable by using other ingredients and became true masters of disguise. Basic products were changed beyond recognition with chefs turning quinces into sea urchins, dough into birds, and carved tuna, mushrooms or blood sausage out of pumpkins. Eventually, the misrepresentations on the table were appreciated so much that cooks whose creations still reminded eaters of the actual ingredients in spite of lush decorations were allegedly given a whipping.

To this day culinary razzle-dazzle can conceal ingredients and constituents. Often enough, the family doctor administers bitter medicines or vaccines on sugar cubes. Many candies were originally devised to camouflage a medical function so as to spare children and unnerved parents troublesome medication intake. "Hustinetten", the market leader among cough drops in Germany, bring relief from coughing – and they are sweet, too.

Masters of Abstraction – Fish Fingers

The idea of manipulating food has culminated in products that use formal abstraction and abstraction of taste as a means to an end. Preparation results in the complete alienation of a basic product and a complete change in aroma and appearance. For about forty years fish fingers have appealed to kids and adults as well as to people who claim they hate fish. The breaded squares are a consumer-friendly pre-portioned instant meal that is easy to prepare and whose abstract form even liked by those who adamantly dislike fish.

Many people are of a similar mind – not only because of the bones and the appearance but also because of the intense taste. But if the fish is cut in a rectangle and covered with a lot of breading, it no longer seems like fish at all. The rigid geometric shape changes the nature of the basic product, concealing its actual composition. Thanks to this design, the processed product – i.e., the fish finger – is ultimately more successful than the natural

Diese Gestaltung bewirkt, dass das verarbeitete Produkt – nämlich das Fischstäbchen – letztlich weitaus mehr Erfolg hat als das Naturprodukt Fisch. Kantig wie ein Bauklotz, überdeckt die knusprige Hülle zudem den meist ohnehin schwachen Fischgeschmack der strahlend weißen Fülle.

Fertige Fischstäbchen erinnern weder in Aussehen noch in Geschmack oder Textur in irgendeiner Weise an Fisch. Dennoch halten sie Millionen von Konsumenten auf der ganzen Welt für die ideale Form der Fischzubereitung. Die Natur wird bewusst unkenntlich gemacht, bevor sie auf den Teller gelangt. Der Durchschnittskonsument assoziiert Fischstäbchen nicht mit Fisch und will an den wahren Inhalt des Tiefkühlfertigproduktes auch nicht erinnert werden. Die Grundsubstanz Fisch wird daher weder auf der Packung noch in Werbespots gezeigt. In der Vorstellung der Konsumenten haben Fischstäbchen „keine Eltern", können weder schwimmen noch fressen und müssen nicht gefangen werden, sondern erblicken in der Fabrik das Licht der Welt. Nach einem ähnlichen Prinzip funktionieren auch die Chicken-Nuggets, deren Aussehen nicht im Entferntesten an die Anatomie eines Huhns erinnert.

Blutrot und süß | Wie viele Kinder verabscheuen Tomaten und lieben Ketchup! Auch die bekannteste Würzsauce der Welt nutzt die Abstraktion geschickt als Designkonzept. Inspiriert von einer asiatischen Gewürzbrühe, wurde die gesüßte Tomatensauce 1878 von Henry Heinz erstmals industriell hergestellt und gilt somit als eines der ersten Fertigprodukte der Lebensmittelgeschichte. Ketchup ist so stark vom Ausgangsprodukt entfremdet, dass selbst erklärte Tomatenhasser ihre Pommes genussvoll darin eintauchen. Weder Geschmack noch Konsistenz erinnern an Tomaten, so dass manche Kinder die unbeliebte Frucht als Grundstoff der roten Sauce nicht einmal zur Kenntnis nehmen.

Zerkleinern, Passieren und Faschieren verändert Zutaten bis zur Unkenntlichkeit. Produkte in ihre Bestandteile zu zerlegen und anschließend neu zusammenzufügen, ist das Grundprinzip jedes Zubereitungsvorganges und bildet die Basis vieler Nahrungsmittel. Die bewusste Entfremdung von Naturprodukten ist ein Erfolgsfaktor für Food Design.

Der Wille zur Abstraktion findet seinen Höhepunkt unter anderem im Anspruch, Lebensmittel völlig richtungslos zu gestalten. Gewachsene Produkte wie Fleisch oder Gemüse haben Fasern, Kerne, Skelette, Häute, Schalen, Knochen oder Stege, doch bewusst Gestaltetes, vom Leber- oder Fleischkäse über diverse Pasteten und Würste bis zum Schmelzkäse, präsentiert sich absolut gleichförmig. Das gleichmäßig sanfte Rosarot eines Wurstrades lässt Konsumenten vergessen, dass sie tierische Bestandteile verzehren. Auch das Spinatblatt büßt als tiefgekühlter Cremespinat sein natürliches Erscheinungsbild völlig ein. Kaum jemand würde wohl ein kräftig grünes Blatt als Ausgangssubstanz hinter dem einheitlich mattgrünen Quader vermuten – und so mancher Kunde weiß wahrscheinlich gar nicht, wie eine Spinatpflanze tatsächlich aussieht.

product. With square edges like a building brick, the crusty coating conceals the fish taste of the radiant white mass which is usually not very striking to begin with.

Ready-to-eat fish fingers do not resemble fish in any way – neither in appearance nor in taste or texture. Still, there are millions of consumers all over the world who think they are the ideal form of fish preparation. Nature is deliberately made unrecognizable before it is served on a plate. The average consumer does not associate fish fingers with fish and does not wish to be reminded of the actual content of the frozen-food product. Fish as a basic substance is not indicated – neither on the package nor in commercials. In the consumers' imagination fish fingers do not have "parents", they cannot swim nor can they eat and they do not have to be caught, they are born in the factory. Chicken nuggets function according to a similar principle – their appearance is not in the faintest related to a chicken's anatomy.

Blood-Red and Sweet | How many children detest tomatoes and love ketchup? Even the world's best-known sauce cleverly makes use of abstraction as a design concept. Inspired by a Chinese spicy broth it evolved into a sweetened tomato sauce. Produced industrially for the first time in 1878 by Henry Heinz, it is regarded as one of the first ready-to-eat products in the history of food. Ketchup has been changed so much that it hardly resembles the product on which it is based. Even declared tomato haters indulgently dip their French fries in ketchup. Neither the taste nor the consistency recalls tomatoes so that some children do not even notice the unpopular fruit as the basic ingredient of the red sauce.

Chopping, straining and mincing can change ingredients to the point of being unrecognizable. To deconstruct products into their components and then reassemble them is the basic principle of any act of preparation and forms the basis of many foods. The deliberate alienation of natural products is a factor that accounts for the success of food design.

The will to abstraction culminates in the attempt to design food so that it lacks all orientation. Time-tested products such as meat or vegetables have fibers, seeds, skeletons, skins, bones or ligaments. Food that is deliberately designed such as Bologna, various patés and sausages or processed cheese, has an absolutely uniform appearance. Thanks to the evenly spread soft pinkish red color of a slice of cold cuts, consumers forget that they are eating animal ingredients. Even a spinach leaf completely loses its natural appearance as frozen creamed spinach. Hardly anyone would imagine an intensive green leaf as being the original substance of the uniformly matte-green squares. And some customers probably do not even know what a spinach plant really looks like.

Der Gabelbissen, auch bekannt unter dem Spitznamen „Tote Oma", stammt ursprünglich aus Schweden, wo man Matjes oder Hering marinierte und mit Sauerrahm, Dill und Gewürzgurken servierte. Seit 1928 werden schwedische Gabelbissen in Deutschland essfertig verpackt angeboten.

The tidbit also known as "Tote Oma" (dead grandma) originally comes from Sweden where matie or herring was marinated and served with sour cream, dill and spiced pickles. Since 1928 Swedish snacks have been offered in Germany in ready-to-eat form.

In Japan liefern die Natur, traditionelle Erzählungen und Gedichte Motive zur Gestaltung von Nahrungsmitteln. So wird etwa der Herbst durch ein kleines Häschen symbolisiert, das jedoch nicht mit dem westlichen Osterhasen verwandt ist, sondern aus einer Kindergeschichte stammt, wonach im herbstlichen Vollmond ein kleiner Hase Reis schlägt (eine traditionelle Art, Reis zuzubereiten).

Motives for designing Japanese food come from nature and traditional stories and poems. Fall, for instance, is symbolized by a little hare which, however, is not related to the western Easter bunny but stems from a children's story according to which a little hare beats rice (a traditional way of preparing rice) under a full moon in the fall.

Schutz vor Kannibalismus | Das Phänomen der Abstraktion hat aber nicht nur mit Geschmacksvorlieben und Bequemlichkeit zu tun, sondern auch mit Psychologie. Mit Ausnahme von Wasser und Salz hat alle Nahrung, die wir verzehren, selbst einmal gelebt. Das heißt, sich ernähren ist immer eine Art von Kannibalismus. Wir müssen anderes Leben zerstören, um uns selbst am Leben zu erhalten. Um diese nicht gerade angenehme Erkenntnis zu verdrängen, arbeitet die Kochkunst seit Jahrtausenden daran, die natürlich gewachsenen Strukturen von Tieren und Pflanzen zu zerstören und unsere Speisen von ihren Grundprodukten zu abstrahieren. Die Entfremdung vom Ausgangsprodukt im Zuge der Zubereitung ist eine Form der kulturellen Verarbeitung, um mit dem grundlegenden Widerspruch des Tötens, um zu leben, fertig zu werden. An der Intensität dieser Abstraktion lässt sich auch der Zivilisationsgrad einer Gesellschaft ablesen. Je komplexer die Entfremdung von Naturprodukten im Zuge der kulinarischen Zubereitung geschieht, desto stärker hebt sich der Mensch von anderen Lebewesen ab – denn Tiere kochen nicht.

Design Food | Einen Extremgrad erfährt die Entfremdung zwischen Gerichten, Zutaten und Inhaltsstoffen heute bei so genanntem Design Food. Design Food bezeichnet Nahrungsmittel, bei deren Herstellung eine Standard-Grundmasse aus günstigen Rohstoffen mittels Aromen, Hydrokolloiden, Farb- und anderen Zusatzstoffen zum gewünschten Nahrungsmittel verarbeitet wird. Design Food setzt Lebensmittelzusatzstoffe – bekannt durch die E-Kennzeichnung auf der Packung – im Vergleich zu herkömmlichen Esswaren in sehr großem Umfang ein. Knabbereien wie Cornflakes, Kartoffelchips, Müsliriegel und extrudierte Produkte wie Erdnussflips zählen zu Design Food. Bekannt ist auch das Krebsfleischimitat Surimi, eine feste Masse aus zerkleinertem, gefärbtem und aromatisiertem Fisch.

Historisch gesehen, entwickelte sich Design Food aus Imitaten und Ersatzprodukten wie Margarine als Butterersatz oder Fisch-, Fleisch- und Wurstersatz aus texturiertem Sojaprotein. Weitere Entwicklungsschübe erfuhr Design Food im Zuge der Gesundheits- und Low-Calorie-Food-Welle ab den 1960er und 70er Jahren. In der Folge wurden Kohlenhydrat- und Fettersatzstoffe sowie künstliche Süßstoffe entwickelt (Fettersatzstoff Z-Trim, unverdaulicher Ölersatz Olestra). Weitere Beispiele für Design Food sind modifizierte Stärke, hydrolysierte Proteine, Glyzeride und Fettaustauschstoffe.[1]

Gegentrend: Reality Food | Ist Entfremdung und Abstraktion ein Erfolgsrezept bei der Gestaltung von Essen, so kann auch das schiere Gegenteil zum besonderen Kick führen. Was im Fernsehen Reality Show heißt, könnte man beim Essen als Reality Food bezeichnen: den Trend zu schonungsloser Wahrheit und Naturnähe. Im Hamburg der 1940er Jahre waren angebrütete Eier mit quasi halbfertigen Küken im Inneren eine Modeerscheinung, die den Genuss durch eine Extraportion Realität garantierte. In Japan werden Fische und Meeresfrüchte nicht nur roh, sondern – als Beweis ultimativer Frische – auch gerne lebend serviert und als Sashimi gegessen. Während man das Fleisch mit den Stäbchen von den Gräten löst, kann man das Tier noch atmen sehen. Eine besondere Spezialität im südjapanischen Fukuoka sind kleine Fische, die lebend und unzerkaut geschluckt werden. Das Vergnügen liefert jenes Kitzeln in der Kehle, welches das Tier beim Zappeln durch die Speiseröhre verursacht. Auch in Europa kennt man solche Realitätsorgien: In Frankreich werden Froschschenkel gerne so serviert, dass der ganze Frosch noch mit dranhängt. Ob derartige Gerichte noch geschmackvoll oder eher schon geschmacklos sind, darüber entscheiden letztlich kulturelles Verständnis und Gewohnheit.

1 | Quelle: Dipl.-Ing. Thomas Birus, (c) Bibliographisches Institut & F. A. Brockhaus AG, 2003

Protection from Cannibalism | The phenomenon of abstraction is not just related to predilections of taste and convenience but also with psychology. With the exception of water and salt all food that we eat has once had a life – so eating is always a form of cannibalism. We must destroy other life to stay alive. In order to suppress this unpleasant insight, culinary arts have been trying to destroy the naturally grown structures of animals and plants for thousands of years and to abstract our foods from their basic products. The alienation of the original product in the process of preparation is a cultural way of coping with the fundamental contradiction of killing to stay alive. The intensity of this abstraction reflects the level of civilization in a given society. The more complex the alienation from natural products in the course of culinary preparation, the stronger human beings stand out from other living creatures – for animals do not cook.

Design Food | So-called design food is subject to todays' extremest degree of alienation. Design food refers to food in whose manufacturing a standard basic mass of inexpensive raw materials is processed by means of aromas, hydrocolloids, color and other additives to produce the desired food. Design food uses a large amount of food additives – familiar to us from the E-label on a package – as opposed to ordinary foods. Nibbles such as cornflakes, potato chips, muesli bars and products created by means of extrusion, such as peanut puffs, are included among design food. Another familiar design food is the crab imitation called surimi, an artificially flavored and dyed mass made of chopped fish.

Historically, design food developed from knockoffs and substitutes such as margarine as an replacement for butter or fish, meat and sausage substitutes made of textured soy protein. Further developments of design food took place in the course of the health and low-calorie food wave from the 1960s and 1970s on. Subsequently, substances replacing carbohydrates and fats as well as artificial sweeteners were developed (Z-Trim, a fat-substituting substance; Olestra, an indigestible oil-substitute). Further examples of design food are modified starch, hydrolyzed proteins, glycerides and fat exchange substances.[1]

Counter-Movement | While alienation and abstraction is a success formula in the design of food, the exact opposite can also lead to a special kick. What is called reality show on television, could be called reality food in eating – the trend towards unrelenting truth and realism. In the 1940s eggs in the process of hatching, with quasi-half-finished chicks inside was all the rage in Hamburg – it guaranteed a thrill with an added dosage of reality. In Japan fish and seafood are not just served raw but also – as evidence of ultimate freshness –live and eaten as sashimi. While using chopsticks to remove the bones from the flesh, one can still see the animal breathing. A special delicacy in southern Japanese Fukuoka is small fish that is still alive and swallowed whole. The pleasure is derived from the tickling sensation in the throat, which is triggered by the animal's flicking its way through the esophagus. Even in Europe such reality orgies are known. In France people like to serve frog legs with the entire frog still attached. Whether such meals are still tasteful or already tasteless, this is something that depends on cultural understanding and habits.

1 | Source: Thomas Birus, © Bibliographisches Institut & F.A. Brockhaus AG, 2003

Zeitgeist
zeitgeist

Zeitgeist - Essen als Modeerscheinung

Was wir essen, signalisiert, wer wir sind. Durch die Auswahl der Speisen definieren wir unseren persönlichen Lebensstil und grenzen uns von anderen Gesellschaftsschichten ab. Essen ist repräsentativ. Gerichte und ihre Gestaltung signalisieren finanzielles Vermögen, Macht und Status.

Schon vor 1500 Jahren demonstrierte der Ostgotenkönig Theoderich seine Allgewalt, indem er mitten im Winter nach frischen Erdbeeren verlangte – und sie auch bekam. Sein Sekretär konstatierte, dass die Macht eines Herrschers darin bestehe, Lebensmittel über weite Strecken hinweg zu transportieren. Im Mittelalter versprach der Konsum von indischem Pfeffer, zur Zeit der Renaissance jener von Zucker Exklusivität und gesellschaftliche Anerkennung. Öffentlich zur Schau gestellter verschwenderischer Umgang mit diesen sündhaft teuren exotischen Zutaten symbolisierte Reichtum und Macht.

Essen und seine Gestaltung unterliegen dem Zeitgeist. Erfolg auf dem Teller hat eben nicht nur, was gut schmeckt, sondern auch was im Trend liegt. Im Mittelalter und in der Renaissance waren detailreich gestaltete Zuckerskulpturen essbare Statussymbole, wie es heute Hummer, Trüffeln oder ein schneller Mercedes sind. Neben der handwerklichen Kunstfertigkeit, die in ihrer Herstellung steckte, bürgte vor allem der teure Rohstoff Zucker für Exklusivität bei Tisch. Die Kristallisation von Zucker aus dem Saft des Zuckerrohrs wurde vermutlich im 4. Jahrhundert nach Christus in Ostindien entdeckt. Über Persien, wo man die Raffination weiterentwickelte, gelangte Zucker zur Zeit der Kreuzzüge ins Abendland. Der lange und gefährliche Transport von Vorderasien nach Europa machte Zucker zu einem Luxusartikel, den sich nur Monarchen und Adelige leisten konnten. Für die breite Masse war das süße Gold unerschwinglich und so kostbar, dass es bis Ende des 16. Jahrhunderts ausschließlich in Apotheken angeboten wurde. Noch um 1770 betrug der jährliche Zuckerverbrauch in Mitteleuropa etwa ein viertel Kilogramm pro Kopf. (Gegenwärtig werden alljährlich etwa dreißig Kilogramm pro Person gegessen.)

Demonstration von Macht und Geld

1747 entdeckte der Berliner Apotheker Sigmund Marggraf, wie man aus der Runkelrübe Zucker gewinnt, und begründete damit die Unabhängigkeit Europas von amerikanischen Zuckerimporten. Die industrielle Herstellung, welche um 1830 einsetzte, verbilligte Zucker innerhalb

Zeitgeist – Eating as a Fad

What we eat signals who we are. By our selection of food we define our own personal lifestyle and delineate ourselves from other social strata. Food is representative, and meals as well as the way they are designed signal wealth, power and status.

Even 1,500 years ago the Ostrogoth king Theoderich already displayed his omnipotence by demanding – and also getting - fresh strawberries in the middle of the winter. His secretary claimed that the power of a ruler was reflected in the ability to transport food over long distances. In the Middle Ages it was the consumption of Indian pepper, in the Renaissance the consumption of sugar that promised exclusivity and social recognition. Profligate use of these prohibitively expensive exotic ingredients in public was a symbol of wealth and power.

Food and its design are dictated by the zeitgeist. For something to be a success on the plate, it does not just have to taste good, it must also reflect a trend. In the Middle Ages and the Renaissance elaborately designed sugar sculptures were edible status symbols, comparable to lobsters, truffles or a fast Mercedes today. In addition to the artisanship that these sculptures displayed it was first and foremost the costly raw material sugar that guaranteed exclusivity at the table. The crystallization of sugar from sugar cane juice was probably discovered in East India the 4th century CE, Via Persia, where the process of refining was further developed, sugar reached the West in the period of the Crusades. The long and perilous transport from Southwest Asia to Europe made sugar a luxury good that only aristocrats and monarchs could afford. For the masses the sweet gold was simply unaffordable; it was so precious that until the end of the sixteenth century it was only sold in pharmacies. Around 1770 annual sugar consumption in Central Europe was still only about a quarter kilo per capita. (At present, the annual per capita consumption of sugar totals about thirty kilos or sixty pounds.)

Demonstration of Power and Money

In 1747 the Berlin pharmacist Sigmund Marggraf discovered how sugar could be derived from beets, thereby establishing Europe's independence from American sugar imports. The industrial production of sugar, which set in around 1830, made the price of sugar drop so dramatically within a short time that it entered all kitchens in all

kurzer Zeit so deutlich, dass er Einzug in die Küchen aller Gesellschaftsschichten hielt. Es ist kein Zufall, dass auch die Entwicklung der klassischen, mitteleuropäischen Mehlspeisküche gerade in diesen Zeitraum fällt. Der nunmehr erschwingliche Rübenzucker erlaubte den verschwenderischen Umgang mit Zucker für die Zubereitung von Torten, Kuchen, Golatschen und Krapfen. Heute ist Zucker zur Massenware geworden, sein Genuss weder zeitgeistig noch trendy.[1]

Repräsentation und Angeberei sind der Ursprung vieler kulinarischer Trends. Der Verzehr von Nahrung, deren Zutaten teuer, schwer erhältlich oder exotisch sind, verspricht Aufmerksamkeit, Anerkennung und Differenzierung. Auch die Schokolade wurde ursprünglich als kulinarische Prahlerei gegessen. Der Kakaobaum ist rund um den Äquator beheimatet, seine Früchte waren bis zur Eroberung Amerikas unbekannt. Die Azteken und Mayas schätzten die belebende Wirkung von Kakaobohnen und genossen Schokolade ausschließlich als ungesüßtes Getränk, das stark rituellen Charakter besaß und mit Gewürzen wie Pfeffer oder Muskatnuss verfeinert wurde. 1492 kam Kakao erstmals nach Spanien, wo er wegen der strengen Geheimhaltung der Rezeptur schnell Furore machte. Als vermeintlicher Förderer der Wollust wurde die Schokolade auch immer wieder Gegenstand religiös motivierter Verbote und Verhetzungen, was ihre Beliebtheit natürlich nur noch weiter steigerte. Im 17. Jahrhundert fand sie dann von Italien ausgehend in ganz Europa Verbreitung und blieb noch bis ins 20. Jahrhundert ein Luxusprodukt. Um Schokolade für die breite Masse erschwinglicher zu machen, verringerte Milton Snavely Hershey 1902 den Kakaoanteil bei der Herstellung und ersetzte ihn durch Milch, die im agrarisch geprägten Pennsylvania günstig zu haben war. Der bis heute geheime „Hershey Process" legte den Grundstein für eine sehr günstige Milchschokolade, die bis heute einen fixen Bestandteil der Standardverpflegung der US-Armee bildet.[2]

Vergoldetes Fleisch | Auch der Vorläufer des Wiener Schnitzels zählt zu jenen Gerichten, die ursprünglich repräsentative Zwecke erfüllten. Um einen gehobenen Lebensstil zu demonstrieren, belegten die am Gewürzhandel reich gewordenen Venezianer zur Zeit der Renaissance Speisen mit Blattgold. Im 16. Jahrhundert breitete sich die Sitte, Bonbons zu vergolden, in ganz Oberitalien aus. Wer es sich leisten konnte, ließ nicht nur kleine Pralinen,

social strata. It is no coincidence that even the development of the classic Central European pastry confections took place in this period. Beet sugar, meanwhile affordable for everyone, made it possible to make profligate use of sugar in the preparation cakes, tarts, and all manner of pastries. Today sugar is a mass product, and its consumption is neither a reflection of the zeitgeist nor any trend.[1]

Representation and bravado mark the beginning of many culinary trends. The consumption of food whose ingredients are expensive, hard to come by or exotic, promises attention, recognition and differentiation. Even chocolate was originally eaten to show off by culinary means. The home of the cocoa tree is found around the Equator, and its fruits were unknown until America was discovered. The Aztecs and Mayas appreciated the invigorating effect of cocoa beans and enjoyed liquid chocolate only in the form of an unsweetened drink which had a marked ritual character and was refined with spices such as pepper or nutmeg. In 1492 cocoa first arrived in Spain where it soon made a splash since its recipe was kept a secret. As something that was claimed to heighten lust, chocolate was repeatedly the object of religiously motivated bans and brainwashing, which of course only served to further increase its popularity. In the seventeenth century it then spread from Italy all over Europe and remained a luxury product up into the twentieth century. To make chocolate more affordable for the masses, Milton Snavely Hershey reduced the percentage of chocolate during production in 1902, replacing it by milk, which was available inexpensively in agrarian Pennsylvania. The "Hershey process", a secret to this very day, laid the foundation for a very inexpensive milk chocolate, which to the present serves as an integral part of the US army's standard diet.[2]

Gilded Meat | Even the precursor of the Wiener Schnitzel (breaded veal cutlet) is one of the dishes that originally had a representative function. In order to demonstrate a more noble lifestyle, Venetians who had accumulated wealth from trading in spices in the Renaissance added gold leaf to their food. In the sixteenth century the custom of gilding candies spread throughout Upper Italy. Whoever could afford it had not only small chocolate truffles but also entire pieces of meat covered with a gilded layer.

Essen als Repräsentation: Um ihren Reichtum zu demonstrieren, belegten die Venezianer zur Zeit der Renaissance Speisen mit Blattgold. Als das Vergolden 1514 verboten wurde, griff man auf ein älteres Zubereitungsverfahren zurück: Die goldgelbe Panier. Ende des 19. Jahrhunderts gelangte das Rezept dann angeblich durch Feldmarschall Radetzky über Mailand nach Wien, wo es leicht adaptiert als Wiener Schnitzel bekannt wurde.

Food as representation: In order to demonstrate their riches, the Venetians covered their foods with leaf gold during the Renaissance. When gilding was prohibited in 1514, they took recourse to an older mode of preparation: the yellow-golden breading. At the end of the nineteenth century, the recipe allegedly reached Vienna via Milan through Field Marshal Radetzky. In Vienna it then became known in a slightly adapted form as a Wiener Schnitzel.

Links | Ebenso wie Mode und Musik unterliegt auch Essen dem Zeitgeist. In den 1970ern waren Shrimpscocktails und Fondues en vogue, um das Jahr 2000 aßen Trendsetter Sushi, Tramezzini und Ingwer.
Rechts | Eine Kalte Platte, wie man sie um 1930 servierte; Flugzeugessen im Spiegel der Zeit: Bratwürste um 1960, Räucherlachs in den 1970ern und Salat in den 1990ern

left | Similar to fashion and music, food is also subject to the zeitgeist. In the 1970's shrimp cocktails and fondues were the rage, and in 2000 trendsetters ate sushi, tramezzini and ginger.
right | A platter of cold cuts as they were served around 1930; Airplane food in the course of time: sausages around 1960, smoked salmon in the 1970's and salad in the 1990's

sondern gleich ganze Fleischstücke mit einer goldenen Hülle versehen. Als man den Ausschweifungen 1514 einen Riegel vorschieben wollte und das Vergolden von Speisen gesetzlich untersagte, besannen sich die italienischen Köche auf ein älteres, alternatives Zubereitungsverfahren: die goldgelbe Panier.

In Weißbrotbrösel gebackenes Fleisch lässt sich bereits im 12. Jahrhundert bei der jüdischen Bevölkerung Konstantinopels und auch bei den Mauren nachweisen, welche die Rezeptur über Andalusien nach Spanien brachten. Der Überlieferung nach gelangte das Rezept dann unter Feldmarschall Joseph Graf Radetzky von Radetz Ende des 19. Jahrhunderts über Mailand nach Wien, wo es leicht adaptiert als Wiener Schnitzel bekannt wurde.[3]

Food Design und Politik |
Auch politische Ereignisse und wirtschaftliche oder militärische Vormachtstellungen beeinflussen die Ausbreitung kulinarischer Vorlieben und dadurch das Aussehen des Essens. So brachte die führende Rolle Englands und des „Commonwealth of Nations" die weltweite Popularität von Tee mit sich. Nach dem Zweiten Weltkrieg wurden amerikanische Nahrungsmittel in Europa trendy. In den 1940er und 1950er Jahren verbreiteten sich Kaugummis und Schokoladen durch die US-Soldaten als begehrte Souvenirs aus den USA und trugen zur Popularität des american way of life bei. Coca-Cola, Popcorn, Cornflakes, Hamburger und anderes Fastfood verdanken ihren weltweiten Erfolg nicht zuletzt dem Umstand, dass sie allesamt Erfindungen einer Supermacht sind und zudem von einer der größten Filmindustrien der Welt permanent beworben werden. Mittlerweile läuft der so genannte mediterrane Lebensstil dem amerikanischen in puncto Essen allerdings den Rang ab. Italien wurde zur globalen Stilikone erhoben, Pasta, Sughi, Olivenöl und getrocknete Tomaten wurden zu gesunden und dennoch genussreichen Delikatessen erklärt. Es ist zum Beispiel beachtlich, dass ein durchschnittliches Brot wie das Ciabatta, das eigentlich gar kein traditionelles italienisches Brot ist, sondern 1982 vom ehemaligen, italienischen Rennfahrer Arnaldo Cavallari erfunden wurde weltweit so erfolgreich ist. Nicht zuletzt durch den englischen TV-Koch Jamie Oliver signalisiert das blasse, grobporige Brötchen aus Italien dem Esser, am Puls der Zeit zu sein.

Nahrung als Zeitgeisterscheinung |
Nahrung und Geschmack unterliegen Modeströmungen. Lebensmittel, Gerichte und Getränke definieren den Zeitgeist genauso wie Kleidung, Autos oder Musik. Die unterschiedlichen Speisen und Produkte der Siebziger-, Achtziger- oder Neunzigerjahre haben ihre Zeit ebenso mitbestimmt wie der Minirock oder die Rolling Stones. Um am Puls der Zeit zu bleiben, durchlaufen manche Esswaren ebenso modische Veränderungen wie Verpackungen oder Firmenlogos. Ähnlich wie die Mickymaus oder andere Comic-Helden, die ihr Äußeres regelmäßig dem jeweiligen Geschmack der Zeit anpassen, modernisierte zum Beispiel auch der Gummibär sein Erscheinungsbild über die Jahrzehnte hinweg mehrmals. Als Hans Riegel 1922 in Bonn begann, kleine Bären aus Fruchtgummi zu gießen, waren diese sowohl deutlich größer als auch schwerer und gingen einzeln über den Ladentisch. Inspiriert vom damals noch geläufigen Tanzbären, schmückte ein zotteliges Fell den

When an attempt was made to put a stop to these excesses in 1514 and gilding food became illegal, the Italian cooks came up with an alternative way: the golden brown breading.

Meat fried in white bread crumbs was already documented in the twelfth century in the Jewish population of Constantinople as well as among the moors who brought the recipe to all of Spain via Andalusia. It has been related that the recipe came to Vienna via Milan thanks to Field Marshal Joseph Count Radetzky of Radetz at the end of the nineteenth century. In Vienna a slightly adapted version of the "Costoletta alla Milanese" then became known as the "Wiener Schnitzel".[3]

Food Design and Politics |
Even political events and economic or military hegemony influenced the growth of culinary predilections and thus also the appearance of food. As a result of the leading role of England and the "Commonwealth of Nations" tea become popular worldwide. In the 1940's and 1950's chewing gum and chocolates spread through the presence of US soldiers as sought-after souvenirs from the USA and contributed to the popularity of the American way of life. Coca-Cola, popcorn, cornflakes, hamburgers and other fast foods also became such a success worldwide thanks to the fact that they were all invented by a superpower and one of the world's largest film industries was constantly advertising them. In the meantime the so-called Mediterranean lifestyle has overtaken the American way of life when it comes to food. Italy has become a global style icon. Pasta, sughi, olive oil and dried tomatoes have been declared healthy yet tasty delicatessen products. It is, for instance, amazing that ordinary bread such as the ciabatta, which is not even traditional Italian bread but was invented by the former Italian rally driver Arnaldo Cavallari in 1982, became so successful worldwide. Thanks to the English TV cook Jamie Oliver the viewer knows that it is trendy to eat the pale, large-pored bread from Italy.

Food as a Manifestation of the Zeitgeist |
Diet and taste are subject to fads. Food products, dishes and drinks define the zeitgeist just as much as clothing, cars or music do. The various foods and products of the 1970's, 1980's and 1990's have shaped their time just as much as the miniskirt or the Rolling Stones did. To be in, many foods also undergo fashion-related changes just like packaging or company logos. Like Mickey Mouse or other comic-book heroes, the gummi bear, for instance, modernized its look several times over the decades. When, in 1922, Hans Riegel from Bonn, Germany, began creating small bears consisting of poured fruit gum they were much larger and heavier and were sold individually. Inspired by the dancing bears that were still common at that time, the body of the original gummi bears was considerably slenderer than today and covered by scraggly fur. Later the teddy bear, named after

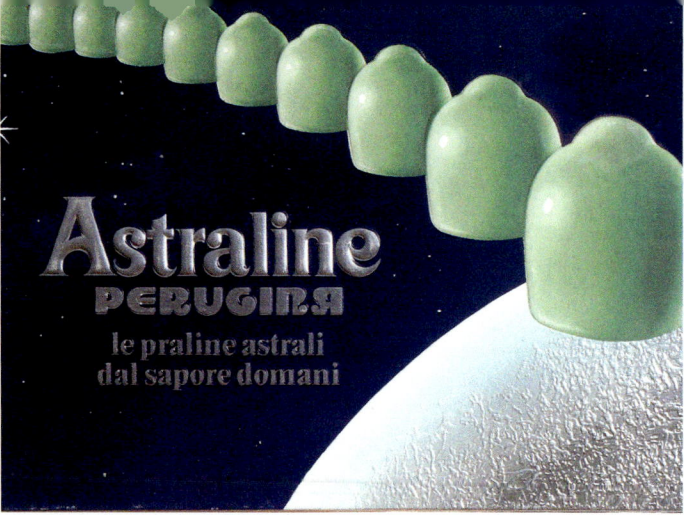

Esswaren unterliegen genauso Modetrends wie Frisuren, Kleidung oder Autos.
Oben | Nach der geglückten Mondlandung 1969 ergriff der Weltraumlook auch das Lebensmitteldesign: Baci als mintgrüne Astraline; vorverpacktes Supermarktgemüse
Rechts | Dreieckssandwich aus Marshmallow
Unten | Auslage der Hofkonditorei Demel in Wien, Flugzeugmenüs heute: Sushi und Schokoladenfondue bei Lufthansa

Food is just as much subject to fashion trends as hairstyles, clothing or cars.
top | After the successful moon landing in 1969 the space look also spread to food design. Baci as mint-green Astraline; pre-packaged supermarket vegetables
right | triangular sandwich made out of marshmallow
bottom | shop window of the court confectioner Demel in Vienna, airplane food today: sushi and chocolate fondue at Lufthansa

wesentlich schlankeren Körper des Urgummibären. Später inspirierte der nach Theodore Roosevelt benannte Teddybär das Design von Packung und Inhalt. Der heutige, stark abstrahierte und etwas klobige Bär mit den ausgestellten Beinen ist schon mehrere Jahrzehnte alt und entspricht mittlerweile auch nicht mehr dem aktuellen ästhetischen Empfinden, passt aber gut in die Retrowelle. Auch die Farbgestaltung der Gummibären veränderte sich im Laufe der Jahre. Im Zuge der Gesundheitswelle wurde in Europa auf pflanzliche und somit mattere Farbstoffe umgestellt.

Knallfarben versus Pastell, Geometrie versus Natur |

Der Zeitgeist prägt die gestalterischen Anforderungen an unser Essen. Dabei schlagen sich Moden nicht nur in Zutaten und Zubereitungsarten, sondern auch in formalen und farblichen Vorlieben nieder. Je nach Dekade regieren gerade Linien und rechte Winkel, verspielte Schnörkel und üppige Dekorationen, knallige oder pastellige Farbtöne. Der Zeitgeist ist ein Gestaltungsparameter von Möbeln und Kleidung, aber auch Nahrungsmittel werden regelmäßig adaptiert, um nicht aus der Mode zu kommen. Der feste Glaube an Fortschritt und Industrie in den 1950er und 1960er Jahren drückte sich in klaren Formen und starken Farben aus: Geometrische Muster und kräftige Primärfarben repräsentierten eine rationale, moderne Gesellschaft. Die Produkte waren einfach, praktisch und preisgünstig. Exakt geschnittene, kreisrunde, quietschgelbe Ananasscheiben kamen aus der Konserve, achteckige Eislutscher in den Kontrastfarben Orange und Grün aus der Gefriertruhe. Tiefkühlgemüse wurde zu kleinen, bunten Einheitswürfeln in Orange, Grün und Weiß geschnitten, so dass man gar nicht mehr erkannte, ob man eigentlich Karotten, Fisolen oder Rüben aß. Geometrisch gestaltete Nahrung stand für die heile, fortschrittsgläubige Welt des Wirtschaftswunders und für die Bezwingung der Natur, die in der geglückten Mondlandung ihren Höhepunkt fand. Mit der bemannten Raumfahrt setzte in den 1970er Jahren eine Weltall-Euphorie ein, die sich in diversen Filmen, in der Mode, aber auch in der Gestaltung von Esswaren niederschlug. Die italienische Firma Perugina beispielsweise brachte ihre bekannte Bacio-Praline im Raumfahrtslook mit mintgrüner Zuckerglasur unter dem Namen Astraline auf den Markt.

Künstlichkeit versus Natürlichkeit |

Mit der Postmoderne kam die Nouvelle Cuisine auf, die Rückbesinnung auf Eigengeschmack und Frische der Zutaten setzte langsam ein. Gemüse wurde kurz gegart, als „Julienne" gestylt und dafür in elegante, längliche Stifte geschnitten. Im Gegenzug zur puristischen, technologieverliebten Küche der Nachkriegsjahrzehnte begann etwa zeitgleich mit der Hippie-Bewegung die Hinterfragung automatisierter Herstellungsprozesse, die Wiederentdeckung der Natur und in weiterer Folge die Biowelle. Zweifel an der gesundheitlichen Unbedenklichkeit industriell gefertigter Produkte begannen die Öffentlichkeit zu beschäftigen. Die Angst vor Schadstoffen, künstlichen Aromen und Gentechnik steigerte die Sehnsucht nach möglichst „natürlicher" Nahrung. Kein Zufall also, dass heute in der Werbung Großmütter in Holzbottichen rühren und Köche Karotten und anderes Wurzelwerk im Ganzen inklusive grüner Blattansatz auf dem Teller drapieren. Der „Zurück zur Natur"-Trend findet allerdings mehr im Design und in den Köpfen der Konsumenten als tatsächlich in der Realität statt. Auch wenn das

Theodore Roosevelt, inspired the design. Today's bear, a radical abstraction of a bear with turned-up feet looking a bit clumsy, has meanwhile been around for several decades. It no longer reflects today's esthetic understanding but it fits nicely with the present retro-wave. Even the color design of the gummi bears changed over the years. As a result of the health wave in Europe there was a shift to plant-based dyes which are rather matte.

Garish Colors Instead of Pastels – Geometry vs. Nature |

The demands made on food design are influenced by the zeitgeist. Fashions are not just reflected in ingredients and modes of preparation but also in preferences regarding shape and color. Depending on the decade, straight lines and right angles, playful curlicues and lavish decorations, garish colors or more pastel shades prevail. Zeitgeist is not just a design parameter of furniture and clothing but foods, too, are regularly adapted in response to fashions. The unfaltering belief in progress and industry in the 1950's and 1960's was reflected in clear forms and strong colors. Geometric patterns and strong primary colors stood for a rational modern society. The products were simple, practical and inexpensive. Precisely cut, circular, bright yellow pineapple slices came out of can. Octagonal popsicles in contrasting colors of orange and green came out of the freezer. Frozen vegetables were cut into standard cubes in orange, green and white so that it was difficult to tell whether one was eating carrots, green beans or beets. Geometrically styled food stood for a wholesome progressive world of the economic boom and for bringing nature under control, a development which culminated in the successful moon landing. With manned space travel, space euphoria emerged in the 1970's, manifestations of which could be found in various films, in fashion and in the design of food. The Italian company Perugina, for instance, launched its well-known Bacio chocolates in a space age look with mint-green sugar glazing under the name "Astraline".

Artificial vs. Natural |

Nouvelle Cuisine appeared with post-modernity. Only gradually did people develop awareness for the intrinsic taste and freshness of ingredients. Vegetables were cooked very briefly, styled as "julienne" and cut into elegant long sticks. As a reaction to the purist, technology-obsessed cooking of postwar decades, around the time when the hippie movement started, people also began questioning automatized production processes, rediscovering nature and then riding the eco wave. People became increasingly skeptical that industrially produced food had no negative effect on one's health. The fear of noxious substances, artificial aromas and genetic engineering resulted in a heightened desire for really "natural" food. It is thus no coincidence that today's commercials show grannies stirring foods in wooden barrels and that cooks drape plates with whole carrots and other root plants including the green leaves. The "back-to-nature" trend, however, is more manifest in design and in the minds of consumers than in reality. Even if consumers are becoming more aware of organic products and regional cuisines are more "in" than ever

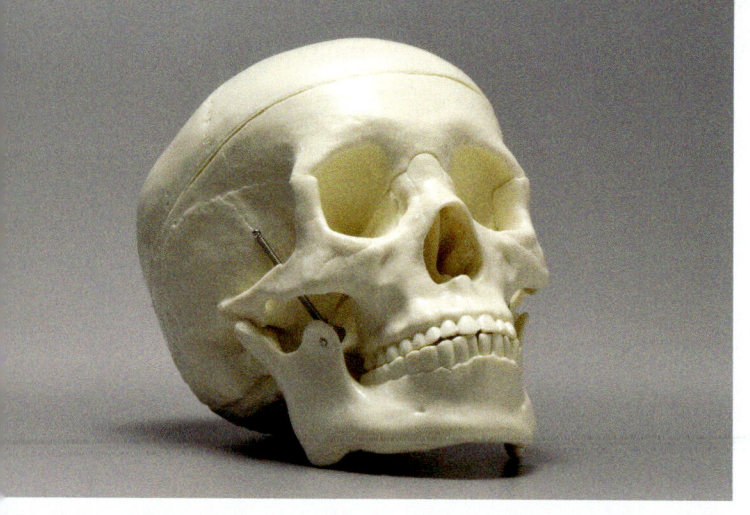

Die Nachahmung nicht essbarer Gegenstände ist eine häufige Methode, um Lebensmitteln Gestalt zu verleihen. Die Motive reichen von positiv belegten Alltagsgegenständen wie dem Teddybären, den wir als Gummibär verspeisen, über Sujets aus der Tier- und Pflanzenwelt (Marzipanrosen, Geleefrüchte, Knusperfische, etc.) bis hin zu essbarem „Spielzeug" und ironisch absurden Darstellungen wie Kaugummizigaretten, Schokoladehandys oder Pommes aus Marzipan.

Links von oben nach unten | Mexikanische Totenköpfe aus Zuckerguss, das Bacio erinnert markant an eine weibliche Brust
Zweite Spalte von oben nach unten | Muscheln aus Marzipan, Herzkeks mit Hello-Kitty-Kopf
Reihe oben | Gummibären, Schokoauto und Croissant als Nachahmung der Mondsichel
Reihe unten | Östereichische Aufschnittwurst mit Tannenbaummotiv und Lebkuchenherzen zum Selbstbemalen (gesehen bei FOA Schwarz in NYC)

The imitation of inedible objects is a method often used to give shape to food. Motives include positively connoted things from every-day life such as teddy bears, which we eat as gummi bears, plants and animals such as marzipan roses, jelly fruit, fish crackers etc., and edible "toys" or ironically absurd copies of real-life objects, including chewing-gum cigarettes, chocolate mobile phones and fries made of marzipan.

First column from top to bottom | Mexican skulls made of sugar icing. The Bacio has a striking resemblance with a female breast.
Second column from top to bottom | Shells made of marzipan, heart cookie with Hello Kitty head
Top row | Gummi bears, chocolate car and croissant imitating the moon crescent
Bottom row | Austrian cold cut with fir tree motive and gingerbread hearts that can be DIY painted (seen at FOA Schwarz in NYC)

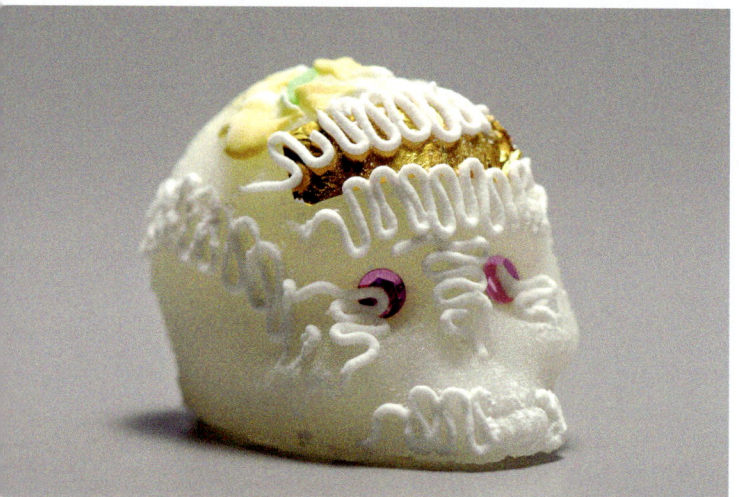

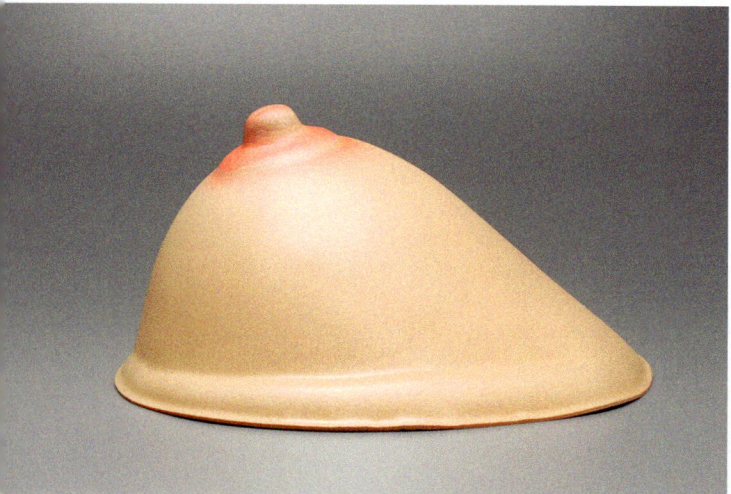

Bekenntnis zu Bioprodukten steigt und Regionalküchen „in" sind wie nie, will der Konsument auf die Annehmlichkeiten standardisierter Produkte wie Hygiene, lange Haltbarkeit oder Lagerfähigkeit nicht verzichten. Wenn Spargel heute in Deutschland als regionstypisches Gemüse beworben wird, dann zeigt die Tatsache, dass 90 Prozent dieses Spargels eine einzige Sorte sind, wie weit Naturromantik und Produktionsrealität voneinander entfernt sind.

Ernährung als Entertainment | Neben der Natürlichkeitseuphorie hat die Industrialisierung der Nahrungsproduktion aber noch andere Zeitgeistphänomene hervorgebracht, beispielsweise die Idee von Essen als Unterhaltung. Die automatisierte Massenproduktion bewirkte in der westlichen Welt, dass Nahrung erstmals in der Geschichte der Menschheit zu einem stetig verfügbaren Gut wurde. Unter Einfluss dieses Bewusstseins setzte nach Auskunft von Elani Raider vom Culinary Institute of America in den vergangenen Jahrzehnten in den USA ein gesellschaftlicher und medialer Prozess ein, der Ernährung primär als Entertainment sieht. Speziell das Angebot für Kinder arbeitet nach dem Konzept der essbaren Unterhaltung: Marshmallows erscheinen als Geister, Fruchtgummis als Vampire und Bonbons beinhalten „magische" Zutaten, die beim Verzehr die Zunge färben. Knallige Frühstückscerealien und lackiertes Popcorn bevölkern die Supermarktregale gemeinsam mit verzehrfähigen Comic-Figuren, die in direkter Zusammenarbeit mit der Trickfilmindustrie entstehen und zu Menüs in Anlehnung an Filmszenen oder Computerspiele zusammengestellt werden: „Monsters versus Aliens" lautet dann etwa die Aufschrift auf der Packung eines Kinderfertiggerichts.

Antike Spielereien | Das Konzept, zu Unterhaltungszwecken Motive aller Art aus Esswaren nachzubilden, ist bei der alltäglichen Nahrung für breite Bevölkerungsschichten relativ neu. Für besondere Anlässe, zu Repräsentationszwecken oder im Bereich von Süßwaren ist es jedoch eine altbekannte Strategie. In der Renaissance präsentierten wohlhabende Gastgeber mit aufwändig gestalteten Speisen, die zuallerletzt der

before, they do not want to part with the conveniences of standard products such as hygiene, long shelf life or easy storage. If asparagus is today advertised as a regional vegetable, then the fact that 90 per cent of asparagus is of one single type shows how distant the romantic view of nature and the reality of production lie from each other.

Food as Entertainment | In addition to the nature craze the industrialization of food production has also brought forth other zeitgeist phenomena, as for instance the idea of food as entertainment. One effect of automatized mass production in the western world was that for the first time in the history of humankind, food has become a constantly available good. According to Elana Raider from the Culinary Institute of America, this awareness has resulted in a social and media-driven process which sees food as being first and foremost entertainment. In particular the food products offered for children work according to the idea of edible communication. Marshmallows appear as ghosts, fruit gummies as vampires and candies contain "magical" ingredients coloring the tongue when they are eaten. Bright-colored breakfast cereals and coated popcorn inhabit the shelves of supermarkets together with edible comic figures that have emerged in direct collaboration with the cartoon industry and are compiled as TV dinners based on film scenes or computer games. "Monsters versus Aliens" is the label on the package of an instant meal made for children.

Ancient Gimmicks | The idea of using food to reconstruct all manners of motives as a form of entertainment is relatively new in everyday food geared to wide segments of the population. It is however an old strategy for special occasions, for representational purposes or in the sweets sector. In the Renaissance wealthy hosts showed off their extravagant life style with lavishly designed foods, which were least conducive to the intake of food. Food styled to reflect the occasion was offered as entertainment for guests, as were silver trees adorned

Kalorienzufuhr dienten, ihren ausschweifenden Lebensstil. Esswaren in Formen, die den Anlass der Feier thematisierten, unterhielten die Gäste ebenso wie Bäumchen aus Silber, an denen aus Teig gebackene Blüten und Früchte hingen. Die Venezianer waren bekannt für die Kunst, pompöse Zuckerskulpturen anzufertigen, die ganze Landschaften und allegorische Szenen darstellten. Figuren aus Buttercreme, manierierte Zuckerspinnereien und mehrstöckige Gelees zierten über die Jahrhunderte hinweg die Tafeln und Feste der Reichen – und sind teilweise noch heute in den Schaufenstern berühmter Konditoreien zu bewundern. Der traditionelle Wiener Hoflieferant Demel beispielsweise präsentiert in seinen Schaufenstern detailgetreue Nachbildungen von Personen des öffentlichen Lebens. Zur Freude der Vorübergehenden lachen Politiker, Fernsehstars und Spitzensportler in Zuckerguss und Schokolade aus den Schaufenstern.

Essen als Spielware findet sich heute als Gestaltungsschema vor allem bei Süßwaren und Naschereien. In Disneyland kann man die Mickymaus nicht nur sehen und angreifen, sondern in Form von Burgern, Brezeln und Keksen auch verspeisen. Auch die japanische Firma Sanrio hat das Potenzial essbarer Merchandising-Artikel schon vor einiger Zeit erkannt. Zu den beliebtesten Comic-Figuren des Fernen Ostens zählt „Hello Kitty", eine kleine, weiße Katze mit rosaroter Masche. Die Auswahl an Fanartikeln reicht von Hausschuhen über Schultaschen bis zu Esswaren. Anhänger verspeisen Kekse, Nudeln und Schlecker in Form des Katzengesichts von Hello-Kitty-Tellern mit Hello-Kitty-Besteck.

1 | vgl.: Christoph Wagner: Süßes Gold, Kultur und Sozialgeschichte des Wiener Zuckers; Brandstätter, Wien, 1999
2 | Wir danken Dr. Floros für diese Information.
3 | Wir danken Christoph Wagner für diese Information.

with baked blossoms and fruits made of dough. The Venetians were known for their art of producing pompous sugar sculptures that depicted entire landscapes and allegorical scenes. Figures made of butter cream, mannerist flights of fancy in sugar and multi-level jellies decorated the banquets and feasts of the rich for centuries – and they can, in part, still be admired today in the shop windows of famous pastry shops. Demel, the traditional Viennese purveyor to the court, for instance, presents in its show windows detailed, elaborate depictions of figures of public life. Passers-by are regaled with laughing politicians, TV stars and top athletes coated with sugar frosting and chocolate behind glass.

Food as a game can primarily be found as a design strategy in sweets and snacks today. In Disneyland, for instance, Mickey Mouse can not only be seen and touched but also eaten in the guise of burgers, pretzels and cookies. The Japanese Sanrio company already recognized the potential of edible merchandising products some time ago. Among the most popular comic figures of the Far East we find "Hello Kitty", a small, white cat with a pink-red bow. The selection of fan articles ranges from slippers to school satchels and food. Fans eat cookies, noodles and lollipops shaped like a cat's face from "Hello Kitty" plates using "Hello Kitty" cutlery.

1 | Christoph Wagner: Süßes Gold, Kultur und Sozialgeschichte des Wiener Zuckers; Brandstätter, Wien, 1999
2 | We thank Dr. Floros for this information.
3 | We thank Christoph Wagner for this piece of information.

Unten | Sushi-Bonbons, Käse und Sandwich aus Marzipan
Rechts | Japanische Konfektschachtel mit Herbstmotiven: Chrysanthemenblüten und –blätter, Matzutakepilze, Ginkonüsse, verfärbtes Ginko- und Ahornblatt.

bottom | Sushi candies, cheese and sandwich made of marzipan
right | Japanese confectionery box with fall motives: chrysanthemum blossoms and leaves, matsutake mushrooms, ginkgo nuts, dyed ginko and maple leaf.

325

Der Mythos-Faktor | „Wir gestalten unser Essen, um uns auszudrücken. Wir essen Werte, Traditionen und Symbole. Ob Modetrend, Kultspeise, Opfergabe oder Fastfood: Die Aufnahme von Nahrungsmitteln hat immer mit Mythos, das heißt mit Unwissen zu tun. Die Überlieferung, die Tradition, eine Erzählung oder Sage bestimmt im Gegensatz zum Logos, also der Vernunft, den Erfolg von Lebensmitteln wesentlich mehr als der sinnliche Genuss oder rationale Argumente. Essen ist kein vernünftiger, sondern ein irrationaler Vorgang." Werner Mlodzianowski

Erfolgreiche Nahrungsmittel schmecken nicht nur, weil sie gut sind, sondern auch weil sie eine Geschichte erzählen. Sie tragen Bedeutungen in sich und sind mehr, als sie zu sein scheinen. Gute Esswaren regen unsere Phantasie an und vermitteln Gefühle. Auch moderne Lebensmittel und Neukreationen spielen mit dem Faktor Mythos. Welterfolge wie Coca-Cola oder Red Bull zeigen, dass die Lebensmittelindustrie immer wieder imstande ist, große Ikonen zu erschaffen. Der Konsument hat eine Schwäche für die Geschichten, die in seiner Nahrung stecken, sei es die Erzählung von Natur, Identität und Gemeinschaft oder von Gesundheit, Schönheit und ewiger Jugend. Wir lieben moderne Zaubertränke, die uns Kraft und Stärke versprechen, wie stark gezuckerte Joghurtdrinks, die (angeblich) die Abwehrkräfte stärken. Mit anderen Lebensmitteln verschlingen wir die Vorstellung von

The Myth Factor | *"We design our food to make a statement. We eat values, traditions and symbols. Be it a fashion trend, sacrifice or fast food. The consumption of food is always related to myth, that is, to ignorance. Traditions, legends, stories or sagas – as opposed to logos, i.e., reason – have much more influence on the success of food than sensual pleasure or rational arguments. Eating is not a rational but an irrational process."* Werner Mlodzianowski

Successful food not only tastes good because it is good but also because it tells a story. It is invested with meaning – more than it appears to be. Good food inspires our imagination and conveys feelings. Even modern foods and new creations play with the myth factor. Worldwide successes such as Coca-Cola or Red Bull show that the food industry is again and again capable of creating big icons. The consumer has a weakness for stories behind his/her food, be it a tale of nature, identity and community or of health, beauty or eternal youth. We love modern magic potions that promise us power and strength, like heavily sweetened yoghurt drinks that (supposedly) strengthen our immune system. With other food we devour the idea of adventure, exotic worlds, as the present trend of Asian foods in Europe shows. Food can also convey moral values and reassure us that we "are doing

Lebensmittel als Kultobjekte: Stonehenge aus Marsriegeln. Essen bedeutet Identität. Was wir essen, definiert unsere Person, unseren Lebensstil, unsere Kultur.

Food as cult objects: Stonehenge made out of Mars bars. Food signifies identity. What we at, defines our personalities, our lifestyle, our culture.

Abenteuer, fremden Welten und Exotik, wie der aktuelle Trend zu asiatischen Lebensmitteln in Europa zeigt. Esswaren können aber auch moralische Werte vermitteln und uns bestätigen, dass wir „das Richtige tun", indem wir zum Beispiel die Umwelt schützen oder dem Erzeuger einen fairen Preis für sein Produkt bezahlen.

Die Illusion, die eine Essware in uns hervorruft, ist bei der Entscheidung, ob wir sie kaufen oder nicht, ausschlaggebend. Die Minderheit der Konsumenten ernährt sich nach Inhaltsstoffen und Nährwerttabellen, die große Mehrheit „isst" den Mythos. Der Kult rund um „Coca-Cola" beispielsweise entstand nicht, weil es so gut schmeckt, sondern infolge seiner Entstehungsgeschichte als kokainhaltigen Rauschgetränks. 1886 brachte der Apotheker Dr. John Stith Pemberton nach dem Vorbild eines französischen Modegetränks, für das Bordeauxwein mit Kokablättern versetzt wurde, „Pemberton's French Wine Coca" heraus. Unter der Prohibition musste Pemberton den Alkohol entfernen und bewarb sein Produkt nunmehr als „alkoholfreies Getränk für Intellektuelle". Coca-Cola galt als „gut für die Gesundheit und lebensverlängernd" und wurde gegen Müdigkeit, Verdauungsstörungen und Nervenkrankheiten eingesetzt. 1902 verschwand auch die kokainhaltige Substanz aus der Rezeptur, womit der Entwicklung zum Kinder- und Jugendgetränk nichts mehr im Wege stand. Doch das leicht rebellische Image blieb dem prickelnden dunklen Drink bis heute erhalten.

the right thing" by, for instance, protecting the environment or paying the producer a fair price for the product.

The illusion that a food triggers in us is decisive in determining whether we buy it or not. The minority of consumers nourishes itself according to ingredients and nutrition charts, while the majority literally "eats" the myth. The "Coca-Cola" myth, for instance, did not evolve because the drink tastes so great but because of its history as an intoxicating drink containing cocaine. In 1886 the pharmacist Dr. John Stith Pemberton modeled his beverage after a popular French drink, in which coca leaves had been added to Bordeaux wine. "Pemberton's French Wine Coca" was the result. Under the prohibition Pemberton had to remove the alcohol and then advertised his product as a "non-alcoholic beverage for intellectuals". Coca-Cola was seen as "beneficial to health and promoting a long life" and was used to fight fatigue, digestive disorders and nervous illnesses. In 1902 the substance containing cocaine also disappeared from the concoction, paving the way for the development of a beverage for children and young people. The slightly rebellious image, however, has remained a part of this sparkling, dark drink to this very day.

Illusionen vom Schlaraffenland | Der Erfolg vieler Produkte resultiert kaum aus ihrem Nähr- oder Genusswert, sondern eher aus ihrer psychologischen Wirkung. Süßigkeiten und Chips beispielsweise verführen mit Visionen vom paradiesischen Überfluss. Der Griff in eine scheinbar unendliche Menge kleiner Knabbereien verschafft ein wohliges Gefühl von Zufriedenheit und Überlegenheit und verleitet dazu, mehr zu essen, als man eigentlich will. Beim Anblick prall gefüllter Schüsseln mit Brezeln, Fischen und Salzstangen in Miniaturform erwachen Urinstinkte von Überfluss und Reichtum. Knabbereien lassen den Traum vom legendären Schlaraffenland für einen kurzen Augenblick wahr werden und bringen so das kleine Glück in salzig-fettiger Form in jedes Wohnzimmer.

Auch der Erfolg von Gummibären beruht unter anderem auf ihrer psychologischen Wirkung. Der Bonbonkocher Hans Riegel schuf 1922 aus Zucker, Gummi arabicum, Säuerungsmitteln und Aromen eine essbare Abbildung des damals noch weit verbreiteten „Tanzbären" und damit eine Essware, der heute von Psychologen aggressionslindernde Eigenschaften nachgesagt werden. Der Verzehr des eigentlich gefährlichen Raubtiers im Miniformat soll dem Esser ein Gefühl von Stärke und Überlegenheit verschaffen und ihn so manchen Ärger vergessen lassen.[1]

Konservierte Erinnerung | Wirklich erfolgreiche Esswaren tragen kulturelle Botschaften in sich, die jeden ansprechen. So genannte „Cross-Border"-Produkte werden quer durch alle sozialen Schichten, Ethnien oder Glaubensgemeinschaften gegessen. Geschmack ist dabei völlig zweitrangig. In den USA zählen zum Beispiel die berühmten „Oreos", dunkle, runde Doppeldeckerkekse mit einer weißen Creme, zu dieser Gruppe von Esswaren. Oreos wurden 1912 von der Firma Nabisco in New York erfunden, heute erweckt ihr Verzehr vor allem sentimentale Gefühle. Oreos stehen für Nostalgie und den Glauben, Teil der großen amerikanischen Familie zu sein. Sie beschwören Erinnerungen an unbeschwerte Kindheitstage und die heile Welt der „goldenen" 50er Jahre herauf. Gerade Esswaren eignen sich hervorragend, um Gefühle zu konservieren, lang vergangene Gerüche, Geschmäcker und Geräusche wieder zu erleben und so den Esser seiner biographischen Kontinuität zu versichern.[2] Die Popularität österreichischer Mannerschnitten, deutscher Leibniz-Kekse oder italienischer Ferrero-Pralinen basiert auf ähnlichen romantischen Verklärungen. Vertraute Nahrungsmittel, ihre wiederkehrenden Formen und Geschmäcker geben Sicherheit in einer Welt stetiger Veränderungen.[3]

Kulinarische Markenprodukte | Mit der Imagination in den Köpfen der Konsumenten arbeiten auch Designs, die Produkte gestalterisch aufwerten, indem sie Firmenlogos direkt auf Esswaren anbringen. Die Marke ist das qualitative Aushängeschild vorgefertigter Waren. Wie einst Pressmodel und Waffeleisen das Familienwappen in die Butter oder den Waffelteig drückten, essen wir heute Schriftzüge auf Schokoladetafeln, Keksen und Bonbons. Der Trend zum Markenprodukt durchdringt mittlerweile alle Lebensbereiche und macht auch vor dem Nahrungssektor nicht Halt. Lebensmittel werden wie Designermode, Autos oder Uhren zu Markenprodukten hochstilisiert, mit denen der Kunde nicht nur sein

Illusions of the Land of Milk and Honey | The success of many products is hardly a result of their nutritional or pleasure-oriented value. It is much more related to their psychological effect. Sweets and chips, for instance, seduce the eater with visions of paradisiacal cornucopia. Reaching for a seemingly infinite amount of snacks creates a pleasant feeling of satisfaction and superiority and induces us to eat more than we actually want to. Looking at bowls full of small hard pretzels, goldfish crackers or pretzel sticks trigger primal instincts of glut and wealth. Snacks let the dream of the legendary land of plenty come alive for a short moment and thus bring a little bit of happiness into everyone's living room in a salty-fatty form.

Even the success of gummi bears is based on their psychological effect. In 1922, candy cook Hans Riegel used sugar, gummi arabicum, acidifiers and aromas to create an edible copy of the "dancing bears" which were then still widespread; thus, he produced food to which psychologists today ascribe qualities that mitigate aggression. The consumption in miniature format of what is actually a dangerous predatory animal was supposed to give the eater a sense of strength and superiority and to thus help him/her forget certain anger.[1]

Preserved Memory | Truly successful foods transport cultural messages which appeal to everyone. So-called "cross-border" products are eaten throughout all social strata, ethnic or denominational groups. Taste is completely secondary here. In the U.S., for instance, the famous "Oreos", the dark round double-decker cookies with white cream, belong to this group. Oreos were invented by the Nabisco company in New York in 1921. Today they trigger a sense of nostalgia and the belief of being part of the great American family when eating them. They conjure up memories of untroubled childhood days and the wholesome world of the "golden" fifties. Foods are particularly well suited for preserving feelings, reawakening long forgotten smells, tastes and sounds and thus reassuring the eater of his/her biographical continuity.[2] The popularity of Austrian "Mannerschnitten" wafers, German Leibniz cookies or Italian Ferrero chocolates is based on similar romantic transfiguration. Familiar foods, their recurring shapes and tastes offer comforting reassurance in a world of constant change.[3]

Culinary Brand Products | Designs that enhance products by applying company logos right on the food in question also work with the consumers' imagination. Just as once upon a time molds and waffle irons imprinted a family's coat of arms into butter or waffle dough, we eat the lettering on chocolate bars, cookies and candies today. The trend towards brand-name products has meanwhile penetrated all realms of life and does not exclude the food sector. Similar to designer fashion, cars or watches, food is being talked up as a brand product. What the consumer buys is more than just that. It stands for an entire feel for life. Each individual bean of the famous product manufactured by the Jelly Beans co., for instance, bears the company's name. The Austrian

Moderner Zaubertrank: Erfolgreiche Nahrungsmittelprodukte schmecken nicht nur, weil sie gut sind, sondern auch weil sie eine Geschichte erzählen. Der Konsument liebt die Legenden rund um sein Essen: die Erzählungen von Gesundheit, Schönheit oder ewiger Jugend. Gute Esswaren regen unsere Phantasie an und vermitteln Gefühle.

Modern magic potion: Successful food products not only taste good because they are good but because they tell a story. Consumers love the legends surrounding their food - tales of health, beauty or eternal youth. Good food products stimulate our imagination and convey feelings.

Kuchenformen, Buttermodel, Butterroller, Keksausstecher und Puddingformen: Food Design ist keine neue Erfindung. Manche Formen, die noch heute gebräuchlich sind, existieren bereits seit 1000, 2000 oder mehr Jahren, unter ihnen das Croissant, die Brezel, der Brotzopf und der Gugelhupf.

Cake pans, butter molds, butter rollers, cookie cutters and pudding forms: Food design is not a new invention. Some forms that are still in use today have been around for 1,000, 2,000 or more years. These include the croissant, the pretzel, the plaited bun and the Bundt cake.

Essen, sondern gleich ein ganzes Lebensgefühl kauft. So trägt jede einzelne der berühmten, süßen Bohnen von Jelly Beans den Firmenschriftzug. Die österreichische Marke Neuburger nimmt bewusst Anleihen bei klassischen Designerprodukten und bringt ihr Logo direkt an der Kruste der 14 kg schweren Fleischblöcke an, die sich im Format bewusst von traditionellen Leberkäse-Produkten unterscheiden. Die Idee des essbaren Logos benutzen mittlerweile auch Kultmarken, um buchstäblich in aller Munde zu bleiben. Italiens Moderiese Giorgio Armani bietet unter dem Namen „Armani dolce" Konfekt an. Ein plastisch hervortretendes A, das Markenzeichen des Mailänder Modeimperiums, ziert die Oberfläche der Pralinen. Wer weiß, vielleicht essen wir schon bald Leberpastete mit dem Logo von Louis Vuitton oder sportgerechte Cornflakes mit dem Swoosh von Nike …

Der Mythos ist der bedeutendste, gleichzeitig aber auch der am schwierigsten zu fassende Faktor von Food Design. Ein Lebensmittel kann nur dann wirklich erfolgreich sein, wenn es die Imagination des Konsumenten anregt und positive Gefühle in ihm weckt. Der Kunde kauft natürlich lieber ein Nahrungsmittel, das ihm irgendeine Erfahrung verspricht, als eines, das seinem Körper gut täte. Tatsächlich will auch kein Entwickler die irrationale Komponente aus dem Essen wegrationalisieren, ganz im Gegenteil: Für viele Hersteller ist sie ein gutes Geschäft. Somit gilt, dass Essen immer ein emotionaler und insofern auch vom Mythos getriebener Vorgang ist, war und sein wird.

Zukunftsszenarien – Technologie versus Natur | *„Essen ist Technologie. Die Vorstellung, dass wir Naturprodukte essen, ist reine Fantasie. Natürliches wächst im Regenwald am Äquator, unsere Kühe dagegen sind über tausende von Jahren hochgezüchtet. Im Regenwald gibt es keine großen Kühe wie bei uns, die Kühe dort sind klein und mager. Auch Karotten oder Gemüse sind das Ergebnis technologischer Prozesse, aber wir vergessen das gerne." Hervé This*

Massenerzeugung und technologische Veränderung von Nahrungsmitteln sind mittlerweile ein selbstverständlicher Bestandteil unseres Alltags und somit ein essenzielles Element unserer Kultur. In der Geschichte gab es immer wieder Bewegungen, die ein „Zurück zur Natur" forderten, tatsächlich ist die Industrialisierung des Essens aber so weit fortgeschritten, dass eine Rückentwicklung ins vorindustrielle Zeitalter hin zu rein handwerklich erzeugten Nahrungsmitteln unmöglich erscheint. Kulturelle Prinzipien wie Sauberkeit, Hygiene oder standardisierte Produkte und die Idee, alle Nahrungsmittel für jeden verfügbar zu machen, spiegeln sich in allen Industrieprodukten wider und sind auch fixer Bestandteil unserer Esskultur geworden. Ändern sich diese gesellschaftlichen Normen, verändern sich auch die Produkte und die Anforderungen an ihre Gestaltung. Die zentrale Frage der Zukunft lautet: Wie können technologische Lebensmittel möglichst nachhaltig, umweltgerecht und fair hergestellt werden, damit sie in 30 Jahren 9 Milliarden Menschen ernähren können?

Jahrzehntelang galten Industrialisierung und rauchende Fabrikschlote als Sinnbild des Wirtschaftswunders und Inbegriff der wohlhabenden westlichen Gesellschaften. Heute, im Zeitalter von Dienstleistungs- und Wissensgesellschaft, wird der Drang zur Individualisierung immer größer. Die automatisierte Erzeugung hat in vielen Bereichen zwar zu einer sprunghaften Vergrößerung der Produktpalette geführt, gleichzeitig aber eine Standardisierung von Verarbeitung und Geschmack mit sich gebracht. So finden sich heute im Kühlregal Dutzende Sorten von Erdbeerjoghurts, die aber alle relativ ähnlich schmecken.

Am Beginn des 21. Jahrhunderts ist industrialisierte Massenware out und Lebensmittelhersteller müssen nach neuen Strategien suchen, um diesen kulturellen Anforderungen gerecht zu werden. „Tailor-made food" ist ein Konzept, das auf die Pluralität der Kundschaft eingeht und unter anderem vom Lebensmittelkonzern Kraft Jacobs verfolgt wird. Dabei handelt es sich um individualisierte Industrienahrung, die vor dem Einkauf im Internet oder an einem Terminal im Supermarkt nach persönlichen Vorlieben zusammengestellt werden kann. Man denkt dabei zum Beispiel an Joghurts oder Müslis, deren Komponenten jeder Kunde selbst

brand Neuburger deliberately borrows from classical designer products and places its logo on the crust of the 28-pound meat blocks whose format is deliberately made to be different from the traditional "Leberkäse" products. Cult brands meanwhile employ the idea of the edible logo to literally pass everyone's lips. Italy's fashion giant Giorgio Armani offers confectionery under the name "Armani dolce". A three-dimensional A, the brand of the Milan fashion empire, adorns the surface of the chocolates. Who knows, perhaps we will soon be eating liver pâté with the logo of Louis Vuitton or sports-oriented cornflakes with the Nike Swoosh…

The myth is the factor in food design which is at the same time most important and most difficult to grasp. Food can only really be successful if it stimulates the consumer's imagination and triggers positive feelings. The customer would, of course, prefer to buy food that promises him/her an experience instead of simply being good for his/her body. In reality, there is also no product developer who would want to completely remove the irrational component of food by rationalization – quite the contrary. For many manufacturers the stories behind their food are also good business. We can thus say that eating has always been and will always be an emotional phenomenon driven by myth.

Future Scenarios – Technology vs. Nature | *"Everything we eat is a product of technology. The idea that we eat natural things is a fantasy. Natural things grow in the equatorial rainforest. Our cows have been selectively bred over thousands of years. Our large cows don't grow in the rainforest, the cows there are small and thin. Our carrots, our vegetables are technological products, but we forget that." (Hervé This)*

Mass production and the technological modification of foodstuffs have meanwhile become a given part of our daily lives and thus an essential element of our culture. In history there have always been movements advocating a "return to nature". In reality, the industrialization of food has reached such an advanced stage that a movement back to the pre-industrial age of hand-made food seems impossible today. Cultural principles such as cleanliness, hygiene or standardized products, and the idea that all food products should be made available to everyone are reflected in all industrial products and have become an integral part of our eating culture. When these social norms change, so will the products and the demands made on their design. The central issue of the future is the following: How can technological food be produced to be as sustainable, environmentally friendly and fair as possible so that it can feed nine billion people 30 years from now?

For decades industrialization and factories with smokestacks were seen as the symbol of the economic boom and the prosperity of western societies. Today, in the age of the service and knowledge society, there is an increasing drive to individualize. In many areas automated production has resulted in a range of products growing by leaps and bounds and at the same time it has led to standardized processing and taste. Today we can thus find a dozen brands of strawberry yoghurt but they all taste pretty much the same.

At the beginning of the twenty-first century industrialized mass goods are on the way out. Food manufacturers now have to find new strategies to respond to these cultural demands. "Tailor-made food" is a concept geared to the plurality of customers and one that the Kraft Jacobs food corporation, for instance, is pursuing. This type of food can be described as individualized, industrially produced nutrition

wählen kann: Fettgehalt, Kalzium, Viskosität, Zuckergehalt, Fruchtzugabe, Vitamine … alles frei wählbar. Die einzelnen Ingredienzien werden dann vor Ort oder im anliefernden Verteilerzentrum gemischt, verpackt und dem Kunden ausgehändigt. In den USA können Kunden im Supermarkt an so genannten „coffee bars" ihren persönlichen Kaffee bereits aus verschiedenen Sorten selbst zusammenmixen.

Was aber wird die Zukunft in Hinblick auf neue Lebensmitteltechnologien und Zubereitungsverfahren bringen und wie werden sich diese auf die Gestaltung unseres Essens auswirken? Für Werner Mlodzianowski, Leiter des Technologietransferzentrums Bremerhaven in Deutschland, liegt die Zukunft unseres Essens vor allem in der Technologie. Eine aktuelle Forderung ist etwa, chemische Zusatzstoffe durch entsprechende physikalische Methoden zu ersetzen. Auch wenn das vielleicht paradox klingt: Man versucht mittels neuer Verarbeitungstechnologien, eine möglichst große Naturnähe zu erreichen.

Der Konsument wünscht sich Lebensmittel, die weniger beziehungsweise überhaupt keine chemischen Zusatzstoffe enthalten, will aber auf die Annehmlichkeiten, die diese mit sich bringen, nicht verzichten. Genau hier setzen heute viele Forschungsprojekte an, um Technologien zu entwickeln, welche die gewünschten Eigenschaften auf rein physikalischem Weg herbeiführen. Das Ergebnis bezeichnet man dann als „Clean-Label"-Produkt. Beispielsweise lässt sich die gewünschte Haltbarkeit von Brot ohne den Zusatz von Konservierungsmitteln durch eine entsprechend hermetische Produktion erreichen: Brot direkt aus dem Backofen ist durch die Hitze steril. Kommt das Brot nahtlos vom Ofen in die Kühlkette, in der es schrittweise abgekühlt und ohne Kontakt zur Außenluft verpackt wird, ist es physikalisch konserviert. Eine Revolution auf diesem Sektor wäre beispielsweise auch eine neue Technologie, die Weißbrot vierundzwanzig Stunden und länger frisch und knusprig hält.

Der Ersatz von chemischen Zusätzen ist ein Forschungsschwerpunkt des TTZ, die Untersuchung gewisser physikalischer Parameter ein anderer. Tiefe Temperaturen zum Beispiel werden heutzutage vor allem zur Konservierung eingesetzt, sie eignen sich aber auch hervorragend, um spezielle, neuartige Texturen herzustellen. Das größte Zukunftspotenzial am Nahrungssektor sieht Werner Mlodzianowski allerdings nicht in der Physik, sondern in Biologie und Biotechnologie. Bis heute ist die mikrobiologische Funktionsweise vieler Herstellungsverfahren, beispielsweise der Sauerteiggärung, nicht geklärt. Sobald man die Wirkungsweise derartiger Vorgänge kennt, ließen sie sich gezielt auch für andere Produkte einsetzen, bislang ungeahnte Technologien und neuartige Lebensmittel wären die Folge...[4]

1 | Helene Karmasin, Die geheime Botschaft unserer Speisen, Bastei Lübbe, 2001, S. 304ff., S. 309f.
2 | Wir danken Elani Raider für diese Information.
3 | Helene Karmasin, Die geheime Botschaft unserer Speisen, Bastei Lübbe, 2001, S. 311
4 | Quelle: Interview mit Werner Mlodzianowski, Leiter des Technologietransferzentrums Bremerhaven, am 18. u. 19.10.2007

In Japan sind Kirschblüten ein Symbol für Schönheit und Vergänglichkeit. Um ihre Farbe und ihren Geruch zu konservieren, werden sie eingesalzen und das restliche Jahr über als Tee getrunken.

In Japan cherry blossoms are a symbol for beauty and transitoriness. To conserve their color and smell, they are salted and served as tea during the rest of the year.

that can be put together on the Internet or at a terminal in the supermarket according to personal preferences before shopping. For instance, customers can select the components of yoghurts or mueslis: fat content, calcium, viscosity, sugar content, fruit, vitamins – it is all your choice The individual ingredients are then mixed, packed and sent to the customer on site or at the supplier's distribution center. In the U.S. customers can already mix their own personal coffee from different flavors at so-called "coffee bars".

What will the future bring in terms of new food technologies and preparation techniques? What effect will these have on the design of our food? For Werner Mlodzianowski, director of the German Bremerhaven technology transfer center TTZ, the future of our food is mainly to be found in technology. A current demand, for instance, is to replace chemical additives by means of physical methods. Even if this may sound paradoxical, there are attempts to use new processing technologies to obtain the highest degree of naturalness.

The consumer wants food with less or no chemical additives at all without having to dispense with the pleasant features that these entail. This is precisely what many research projects target in developing technologies which bring about the desired qualities by purely physical means. The result is then described as a "clean-label" product. For instance, the desired shelf life of bread is achieved by means of adequately hermetic production conditions without needing to add preservatives. Bread that comes straight out of the oven is sterile because of the heat. If the bread goes right from the oven to the cooling chain where it is gradually cooled and packed without coming into contact with the outside air, it is preserved by physical means. A new technology that keeps white bread fresh and crispy for twenty-four hours and longer would also be a revolutionary development in this sector.

One research focus of the TTZ (Technology Transfer Center) is to replace chemical additives, another one is to study certain physical parameters. Today, low temperatures are mainly used to preserve food products but they are also excellently suited to create special, novel textures. Werner Mlodzianowski sees the greatest future potential in the food sector not in physics but in biology and biotechnology. To this day, the macrobiological functioning of many manufacturing techniques, such as the fermentation of sourdough, has yet to be explained. As soon as we know how these processes work down to every detail, we could deliberately use them for other products, and new, hitherto unknown technologies and food could emerge from this.[4]

1 | Helene Karmasin, Die geheime Botschaft unserer Speisen, Bastei Lübbe, 2001, pp. 304, p. 309 ff.
2 | We thank Elana Raider for this information.
3 | Helene Karmasin, Die geheime Botschaft unserer Speisen, Bastei Lübbe, 2001, p. 311
4 | Source: Interview with Werner Mlodzianowski, Director of the Technologie Transfer Zentrum Bremerhaven, 18 and 19 October 2007

1756 Erfindung der **Mayonnaise** / Frankreich (Legende) | **1760** Herstellung von **Lakritze** / England | **1762** Erfin
1756 mayonnaise invented / France (legend) | **1760 licorice** produced / England | 1762 **sandwich** invented

1778 Pasta mit Tomatensauce wird erstmals erwähnt | **1805** Erfindung der **Frankfurter** / Wien | **1812** Eröffnung d
1778 pasta with tomato sauce mentioned for the first time | **1805 Frankfurter** sausage invented /Vien

die **Sachertorte** / Wien
Sacher invented the **Sacher cake** / Vienna

1835 Erfindung der handbetriebenen **Eismaschine**
1835 Hand-operated ice-cream m

Schokolade in festem Zustand / Bristol | **1847** Justus von Liebig erfindet den **Fleischextrakt** / Deutschland | 18
developed the first **solid chocolate** / Bristol | **1847** Justus von Liebig invented **meat extract** / Germany | 1853

(Legende) | **1865** Küfferle bringt die Wiener **Katzenzungen** heraus / Wien | **1866** die **Suppeneinlage** kommt
Küfferle launched the **Viennese Katzenzungen** (cat tongues) /Vienna | **1866 Soup additions** became fash

Adams erfindet den **Kaugummi** / USA | **1875** Erfindung der **Milchschokolade** / Schweiz | **1875** die Produktio
Adams invented **chewing gum** / U.S.A. | **1875 Milk chocolate** invented / Switzerland | **1875** the production o

hergestellt / USA |
/U.S.A. |

1880 Philadelphia Brand Cream Cheese wird erstmals produziert / New
1880 Philadelphia Brand Cream Cheese produced for the first tim

Erfindung der **Prinzenrolle** / Belgien | **1886** Dr. John Stith Pemberton erfindet das **Coca-Cola** / USA
the **Prinzenrolle** invented / Belgium | **1886** Dr. John Smith Pemberton invented **Coca-Cola** /U.S.A. |

gegründet / USA | **1889** die **Erbswurst** von Knorr geht in Produktion | **1890** Paul Fürst erfindet die **Mozartkugel**
founded /USA | **1889** Knorr launched production of **dried pea soup** shaped like a sausage | **1890** Paul Fürst invented

Deutschland | **1893** William Wrigley bringt den Spearmint und den Juicy Fruit **Kaugummi** auf den Markt / Chicago | 189
William Wrigley launched **spearmint** and **juicy fruit** chewing gum / Chicago | **1894** Dr. John Harvey Kellogg invented

Suchard bringt die **Milka** auf den Markt / Österreich, Schweiz, Deutschland
Switzerland, Germany

Carte-Restaurant / London | **1908** Theodor Tobler erfindet die **Toblerone** / Schweiz |
first **à-la-carte restaurant** / London | **1908** Theodor Tobler invented **Toblerone** /Switzerland

s Sandwiches / England (Legende) | ca. 1790 Erfindung der Kaisersemmel / Wien
d (legend) | ca. 1790 "Kaiser" (imperor) roll invented (Vienna)

Konservenfabrik / England | 1827 Erfindung der Linzertorte / Linz | 1832 Franz Sacher erfindet
12 first food can factory opened / England | 1827 the Linzertorte (cake) invented / Linz | 1832 Franz

1843 Jakob Christoph Rad erfindet den Zuckerwürfel / Böhmen | 1847 Firma Fry entwickelt die erste
ne invented / U.S.A. | 1843 Jakob Christoph Rad invented the sugar cube / Bohemia | 1847 Fry company

dung der Pommes frites / New York (Legende) | 1855 Erfindung der Malakofftorte / Halbinsel Krim
ch fries invented / New York (legend) | 1855 Malakoff cake invented / Krim peninsula (legend) | 1865

e / Bayern | 1867 Erfindung der Erbswurst / Berlin   ca. 1870 Thomas
/Bavaria | 1867 dried pea soup shaped like a sausage invented / Berlin   ca. 1870 Thoman

Knorr-Suppen beginnt   1878 Ketchup wird von Harry Heinz erstmals industriell
rr-soups starts   1878 Ketchup mass-produced for the first time by Henry Heinz

a. 1880 Erfindung des Hot Dog / USA | 1880 Victor Schmidt erfindet das Ildefonso / Wien | ca. 1885
w York | ca. 1880 Hot dog invented / U.S.A. | 1180 Victor Schmidt invented Ildefonso / Vienna | ca. 1885

1886 Maggi's flüssige Speisen- und Suppenwürze kommt auf den Markt / Schweiz | 1888 Pepsi wird
1886 Maggi's liquid food and soup seasoning launched /Switzerland | 1888 Pepsi

rg   1891 Bahlsen bringt den Leibniz-Keks auf den Markt /
ozartkugel / Salzburg   1891 Bahlsen launched the Leibniz cookie /Germany | 1893

ohn Harvey Kellogg erfindet die Corn Flakes / USA | 1898 Erfindung der Mannerschnitte / Wien | 1901
n Flakes / U.S.A. | 1898 Mannerschnitte invented / Vienna | 1901 Suchard launched Milka /Austria,

1902 Eröffnung des ersten Automatenbuffet / Philadelphia | 1903 Escoffier eröffnet das erste À-la-
1902 The first snack vending machine opened / Philadelphia | 1903 Escoffier opened the

1908 Maggi's Rindsuppenwürfel kommt auf den Markt / Schweiz | 1913 Rumkugeln
1908 Maggi's beef stock cubes launched /Switzerland | 1913 Casali rum balls |

1914 Wrigley's bringt den Doublemint auf den Markt / Chicago | 1915 das deutsche Wort Keks wird in den Duden a
1914 Wrigley's launched doublemint / Chicago | 1915 the German word "Keks" is entered in the Ge

der erster elektrische Toaster mit Zeitschaltung wird patentiert / USA | 1920 Franklin Clarence Mars erfindet den Schokorieg
the first electrical toaster with timer patented /U.S.A. | 1920 Franklin Clarence Mars invented the chocolate bar Mars |

als Haribo „Tanzbär" erfunden / Bonn | 1922 C.K. Nelson erfindet die Schokoglasur für Speiseeis / USA | 1923 das Eis
is invented as the Haribo "dancing bear"/Bonn | 1922 C.K. Nelson invented the chocolate glazing for ice cream /U.S.A.

Krispies heraus / USA | 1930 das Kaubonbon Maoam wird patentiert / Düsseldorf | 1930er Franz Wiesbauer erfinde
Krispies launched by Kellogg's /U.S.A. | 1930 the Maoam chewing candy patented /Düsseldorf | 1930's Franz Wiesba

erfindet die Alete Babynahrung / München | 1938 Nestlé führt den Nescafé ein / Schweiz
food invented by Prof. Dr. Günther Malyoth /Munich | 1938 Nescafé introduced by Nestlé/ Switzerland

USA | 1941 das Eisstanitzel wird patentiert / USA | 1944 Wrigley bringt Orbit heraus / USA | 1945 das Automa
McDonald | 1941 the ice-cream cone patented / U.S.A. | 1944 Orbit launched by Wrigley /U.S.A. | 1945 Soft

erfunden / Feldbach, Österreich |  1950 Kraft Deluxe process cheese slices: 1. S
Soletti salt sticks invented/ Feldbach, Austria | 1950 Kraft Deluxe process cheese :

/ USA | 1953 Bahlsen führt Messino unter dem Namen „Messina" ein / Deutschland | 1953-54 Burger King wird ge
launched Messino under the name "Messina"/ Germany | 1953-54 Burger King founded /U.S.A. | 1955 1st M

1957 Ferrero führt Mon Chéri in Deutschland ein  1958 Pizza Hut wird geg
1957 Mon Chéri introduced by Ferrero in Germany 1958 Pizza Hut fo

ein | 1966 Cornetto kommt auf den Markt | 1967 Jolly kommt auf den Markt | 1967 Ferrero führt Kinder Schokola
Cornetto launched | 1967 Jolly launched | 1967 Children's chocolate introduced by Ferrero in Germany |

die Kinderüberraschung ein / Deutschland | 1979 Ferrero führt die Milch-Schnitte in Deutschland ein | 19
a surprise for children / Germany | 1979 Ferrero launched "Milchschnitte" in Germany | 1983 Coca-Cola Ligh

1990 Magnum kommt auf den Markt  1996 probiotisches Produkt LC1 kommt von Nest
1990 Magnum launched 1996 probiotic product LC1 launched by Nestlé |

mmen | 1917 Kikkoman wird gegründet / Japan  1919
ctionary Duden   1917 Kikkoman founded/ Japan   1919

   1922 Luisa Spangoli erfindet für Perugina das Bacio / Italien | 1922 Haribo Goldbär wird
   1922 Luisa Spangoli invented the Bacio for Perugina /Italy | 1922 Haribo golden bear

Stiel wird erfunden / USA | 1927 Ed Haas III erfindet das Pez Bonbon / Wien | 1928 Kellogg's bringt die Rice
3 The popsicle (ice cream on a stick) invented /U.S.A. | 1927 Ed Haas invented the Pez candy /Vienna | 1928

ergsteiger / Wien | 1930er Erfindung der Schwedenbombe / Wien | 1935 Prof. Dr. Günther Malyoth
nted "Bergsteiger" /Vienna | 1930's Schwedenbombe invented / Vienna | 1935 Alete baby

1940 die Gebrüder Dick und Mac Mc-Donald eröffnen ein Schnellservice-Restaurant in San Bernardino /
1940 the fastfood restaurant in San Bernardino/U.S.A. opened by the brothers Dick and Mac

Softeis wird erfunden | 1949 die Curry-Wurst wird erfunden / Berlin | 1949 Soletti-Salzstangen werden
cream dispenser AUTOMATEN-SOFTEIS invented | 1949 the curry sausage invented /Berlin | 1949

blettenkäse / USA | 1950 das Carpaccio wird erfunden / Venedig | 1952 Kellog's führt die Frosties ein
/U.S.A. | 1950 Carpaccio invented / Venice | 1952 Frosties launched by Kellog's /U.S.A. | 1953 Bahlsen

t / USA | 1955 die 1. McDonald´s Restaurant eröffnet / Chicago | 1957 Whopper wird erfunden |
nalds restaurant opened /Chicago | 1957 Whopper invented |

| ca. 1962 das Fischstäbchen wird eingeführt / Großbritannien | 1964 Ferrrero führt Nutella in Deutschland
| ca. 1962 Fish sticks introduced /Great Britain | 1964 Nutella launched by Ferrero in Germany | 1966

Deutschland ein | 1968 Twinni wird eingeführt | 1968 der Big Mäc wird eingeführt | 1974 Ferrero führt
3 Twinni introduced | 1968 Big Mac introduced | 1974 Ferrero launched chocolate egg containg

oca-Cola Light / Diet Coke wird eingeführt / Europa | 1990-91 Zitronenpresse „Juicy Salif" von Starck |
Diet Coke introduced / Europe | 1990-91 Lemon press "Juicy Salif" by Starck |

den Markt | 2001 grünes Ketchup „Heinz Blastin' Green" kommt auf den Markt
1 green ketchup "Heinz Blastin Green" launched

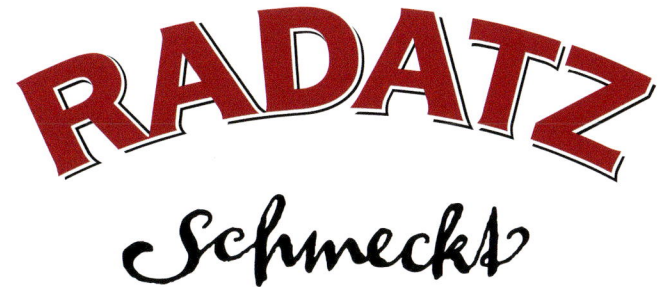

Danksagung | acknowledgment

Herzlichst bedanken wir uns bei folgenden Personen für Ihre Unterstützung:

für ihr Wissen (in alphabetischer Reihenfolge) | Albert Adria, Koki Ando, Elisabeth Andoh, Eva Bakos, Benedict Beaugé, Emmerich Berghofer, Friedrich Blutner, Marc Brétillot, Manfred Buchinger, Stéphane Bureaux, Angelika Deutsch, Klaus Dürrschmid, Martina Kaller Dietrich, Wolfhard Fechner, John Floros, Frank Förster, Georg Friedl, Peter Gnaiger, Fritz Gnan, Gisla Gniech, Marti Guixé, Stephan Gruber, Hanss Hatt, Heinz Hanner, Dave Hardt, Nico Heukels, Josef Honeder, Florian Holzer, Jeannie Houchins, Dieter Isakeit, Christina Lund Jakobsen, Jelly Mongers, Raja Murugan, Alexander Knakal, Peter Krammer, Ulrich Kulozik, Paul Levy, Konrad Paul Liessmann, Cathie Martin David Julian McClements, Barbara van Melle, Carita Merenmies, Achim Meyrhofer, Richard Mithen, Werner Mlodzianowski, Dietmar Muthenthaler, Keiko Nakayama, Natias Neutert, Rudolf Kellner, Franz Kiesenhofer, Paul Renner, Georg Reier, Gerhard Riess, Hanni Rützler, Artemio Sanzio, Wolfgang Schlüter, Vivek V. Shankar, Kantha Shelke, Christophe Spotti, Thomas Strini, Jake Tilson, Hervé This, Kim Thornton, Herr Touzimsky, Steven Twining, Barbara van Melle, Renate und Christoph Wagner, Lojze Wieser, Peter Wilde, Christian Wrenkh, Margit Zeiler

für ihre Unterstützung bei organisatorischen Angelegenheiten (in alphabetischer Reihenfolge) | Stella Avallone, Nadja Benndorf, David Cobbold, Barbara Fuchs-Puchner, Ingeborg Gasser-Kriss, Markus Glaser, Eva Gombocz, Cesare Griffa, Martha Hablesreiter, Roland Hablesreiter, Monika Hirsch, Hans Jörg Hörmann, Anke Jansen, Wencke Lemmes Pireaux, Barara Knapp, Franz Kiesenhofer, Silvia Kolenz, Jutta Meyer, Werner Mlodzianowski, Aiko Morise, Takat Nakamura, Jill Norman, Svend Pedersen, Britta Rollert, Joanna Savill, Wolfgang Schlüter, Franz Schubert, Barbara Taussig, Barbara Taxacher, Nils This, Akira Uehara, Viktor Vanicek, Anne Wendelin,

für Informationen und Werksbesichtigungen (in alphabetischer Reihenfolge) | Liz Brenna, Ben & Jerry's, South Burlington VT; Gernot Bartussek, Chocolaterie & Großkonditorei Aida-Prousek & Co., Wien; Corinna Casale, Nestlé Italiana S.p.A., Perugina Historical Museum, Perugia; Dipl.-Ing. Martin Doppler, Zuckerraffinerie Tulln, Agrana; Ing. Karl Peter Egger, Grand Cru; Petra Franzmeier, PEZ International AG; Kenny k. Fujimoto, Director, Nadaman Restaurants Tokyo; Herta und Alois Gölles, Gölles Schnapsbrennerei & Essigmanufaktur GmbH, Riegersburg; Giancarlo Gonizzi, Archivo storico de Barilla, Parma; Gerhard Gugg, Restaurant Mesnerhaus, Mauterndorf; Reiner Heilmann, Hotel Sacher, Wien; Hiroyasu Kawai & Keiji Toyizumi, Kakiyama, Tokio; Yasuhiro Yamada, Nissin Foos, Kanto Plant; John King & Nanci Kah, P&G Global Snacks, Cincinnati, OH; Hiroaki Konishi, Snow Brand Milk Products, Tokyo; Simon Lindenthaler, QimiQ Handels GmbH, Hof bei Salzburg; Dietmar F. Muthenthaler & Christian Gahn, k.u.k. Hofzuckerbäcker Ch. Demel's Söhne; Mutti S.p.A., Montechiarugolo; Keiko Nakayama, Toraya Gallery, Toraya Confectionery LTD, Tokyo; Dipl.-Volksw. Andreas Nickenig, Haribo Linz; Markus Pois, A. Egger's Sohn Süßwaren; Edgar Petry, Lufthansa; Gerfried Pichler, Frisch und Frost Nahrungsmittel Ges.m.b.H., Hollabrunn; Elisabeth Radatz, Radatz GmbH., Wien; Herr Reichert, Simpert Reiter GmbH., Augsburg; Otto Wilhelm Riedl, B.A., Manner; Franz Rohringer, Neuburger GmbH & Co.KG. Ulrichsberg; Gigi Schoeller, Wien - Kikkoman Trading Europe; Dr. Paulus Stuller, k & k Hofzuckerbäcker L. Heiner; Gulliver Wagner, Felix Austria GmbH, Mattersburg; Thomas Wieder, Nestlé, Mailand; Restaurant Zum schwarzen Kameel

besonderer Dank an | special thanks to
Barbara Daser, David Groiss, Roland Hablesreiter, Markus Hiden, die Mäuse, Jan Schlink

Model: Tom Hanslmaier

Bildverzeichnis | Photographic Credits

Vorwort | foreword
S1: Render Farfalle / Michael Mähring und Philipp Kratzer; S2- 3: Tiefkühlkarotten / Studio Köb; S5: Sandkasten / Ludwig Löckinger; S6- 7: Fisch-Lastauto / Studio Köb; S9: Melone / Sonja Stummerer; S10: Artischocke / Studio Köb; S11: Pillen / Ludwig Löckinger;
S12- 13 Imagefotos / Sonja Stummerer; S15: Geschmackscollage / Studio Köb; S16: Szene in blau / Studio Köb

5 Sinne | 5 senses
S18- 19: Farbkreis / Studio Köb; S20- 21: Infusionsbeutel / Studio Köb; S22- 23: Kirschen im Glas / Ludwig Löckinger; S24-25: Feige und Croissant / Studio Köb; S26- 27: 4 Teller / Studio Köb; S28: v.o.n.u. Bild 1-4 / Studio Köb, Bild 5 / Kikkoman; S29: Blumenchips / Ludwig Löckinger; S30: Blumenwurst / Ludwig Löckinger; S32- 33: -Gewürzecollage / Ludwig Löckinger; S34- 35: Flaschen / Ludwig Löckinger; S36- 37: Konsistenz / Studio Köb; S38: Tomaten quetschen / Studio Köb; S39: Glovebox / Martin Hablesreiter; String Cheese / Studio Köb; S40- 41: Erdbeeren / Ludwig Löckinger; S42- 43: Schlagobers / Ludwig Löckinger; S44- 45: Knabbergebäck / Ludwig Löckinger; S46-47: Zuckerwatte / Ulrike Köb, Schraubzwingen / Studio Köb; S48-49: Saucen / Studio Köb; S50: Collage / Ludwig Löckinger, Waffelcollage / Ludwig Löckinger, Crumpet / Studio Köb; S51: Reiscracker / Studio Köb, Hohlhippe / Studio Köb, Makro / Ludwig Löckinger, Waffelproduktion / Fa Manner; S52: Türkischer Honig / Studio Köb; S52: Bonbon / Studio Köb, Jogurt / Studio Köb; S54- 55 alle Fotos von Studio Köb und Ludwig Löckinger; S 56 alle Fotos von Jellymongers; S57: Kaugummi / Studio Köb; S58: Erdbeerstrohhalm / Studio Köb; S61: Toffee / Studio Köb; S62-63: Praline / Ludwig Löckinger, S64- 65: Japanischer Cracker / Studio Köb; S66: Sojasauceflasche / Fa. Kikkoman; S67: Essigflasche / Fa. Gölles; S68: Käsekrainer / Studio Köb; S69: Käsekrainer Makro / Ludwig Löckinger; S70-71: Getränke / Ludwig Löckinger; S72: Mineralwasser / Fa. Römerquelle; S73: AloeVera Saft / Studio Köb; S74: Imagefoto / Sonja Stummerer; S76-77: Brötchen / Studio Köb; Brötchen / Restaurant Schwarzes Kamel Wien, Gemüse / Michel Bras, Fruchtnougattorte / Sonja Stummerer, japanischer Schlecker-Verkäufer, japanische Kellnerin u. japanische Süßigkeiten / Sonja Stummerer; S81: Gugelhupf / Ludwig Löckinger, Grafik / Martin Hablesreiter; S82- 83: Japanische Suppeneinlage / Studio Köb; S84- 85: Gummibär / Studio Köb; S86- 87: Farbkreis / Ludwig Löckinger, Geyerhalter Film; S88- 89: Pizza, Käse / Ludwig Löckinger, Geyerhalter Film; S90- 91: Collage Rot/Schwarz/ Ludwig Löckinger, Geyerhalter Film; S93: Reis / Studio Köb, Seidenbonbons, japanische Cracker, Zuckerstange, sauer panierte Fruchtgummizungen, zweifärbige Farfalle rigate, Lakritze-Röllchen, transparenter Lollipop mit Kaugummifülle, Farbkreis nach dem Schweizer Maler Johannes Itten, geschichtetes Lakritze-Bonbon / Studio Köb; S94- 95: Getränke / Ludwig Löckinger; S96: Popcorn / Sonja Stummerer; S99: Cup Noodles / Nissin Foods Holdings Co., Ltd.; S100- 101: Plattenspieler / Ludwig Löckinger; S102- 103: Würstchenverkostung/ Fa. Synotech Geyer im Erzgebirge, Mikrofon mit Keks / Ludwig Löckinger, Cornflakes / Fa Kellogg Wien; S104- 105: Knackwurst / Ludwig Löckinger, Bier / Ludwig Löckinger; S106- 107: Schnitzel / Ludwig Löckinger , Frequenzdiagramme / Fa. Synotech Geyer, Friedrich Blutner / Roland Hablesreiter; S109: Schokoladehandy / Studio Köb; S110- 111: Geschmacksforschung Titel / Nils This/ Techologietransferzentrum Bremerhaven, Artischocke, Himbeere u. Feige / Studio Köb, Krapfen / Ludwig Löckinger, Geyerhalter Film; S112- 113: Testgeräte / Nils This/ Techologietransferzentrum Bremerhaven; S114- 115: Labor / Nils This Techologietransferzentrum Bremerhaven, Gaschromatograph u. Schokoladearomen / Barbara Daser; S116- 117: Riechkolben / Ludwig Löckinger; S118- 119: Sensoriklabor / Nils This Techologietransferzentrum Bremerhaven & Sonja Stummerer; S120- 121: Texture Analyzer / Barbara Daser, Sonja Stummerer

Funktion | function
S122- 133: Titelbild / Studio Köb; S124- 125: Pasta / Studio Köb; S126- 127: Einmachgläser / Ludwig Löckinger; S128: Stockfischproduktion / Norwegian Seafood Export Council, Apfelspalten / Studio Köb; S129: Nudeltrocknung / Barilla Historical Archive; S130- 131: Fisch und Fleisch auf Salz / Studio Köb; S132: Marmeladeproduktion Firma Stauds Wien / Petra Schmidt; S133: Parmaschinken / Sonja Stummerer; S134: Sauerkraut, Essiggurken, Gimchi / Studio Köb; S135: Käse / Studio Köb; S136- 137: Tiefkühlgemüse / Studio Köb, Patentzeichnung v. Birdseye / offenes Bild Internet; S138- 139: Gefriergetrocknetes Gemüse / Ludwig Löckinger; S140: Erdbeercombino / Studio Köb; S141: Schwedenbombe / Studio Köb, Render / David Groiss/ Grafik Martin Hablesreiter; S142-143: Japanische Suppe / Studio Köb; S144- 145: Produktion / Sonja Stummerer, Haribo, Barilla, Stephan Gruber, Kikkoman, Rupp, Radatz, Muti, Nissin, Unilever, Bahlsen, Aida; S146: Produktion / Manner, Masterfoods, Unilever; S148: Nudeln / Ludwig Löckinger; S149: Ungarische Schaumrolle / Martin Hablesreiter; S150- 151: Ildefonso u. aufgestellte Schokolade / Ludwig Löckinger; S152- 153: Pizza / Martin Hablesreiter; S154- 155: eckige Pizza, eckige Knödel / Sudio Köb; S157: Gugelhupf / Studio Köb, Render / David Groiss, Grafik / Martin Hablesreiter; S158- 159: Stapelbares / Studio Köb; S160- 161: Fladenbrot / Studio Köb; S162: Hamburger / Michael Mähring & Philipp Kratzer, Geyerhalter Film; S164-165: Qimiq / Martin Hablesreiter nach einem Vorbild von QimiQ, Nudelsuppe mit Ei / Studio Köb; S166: Pringles / Ludwig Löckinger; S168- 169: Transport / Martin Hablesreiter; S170: Brothandtasche / Stephane Bureaux; S171: Transport Indien und Japan / Sonja Stummerer; S172: Snackverkäufer / Sonja Stummerer, Käselaib, Fass, Reifen / Sonja Stummerer; S173: Tom Hanslmaier / Ludwig Löckinger; S174- 175: Bagel / Ludwig Löckinger; S176: Simitverkäufer / Sonja Stummerer; S177: Markt u. LKW in Indien / Sonja Stummerer; S178- 179: Spielzeuglastwägen / Roland Hablesreiter, Europalette / David Groiss; S180- 181: Fischstäbchenturm / Ludwig Löckinger; S183: Schwarzwälderkirschtorte / Studio Köb; S185: Sachertorte / Studio Köb; S186- 187: Zertrümmertes Keks / Sonja

Stummerer, Geyerhalter Film; S188- 189: Verzehrsituation / Studio Köb; S191: Space food / NASA; S192-193: Tom / Ludwig Löckinger, Hot Dog / Studio Köb, Hamburger, BigMac / McDonalds; S194- 195: Dürum / Ludwig Löckinger, Geyerhalter Film; S196: Kebab-Verkäufer / Sonja Stummerer; S197: Reisbälle / Studio Köb; S198- 199: Strudelerzeugung / Martin Hablesreiter; S201: essbares Gedeck / Studio Köb; S202- 203: Eisstanitzel, Hand mit Eis, Löffel, Hohlhippe, Soletti, Pommes, Brot / Studio Köb; S204- 205: Carotte Rapeuse / Stéphane Bureaux; S206- 207: Space Food / NASA; S209: Waage / Ludwig Löckinger; S210- 211: Mikrowelle u. Fertiggericht / Ludwig Löckinger; S212- 213: Erbswurstsuppe / Studio Köb; S215: Suppenwürfel / Unilever; S216- 217: Mundgerechtigkeit / Studio Köb; S218- 219: Puppe (2), Pralinen / Studio Köb, Kind / Sonja Stummerer; S220- 221: Banane, Würstchen, Schokobanane, Fischstäbchen, Dreiecke / Studio Köb, Render Pringle / David Groiss, Soletti / Kellys; S223: Wursterzeugung / Fa. Radatz Wien; S225: Punschkrapfen / Studio Köb; S226- 227: Abhängigkeit / Ludwig Löckinger; S228- 229: Toast, Wurstsemmel / Studio Köb; S230- 231: Wurstsemmeln / Studio Köb, Salzstangerl, Kipferl, Cracker / Studio Köb; S233: Sandwich / Sonja Stummerer; S234- 235: Oberflächendesign / Ludwig Löckinger; S236- 237: Fußballschuh / Ludwig Löckinger; S238- 239: Nudel-Render / Michael Mähring & Philipp Kratzer, Pasta (3) / Ludwig Löckinger, Geyerhalter Film; S240- 241: Teilen / Studio Köb (2); S242- 243: Apfel / Martin Hablesreiter; S244-245: Terry's Chocolate Orange, Orange / Studio Köb; S246- 247: Käse, Tabs und Suppenwürfel / Studio Köb; S249: Schokolade mit Bohrer / Martin Hablesreiter; S250- 251: Schokoladetafel / Studio Köb; S252- 253: Twinniherstellung / Unilever; S254- 256: Collage Teilen / Ludwig Löckinger; S259: Tomaten / John Innes Centre Photography

Designprozess | design process
S262- 263: Renderings / Michael Mähring und Philipp Kratzer; S264: Ahnengalerie / Nestlé, Coca Cola, Bahlsen, Nestlé, Sacher, Kraft Foods, Unilever, Manner; S266- 269: Renderings / Michael Mähring und Philipp Kratzer, Unilever; S271- 273: Portraits / Sonja Stummerer, Martin Hablesreiter, the Jelly Mongers

Kultur | culture
S274- 275: Brote / Studio Köb; S276- 277: No-Masken / Studio Köb; S278- 279: Hirn und Herz / Ludwig Löckinger; S:280- 281: Zöpfe u. Puppen / Ludwig Löckinger; S282- 283: Kipferlkopf / Ludwig Löckinger, Kipferlflagge / Studio Köb; S28- 285: Schokokruzifix / Studio Köb; S286- 287: Monstranz / Ludwig Löckinger, Geyerhalter Film; S288- 289: Geburtstagsbrot / Ludwig Löckinger, Hostie, Buddha-Lolli / Studio Köb; S290- 291: Wagashi / Yasumuro Hisamitsu/ Toraya; Toblerone / Studio Köb; S292- 293: Thunfisch mit Reis / Ludwig Löckinger; S294- 295: Stillleben mit Blutwurst / Studio Köb; S296- 297: Ameisen-Lolli / Studio Köb; S298- 299: Wagashiherstellung, Toraya / Sonja Stummerer (3); S300- 301: Japanische Süßigkeiten / Studio Köb; S302- 303: Kücken / Ludwig Löckinger, Geyerhalter Film; S304- 305: Überraschungsei / Studio Köb; S306- 307: Abstraktion / Ludwig Löckinger, Geyerhalter Film; S308- 309: Stillleben mit Kalbskopf/ Ludwig Löckinger, Geyerhalter Film; S310- 311: Tote Oma / Studio Köb; S312- 313: Hasenkonfekt / Studio; S314- 315: Pfeffer-Mercedesstern / Ludwig Löckinger; S316- 317: Blattgold u. Schnitzel / Studio Köb; S318- 319: Diagramm Modeerscheinungen / Studio Köb, Lufthansa, Kalte Platte / Restaurant Schwarzes Kamel Wien, Flugzeugessen / Lufthansa, Lauda Air; S320: Astraline / Perugina, Gemüse, Marshmallow-Sandwich / Studio Köb, Auslage / Konditorei Demel, Flugzeugessen, Schokofondue / Lufthansa; S322- 323: Totenköpfe, Brust und Bacio/ Ludwig Löckinger, Geyerhalter Film, Teddybär, Auto und Croissant / Ludwig Löckinger, Marzipanmuscheln, Hellokitty-Herz, Weihnachtsbaumwurst u. Lebkuchenherzen / Studio Köb; S324- 325: Sushi-Bonbons u. Marzipankäse / Studio Köb, Japanisches Konfekt / Studio Köb; S326- 327: Mars-Stonehenge / Ludwig Löckinger; S328- 329: Kelch / Ludwig Löckinger; S330- 331: Model / Ludwig Löckinger; S332: Neuburgerblock / Ludwig Löckinger

Zeitdiagramm und Danksagung | timeline and acknowledgment
S336: Sachertorte / Hotel Sacher, Wien; Philadelphia / Kraft Foods - Philadelphia is a registered trademark of Kraft Foods in numerous countries worldwide including e.g. in Germany, Austria and Switzerland and is reproduced by permission of Kraft Foods., Milka / Kraft Foods - Milka is a registered trademark of Kraft Foods in numerous countries worldwide including e.g. in Germany, Austria and Switzerland and is reproduced by permission of Kraft Foods; Toblerone / Kraft Foods - Toblerone is a registered trademark of Kraft Foods in numerous countries worldwide including e.g. in Germany, Austria and Switzerland and is reproduced by permission of Kraft Foods; S337: Semmel / Studio Köb; Erbswurst / Unilever Austria; Goldaugen / Unilever Austria; Coca Cola / Coca Cola Austria; Mozartkugel / Kraft Foods - Mirabell is a registered trademark of Kraft Foods in numerous countries worldwide including e.g. in Germany, Austria and Switzerland and is reproduced by permission of Kraft Foods; S338: Nescafé / Nestlé Austria; Soletti / Kelly s Austria; Mon Chéri / Ferrero; Magnum / Unilever; S339: Sojasauce / Kikkoman; Mars / Masterfoods Austria; S342- 343: Mozartkugel / Studio Köb; S345: japanische Collage / Studio Köb; S348: Portrait Stummerer & Hablesreiter / Studio Köb, Portrait Ulrike Köb / Studio Köb, Portrait Ludwig Löckinger / Sonja Stummerer

Bibliography:

Adbusters No.44: Apetito/Appétit/Appetite, Vancouver, 2002
Ferran Adrià, Joli Soler, Albert Adrià: elBulli 1998-2002, ElBulli, Roses, 2003
Mark Ainsworth: Fish and Seafood, Identification, Fabrication, Utilization, Delmar, Cengage, Learning, NY, 2009
Elisabeth Andoh: Washoku, Ten Speed Press, 2005
Archplus 173, Berlin 2005
Archplus 176/ 177, Berlin 2006
Michael Ashkenazi / Jeanne Jacob: Food Culture in Japan, Greenwood Press, Westport, 2003
BankART 1929, 2006
Era Bakos: Gaumenschmaus und Seelenfutter. Tausend Jahre Wiener Küche, Heyne, München, 1996
Peter Barham: Die letzten Geheimnisse der Koch-kunst, Springer Verlag, Heidelberg, 2001
Barilla: Pasta, History, Technologies and Secrets of Italien Tradition, Barilla, Milano, 2000
Eva Barlösius: Soziologie des Essens, Juventa Verlag, Weinheim und München, 1999
Ralf Beil: Künstlerküche Lebensmittel als Kunstmaterial – von Schiele bis Jason Rhoades, Dumont, Köln, 2002
Ramesh K. Biswas, Siegfried Mattl und Ulrike Davis-Sulikowski: Götterspeisen, Springer, Wien, 1997
Peter Blenke, Thomas Wieke: Kaiser, Köche, Caterer; Kulinarische Gastlichkeit von der Antike bis heute, Mosaik, München, 2001
Marc Bretillot: Design Culinaire, l'Atelier, Ecole Supérieur d'Art et de Design, Reims, 2006
Renate Breuss: Das Maß im Kochen, Haymon, Innsbruck, 1999
Renate Breuss: Die Frauen und das Kochen, Studienverlag, 2001
Renate Breuss: Brennpunkt Küche: planen, ausstatten, nutzen, Ausstellungskatalog, Feldkirch, 2001
Stephane Bureaux: Tool's Food, 2007
Piero Camporesi: Der feine Geschmack. Luxus und Moden im 18. Jahrhundert. Campus Verlag Frankfurt am Main. 1992
Claire Catterall (ed.): Food, Design & Culture, Laurence King, London, 1999
Create eating design and future food, Gestalten, Berlin, 2008
David E. Carter: Branding, Hearst Books, New York, 1999
Alan Davidson: The Oxford Companion to FOOD, Second edition, Oxford University Press Inc., New York, 2006
Ulrike Davis-Sulikowski: Götterspeisen, Springer Wien, 1997
Rotraud Degner: Die Welt der Pasta, 150 Pastasorten in Text und Bild, Collection Rolf Heyne, München, 2000
Edition DETAIL: What Architects cook up, Architekten kochen, München, 2007
Hannsferdinand Döbler: Kochkünste und Tafelfreuden, Orbis, 2002
Jürgen Dollase: Geschmacksschule, Tre Torri Verlag, Wiesbaden, 2005

DOMUS 913: Special Food, Rozzano, 2008
Wiglaf Droste / Nikolaus Heidelbach / Vincent Klink: Wurst, DuMont, Köln, 2006
Klaus Dürrschmid: Gustatorische Wahrnehmungen gezielt abwandeln, Behr's Verlag, Hamburg, 2009
Annette Epp: Gerichte und ihre Geschichte, Kulinarische Zeitreisen, Collection Rolf Heyne, München, 2005
Auguste Escoffier, Walter Bickel: Kochkunstführer Hand- und Nachschlagebuch der klassischen französischen Küche, Pfanneberg, Gießen, 1993
Ullrich Fellmeth: "Brot und Politik", Verlag J.B. Metzler, Stuttgart und Weimar 2001
Karen E. Frank: Food + Architecture, Architectural Design, London 2002
Peter Gnaiger, Wolfgang Hoffmann: In die Suppe gespuckt. Von Sternen, Hauben und anderen Geschäften, Ecowin, Salzburg, 2006
Gisla Gniech: Essen und Psyche Über Hunger und Sattheit, Genuss und Kultur, Springer, Berlin, 2002
Hans-Ulrich Grimm: Die Suppe lügt, Die schöne neue Welt des Essens, Droemer-Knaur, München, 1999
Marti Guixé: 1:1, 010 Publishers Rotterdam, 2002
GUSTO #2: Les influences de l'ailleurs, printemps, Paris, 2007
Hans Jürgen Hansen: Kunstgeschichte des Backwerks, Gerhard Stalling Verlag, Hamburg, 1968
Andrea Heistinger: Die Saat der Bäuerinnen, Loewenzahnverlag
Jost Herbig: Nahrung für die Götter, Hanser Verlag, München Wien, 1988
Gunther Hirschfelder: Europäische Esskultur Geschichte der Ernährung von der Steinzeit bis heute, Campus, Frankfurt/ Main 2001
Kurt Imfeld, Lukas Bidinger: Die molekulare Küche, Rezepte mit Anleitung für Einsteiger und Fortgeschrittene, Foto Plus Schweis GmbH, Luzern, 2008
Martina Kaller-Dietrich: Über den Teller hinaus. Zur Globalisierung der Ernährung, In: iz3w.278/279, 2004
Helene Karmasin: Die geheime Botschaft unserer Speisen, Kunstmann, München, 2000
Jean-Claude Kaufmann: Kochende Leidenschaft: Soziologie vom Kochen und Essen, UVK Verlagsgesellschaft mbH, Konstanz, 2006
Emi Kazuko: Japanese food and cooking. A timeless cuisine: the traditions, techniques, ingredients and recipes, Lorenz Books, London, 2001/2007
Ruth Keenau: Festliche Menüs aus 2000 Jahren, Esskultur und Tafelfreuden vom alten Rom bis heute, Christian Verlag, München, 2001
Kelly: United Snacks of Kelly's, Kelly GmbH, Wien, 2002
Margareta Lehrbaumer: Womit kann ich dienen? Julius Meinl, Auf den Spuren einer großen Marke, Pichler Verlag, Wien, 2000

Carola Lentz: The Porrigde debate: Grain, Nutrition and forgotten Food Preperation Techniques, In Changing Food Habits, Harwood Academic Publishers, Australia, 1999

Uliano Lucas: Italiani a tarola, Mazzotta, Milano, 2003

Erich Lukas: Verdi: kalt/warm, oder die lizenz zum kochen, Linz

Elisabeth Meyer-Renschhausen: Bauern und Bäuerinnen aus städtischer Sicht In Günter Lorenzls Urbane Naturaneignung als agrarische Marktchance?, Köster, Berlin,1996

Mi-kitchen, Miki Torigoe, Pie books, 2007

Momofuku Ando: The Story of the Invention of Instant Ramen, My Resume, Nissin Food Products, Osaka, 2002

Momofuku Ando: Human beings are noodle beings, Nissin Food Products, Tokyo, 2007

Massimo Montanari: Der Hunger und der Überfluss, Kulturgeschichte der Ernährung in Europa, C. H. Beck, München, 1993

Andreas Morel: Der gedeckte Tisch Zur Geschichte der Tafelkultur, Gva-Vertriebsgemeinschaft, 2001

Leo Moulin: Augenlust & Tafelfreude Essen und Trinken in Europa – eine Kulturgeschichte, Zabert Sandmann, München, 2002

Klaus E. Müller: Nektar und Ambrosia, C.H. Beck, München, 2003

Kristiane Müller-Urban: Toast Hawaii und Entenbrust, Die Trendrezepte aus fünf Jahrzehnten, Dumont, Köln, 1999

Yoshihiro Murata: Kaiseki, The Exquisite Cuisine of Kyoto's Kikunoi Restaurant, Kodanska International, Tokyo, 2006

Roswitha Mutterthaler, Elisabeth Limbeck-Lilienau und Gabriele Zuna-Kratky im Auftrag des Technischen Museums Wien: Geschmacksache. Was Essen zum Genuss macht, Wien, 2008

Athenaios von Naukratis: Das Gelehrtenmahl, Dieterich'sche Verlagsbuchhandlung Leipzig, 1985

Peter Noever: Die Frankfurter Küche von Margarete Schütte-Lihotzky, Ernts & Sohn, Brelin, 1997

Diane Norman & Michelle Cornell: Ikebana, A fresh look at Japanese Flower Arranging, Conran Octopus, London, 2002

NZZ Folio: Zucker, Der süße Treibstoff, NZZ, 2006

Catarina Ossuth-Wolkenstein: Die Marke Eskimo, Eine Erfolgsgeschichte, Signum, Wien, 2000

Gert von Paczensky und Anna Dünnebier: Kulturgeschichte des Essens und Trinkens, Orbis Verlag München 1999

Dr. Ute Paul-Prössler: Käse, Entdecken & Genießen, Sigloch Edition, Künzelsau, 2002

Dolly Peters: Kalte Happen und Partysnacks, Falken-Verlag, Niederhausen/ Ts, 1997

Perugina, Una storia d'Azienda, ingegno e passione, Silvana Editoriale, Milano, 1997

Ewald Plachutta, Christoph Wagner: Die gute Küche, Das österreichische Jahrhundertkochbuch, Orac Verlag, Wien, 1993

Karl-Heinz Plattig: Spürnasen und Feinschmecker, Die chemischen Sinne des Menschen, Springer Verlag, Berlin Heidelberg, 1995

Hans-Werner Prahl und Monika Setzwein: Soziologie der Ernährung, Leske und Budrich Verlag, 1999

Gabriele Praschl-Bichler: Kaiserliche Mehlspeisen, Die besten Backrezepte aus dem privaten Kochbuch der österreichischen Kaiserfamilie, Amalthea ,München, 2004

John Robbins: Food Revolution, Hans-Nietsch-Verlag, Freiburg, 2003

Hanni Rützler: Was essen wir morgen? Springer Wien New York, Wien, 2004

Brillat-Savarin: Physiologie des Geschmacks, Insel Verlag, Frankfurt am Main und Leipzig, 1913

Thomas Schueller: Meat, Identification, Fabrication, Utilization, Delmar, Cengage Learning, NY, 2009

Daniel Spoerri: Der Zufall als Meister, Kerber Verlag, Bielefeld, 2003

Clarissa Stadler, Wolfgang Thaler: Aida, Mit reiner Butter, Verlag Christian Brandstätter, Wien, 1995

Jakob Tanner: Fabrikmahlzeit. Ernährungswissenschaft, Industriearbeit und Volksernährung in der Schweiz 1890-1950, Chronos Verlag, Zürich, 1999

Utz Thimm und Karl-Heinz Wellmann (Hrsg.): Essen ist menschlich, Zur Nahrungskultur der Gegenwart, Suhrkamp, Frankfurt, 2003

Yoshio Tsuchiya: The fine art of Japanese Food Arrangement, Kodansha, Tokyo, 1985/2002

Hanne Tügel: Experiment Schlaraffenland, Geo Magazin 6/98

Alice Vollenweider: Italiens Provinzen und ihre Küche, Eine Reise und 88 Rezepte, Wagenbach, Berlin, 1990

Christoph Wagner: Fast schon Food, Campus, Frankfurt, 1995

Christoph Wagner: Das Lexikon der Wiener Küche, Deuticke, Wien, 1996

Christoph Wagner: Prato, Die gute alte Küche, Pichler Verlag, Wien, 2006

Christoph Wagner: Süßes Gold, Kultur und Sozialgeschichte des Wiener Zuckers, Verlag Christian Brandstätter, Wien, 1996

Loukie Werle, Jill Cox: Ingredienzen, Das große Buch der Zutaten, Könemann, Köln, 2000

Reinhart Wolf: Japan, Kultur des Essens, Collection Rolf Heyne, München, 2001

Sonja Stummerer
Geb.: 22.02.1973 in Wien
1991- 98   Studium der Architektur an der Universität für Angewandte Kunst, Mkl.: Prof. Hans Hollein in Wien und an der Designschule Elisava in Barcelona
1999- 01   Ausbildung zum „Master of Architecture" an der „Architectural Association" in London/ GB
2001- 2003 Projektarchitektin unter anderem im Atelier Arata Isozaki & Associates/ Tokyo/ Japan
Seit 2007 staatlich befugte und beeidete Ziviltechnikerin und Mitglied der Wiener Architektenkammer

Martin Hablesreiter
Geb.: 05.02.1974 in Freistadt/ Österreich
1993- 00   Studium der Architektur bei Prof. Hans Hollein an der Universität für Angewandte Kunst in Wien und an der Bartlett School of Architecture in London
2001- 03   Projekt Architekt bei Arata Isozaki & Associates in Tokyo/ Japan und bei Prof. Helmut Richter/ Wien

Sonja Stummerer und Martin Hablesreiter gründeten 2003 das interdisziplinäre Architekturatelier honey & bunny productions in Wien. Sie entwickelten und bauten mehrere Dachausbauten, führten Regie bei „food design – der Film", kuratierten die Ausstellung „food design" für Wien und Graz und nahmen an zahlreichen Einzel- und Gemeinschaftsausstellungen teil. Stummerer und Hablesreiter hielten zahlreiche internationale Vorträge und lehrten unter anderem in Bukarest RU, Istanbul TU und in Chennai IN.

www.honeyandbunny.com

Sonja Stummerer
Born: Feb. 22, 1973 in Vienna
1991-98   Studied architecture at the University of Applied Arts in Vienna under Prof. Hans Hollein and at the Design School Elisava in Barcelona
1999-2001 Studied for "Master of Architecture" at the "Architectural Association" in London/GB
2001-2003 Work as project architect at firms such as Arata Isozaki & Associates / Tokyo / Japan
Since 2007 licensed civil engineer and member of the Vienna Chamber of Architects

Martin Hablesreiter
Born: Feb. 5, 1974 in Freistadt, Austria
1993-2000 Studied architecture under Prof. Hans Hollein at the University of Applied Arts in Vienna and at the Bartlett School of Architecture in London
2001-2003 Project architect for Arata Isozaki & Associates in Tokyo/Japan and for Prof. Helmut Richter in Vienna

In 2003 Sonja Stummerer and Martin Hablesreiter founded the interdisciplinary architecture studio honey & bunny productions in Vienna. They have developed and constructed roof expansions, have directed a film ("Food design – the film"), curated the exhibition "food design" for Vienna and Graz and taken part in numerous single and group shows. Stummerer and Hablesreiter have given a number of international lectures and taught at schools including Bukarest RU, Istanbul TU and Chennai IN.

www.honeyandbunny.com

Ulrike Köb
Geb.: 1959
1988 Gründung eines Fotostudios in Wien mit Schwerpunkt Food- und Stillife.

Ulrike Köb macht Fotografie für diverse Koch- und Fachbücher wie „Die süße Küche", „Pilze", „Wildkräuter-Delikatessen", „Soja", „Genussland Österreich" „Faszination Gemüse", Werbefotografie für Lebensmittelkunden und Fotos für Fachmagazine mit Schwerpunkt Essen und Trinken.

www.koeb.at

Ulrike Köb
Born: 1959
Founded 1988 a photo studio in Vienna with a focus on food and still life.

Ulrike Köb photographs for various cooking and specialized books (e.g.,„Die süße Küche", „Pilze", „Wildkräuter-Delikatessen", „Soja", „Genussland Österreich" „Faszination Gemüse")

www.koeb.at

Ludwig Löckinger

Geb.:   1973 in Linz
Ab 1992   diverse Kurzfilme als Regisseur
1994- 2001 Studium: Kamera an der Filmakademie Wien
Ab 2001   Tätigkeit als Kameramann
Ab 2008   Lehrtätigkeit an der Akademie der bildenden Künste in Wien

www.loeckinger.com

Ludwig Löckinger
Born: 1973 in Linz
From 1992 directed various short films
1994-2001 Studied camera shooting at the Vienna Film Academy
From 2001 Work as camera man
From 2008 Teaching at the Academy of Fine Arts in Vienna

www.loeckinger.com